Even Distribution and Spherical Ball-Packing

Even Distribution and Spherical Ball-Packing

Ying-chien Chang

Edited by: Andy Wen

ISBN: Softcover 978-1-5144-5117-5
eBook 978-1-5144-5116-8

Print information available on the last page.

Rev. date: 02/29/2016

To order additional copies of this book, contact:
Xlibris
1-888-795-4274
www.Xlibris.com
Orders@Xlibris.com
727577

CONTENTS

DEDICATION

To my parents and the Earth

They inspired me with their love of nature.
May our spirits emerge and go on forever.

INTRODUCTION

A set of n points evenly distributed on a surface of a sphere is called an **Even Distribution (E).** There are five distinctive nEs while **n = 4, 6, 8, 12, and 20,** so every individual E is 4E, 6E, 8E, 12E, or 20E**.** Every E has a lattice of polyhedron with *all identical regular faces*. Every point of an E is a vertex of its polyhedron.

The configurations of five nEs

Names by Chang (by no. of vertices)	4E	6E	8E	12E	20E
Lattices of Polyhedron		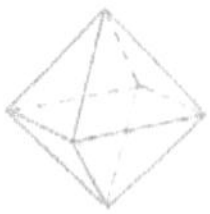	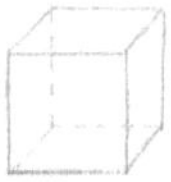	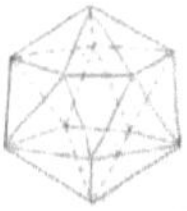	
Archimedean Names (by no. of faces)	Tetrahedron 4 △	Octahedron 8 △	Hexahedron 6 □	Icosahedron 20 △	Dodecahedron 12 ⬠
No. of edges	6	12	12	30	30
Radius R	$\sqrt{\frac{3}{2}} \doteq 1.2247448$	$\sqrt{2} \doteq 1.41421356$	$\sqrt{3} \doteq 1.7320508$	$\sqrt{\frac{1}{2}(5+\sqrt{5})}$ $\doteq 1.902113$	$\frac{\sqrt{3}}{2}(1+\sqrt{5})$ $\doteq 2.802517$

To generate E-extensions and to analyze them is the main goal of this study. This study started with the five basic* configurations and extended from them to establish five new categories of extensions. The extension process and analysis are based on three-dimensional Euclidean geometry. The name assigned to every new configuration is based upon how it is constructed from the mother configuration and how many points it contains. (* The word "basic" indicates that the configuration has all edges of lengths 2. Sometimes, the word "original" is used instead of "basic." The word "configuration" means the entire system of the E pack or E-extension pack.)

The polyhedron of a nE can be inscribed in a spherical sphere of radius R. An nE is a spherical pack system incorporated with its polyhedron. All points of a nE are floating on the surface. Such a surface is called the floating surface, or just the surface. There are two kinds of angles formed in the incorporated configuration: the ***straight angle*** and the ***surface angle.*** The straight angle is an angle on a **plane** face of a polyhedron. The **surface angle** is an angle on the spherical surface corresponding to its straight angle.

The straight angle $\bar{\phi}$, or just ϕ, is the angle formed from a certain point on a surface with two straight lines in space to another two points on the same surface. The **surface angle** $\widehat{\phi}$ is the angle formed from a certain point on the surface with two shortest curved **surface lines** to another two points on the same surface. For instance, the North Pole of the Earth to two locations on the equator 90° apart corresponds to a straight angle of 60° and a surface angle of 90°. The angle measurement of a surface angle is greater than that of its corresponding straight angle while the two edges of the straight angle are equal in length. The total surface angle of a point (the sum of all individual surface angles of a point) is 2π or 360°, and the total straight angle is less than that. (On a plane, the total straight angle is 2π.)

The practical way in studying E-packing is to assume that every point of a configuration is a small ball with a radius of 1 unit in length packed in a spherical container. This induces the study of Es and E-extensions into spherical ball packing. The essential rule of ball packing is to pack all balls in a container under the criterion that ***every ball should be in a state of having absolute 0 degree of freedom in motion (in all directions and distances).*** This regulates the structure of the pack. A pack with every ball under such a state is called a **well-bound (WB)** pack or a good pack.

Generally, the way to construct a new configuration is to add some balls systematically and symmetrically on some faces of a mother configuration, then press these added balls vertically downward the faces. At the mean time, the original balls are forced to go outward from the center of the system. The press stops while the added balls and the original balls are reached on a common surface. It is a new configuration containing all the added and original balls. And it is a good pack if it can pass all WB criteria, otherwise it cannot exist.

There are five WB criteria that a good pack should satisfy. They are:

(1) ***Criterion-A*: Every surface *angle* of a ball with two straight edges of lengths 2 should be smaller than a π.**
(2) ***Criterion-B*: Every ball should have at least three but no more than five *bindings*. (The number of bindings is the number of immediate neighbor balls. "Immediate" means "two balls are in touch and the distance between them is 2).**
(3) ***Criterion-C*: Every layer of added balls should be *confined* in the range of its base layer.**
(4) ***Criterion-D*: Every ball to any other ball, added or original, should have a *distance* of at least 2.**
(5) ***Criterion-E*: For an M-A constructed E-extension (to be discussed in Chapter II), the associate face should have all its associated *edge-lengths* of at least 2.**

The detailed analysis and calculation of these five criteria are in **Chapter II – 2**. The most important one is **criterion-A.** It plays a very important role in this study. It concerns the straight angle and the corresponding surface angle. The outline about the straight angle $\bar{\phi}$ and surface angle $\hat{\phi}$ is as follows:

The formula of straight angle $\bar{\phi}$ of a point on a surface to another two points on the same surface with two straight edge-lengths $\mathbf{2\,d_1}$, $\mathbf{2\,d_2}$, and the opposite edge-length $\mathbf{2\,d_0}$ is

$$\mathrm{Cos}\bar{\phi} = \frac{d_1^2+d_2^2-d_o^2}{2d_1d_2}$$

The formula of surface angle $\hat{\phi}$ of the corresponding straight angle $\bar{\phi}$ developed in this study is

$$\mathrm{Cos}\hat{\phi} = \frac{(d_1^2+d_2^2-d_o^2)R^2-2d_1^2d_2^2}{2d_1d_2\sqrt{R^2-d_1^2}\sqrt{R^2-d_2^2}}$$

While $d_1 = d_2 = \mathbf{d}$, we have $\mathrm{Sin}\frac{\hat{\phi}}{2} = \alpha\mathrm{Sin}\frac{\bar{\phi}}{2}$ where $\alpha = \frac{R}{\sqrt{R^2-d^2}}$, is called the ***surface index***.

The total surface angle of a point on a surface is $\sum\hat{\phi}_i = 2\pi$.

The total straight angle of a point on a surface is always less than 2π. *(**Note:** The total straight angle of a point on a plane is always* 2π*.)*

Theoretically speaking, using the above formula, it is impossible to have three hexagons, or six triangles, intersect at one point on a spherical surface (no matter how large the radius of the sphere is, or how relatively small the hexagons or the triangles are).

The wall of container of the floating E-packing is the boundary of the pack system. It restricts all balls from flying out. Actually, for the ball-packing, the wall is imaginary, so it may be considered that the pack is bounded by its own gravity at the center. Without this gravitational bound, even the pack meets all above five criteria; all balls may still fly out freely. While analyzing the WB criteria, it is assumed that the essential gravity is always there, and so there's no need to put it in further consideration.

The five categories of E-extensions **developed from the five basic Es under the WB criteria are:**

Category I **The First Generation of Even distribution (E') (The analysis is in Chapter II) There are eleven distinctive E's. They are generated from the five basic nEs. Every E' has a lattice of polyhedron with two or three sets of different regular faces. The configurations of the eleven E's are:** $12E_1'$, $12E_2'$, $24E_1'$, $24E_2'$, $24E_3'$, $30E'$, $48E'$, $60E_1'$, $60E_2'$, $60E_3'$, $120E'$**. Their drawings are below. (The sizes of the individual drawings are not in proportion.)**

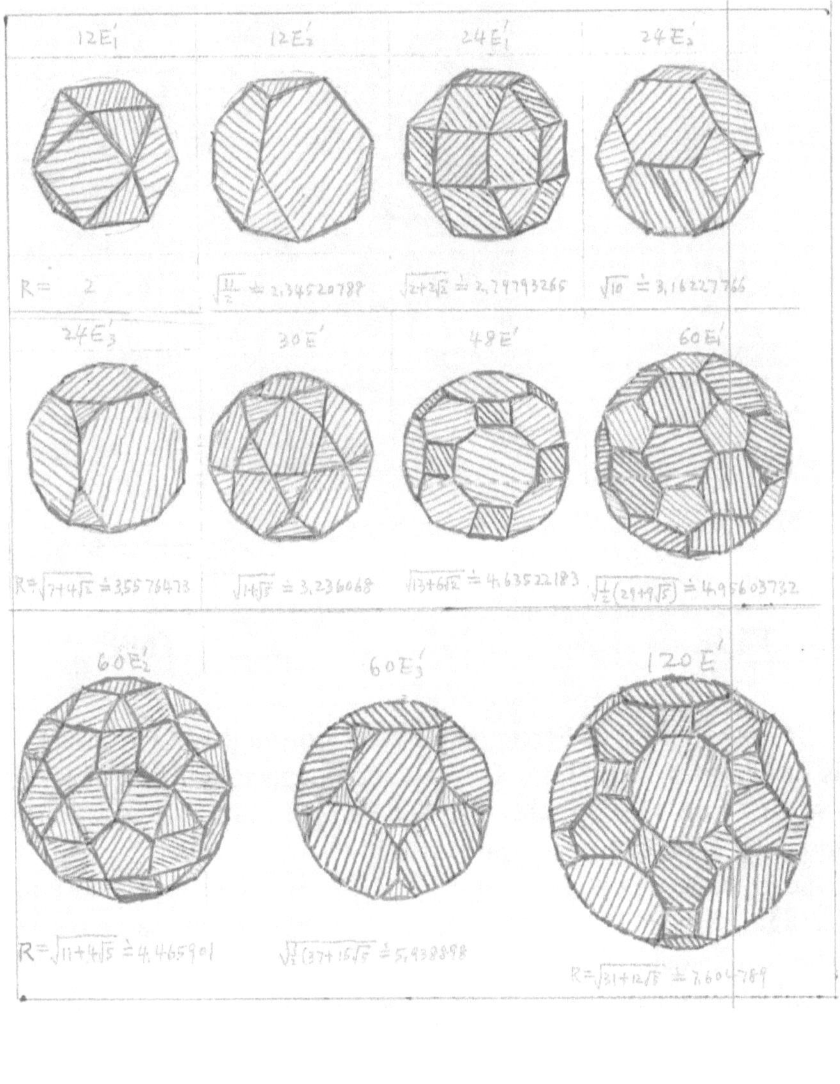

Category II **The Twisted Even-distribution (TE). (The analysis is in Chapter III) There are three distinctive TEs. Each is generated by twisting a certain E'. The candidate E' must have three sets of faces, with one set of all squares and one set of all triangles. The drawings of the three candidate mother E's and three daughter TEs are below. (Also their sizes are not in proportion.)**

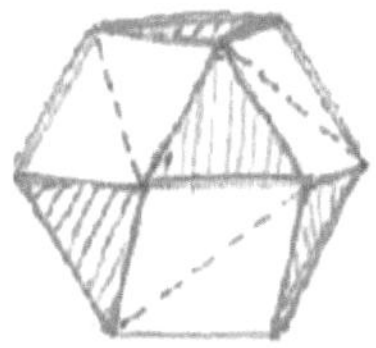

$12E_1'$

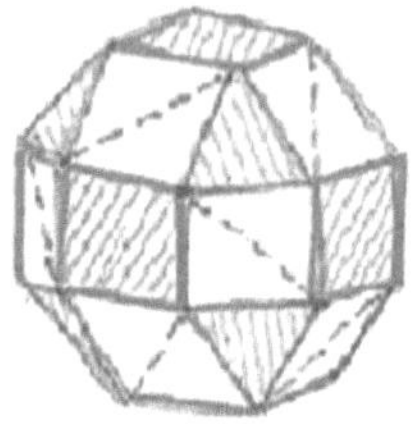

$24E_1'$

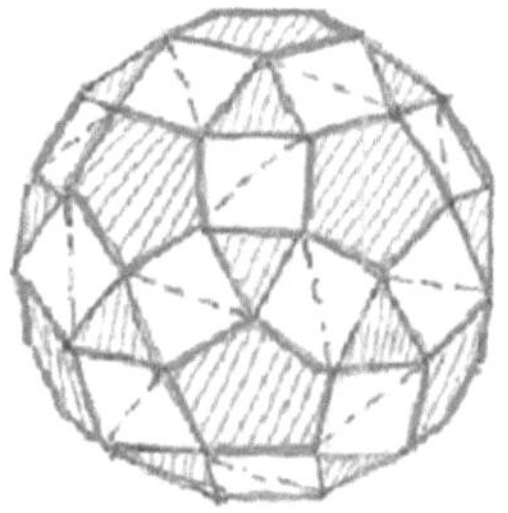

$60E_2'$

T12E

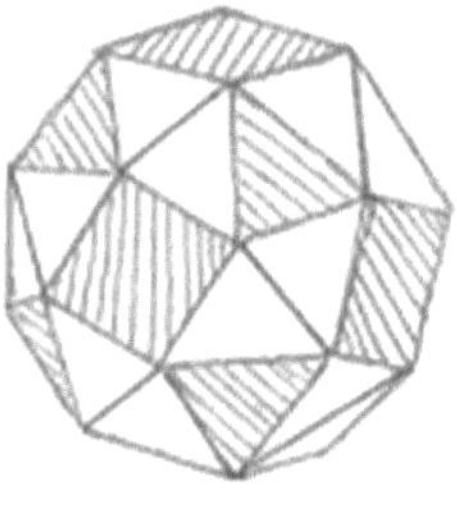

T24E

T60E

Category III **The String Even-distribution (SE) (The analysis is in Chapter IV.) There are two subcategories: the Square-SE (Sq-SE) and the Triangle-SE (Tr-SE).** The SE is generated by inserting a string of squares or paired triangles, called sq-pearl or tr-pearl, between two faces of an E or E'. All strings in a certain SE configuration should contain the same number of the same pearls. The string could contain an infinitely large number of pearls, so it could be infinitely long, and the SE packing could be an infinite ball packing. While inserting strings between faces, some open faces would appear. A long string SE pack would have large open faces, and the original faces become relatively small, even could be condensed as a point. In such a case, the very long string configuration, no matter if it is Sq-SE or Tr-SE, will resemble to another E or an AE (See **Category V**).

The notation of string configurations: For instance, to insert a string containing one sq-pearl in a mother **12E'1** between every square face (4-gon face) and a triangle face (3-gon face), the notation of such configuration is **Sq-S12E'1(4-3, 1).** If the mother **12E'1** is inserted between its 4-gon and 3-gon faces with a string containing one pair of tr-pearl, then the notation of the configuration would be **Tr-S12E'1(4-3, 1)**. (See the illustrations below.)

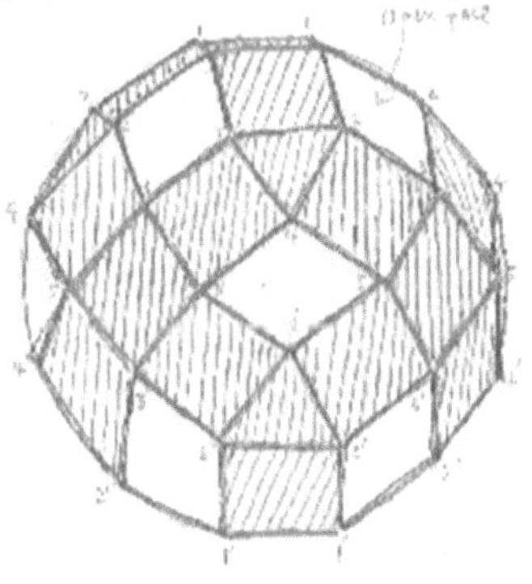

The Sq-S12E'1(4-3, 1)

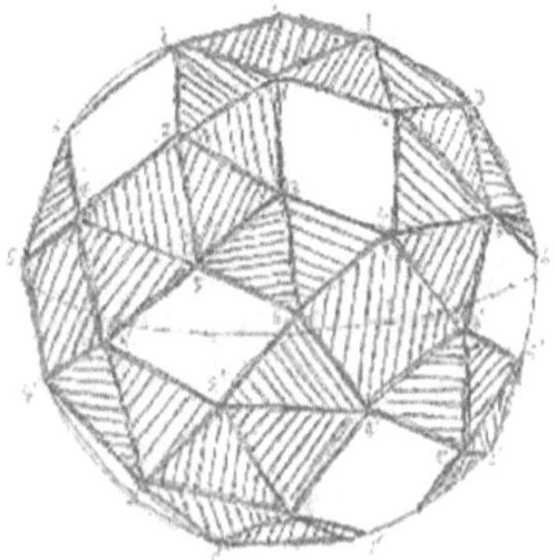

The Tr-S12E'1(4-3, 1)

If the string contains m pearls, then the notation would be **Sq-S12E'1(4-3, m) or Tr-S12E'1(4-3, m).**

***Category IV* The Second Generation of Even Distribution (E"):** The **E"**s are generated by adding layers of points (balls) systematically and symmetrically on one or two sets of faces of an E or E-extension. The arrangement of added points on the faces in the same set of the mother configuration should be identical. The lattice of an E" contains some non-regular faces. An E" formed from a certain mother configuration is called the family member of that mother. Every E or E' or TE or Sq-SE or Tr-SE can have its own family. After a new E" is constructed, filtering it out through the five criteria is a very significant procedure. Only those which passed all five are to become an existing member of the mother's family. Some families are very large, such as the family of **48E'** and **120E'** (see their data in DATAFILE).

There are many different ways to add balls on a certain mother configuration. These include: (1) Single arrangement (S)—add balls only on one certain set of faces; (2) combination arrangement—add balls on two different sets of faces with two different arrangements. There are two different combination arrangements, namely, Main-Associate arrangement (M-A) and Single-Single arrangement (S-S). (For details, see Chapter II.) It is impossible to add balls on all three sets of a mother configuration even if the mother has three sets. The criteria in constructing a new E" with some combination arrangements, such as the M-A arrangement, is very complicated.

Category V **The Amicable Even Distribution (AE):** There are only three **AE**s. The first one is a combination of two 4Es. Called **A8E,** this combination has 8 points. (Actually, **A8E** is the same in construction as **8E)**. The second is the combination of **6E** and **8E**, and is called **A14E**. The third is the combination of **12E** and **20E,** and is called **A32E.** They are the only ones combined with two nEs. The two component nEs are called the amicable pairs. The AEs are actually E"s of the family of either component nE. They have very special characters and are listed as in an individual category. (See analysis in section **I-5** and **IV-8.**)

The Head Signs **To create E"s,** while adding balls on faces of a mother E or E-extension, it is necessary to use some head signs to show how the different arrangements are in the construction. The small head signs (such as "$_\sim$", "$_-$", "$_\vee$", and "$_2$", plus the head signs of surface angle ($\widehat{\phi}$) and straight angle ($\bar{\phi}$), and some more), invented for use in this study, play very significant roles. The developing equations and analyzing the structure of the entire configuration are totally relying on them. Without these signs, the structure of a configuration would be very confusing, no equations can be developed, and thus, no analysis can be performed. The consequence is simply that this study will be jeopardized.

For creating new configurations, the very first primary step is to build up a list of good stacks which can pass all five WB criteria as to form the building blocks. To use these blocks, it is possible to add balls layer by layer on the sets of faces of a mother configuration to create a new E". In the process of constructing, it is required in almost every step to check through the WB criteria due to individual stacks of balls added on it. The existing stacks in the APPENDIX are good stacks which are already filtered out step by step through the WB criteria, and so they are ready to serve as building blocks.

The Archimedean Solids with respect to the names of configurations by Chang:

The very ancient names of the configurations of Es and E-extensions are lost. In modern science, the five basic Es are called the platonic solids, and the eleven E's are called the Archimedean solids. In 1611, Johannes Kepler discovered the TEs called the Kepler's solids. All these three solids are generally called the Archimedean solids. The Archimedean solids do not cover the **T12E**, threc **AEs**, all **SEs,** and all second generations, the **E"s.** The names of Archimedean Solids are based upon the number of faces and their shapes The names of Chang's configurations are based upon how many points the configuration has and how they are constituted. The list below does not include the ones which are missing in Archimedean Solids.

Names of Archimedean Solids	Names in this study by Chang
Tetrahedron	4E
Octahedron	6E
Hexahedron	8E
Icosahedron	12E
Dodecahedron	20E

Cuboctahedron	**$12E_1'$**
Truncated tetrahedron	**$12E_2'$**
Rhombicuboctahedron	**$24E_1'$**
Truncated Octahedron	**$24E_2'$**
Truncated Hexahedron	**$24E_3'$**
Icosidodecahedron	**$30E'$**
Truncated Cuboctahedron (or Great Rhombicoboctahedron)	**$48E'$**
Truncated Icosahedron	**$60E_1'$**
Rhombicusidodecahedron	**$60E_2'$**
Truncated Dodecahedron	**$60E_3'$**
Icosidodecahedron (or Great Rhombicosidodecahedron)	**$120E'$**
Snub hexahedron (or Snub Cuboctahedron)	**T24E**
Snub Dodecahedron (or Snub Icosidodecahedron)	**T60E**

The Application of the E and E-extensions (The Connection between This Study and Other Scientific Areas)

In 1962, Aaron Klug and Donald Kasper discovered a class of viruses shaped as 12E.

In 1985, Richard Smalley, Harold Kroto, Robert Curl, and colleagues discovered the 60-atom carbon molecule, called buckminsterfullerene, constructed as the construction of $60E_1'$. Their discoveries in the field of chemistry initiated this study of even distribution and spherical ball-packing in the field of mathematics. Currently, the fullerene (constituted with twelve 5-gons each surrounded by five 6-gons and many other 6-gons, or a combination construction of 12E and some 6-gons) and the nano-carbon-tube (constituted with all 6-gons in various constructions) open a new high-tech scientific field.

Stan Schein, UCLA neurologist, studies a protein, called clathrin, which is a kind of fullerene with special biological characters.

Luann Becker, UCSB astro-geologist, discovered some helium isotopes caged in the buckminsterfullerenes for billions of years in the asteroid bombarded sites on Earth. Also many scientists found buckminsterfullerenes existed in space.

The constitution of crystals is directly connected with the constructions of many E and E-extensions.

The early life evolution wherein small molecules combined together to form larger ones or complicated ones could be in some ways similar to the process of creating E"s with stacks one layer upon one layer or stacks on different faces and connected by strings. The development of embryo could be concerned with the process of creating some highly condensed E"s (the E"s with large complicated stacks or with long strings of an SE, or both).

CHAPTER I

EVEN DISTRIBUTION AND FIRST GENERATION OF EVEN DISTRIBUTION

[I – 1] The Even Distribution (E):

A set of n points evenly distributed on a spherical surface is called an even distribution (E) which requires:

(1) Every point should have at least three but no more than five immediate neighbor points.
(2) Every point is a vertex of a polyhedron with all identical regular polygon faces.

To meet these two criteria, the number of points of an E is limited to **4**, **6**, **8**, **12**, and **20** only. A certain n-point even distribution is denoted as nE. The five even distributions are **4E, 6E, 8E, 12E, and 20E.**

The polyhedron of an nE can be inscribed in a spherical surface of radius R. All points are floating on the surface. Such a surface is called the floating surface, or just the surface. The nE is a spherical pack system and is incorporated with its polyhedron. There are two kinds of angles formed in such configuration: **the *straight angle*** ($\bar{\phi}$) is the angle appears on the polyhedron, and the ***surface angle*** ( ) appears on the spherical surface corresponding to the straight angle.

The straight angle $\bar{\phi}$, or just ϕ, is the angle from a certain point on the surface with two edges of straight lines in space to another two points on the same surface. The surface angle $\hat{\phi}$ is the angle from a certain point on the surface with two edges of shortest curved lines on the surface to another two points on the same surface. These two corresponding angles are of different angle measurements. For example, the North Pole of the Earth to two locations on the equator with 90° apart, the straight angle is 60° and the corresponding surface angle is 90°. **The surface angle is greater than its corresponding straight angle while the two edges are equal in length. The total surface angle (the sum of all individual surface angles of a point) is 2π or 360°, and the total straight angle is less than 2π.** The equations of the surface angle and its corresponding straight angle are developed in **[II – 2a]**.

To solve an E-pack, mainly is to find the radius R of the surface. And the volume V, the volume density Dv, and the surface density Ds of the pack can be deduced from the R.

Since every ball is of radius 1, the container's radius is R + 1. And the container's volume V is

$$V=\frac{4}{3}\pi\left(R+1\right)^3 \qquad \textbf{(1)}$$

The volume density Dv of a pack is the percentage of occupied space by all **N** balls in the container with the container's volume, which is

$$\mathbf{D_V = \frac{100N}{(R+1)^3}\,\%} \qquad \textbf{(2)}$$

The floating packing has all balls "floating" on the surface, so the surface density Ds is more important than the volume density Dv. The Ds of a pack is defined as the percentage of the area that all **N** balls occupied on the surface with the total area of the surface.

$$\mathbf{D_S = \frac{100N}{4R^2}\%} \qquad \textbf{(3)}$$

Note that if B is the center of a ball floating on the surface S as in **Fig. I-1**. The equator plane of this ball containing D_1 and D_2 is vertical to this paper and has an area measurement of π. But the virtual area of this ball covered on S is the cone area containing E_1 and E_2. These two areas are different in shape but their area measurements are equal. Therefore, the formula **(3)** is true.

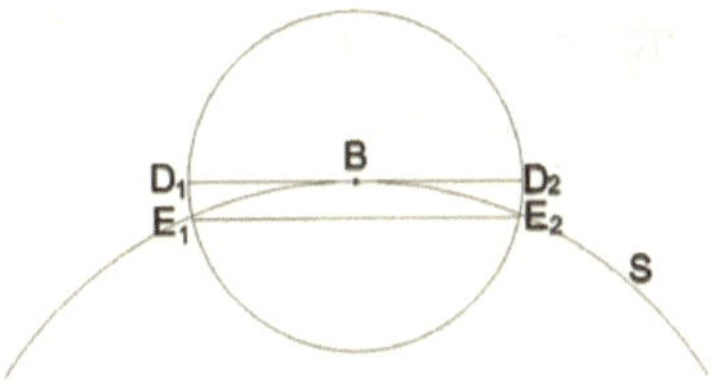

Fig. I-1

The shape of a k-gon: The term "k-gon" used throughout this paper is for k = 3, 4, 5, 6, 8, or 10. While k = 3 or 5, the k-gon is a regular polygon. While k is an even number, the k-gon could be a regular polygon or a half even k-gon, with every next edge of length $\mathbf{2e_1}$ and the other next edges of length $\mathbf{2e_2}$, with all equal inner angle measurement. **[See Fig. I-2a and I-2b.]**

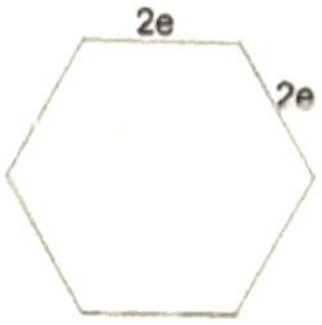

Fig. I-2a

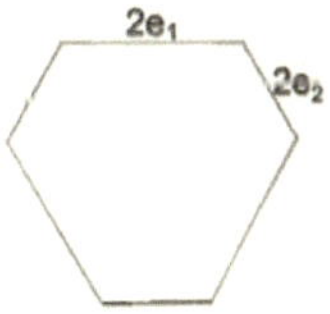

Fig. I-2b

Another case is that a 2k-gon could be a near **k-gon.** For instance, a 4-gon could be a diamond with all 4 edges of same length, and one diagonal is longer than the other one **[Fig. I-2c]**. For a 6-gon, it can have all 6 edges of same length and two kinds of different angles **[Fig. I-2d]. The same goes for 8-gons and 10-gons**—with all same edge-length and two different kinds of angles.

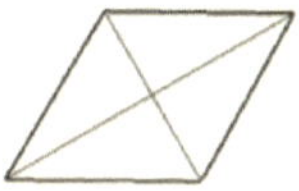

Fig. I-2c

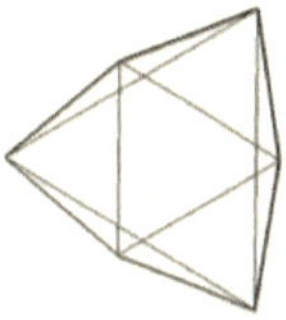

Fig. I-2d

All equations in this paper are expressed in traditional style, for example: $A=\frac{(2BC+1)R^4}{(R^3+1)^2}$

While it is for the computer to work, it should be expressed as: $A=(((2*B*C+1)*(R\wedge 4))/((R*R+1)\wedge 2))*\pi$

[I – 2] The First Generation of Even Distribution (E')

There are two ways to generate an E' from a mother E:

The first way: This is to add points on every face or on every edge of a mother E. The added points form a new E'. (The mother points are not counted). The following signs show the way of adding points on the mother E to create an E':

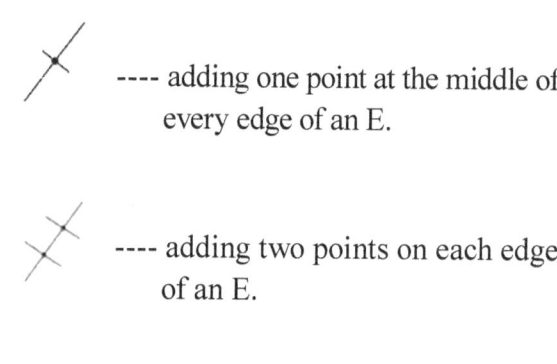

---- adding one point at the middle of every edge of an E.

---- adding two points on each edge of an E.

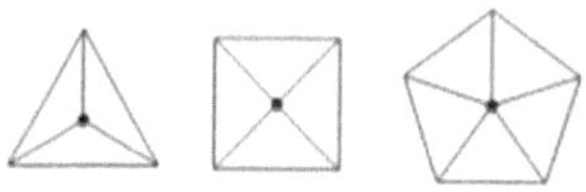

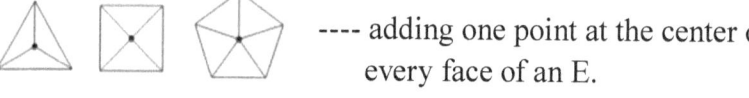

---- adding one point at the center of every face of an E.

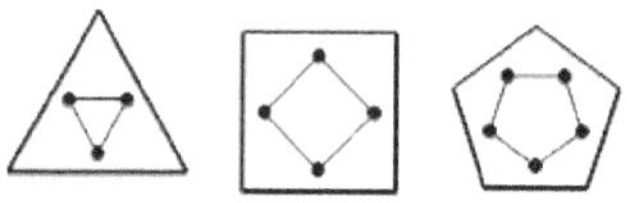

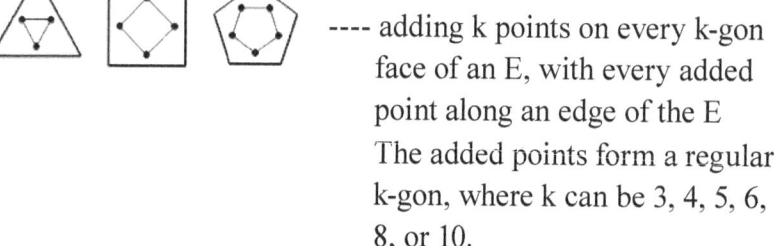

---- adding k points on every k-gon face of an E, with every added point along an edge of the E The added points form a regular k-gon, where k can be 3, 4, 5, 6, 8, or 10.

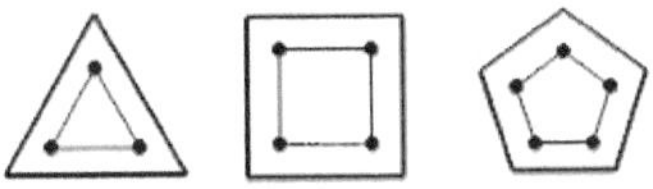

---- adding k points on every k-gon face of an E, with every added point along the corner of the face. The added points form a regular k-gon, where k can be 3, 4, 5, 6, 8, or 10.

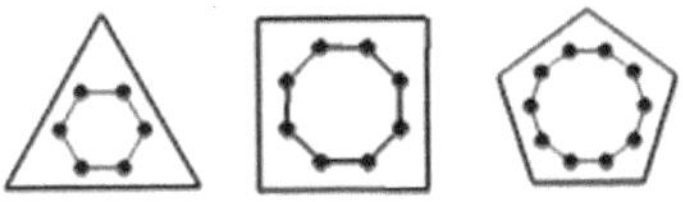

---- adding 2k points on every k-gon face of an E, with every two added points along an edge of the face, (which is the same as along every corner of the face.) The added 2k points form a regular 2k-gon, where k can be 3, 4, 5, 6, 8, or 10.

The result E's from the mother Es:

Way of generating					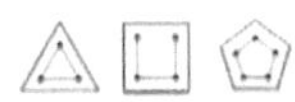	
From the mother E	The resulting configurations					
4E	6E	$12E_2'$	4E	$12E_1'$	$12E_2'$	$24E_2'$
6E	$12E_1'$	$24E_2'$	8E	$24E_1'$	$24E_2'$	48E’
8E	$12E_1'$	$24E_3'$	6E	$24E_1'$	$24E_2'$	48E’
12E	30E’	$60E_1'$	20E	$60E_2'$	$60E_3'$	120E’
20E	30E’	$60E_3'$	12E	$60E_2'$	$60E_1'$	120E’

The second way: This involves cutting away all corners at one-half or cutting away the corners and leaving the central portion of the edge as of length 2. The result configuration is an E' with two sets of faces. To generate an E' with three sets of faces, the way is to cut the corners away then remove all of the original edges. The leftover points form a configuration with three sets of polygons. (This second way of generating the individual E's is not illustrated.)

The General Characters of the First Generation

(1) Every lattice of an E' has two or three different sets of polygon faces. All faces in the same set are identical.

(2) Every point of an E' is a vertex of a lattice. Every vertex is an intersection of a same arrangement of faces from two or three different sets. Such a vertex with its identical arrangement of faces is called a "**Village.**" (See chart below.)

(3) The centers of all faces in the same set are distributed as a form of an E or another E'. For example, the **$60E'_2$** has 12 pentagons, 20△, and 30□. The 12 centers of the 12 pentagons are distributed as 12E. The 20 centers of the 20□ are distributed as 20E. And the 30 centers of the 30□ are distributed as 30E'.

The Villages and R of the Eleven E's:

E'	Village	Faces contained in a village	R
$12E'_1$		two (4-gons) + two (3-gons)	2
$12E'_2$		one (3-gon) + two (6-gons)	$\sqrt{\frac{11}{2}} = 2.34520788$
$24E'_1$		One 4-gon from 6 (4-gon) set, one 3-gon from 8 (3-gon) set, and two 4-gons from 12 (4-gon) set	$\sqrt{5+2\sqrt{2}} = 2.79793265$
$24E'_2$	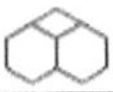	One (4-gon), and two (6-gons)	$\sqrt{10} = 3.16227766$
$24E'_3$		one (3-gon), and two (8-gons)	$\sqrt{7+4\sqrt{2}} = 3\ 65764729$
$30E'$		two (5-gons), and two (3-gons)	$1+\sqrt{5} = 3.236068$
$48E'$		One 8-gon, one 6-gon, and one 4-gon	$\sqrt{13+6\sqrt{2}} = 4.635221825$
$60E'_1$		one (5-gon), and two(6-gons)	$\sqrt{\frac{1}{2}(29+9\sqrt{5})} = 4.9560373$
$60E'_2$		One 5-gon, one 3-gon, and two 4-gons	$\sqrt{11+4\sqrt{5}} = 4.465901$
$60E'_3$		One 3-gon and two 10-gons	$\sqrt{\frac{1}{2}(37+15\sqrt{5}} = 5.038898$
$120E'$		One 10-gon, one 6-gon, and one 4-gon	$\sqrt{31+12\sqrt{5}} = 7.604789$

[I – 3] Layered Diagram and Top-view Diagram

Layered Diagram ---- The layered diagram is a diagram which shows the layers of balls on the surface of a pack. Each layer is a plane perpendicular to this paper and shows as a line perpendicular to the N-S axis of the pack system. The **Fig. I-3a and I-3b** are two typical layered diagrams. **Fig. I-3a** has three layers at the northern hemisphere and three at the southern. **Fig. I-3b** has a single ball B_N at the N-pole, and a single ball B_S at the S-pole, and has two layers at the northern hemisphere, plus a layer at the equator plane. The layers distributed at the southern hemisphere and northern hemisphere are symmetric with respect to the equator plane containing point O, the center of the pack system. The equator plane is also perpendicular to this paper. The r_i is the radius of the i-th layer, and **hi** is the vertical distance between (i-1)-th and i-th layers. There are k_i balls of B_i on the circumference of i-th layer and the same number of k_i balls of B_i' at the i-th layer from S-pole. The layers show only their relative positions.

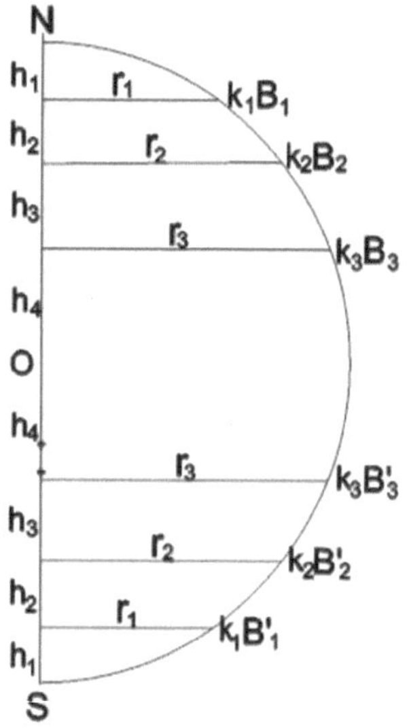

Fig. I-3a

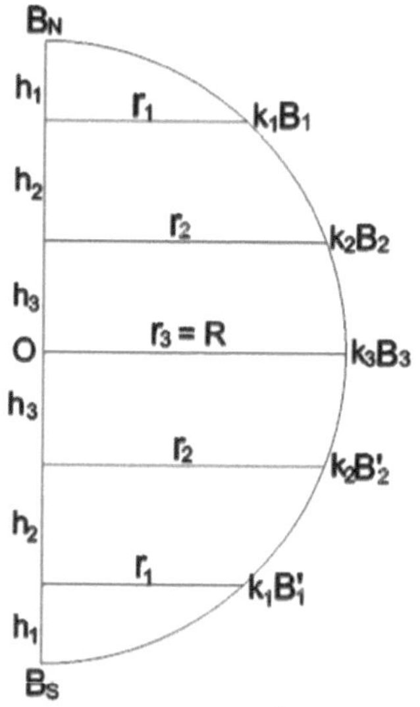

Fig. I-3b

Top-view Diagram ---- The top-view diagram is a bird's eye view diagram from the zenith above the North Pole of a setting.

The Fig. I-4a is the setting of $\mathbf{24E_1'}$ and **Fig. I-4b** is its top-view diagram. In the top-view diagrams, it is clear to see the distances among balls on the same layer or on different layers. The angles between two balls on the same layer as well as on two different layers with respect to N-S axis are also clearly shown. These distances and angles play vital roles in developing equations.

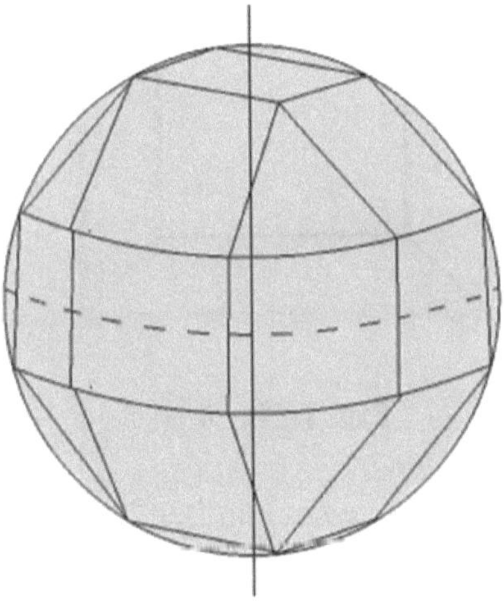

Fig. I-4a. The setting

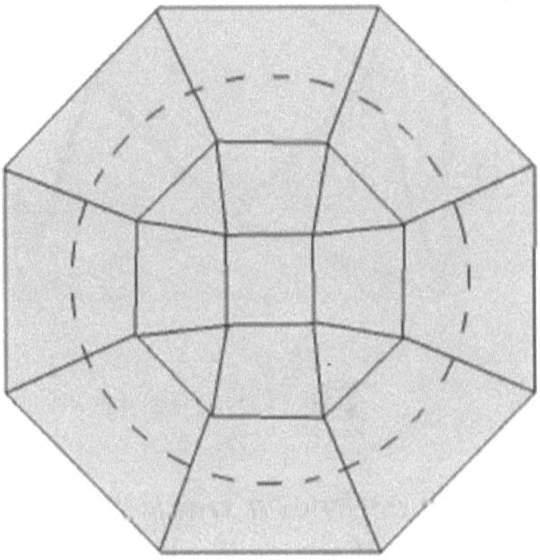

Fig. I-4b. The top-view diagram

In solving an E-pack, the layered diagram and top-view diagram are two very useful tools for developing equations. Sometimes, just draw a village and the layered diagram, and this is enough for setting equations, so the top-view diagram could be eliminated.

The Head Signs ---- Associated with the two diagrams, there are some important head signs for showing the stack arrangements.

The sign "$\sim$", **reads as "switch fitted,"** which means in a Stack, every ball at the upper layer is above the mid-edge of the lower layer. For example, the arrangement $(4 \overset{\sim}{\ } 4)$ as shown in **Fig. I-5a.**

***The sign* "$\_$ ", reads as "straight fitted,"** which means every ball at the upper layer is above a corner ball at the lower layer. For example, the arrangement $(4 \overset{-}{\ } 4)$ as shown in **Fig. I-5b.**

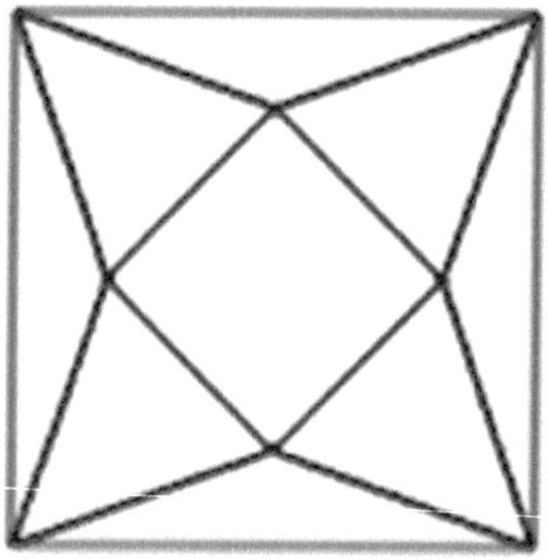

Fig. I-5a. $(4 \overset{\sim}{\ } 4)$

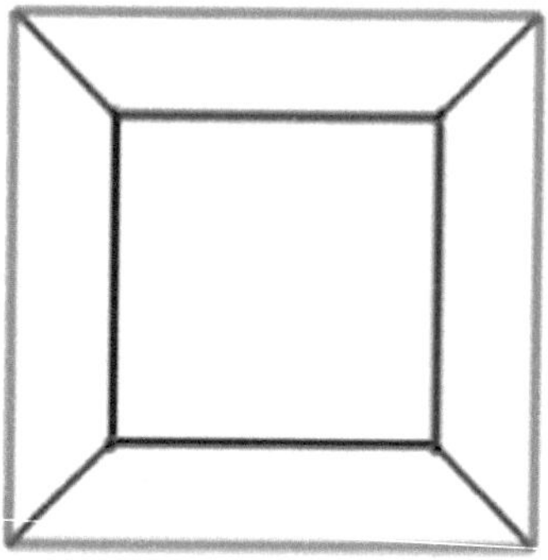

Fig. I-5b. $(4 \overset{-}{\ } 4)$

Sometimes a small "$_2$" appears above an even number 2k, **read as "2-ball united"**. The meaning is this: the layer contains 2k balls united between (in touch with) distance **2**, and they appear as one ball. Such united pairs can be arranged in a switch-fitted or straight-fitted way with its upper or lower layered balls. The following figures, **Fig. I-6a** through **I-6f,** show the different arrangements with the united pairs. (The red edge is enlarged, and the dark one is of length **2.**)

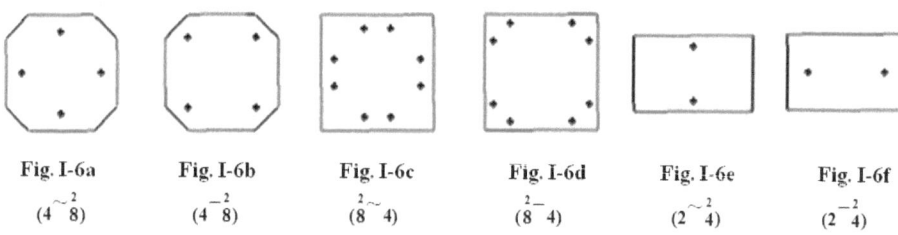

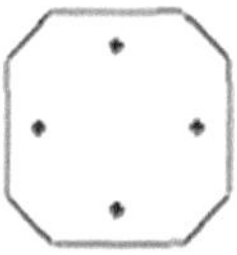

Fig. I-6a

$(4 \overset{\sim 2}{\ } 8)$

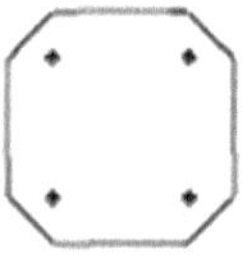

Fig. I-6b

$(4 \overset{-2}{\ } 8)$

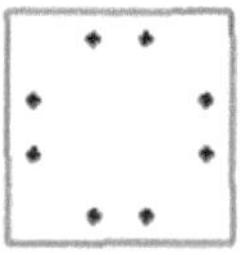

Fig. I-6c

$(8 \overset{2 \sim}{\ } 4)$

Fig. I-6d

$(8 \overset{2 -}{\ } 4)$

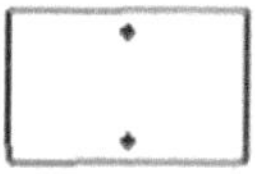

Fig. I-6e

$(2 \overset{\sim 2}{\ } 4)$

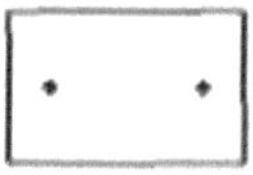

Fig. I-6f

$(2 \overset{-2}{\ } 4)$

In some cases, a small **"$\vee$"** appears above a number, **read as "equal distance apart,"** such as $(2\overset{\sim}{\ }\overset{\vee}{4})$. That is the lower layered four balls are of ***equal distance apart***. The **Fig. I-7a, b, c** show the differences among the arrangements $(2\overset{\sim}{\ }\overset{\vee}{4})$, $(2\overset{\sim}{\ }\overset{2}{4})$, **and** $(2\overset{-}{\ }\overset{2}{4})$.

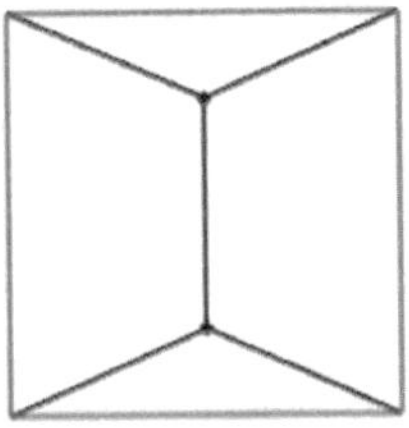

Fig. I-7a

$(2\overset{\sim}{\ }\overset{\vee}{4})$

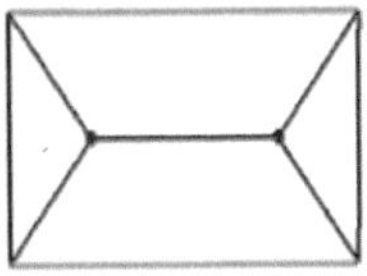

Fig. I-7b

$(2\overset{-}{\ }\overset{2}{4})$

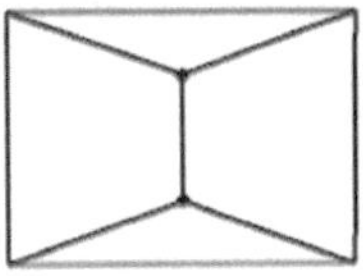

Fig. I-7c

$(2\overset{\sim}{\ }\overset{2}{4})$

The sign "$\vee$" is only to emphasize the case of equal-distance. In some cases, the equal distance is obvious, and this sign can be eliminated.

The Layered Expression The layered and the top-view diagrams are all diagramed depending on how the lattice of the configuration is set. For instance, the 6E can be set as one point at the N-pole, 4 points at the equator plane, and one point at the S-pole. Such a setting can be expressed as the **layered expression** $[1\ \overset{\vee}{4}\ 1]$. Also, the 6E can be set with one triangle face at the top and the opposite triangle face at the bottom. The layered expression would be $[3 \sim 3]$. These two different settings have different layered- and top-view diagrams, and so have different equations. Nevertheless, the final solution of **R, V, Dv, and Ds** would be consistent (except the **ri** and **hi**).

Not only 6E can be expressed as several different layered expressions, actually all E packs and E-extensions can be expressed in some different layered expressions upon different settings. The **[]** of a stack is used to mean the layered expression of the entire system, and the () as the layered stack arrangement on a certain set of base faces of a configuration system. The base face with a stack of all layers of balls on it is called **the stack.** While constructing an E", for instance, add two layers of balls each containing 3 and 6 balls, respectively, on a base of a 6-gon face to form a three-layered stack. There are four possible different arrangements for such 3-6-6 stack. Their layered diagrams are same and thus are eliminated here. Their top-view diagrams are shown below in **Fig. I-8a, b, c, d.** (The colored edges are enlarged and the dark edges are of length 2.)

Fig. I-8a

$(3\,\overset{\sim 2}{}\,6\,\overset{\sim 2}{}\,6)$

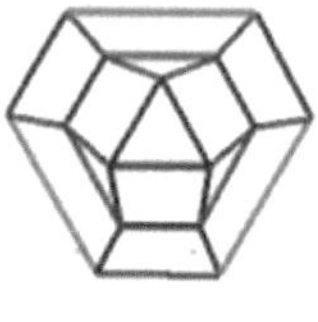

Fig. I-8b

$(3\,\overset{\sim 2}{}\,6\,\overset{-2}{}\,6)$

Fig. I-8c

$(3\,\overset{-2}{}\,6\,\overset{\sim 2}{}\,6)$

Fig. I-8d

$(3\,\overset{-2}{}\,6\,\overset{-2}{}\,6)$

*** For all diagrams throughout this paper, the inner face is the upper layer and the outer face is the lower layer.***

[I – 4] The General Method for Solving E-Packs

Let r_i be the radius of the i-th layer and k_i be the number of balls B_i on i-th layer **(see Fig. I-9a).** If there is only one ball at the very top layer, then it is at the North Pole and the layer is the **0**-th layer with r_0 = **0**. If there is no ball at the N-pole, then the first layer k_1 balls always form a regular k_1-gon with every edge of length **2**. Let h_i be the vertical distance between (i-1)-th and i-th layers, **2d** be the distance between two certain balls on two different layers, and $è_{(i-1),i}$ be the angle of B_{i-1} and B_i with respect to **N-S** axis. Assume the half- edge-length of two immediate layers are $e_{(i-1)1}$, $e_{(i-1)2}$, e_{i1} , and e_{i2} **(see Fig. I-9b).** The half-edge length can be found from the stack formulae as a function of R. (The stack formula will be discussed later.).

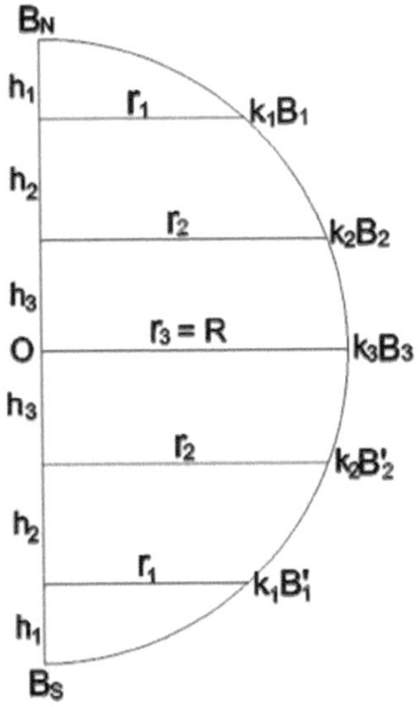

Fig. I-9a. The layered diagram

Calculation:

$$S_{si} = \text{Sin}\left(\frac{\pi}{k_{si}}\right) \quad C_{si} = \text{Cos}\left(\frac{\pi}{k_{si}}\right) \qquad \textbf{(1)}$$

Where $= k_i$ while **k = 3, 4, or 5**; and $k_{si} = k_i /2$ while **k = 6, 8, or 10**.

$$r_i = \left(\frac{1}{S_{si}}\right)\sqrt{e_{i1}^2 + e_{i2}^2 + 2C_{si}e_{i1}e_{i2}} \qquad \textbf{(2)}$$

$$h_i^2 = 4d^2 - r_{i-1}^2 - r_i^2 + 2r_{i-1}r_i\text{Cos}\left(\theta_{(i-1),i}\right) \qquad \textbf{(3)}$$

where

$$\mathrm{Cos}(\theta_{i-1,i}) = \frac{1}{r_{i-1}r_i}\left(\sqrt{r_{i-1}^2 - e_{i-1}^2}\sqrt{r_i^2 - e_i^2} + e_{i-1}e_i\right) \quad (4)$$

According to the layered diagram, that

$$h_i^2 = \left[\sqrt{R^2 - r_{i-1}^2} - \sqrt{R^2 - r_i^2}\right]^2 \quad (5)$$

The "**Net Equation**" of the system is

$$\sum_1^m h_i = R \quad (6)$$

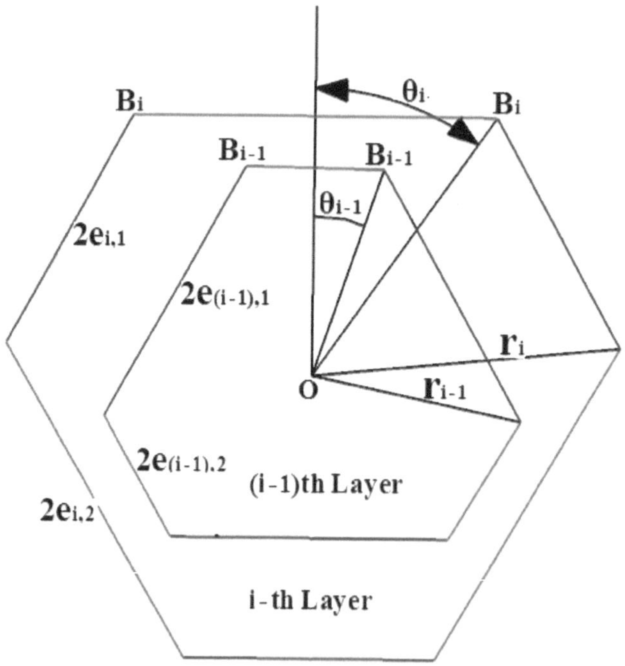

Fig. I-9b. The Top-view Diagram

The Eq. (6) is called the *Net Equation*. While applying all h_i into the **Net Eq. (6)**, the **Eq. (3)** of a certain i should be used at least once.

Solving basic Es and E's

To solve the basic E or E', the first step is to draw its lattice to show how it is set. Next, draw the layered diagram and top-view diagram according to the setting. In some individuals, the lattice already clearly reveals the distances and angles among balls and layers, so just drawing the village with one of the two diagrams will be enough, or even both diagrams can be eliminated. The third step is to set equations according to the diagrams (or the setting, or the village). The final step is to solve them by the general method. All the basic Es, E's, AEs, and TEs are solved by manual calculation for their virtual value of R and the rest of the solution.

Solving most E"s would have to depend on the computer. The process that the computer works on the Net Equation (6) is **"To adjust the value of R in Equation (6), until the left side of the equation is *satisfactorily equal* to the right side, such a value of R with all r_i, h_i, Dv and Ds is the *acceptable solution*."** The notation of such process of finding the value of R is:

$$R \rightarrow Eq.\ (6) : LT = RT$$

(* All basic Es and E's can also be solved by using the total surface angle method, which will be discussed in the next chapter.)

The data of basic Es and E's are in DATAFILE: [DATA 1-1]

[I – 5] Upper Limit of Surface Density and Volume Density

The Upper Limit of Surface Density Ds:

The surface of a container becomes a plane while the container's radius R is infinitely large. Every ball on such a surface can have six immediate neighbor balls around, and all seven balls touch each other (see **Fig. I-10**). The area of $\triangle ABC$ is $\sqrt{3}$. And the occupied area by the three balls inside the triangle $\triangle$ABC is $\frac{\pi}{2}$. So the **upper limit of the surface density** is

$$\mathbf{Ds} = \frac{\pi}{2\sqrt{3}} \doteq \mathbf{90.68996\%}$$

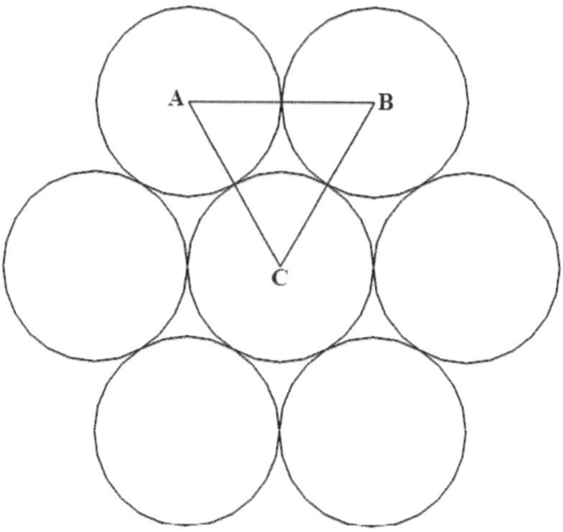

Fig. I-10

The Upper Limit of Volume Density Dv:

The method used here is called the **Cutting-12E$'_1$** to find the upper limit of the volume density of the E-pack. Here, $\mathbf{12\hat{E}'_1}$ means the spherical pack of **12E$'_1$**, and $\mathbf{12\bar{E}'_1}$ means the lattice of **12E$'_1$**. For a large number of balls in a huge container (it does not matter what the shape of the container is), these are naturally arranged in a well-packed manner as an arrangement of **1-on-3** or as a **4E**-pack. In such a pack, every ball is a *central ball* surrounded by 12 balls which are distributed as **12E$'_1$** of radius **R = 2**. The $\mathbf{12\hat{E}'_1}$ contains six □-cone-face and eight △-cone-face, so $\mathbf{12\hat{E}'_1}$ can be cut from the edges of the surface to its center to get 6 □-cone-pyramids and 8 △-cone pyramids (see **Fig. I-11a, b, c and d**). All edges of these pyramids are of lengths 2.

Let $\widehat{V}_\square$ be the volume of the $\square$-cone-pyramid **(See Fig. I-11a)**

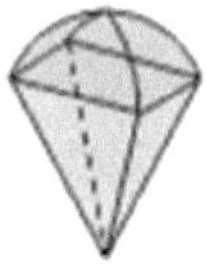

Fig. I-11a

$\overline{V}_\square$ be the volume of the $\square$-flat-pyramid, $V_\square = \frac{4}{3}\sqrt{2}$ **(See Fig. I-11b)**

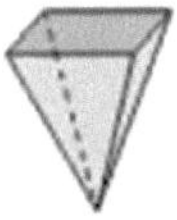

Fig. I-11b

$\widehat{V}_\triangle$ be the volume of the $\triangle$-cone pyramid **(See Fig. I-11c)**

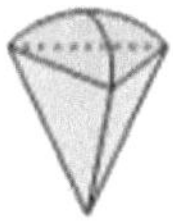

Fig. I-11c

$\overline{V}_\triangle$ be the volume of the $\triangle$-flat-pyramid, $V_\triangle = \frac{2}{3}\sqrt{2}$ **(See Fig. I-11d)**

Fig. I-11d

$\widehat{V}_{12E_1'}$ be the volume of the spherical pack of the $\mathbf{12E_1'}$

$\overline{V}_{12E_1'}$ be the volume of the lattice pack of $\mathbf{12E_1'}$, $V_{12E_1'} = \frac{40}{3}\sqrt{2}$

V_{4b} be the volume of the void apace among 4 balls which stacked as 4E

V_{6b} be the volume of the void space among 6 balls which stacked as 6E

$A_{\square}$ be the plane area of the base of the $\square$-pyramid, (which is the area of a $\square$), $A_{\square} = \mathbf{4}$

$A_{\triangle}$ be the plane area of the base of the $\triangle$-pyramid, (which is the area of a $\triangle$), $A_{\triangle} = \sqrt{3}$

The ratio of $\widehat{V}_{\square}$ and $\widehat{V}_{\triangle}$ is the same as that of $A_{\square}$ and $A_{\triangle}$. That is

$$\frac{\widehat{V}_{\square}}{\widehat{V}_{\triangle}} = \frac{A_{\square}}{A_{\triangle}} = \frac{4}{\sqrt{3}}$$

And

$$\widehat{V}_{12E_1'} = 6\,\widehat{V}_{\square} + 8\,\widehat{V}_{\triangle} = \frac{32}{3}\pi$$

So we have

$$\widehat{V}_{\square} = \frac{8}{9}\left(3 - \sqrt{3}\right)\pi$$

$$\widehat{V}_{\triangle} = \frac{2}{3}\left(\sqrt{3} - 1\right)\pi$$

By definition, the

$$V_{6b} = \bar{V}_{6E} - 6\left({}_1\hat{V}_\square\right) \qquad \text{where} \quad \bar{V}_{6E} = 2\,\bar{V}_\square$$

and

$$\hat{V}_\square = \frac{1}{8}\hat{V}_\square, \qquad \{\text{ that is } \left({}_1\hat{V}_\square\right) = \frac{1}{8}\left({}_2\hat{V}_\square\right)\}$$

Therefore,

$$V_{6b} = 2\bar{V}_\square - 6\left(\frac{1}{8}\hat{V}_\square\right) = \frac{2}{3}\left[4\sqrt{2} - \left(3 - \sqrt{3}\right)\pi\right]$$

The same that

$$V_{4b} = \bar{V}_{4E} - 4\left({}_1\hat{V}_\Delta\right) \qquad \text{where} \quad \bar{V}_{4E} = \bar{V}_\Delta \quad \text{and} \quad {}_1\hat{V}_\Delta = \frac{1}{8}\hat{V}_\Delta$$

So

$$V_{4b} = \bar{V}_\Delta - 4\left(\frac{1}{8}\hat{V}_\Delta\right) = \frac{1}{3}\left[2\sqrt{2} - \left(\sqrt{3} - 1\right)\pi\right]$$

The central ball in $12\hat{E}_1'$ occupies a volume territory **T**, which is its own volume plus 1/6 of six V_{6b} and 1/4 of eight V_{4b}.

So that

$$T = \frac{4}{3}\pi + \frac{1}{6}\left(6V_{6b}\right) + \frac{1}{4}\left(8V_{4b}\right) = 4\sqrt{2}.$$

And the upper limit of Dv is

$$\mathbf{U(D_v) = (\tfrac{4}{3}\pi) / T}$$

which yields

The Upper Limit of the Volume Density $\boxed{U(D_v) = (\frac{\pi}{3\sqrt{2}}) \doteq 74.04804\%}$

(* The upper limit of Dv calculated here agrees with that calculated by Dr. Thomas C. Hales at the University of Michigan at Ann Arbor.)

CHAPTER II

SECOND GENERATION OF EVEN DISTRIBUTION

[II – 1] Second Generation of Even Distribution (E")

The second generations of even distribution (E") are extended by adding some balls on every face of one or two sets of an E or E-extension, then pressing all added balls downward vertically to the base face. In the mean time, the balls of the mother configuration are forced to go outward from the center of the system until all added balls and mother balls reach one surface, which is the surface of the newly formed E" configuration. The arrangements on the same set of faces of the mother configuration should be identical. The E" may contain some non-regular faces. The edges of the base face with loaded balls are enlarged from 2 to 2e. The e is called the enlargement factor, and its length depends on how the added balls are arranged. On a certain base face, there are many ways to add balls on, and the added ball stack can be multi-layered. The newly formed E", if it passes all WB criteria, is a family member of the mother configuration.

The General Characters of the Second Generations

(1) **All balls of an E" are floating on the surface of a spherical container.**
(2) **The lattice of an E" may contain some non-regular faces.**
(3) **All arrangements on a certain set of faces of the mother configuration should be identical.**
(4) **All Villages with their stacks are identical. The village arrangement is the same as that of the mother's.**

The Construction Status of E"

There are three construction status of constructing an E". When balls are only added on one set of faces, it is called the **Single status of construction (S).** If the added balls are on two sets of faces, it is called the combination status of construction, which include the **M-A** status and **S-S** status. It is impossible to add balls on all three sets of faces even if the mother has three.

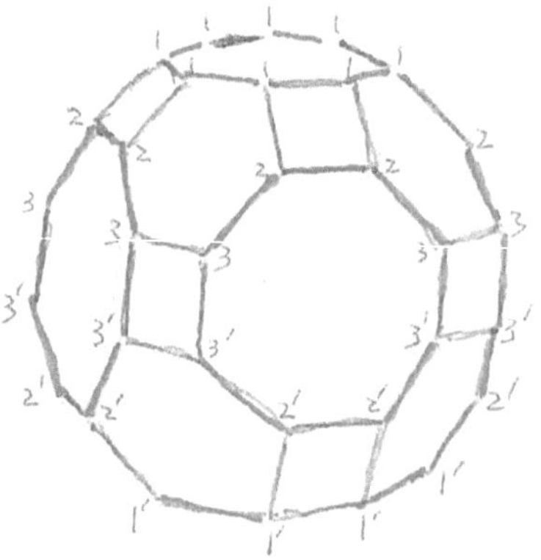

Fig. II-1. The setting of 48E'

Here, adding balls on 48E' was used to illustrate the different status of constructions. Every village of **48E'** contains an 8-gon, a 6-gon, and a 4-gon. The three different edges are colored differently.

(See **Fig. II-1** at right)

(1) ***S-Status of Construction (Single status of construction):*** While balls are added only on the set of six 8-gons, it is a **single arrangement (S status)**. It may be arranged as all edges are equally enlarged or four edges enlarged, and the other four remained as 2. If all edges are equally enlarged, the notation is **(8 S)e** (read as "all 8-gons are arranged as single status of construction with all edges equally enlarged"). If four edges are enlarged and they are adjacent with 6-gons, the

notation is $(8S)_{6\text{-}e}$. If the four enlarged edges are adjacent with 4-gons, the notation is $(8S)_{4\text{-}e}$. **(See Fig. II-2a, 2b, and -2c.)**

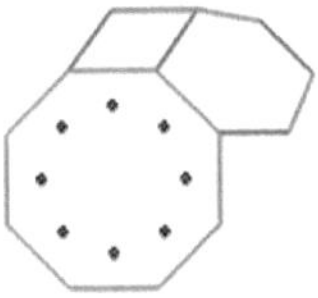

Fig. II-2a. (8 S)e

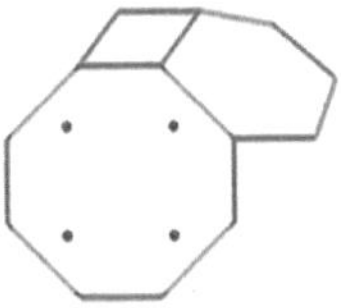

Fig. II-2b. (8 S)6-e

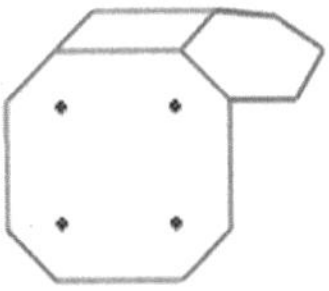

Fig. II-2c. (8 S)4-e

(2) ***M-A status of Construction (Main-Associate status of construction):*** Suppose to add balls on the set of 8-gons to make four edges which are common with 6-gons enlarged to be **2e**, then add some balls on every 6-gon. Since every 6-gon has already had its three edges which adjacent with the 8-gons enlarged to be **2e**, so there are only other three edges left to be enlarged by its own added balls. In such a case, the 8-gons are the **"Main" (M)** faces and the 6-gons are the **"Associate" (A)** faces. The notation is **(8 M)e,1 + (6 A)**. The footnote **"e,1"** means half number of the edges are enlarged, and the other half remained as **1** (the half-edge-length of **2**). Since the A-face is always with half number of edges enlarged by its **M**-neighbors (those edges are called the **M**-edges) and the other half number of edges enlarged by its own added balls (**A**-edges), hence, for an A-face, no footnote is needed.

For an **M-A** arrangement, if the edges of **M**-faces are all equally enlarged and the A-faces are 6-gons **(Fig. II-3a),** the notation is **(8 M)e + (6 A). The Fig. II-3a, 3b, and 3c** show the different cases of **M-A** arrangements.

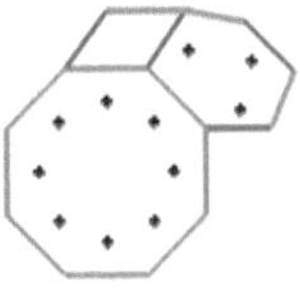

Fig. II-3a
(8 M)e + (6 A)

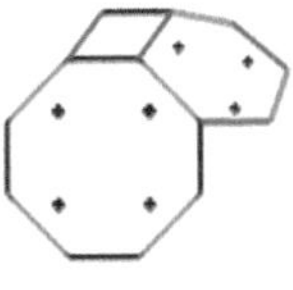

Fig. II-3b
(8 M)e,1 + (6 A)

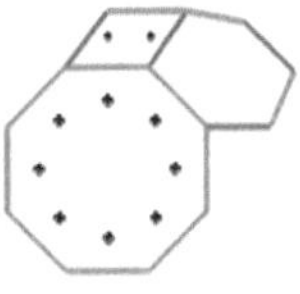

Fig. II-3c
(8 M)e + (4 A).

(3) S-S status of Construction (Single-Single status of construction): For this status, some balls on the faces of the set of 8-gons and some balls on the set of 6-gons are added in a way that their common edges remained as **2,** while the other half number of edges of both 8-gons and 6-gons are enlarged independently by their own added balls. **(See Fig II-4a, b, c.)** In such an arrangement, both the 8-gons and 6-gons are of single status, and the notation of such arrangement would be **(8 S1) + (6 S2)**.

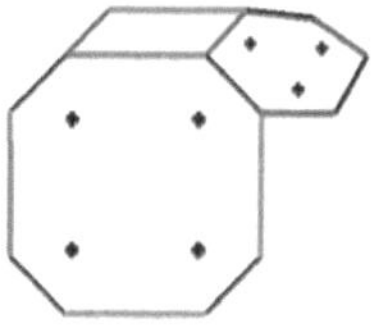

Fig II-4a (8 S1) + (6 S2)

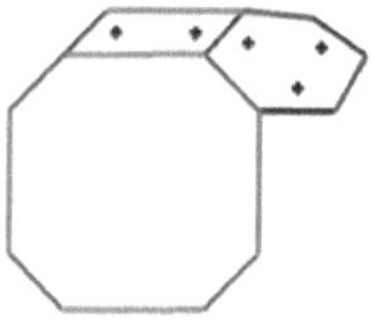

Fig II-4b (6 S1) + (4 S2)

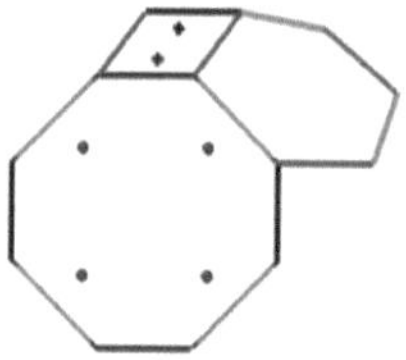

Fig II-4c (4 S1) + (8 S2)

* For all the above figures, the red edges are M-edges, the blue edges are A-edges, and the dark edges are of length 2.
* The dots on some faces of the above figures are just to mean that those faces are loaded with balls.
* Some faces have no balls added on, but their edges are enlarged; that is due to their adjacent faces' enlargement.

[II – 2] Well-Bound Criteria

A good pack should have its every member pass in all of the following five well-bound (WB) criteria.

(1) ***Criterion-a*: Every surface *angle* of a ball with two straight sides of lengths 2 should be smaller than a π.**
(2) ***Criterion-b*: Every ball should have at least three but no more than five *bindings* (the binding is the immediate neighbor ball of distance 2).**

(3) ***Criterion-c*: Every layer of added balls should be *confined* in the range of its base layer.**
(4) ***Criterion-d*: Every ball to any other ball, added or original, should have a *distance* of at least 2.**
(5) ***Criterion-e*: The length of an *edge* should be at least 2. For an enlarged edge, its length will be limited by Criterion-a, -c, or d.**

***Criterion-a (angle)*:**

If a ball with its two immediate balls has a surface angle equal or greater than **180** degrees, then this ball can slip through this large angle. Hence, every ball should have its every surface angle less than 180 degrees. The analyzing of the surface angle with respect to its corresponding straight angle is in **[II – 2a]**

***Criterion-b (binding)*:**

It is obvious that every ball should have a certain number **b** of binding lines to bind it. The number of binding lines is the number of immediate neighbor balls (of distance 2). The number **b** should not be less than **3** or exceeding **5**. If a ball has less than three immediate neighbor balls, say two, then it will be able to go left or right freely. If a ball has **b** > **5**, say **b** = 6, then this ball with its 6 surrounding balls are on a plane, not on a surface, so the central ball can go straight upward or downward freely and cause the entire system to collapse. This criterion usually is obvious for all constructions.

***Criterion-c (confinement)*:**

Every layer of added balls should be **confined** within the range of its base layer. The boundary of the range of the base face is an invisible fence that extends from the center of the system to the edges of the base face, and extending straight outward to the surface. (Note that the fence is not vertical to the base face.) The ground edge of the boundary of the base face could be smaller than that of the added layer of balls, because the base face is closer to the center of the system. The analysis is in section **[II-2c]**.

***Criterion-d (distance)*:**

Every ball on the surface should have distance at least **2** to any other ball. In other words, every ball should have all its straight angles at least **60** degrees. Calculating the distance between two balls will be discussed in **[II – 2d]**.

***Criterion-e (edge length)*:**

While constructing an **M-A** constitution, in some cases, the **M**-edges are too long; so the **A**-edges would be too short, shorter than **2** or even not existed. Or, if the **M**-edges are too short, then the stack on **A**-face will be too long and not well confined in the range above its base face. **Therefore, for an M-A construction, the associate faces should have their associate *edges* of lengths at least 2.** Also for the string -E configurations, the edge length of the P-face is limited for the strings' existence due to Criterion-a, -c, or -d. (The string-E will be discussed in Chapter IV.)

[II – 2a] The Surface Angle and Its Corresponding Straight Angle

Let B_0, B_1, and B_2 be the three points on the surface of a pack with B_0 at the North Pole as shown in **Fig. II-5**. Let B_1' and B_2' be the vertically projected points of B_1 and B_2 on the equator plane of axis OB_0.

The surface angle of a point is the angle of its corresponding straight angle projected vertically onto its equator plane.

The straight angle of B_0 is $\bar{\phi} = \angle B_1B_0B_2$, and its corresponding surface angle is $\hat{\phi} = \angle B_1'B_0B_2'$.

Calculation:

Let $\overline{B_0B_1} = 2d_1$, $\overline{B_0B_2} = 2d_2$, $\overline{B_1B_2} = 2d_0$.

Let C_1 and C_2 be the centers of the layers containing B_1 and B_2 respectively. So C_1 and C_2 are on the axis OB_0, and

Let $r_1 = C_1B_1$, and $r_2 = C_2B_2$; $h_1 = B_0C_1$ and $h_2 = C_1C_2$; **and** $\angle B_0OB_1 = 2\theta_1$ and $\angle B_0OB_2 = 2\theta_2$.

$$\text{Sin}(\theta_1) = \frac{d_1}{R},\ \text{Sin}(\theta_2) = \frac{d_2}{R} \tag{1}$$

$$r_1 = R \times \text{Sin}(2\theta_1) = \frac{2d_1\sqrt{R^2 - d_1^2}}{R} \tag{2}$$

$$r_2 = \frac{2d_2\sqrt{R^2 - d_2^2}}{R} \tag{3}$$

$$h_1 = R - R \times \text{Cos}(2\theta_1) = \frac{2d_1^2}{R} \tag{4}$$

and $h_1 + h_2 = \dfrac{2d_2^2}{R}$ $\quad\therefore h_2 = \dfrac{2}{R}\left(d_2^2 - d_1^2\right)$ (5)

Also $h_2^2 = 4d_0^2 - r_1^2 - r_2^2 + 2r_1r_2\text{Cos}(\widehat{\phi})$ (6)

Since $\text{Cos}(\overline{\phi}) = \dfrac{d_1^2 + d_2^2 - d_0^2}{2d_1d_2}$ (7)

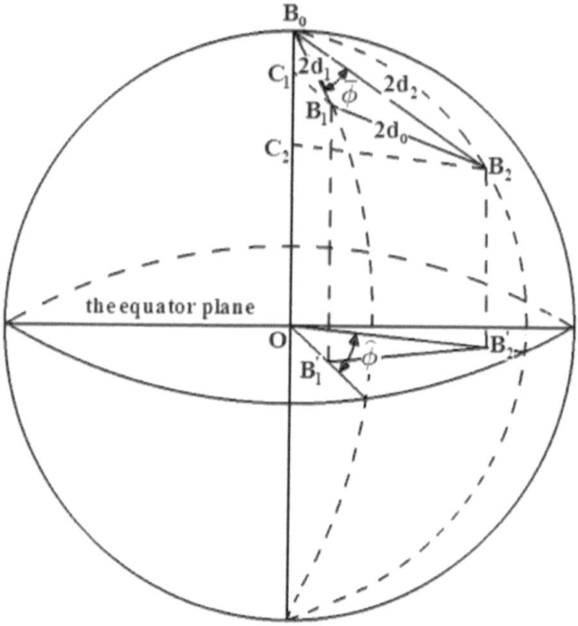

Fig. II-5

From all equations above, it yields

$$\boxed{\mathrm{Cos}\left(\hat{\phi}\right)=\frac{\left(d_1^2+d_2^2-d_0^2\right)R^2-2d_1^2d_2^2}{2d_1d_2\sqrt{\left(R^2-d_1^2\right)\left(R^2-d_2^2\right)}}} \qquad (8)$$

The Eq. (8), can be represented as $\hat{\phi}=[d_1, d_2, d_0]$

In many cases, $d_1=d_2=d$**. Combining Eq. (7) and (8) yields**

$$\boxed{\mathrm{Sin}\left(\frac{\hat{\phi}}{2}\right)=\alpha\times\mathrm{Sin}\left(\frac{\bar{\phi}}{2}\right) \quad \text{where} \quad \alpha=\frac{R}{\sqrt{R^2-d^2}}} \qquad (9)$$

The α is called the "surface index". The Equations (8) and (9) are called the *Surface Angle formulae.*

And correspondingly, the Eq (7) is called the *straight angle formula.*

[II – 2b] *Criterion-b*:

On a surface, a ball Bo has b, (say, b = 3), binding balls B1, B2, and B3. The three binding balls form a 3-corner for the central ball Bo, which gives the name of the location of Bo as **3C.** Obviously, a 3C has three angles called the **three 3C angles** (usually they refer to straight angles if not specified). Every angle of the three 3C angles should be < 180 deg. and > 60 deg. **The total angle measurement of the three 3C angles is always smaller than 2π. [For b = 4, the total angle measurement of four 4C angles should be < 2π. For b = 5, the total angle measurement of five 5C angles should be < 2π.]**

[II – 2c] *Criterion-c*:

The added layer of balls should be above and confined in the range of its base face.

Suppose adding a layer **La** of 6 balls on a base layer **Lb**, also of 6-balls **(see Fig. II-6a, the top-view diagram).** Let **O** be the center of the pack, o_1 and o_2 be the center of the layers **La** and **Lb,** respectively. (On the top-view diagram, **Fig. II-6a,** these three points are merged as one point on point **O**. On the layered diagram **(Fig. II-6b,** these three points form a straight line.) Let e_1, e_2, e_3 and e_4 be the half length of each edge as shown in **Fig. II-6a**; and let M_1, M_2, M_3, and M_4 be the midpoints on each edge. Let r_1 and r_2 be the radius of **La** and **Lb.**

Calculation:

$$S = \mathrm{Sin}(\frac{\pi}{k_s}) \quad C = \mathrm{Cos}(\frac{\pi}{k_s}) \quad [\text{while } k = 3, 4, 5,\ k_s = k; \quad \text{while } k = 6, 8, 10,\ k_s = \frac{k}{2}]$$

$$r_1 = \frac{1}{S}\sqrt{e_1^2 + e_2^2 + 2Ce_1e_2} \qquad r_2 = \frac{1}{S}\sqrt{e_3^2 + e_4^2 + 2Ce_3e_4} \quad (1)$$

Let $\beta_1 = \angle M_1 O O_1$, $\beta_2 = \angle M_2 O O_1$, $\beta_3 = \angle M_3 O O_2$ and

$\beta_4 = \angle M_4 O O_2$ (See Fig. II-6b. the side view)

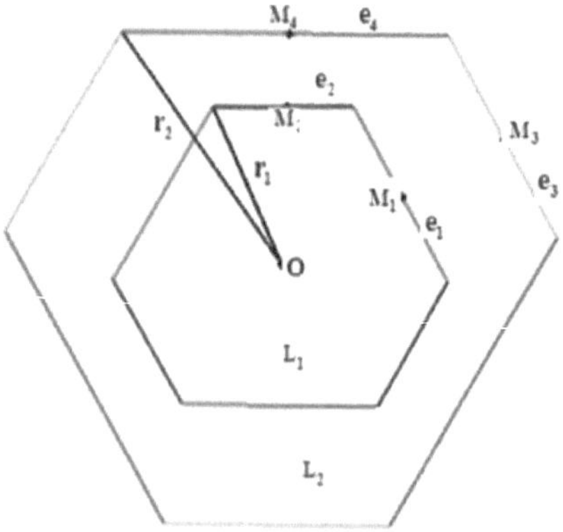

Fig. II-6a. Top-view

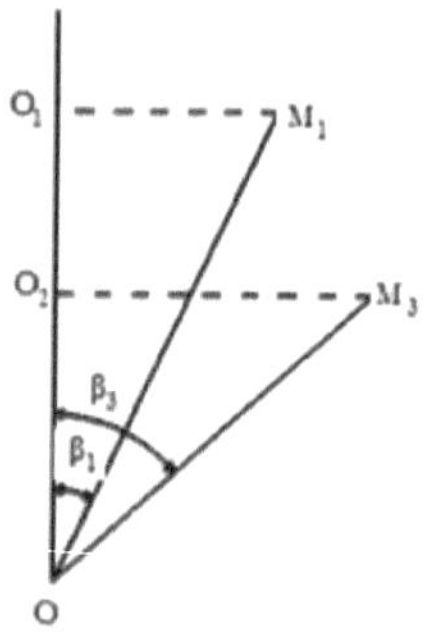

Fig. II-6b. Side view

It is obvious that:

If $\beta_1 < \beta_3$, the M_1 is inside the border of the edge of M_3.

If $\beta_1 = \beta_3$, the three points M_1, M_3, and O form a straight line.

The two end balls of the line with midpoint M_1 are on the *boundary* of the line with midpoint M_3. In such a case, the two end balls can run out freely.

If $\beta_1 > \beta_3$ the two balls on the added layer are out of range.

Hence, for confinement situation, it needs both

$$\beta_1 < \beta_3 \text{ and } \beta_2 < \beta_4 \text{, or}$$

$$\mathrm{Sin}(\beta_1) < \mathrm{Sin}(\beta_3) \quad \text{and} \quad \mathrm{Sin}(\beta_2) < \mathrm{Sin}(\beta_4) \qquad (2)$$

***Therefore, for the added layer of balls are confined in the range above its base layer, iff* Fig. II-6b. Side view**

$$\boxed{\frac{r_1^2 - e_1^2}{R^2 - e_1^2} < \frac{r_2^2 - e_3^2}{R^2 - e_3^2} \quad \text{and} \quad \frac{r_1^2 - e_2^2}{R^2 - e_2^2} < \frac{r_2^2 - e_4^2}{R^2 - e_4^2}} \qquad (3)$$

Eq. (3) is called the *formula of confinement.*

*** Note that O_1M_1 could be > O_2M_3, even $\beta_1 > \beta_3$. Or $O_1M_2 > O_2M_4$, even $\beta_2 < \beta_4$. But as far as both $\beta_1 > \beta_3$, and $\beta_2 < \beta_4$, then all balls on the upper layer are confined. The stacks $(3\tilde{\ }3)$ and $(2\tilde{\ }\overset{2}{4})$, in Fig. II-6c, 6d are two obvious examples in this case.**

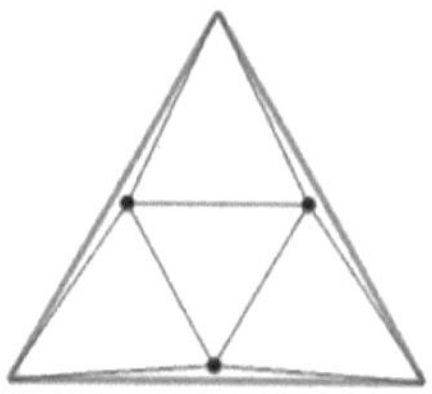

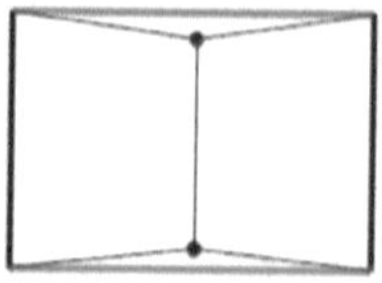

Fig. II-6c. $(3 \overset{\sim}{} 3)$ Fig. II-6d. $(2 \overset{\sim}{} \overset{2}{4})$

[II – 2d] *Criterion-d*:

To find the distance **2d** between two balls on the surface, while d_1, d_2, and $\widehat{\phi}$ are known, it can be directly deduced from the surface angle formula developed in **[II -- 2a]** as

$$\boxed{d = \frac{1}{R}\sqrt{(d_1^2 + d_2^2)R^2 - 2d_1^2 d_2^2 - 2d_1 d_2 \sqrt{R^2 - d_1^2}\sqrt{R^2 - d_2^2}\,\mathrm{Cos}\widehat{\phi}}} \qquad (10)$$

Eq. (10) is called the distance formula and can be represented as $d = [d_1, d_2, \widehat{\phi}]$

$$\textbf{If } d_1 = d_2 = 1\textbf{, then } d = \frac{\sqrt{2(R^2 - 1)}}{R}\sqrt{1 - \mathrm{Cos}\widehat{\phi}} \qquad (11)$$

[II – 3] Stack Formulae

While adding layers of balls on a base face, the edges of the base face are enlarged. To find the enlarged half-edge-length, the first step is to develop equations for every kind of arrangement between two immediate layers of balls. Let the edges of top layer be **E1** and **E2,** and the bottom layer edges be **E3** and **E,** where **E1//E3** and **E2//E.** Their half-edge-lengths are e_1, e_2, e_3, and **e**, respectively. The e_1, e_2, and e_3 are known, and **e** is the length to be found. The **stack formulae** for the length **e** of all kinds of arrangements are developed as following:

The C and S are defined as $C = \cos(\frac{\pi}{k})$ $S = \sin(\frac{\pi}{k})$ **for all following cases**

Case 1: (1 k) **---- k can be 3, 4, or 5 only.**

$$e = \frac{2S\sqrt{R^2-1}}{R}$$

Special Case:  $(1\ \overset{2}{4})$: $e = \frac{1}{R}\sqrt{3R^2-4}$

Case 2: **(k $\overset{\sim}{}$ k) ---- k = 3, 4, 5, 6, 8, or 10**

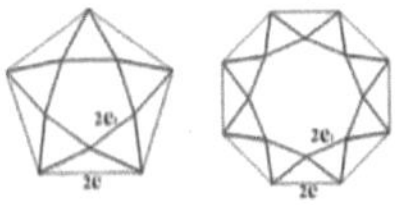

$$e = \frac{1}{R^2 - e_1^2}\left[Ce_1\left(R^2-2\right) + \sqrt{\left(S^2R^2 - e_1^2\right)\left[\left(4-e_1^2\right)R^2 - 4\right]}\right]$$

Special case: $(3 \sim 3)$:  $e = \frac{2R^2-3}{R^2-1}$

Case 3 **(k $\overline{\ }$ k) ---- k = 3, 4, 5, 6, 8, or 10.**

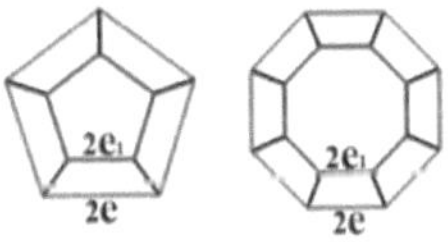

$$e = \frac{1}{R}\left[\left(R^2-2\right)e_2 + \sqrt{\left(S^2R^2 - e_2^2\right)\left(R^2-1\right)}\right]$$

Case 4 **$(k \sim \overset{2}{2k})$ ---- k = 2, 3, 4, or 5.**

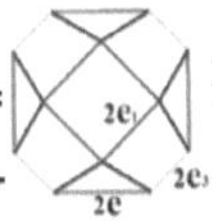

$$e = \frac{1}{-e_1^2}\left[Ce_1(R^2-2-e_1e_3) + \sqrt{(S^2R^2-e_1^2)\{[4-(e_1-e_3)^2]R^2 - 4(e_1e_3+1)\}}\right]$$

While $e_1 = 1,\ e_2 = 0,\ e_3 = 1$: $e = \frac{2}{R^2-1}\left[-C + \sqrt{\left(R^2-2\right)\left(S^2R^2-1\right)}\right]$

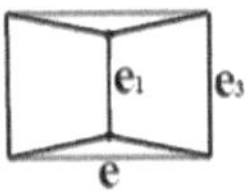

Special Case $(2 \sim \overset{2}{4})$ **:** $e = 2\sqrt{\frac{R^2-2}{R^2-1}}$

Case 5 $(k\ \bar{}\ \overset{2}{2}k)$---- k = 2, 3, 4, or 5.

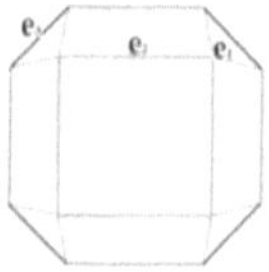

$$e = -Ce_3 + \frac{1}{R}\left[\left(R^2-2\right)e_2 + \sqrt{\left(S^2R^2-e_2^2\right)\left[\left(4-e_3^2\right)R^2-4\right]}\right]$$

While: $e_1 = 0,\ e_2 = e_3 = 1$: $e = -C + \left[\left(R^2-2\right) + \sqrt{\left(S^2R^2-1\right)\left(3R^2-4\right)}\right]$

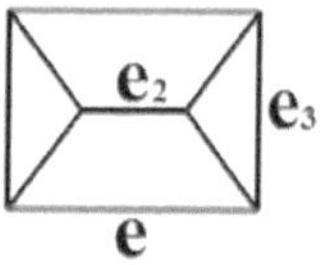

Special Case: $(2\ \bar{}\ \overset{2}{4})$ $e = -C + \left[\left(R^2-2\right) + \sqrt{\left(R^2-1\right)\left(3R^2-4\right)}\right]$

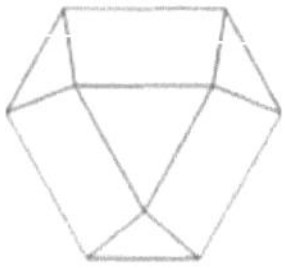

Special Case: $(3\ \bar{}\ \overset{2}{6})$ $e = \frac{2\left(R^2-2\right)}{R^2}$ (e1=0, e2=1, e3=1)

Case 6 $(\check{2}k\ k)$ ---- k = 3, 4, 5, 6, 8 or 10

$e_1 = e_2 = 1$ *for all k in this case*

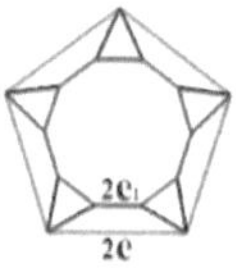

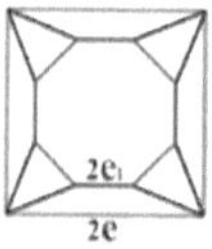

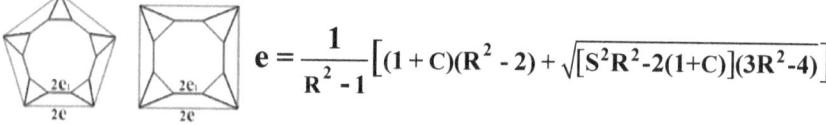

$$e = \frac{1}{R^2-1}\left[(1+C)(R^2-2) + \sqrt{[S^2R^2-2(1+C)](3R^2-4)}\right]$$

Case 7 **(2k k) ---- k = 3, 4, 5, 6, 8, or 10.**

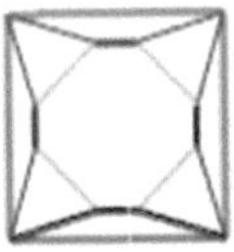

$$e = \frac{1}{R^2 - e_1^2}\left[(ce_1 + 1)(R^2 - 2) + s\sqrt{(R^2 - r_1^2)\left[(4 - e_1^2)R^2 - 4\right]}\right]$$

$$\text{where } r_1 = \frac{1}{s}\sqrt{e_1^2 + 1 + 2ce_1}$$

Case 8 **($2\overset{2}{k}\ \bar{}\ k$) ---- k = 3, 4, 5, 6, 8 or 10.**

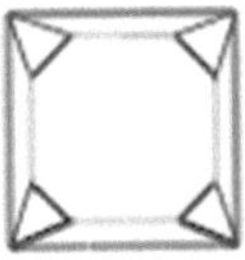

$$e = \frac{1}{R^2 - 1}\left[(c + e_2)(R^2 - 2) + s\sqrt{(R^2 - r_1^2)(3R^2 - 4)}\right]$$

$$\text{where } r_1 = \frac{1}{s}\sqrt{1 + e_2^2 + 2ce_2}$$

Case 9 **($k^{\sim}\ \check{2}k$) ---- k = 2, 3, 4, or 5**

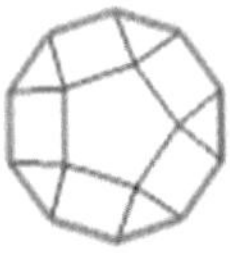

$$e = \frac{1}{2R^2 - e_1^2}\left[\sqrt{1 + c^2}\,(R^2 - 2)e_1 + \sqrt{(s^2R^2 - e_1^2)\left[(8 - e_1^2)R^2 - 8\right]}\right]$$

** ***For this case, actually only*** ($2^{\sim}\ \check{4}$) ***is possible. Its stack formula is***

$$e = \frac{1}{2R^2 - 1}\left[(R^2 - 2) + \sqrt{(R^2 - 1)(7R^2 - 8)}\right] \text{ ------------ (9)}$$

The Universal Stack Formula

The first eight stack formulae can be generalized as one formula, (see Fig. II-7).

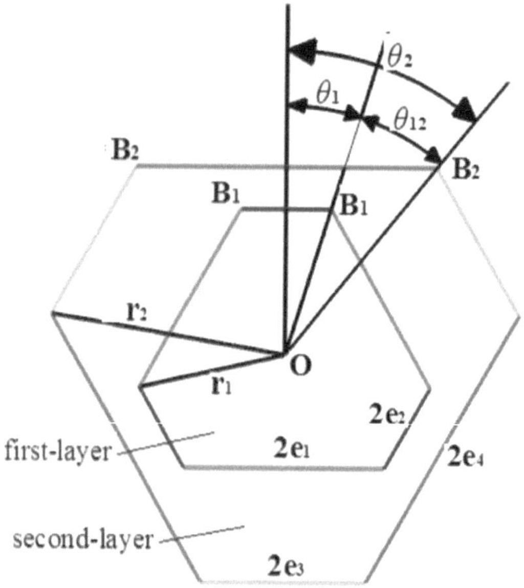

Fig. II-7

While k = 3, 4, 5: $k_s = k$ **; while k = 6, 8, 10:** $k_s = \frac{k}{2}$

$$S = \mathrm{Sin}(\frac{\pi}{k_s}) \quad C = \mathrm{Cos}(\frac{\pi}{k_s}) \tag{1}$$

$$r_1 = \frac{1}{S}\sqrt{e_1^2 + e_2^2 + 2Ce_1e_2}$$

$$r_2 = \frac{1}{S}\sqrt{e_3^2 + e_4^2 + 2Ce_3e_4} \tag{2}$$

The e_1, e_2, and e_3 are known, and **e** (e_4) is to be found.

The angle functions are

$$\mathrm{Sin}\left(\theta_1\right)=\frac{e_1}{r_1} \quad \text{and} \quad \mathrm{Sin}\left(\theta_2\right)=\frac{e_3}{r_2} \qquad (3)$$

Let **2d** be the length of B_1B_2 and **h** be the distance between the two layers.

$$h^2=4d^2-r_1^2-r_2^2+2\mathrm{Cos}\left(\theta_1-\theta_2\right)\equiv\left(\sqrt{R^2-r_1^2}-\sqrt{R^2-r_2^2}\right)^2 \qquad (4)$$

Solving above equations, we have the formula for the enlargement factor e (or e_4) as

$$e=-Ce_3+\frac{S}{A}\left[B+\sqrt{B^2-AD}\right]$$
$$\text{where } W=R^2-2d^2-e_1e_3$$
$$A=R^2-e_1^2 \qquad (5)$$
$$B=W\sqrt{r_1^2-e_1^2}$$
$$D=W^2-\left(R^2-r_1^2\right)\left(R^2-e_3^2\right)$$

The Formula (5) is called the *Universal Stack Formula*, which covers all the first eight formulae except the special case $(2^{\sim}\check{4})$. The e of the special case $(2^{\sim}\check{4})$ is

$$=\frac{}{2R \quad 1}\left[\left(R^2-2\right)+\sqrt{\left(R^2-1\right)\left(7R^2-8\right)}\right] \qquad (9)$$

Examples of How to Use the Universal Stack Formula

To find the half-edge-length of a k-gon or 2k-gon base face, the first step is drawing the diagram of the stack—the half-edge-length e is colored red which is the very base layer and is to be found. All other colored edges are > or = 1; dark colored is of length 2. The second step: to set (e_1, e_2, e_3) such that $e_1 // e_3$ and $e_2 // e$. The third step: use the stack formulae or the universal formula to solve layer by layer until the very base layer's half-edge-length e is found.

To set C and S as: $C = \mathrm{Cos}\left(\frac{\pi}{k}\right)$ and $S = \mathrm{Sin}\left(\frac{\pi}{k}\right)$.

Example 1: $(4\ \tilde{}\ 4)$

Here $\mathbf{k = 4}$, and $(e_1, e_2, e_3) = \mathbf{(1, 0, 0)}$ using the stack formula or universal formula to find **e**.

Example 2: $(3\ \bar{}\ 3)$

k = 3, and $(e_1, e_2, e_3) = \mathbf{(0, 1, 0)}$ using the stack formula or universal formula to find **e**.

Example 3: $(1\ 4\ \overset{-\,2}{8})$

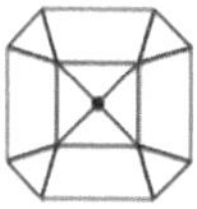

First step: **k = 4**, and $(e_1, e_2, e_3) = \mathbf{(0, 0, 0)}$ to find **e** of the second layer.
Second step: Set $e_2 = e$, $k = 4$, and $(e_1, e_2, e_3) = (0, e_2, 1)$ to find **e** of the third layer.

Example 4: $(\underline{10\ 5}\,\bar{}\,5\,\tilde{}\,\overset{2}{10})$

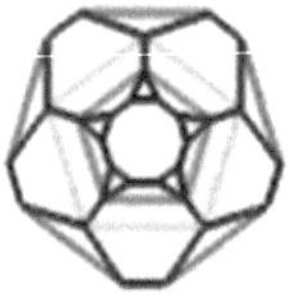

Here **k = 5**, and $(e_1, e_2, e_3) = \mathbf{(1, 1, 0)}$ to find **e** of the 2nd layer.

Next to set $e_2 = e$ and $(e_1, e_2, e_3) = (0, e, 0)$ to find **e** of the 3rd layer.

Finally, to set $\mathbf{e}_1 = \mathbf{e}$ and $(\mathbf{e}_1, \mathbf{e}_2, \mathbf{e}_3) = (\mathbf{e}, 0, 1)$ to find **e** of the very base layer.

Example 5: $(2\ \check{4}^{\sim}\overset{2}{8})$

First step, use the formula **(5)** of the special case of $(2^{\sim}\check{4})$, which is

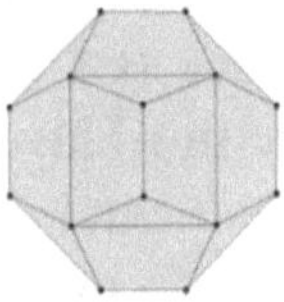

$$\mathbf{e} = \frac{1}{2R^2 - 1}\left[\left(R^2 - 2\right) + \sqrt{\left(R^2 - 1\right)\left(7R^2 - 8\right)}\right]$$

Second step: set $\mathbf{e}_1 = \mathbf{e}$, $\mathbf{k} = 4$, and $(\mathbf{e}_1, \mathbf{e}_2, \mathbf{e}_3) = (\mathbf{e}, 0, 1)$, to use the universal formula to find the final **e**.

[II – 4] Existing Stacks

For the purpose of constructing E"s of the families of Es and E-Extensions, it is necessary to establish a list of existing stacks with base k-gons for all k, where k = 3, 4, 5, 6, 8, and 10. The first step is to create a list for all possible stacks (up to four layers in this study) for a certain k. For example, the very original list of possible stacks for k = 3 are:

1. $(1\ 3)$ 2. $((3^{\sim}3)$ 3. $(3^{-}3)$ 4. $(\underline{6\ 3})^{*1}$ 5. $(1\ 3^{\sim}3)$ 6. $(1\ 3^{-}3)$ 7. $(3^{\sim}3^{\sim}3)$ 8. $(3^{\sim}3^{-}3)$ 9. $(3^{-}3^{\sim}3)$, 10. $(3^{-}3^{-}3)$ 11. $(3^{\sim}\overset{2}{6}^{\sim}3)$ 12. $(3^{\sim}\overset{2}{6}^{-}3)$ 13. $(3^{-}\overset{2}{6}^{\sim}3)$ 14. $(3^{-}\overset{2}{6}^{-}3)$ 15. $(\underline{6\ 3}^{\sim}3)$ 16. $(\underline{6\ 3}^{-}3)$, 17. $(\underline{6\ 3}^{\sim}\overset{2}{6})$ 18. $(\underline{6\ 3}^{-}\overset{2}{6})$ 19. $(\underline{12\ 6\ 3})^{*2}$ 20. $(1\ 3^{\sim}3\ \ 3)$…

*1. A stack of $(\underline{2k\ k})$ with underline means the top layer of 2k balls formed a 2k-gon with every edge equal in length. It can be considered as switch fitted or straight fitted way of stack between two layers.

*2. For all k, it is impossible to stack balls in layers of the form $(\underline{4k\ 2k\ k})$.

After the long list of stacks of a certain base k-gon is made, use the stack formulae or universal stack formula to find every enlarged half-edge- length $\mathbf{e}_i$, one layer by one layer, and check each $\mathbf{e}_i$ through all five WB criteria. For a multi-layered stack, check it one layer by one layer. Any one that does not pass any of the five criteria will be thrown away. Sometimes, this enables a shortcut while checking. Say, you found that $(3\,\tilde{}\,3)$ is a good stack, but if you just add one ball atop of the first layer, as a new stack $(1\ 3\,\tilde{}\,3)$, but you found that the second layer is out of range of the third layer, so you throw $(1\ 3\,\tilde{}\,3)$ away from the possible list. And you can be sure that all three layered stacks with last two layers as $(\ldots 3\,\tilde{}\,3)$ are no good. You can save a lot of time. After filtering out all bad stacks, there are only six existing for **k = 3**. **They are ## 1, 2, 3, 4, 8, and 16.** For **k = 3**, there is no good stacks for more than three layers (include the base layer). And for **k = 3**, the number of possible stacks is the least among all k.

Some stacks can only serve as one or two status of the three construction status, **(S, M, A)**, so it should be considered for the availability of every existing stack while serving in different status. For a prime numbered k-gon face, while balls are added on, all edges are equally enlarged, so it cannot serve as an A-face; it can only serve as S- or M-face. For the stacks $(k\,\tilde{}\,k)$ that all k edges of the base layer are also equally enlarged; so this kind of switch-fitted stacks also cannot serve in **A-status.**

Some stacks, such as $(3\,\overset{-2}{6}\,\overset{-2}{6})$, cannot pass the **criterion-A** (see **Fig. II-8a**). In such case, it does not need to use the surface angle formula to check criterion-A, because the balls in the mid-layer, each has three straight angles of 60°, 90°, and a θ, where θ is obviously $> 180^{\circ}$, so, its corresponding surface angle is even more greater than 180°. It tells that the stack $(3\,\overset{-2}{6}\,\overset{-2}{6})$ cannot serve as **S**- or **M**-status. It can serve as **A**-status only if the blue edge is lengthened enough to have the surface angle $\widehat{\theta}$ lessened to be $< 180^{\circ}$ **(see Fig. II-8b)**. Thus, the stack $(3\,\overset{-2}{6}\,\overset{-2}{6})$ is only qualified to be listed in the existing list as "possible" for serving as **A**-status.

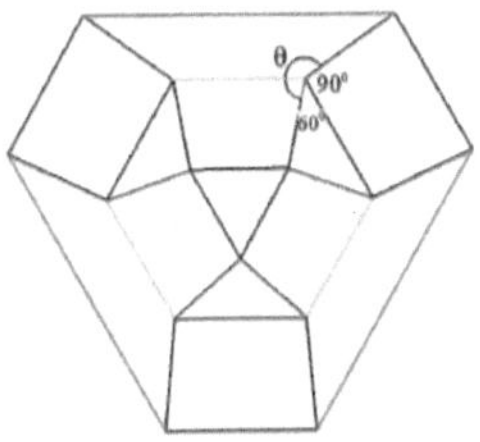

Fig. II-8a. The stack

$(3\ \overline{6}^{2}\ \overline{6}^{2})$ as S- or M-status

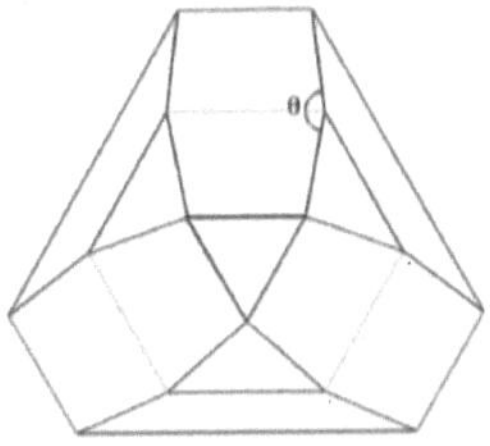

Fig. II-8b. The stack

$(3\ \overline{6}^{2}\ \overline{6}^{2})$ as A-status

There is a certain style of multi-layered stacks as

$\mathbf{(k^{\sim}\ 2k^{2}\ \overline{2k}^{2}\ \overline{2k}^{2} \ldots\ldots \overline{2k}^{2}\ \overline{2k}^{2})}$ **; or** $\mathbf{(2k^{2}\ \overline{2k}^{2}\ \overline{2k}^{2}\ \overline{2k}^{2} \ldots\ldots \overline{2k}^{2}\ \overline{2k}^{2})}$**, where k = 3, 4, 5**

This style of stacks that can be stacked one layer upon one layer *endlessly.* And so, the pack carrying such stack can be an infinite ball-packing. Also, this is a style of arrangement in the category of string-E packing (SE), which will be discussed in Chapter IV.

To establish the lists of existing stacks for all k, where k = 3, 4, 5, 6, 8, and 10: The procedures of making a list of existing stacks for all k are same, except the possible list, while k is greater than 3, will be much longer, and so needs more patience to do the filter job. The one passed all criteria will be recorded in the list of existing-stacks. Some stacks can only survive under certain conditions, such as in a certain R range, or can only serve in a certain construction status, then the existing condition should be remarked.

The existing stacks for k = 3, 4, 5, 6, 8, and 10, up to four layers (including the base layer), are listed in **APPENDIX.**

[II – 5] Construction of a New E"

In order to construct a new E" for a family of a certain mother configuration, there are several steps shown in the examples in the next section. The most important step is to check a newly constructed system through all WB criteria for its existence. There are only five WB criteria, but for some construction, there are many different surface angles or distances need to be checked. Also, the existing stacks of M- or S-status are already passed the five WB criteria, but for the A-edges in a M-A construction or the two M1- and M2-edges in S1-S2 construction, due to their individual situation of enlarged edges, there would be more angles or distances needed to be checked. The following is a checklist for these two constructions:

The Criteria for M-A Status:

(The definitions of many different edges and angles are shown in II-9a and -9b below.)

(1) $E_a \geq 1$
(2) A-face confined
(3) $\hat{U}_m < \pi$
(4) $\hat{U}_a < \pi$
(5) $\hat{V}_a < \pi$ **if** $E_{a1} > 1$
(6) $\hat{V}_m < \pi$ **if** $E_{m1} > 1$
(7) $d_a \geq 1$ **(da is the distance of two balls across the edge Ea)**
(8) $d_m \geq 1$ **(dm is the distance of two balls across the edge Em)**

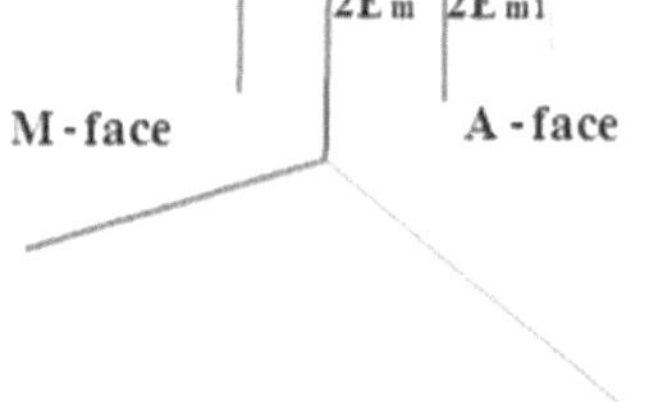

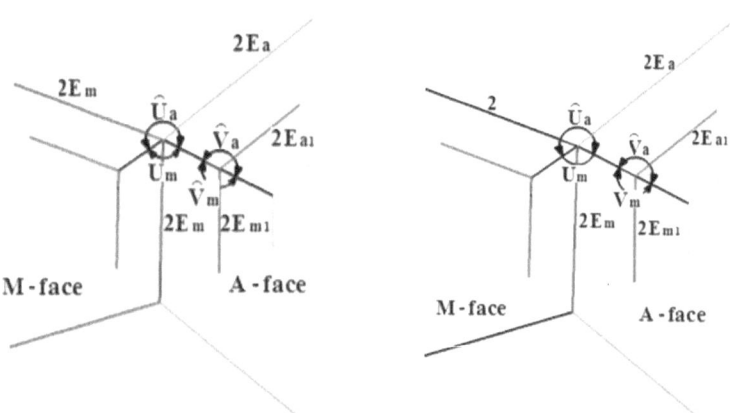

Fig. II-9a. While all edges of M-faces are equally enlarged.

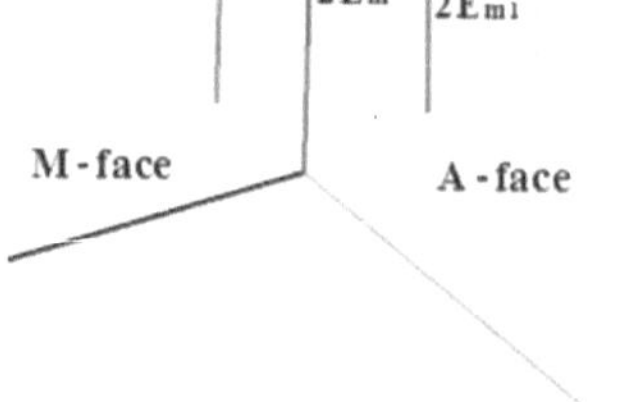

Fig. II-9b. The M-face has half number of edges enlarged, and the others remained as 2.

The Criteria for S1-S2 Status:

(1) $\hat{U}_a < \pi$
(2) $\hat{V}_{11} < \pi$ **if** $E_{11} \geq 1$
(3) $\hat{V}_{12} < \pi$ **if** $E_{12} \geq 1$
(4) $\hat{V}_{21} < \pi$ **if** $E_{21} \geq 1$
(5) $\hat{V}_{22} < \pi$ **if** $E_{22} \geq 1$

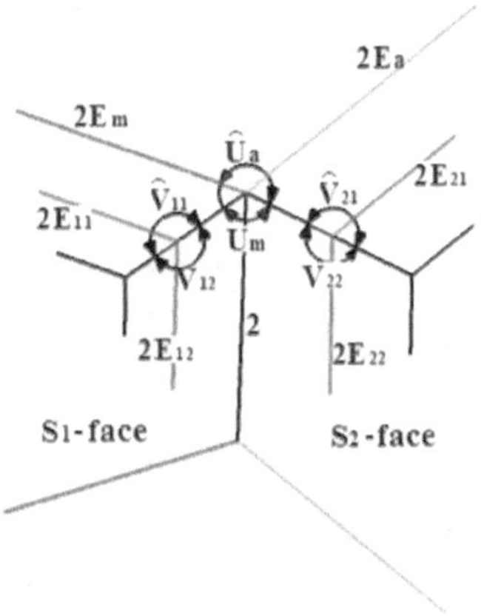

Fig. II-9c.
The adjacent edge between S1 and S2 is always of length 2.

*** For M-status or S-status, the criteria (2), (3), (4), and (5) should be passed. Since all existing stacks for serving as M- or S-status have already been checked under these criteria, so while constructing S-S configurations, only criteria (1) need to be checked.**

Examples of Creating a New E":

***Example 1*: To Create E"s for the Family of 48E'**

***(Step 1)* Draw the setting of mother configuration 48E' and its village (Fig. II-10a and -10b). The three edges, Edge-X, -Y, and -Z, each of length 2x, 2y, and 2z, are colored red, green, and purple, respectively. After adding balls on the faces, the lengths of x, y,**

and z are determined by the stack formulae according to their arrangements.

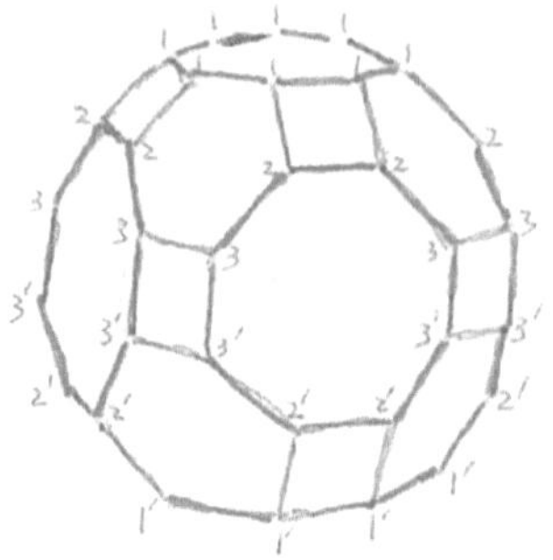

Fig. II-10a. The setting of 48E'

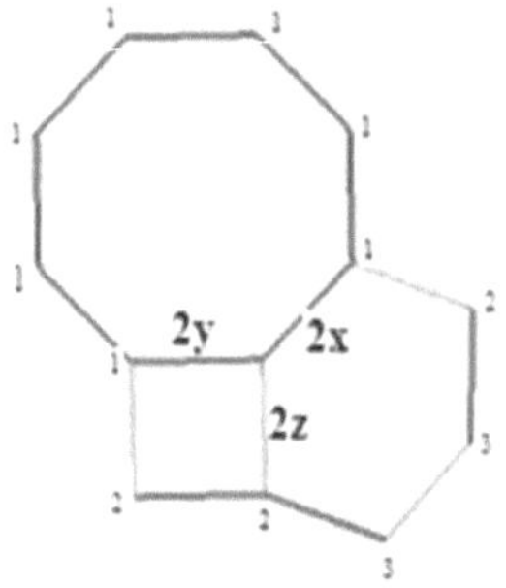

Fig. II-10b. The Village of 48E'

(Step 2) **Develop equations according to the setting and the village:**

$$\mathbf{k = 4,}\ S = \operatorname{Sin}\left(\frac{\pi}{k}\right),\ C = \operatorname{Cos}\left(\frac{\pi}{k}\right) \tag{1}$$

$$r_1 = \frac{1}{S}\sqrt{x^2+y^2+2Cxy} \tag{2}$$

$$r_2 = \frac{1}{S}\sqrt{z^2+(x+y)^2+2Cz(x+y)} \tag{3}$$

$$r_3 = \frac{1}{S}\sqrt{y^2 + \left(z + \sqrt{2}x\right)^2 + 2Cy\left(z + \sqrt{2}x\right)} \quad (4)$$

$$h_1 = R - \sqrt{R^2 - r_1^2} \quad (5)$$

$$h_2^2 = 4y^2 - r_1^2 - r_2^2 + 2r_1r_2\mathrm{Cos}\left(\theta_1 - \theta_2\right) \quad (6)$$

where $\mathrm{Sin}\left(\theta_1\right) = \frac{z}{r_1}$, $\mathrm{Sin}\left(\theta_2\right) = \frac{z}{r_2}$ (7)

$$h_3 = \sqrt{R^2 - r_2^2} - \sqrt{R^2 - r_3^2} \quad (8)$$

$$h_4 = z \quad (9)$$

$$R = \sum_{i=1}^{4} h_i \quad (10)$$

Solve R by R à Eq. (10): LT = RT

While x = y = z = 1, we have the solution (by manual calculation) of the mother 48E' as:

$$R = \sqrt{13 + 6\sqrt{2}} \doteq 4.63522183$$

$$r_1 = \sqrt{2\left(2 + \sqrt{2}\right)} \doteq 2.6131259, \quad h_1 = \sqrt{13 + 6\sqrt{2}} - \sqrt{9 + 4\sqrt{2}} \doteq 0.8068$$

$$r_2 = \sqrt{2\left(5 + \sqrt{8}\right)} \doteq 3.9568742, \quad h_2 = \sqrt{2} \doteq 1.41421356$$

$r_3 = \sqrt{6\left(2+\sqrt{2}\right)} \doteq 4.5260668$ $h_3 = \sqrt{2} \doteq 1.41421356$

$h_4 = 1$

Dv = 26 823 % Ds = 55.8552 %

(*Step 3)* Make a list of all possible single and combination arrangements for 48E':

There are nine groups for a single arrangement:

(1) 6 (8 S)e
(2) 6 (8 S)6-e
(3) 6 (8 S)4-e
(4) 8 (6 S)e
(5) 8 (6 S)8-e
(6) 8 (6 S)4-e
(7) 12 (4 S)e
(8) 12(4 S)8-e
(9) 12(4 S)6-e

There are twelve groups of M-A constitutions:

(10) 6 (8 M)e + 8 (6 A)
(11) 6 (8 M)e + 12 (4 A)
(12) 6 (8 M)e,1 + 8 (6 A)
(13) 6 (8 M)e,1 + 12(4 A)
(14) 8 (6 M)e + 12(4 A)
(15) 8 (6 M)e + 6 (8 A)
(16) 8 (6 M)e,1 + 12(4 A)
(17) 8 (6 M)e,1 + 6 (8 A)
(18) 12 (4 M)e + 6 (8 A)
(19) 12 (4 M)e + 8 (6 A)
(20) 12 (4 M)e,1 + 6 (8 A)
(21) 12 (4 M)e,1 + 8 (6 A)

There are three groups of S1-S2 constitutions (the adjacent edge is of length 2):

(22) 6 (8 S1) + 8 (6 S2) **(23) 8 (6 S1) + 12 (4 S2)** **(24) 12(4 S1) + 6 (8 S2)**

Totally, there are twenty-four cases for creating E" for the 48E' family. Each group may contain a number of from a few to more than 3,000 different arrangements of E"s.

(Step 4) **Choose a certain arrangement from above list and find the enlarged edge-lengths x, y, z**

Suppose the single arrangement of #1, 6(8S)e, is chosen, so look through the list of 8-gon in Existing Stacks. There are seven possible stacks for serving in **S**-status with all edges equally enlarged. They are

3. (8 8)e 6. (16 8)e 21. (8 8 8)e 23. (8 8 8)e 34. (16 8 8)e 38. (16 16 8)e 184. (16 16 16 8)e

Take them one by one to use the stack formulae (or universal formula) to find the three enlarged edges **x, y, z.**

Suppose the arrangement # 17, 8 (6M)e,1 + 6 (8A), is chosen, and there are 22 possible M-stacks in the list of 6-Gon, and 61 possible A-stacks in the List of 8-Gon are found. So to combine every M-stack from the 22 with every A-stack from the 61, there are (22 × 61) = 1342 possible combined constitutions. For each one, use the stack formula or universal formula to find the three enlarged edge-lengths **x, y,** and **z.**

(Step 5) **Solving:**

For every individual constitution, put x, y, and z into the equations developed in step 1 and calculate the value of R by:

R ➔ Eq. (9): LT = RT

(Step 6) **Check through all WB Criteria:**

This is the most important and most complicated step, especially while the configuration is an M-A arrangement, or S-S arrangement.

(Step 7) **Register the good one as an existing member in the family of 48E' in the DATAFILE "DATA [2B-7] 48E' Family."**

Do every individual arrangement **of all twenty-four groups for 48E' in this way from step 1 to step 7.**

Example 2: To Create E"s for the Family of 6E

To create the E"-members for the family of **6E** is very easy and simple. The **6E** has only one set of eight 3-gons, and it can only make an S- arrangement on every 3-gon. The list of 3-gon contains the least number of existing stacks, only six. All six are qualified for **S** constructions. So do the seven steps as shown above for each one. Establishing the whole family can be done in five minutes. **(See DATAFILE "[DATA 2A] The Es and Families")**

To make families for all five nEs is just as simple as that.

Using TSA Method in Calculation

Using the general method to solve basic Es, E's, and E"s is generally applied. Using the **total surface angle (TSA)** method, upon the fact that the total surface angle of a point on a surface is always 2π, to solve the basic Es and E's would be easier than using the general method. However, it requires that all straight angles in a village are known and are special angles. This method has another advantage that while in solving some complex structures, such as T24E and T60E (to be discussed in the next chapter), it is the only useful tool for finding the virtual value of R, even if it is a very time-consuming task.

The following is an example of using TSA method to solve 24E'2 : The village of **24E'2** contains two hexagons and one square intersected at the central point as shown in **Fig. II-11.** The three straight angles of the central ball are 120°, 120°, and 90°. Let their corresponding surface angles be $2\widehat{\theta_1}$, $2\widehat{\theta_1}$ and $2\widehat{\theta_2}$ respectively.

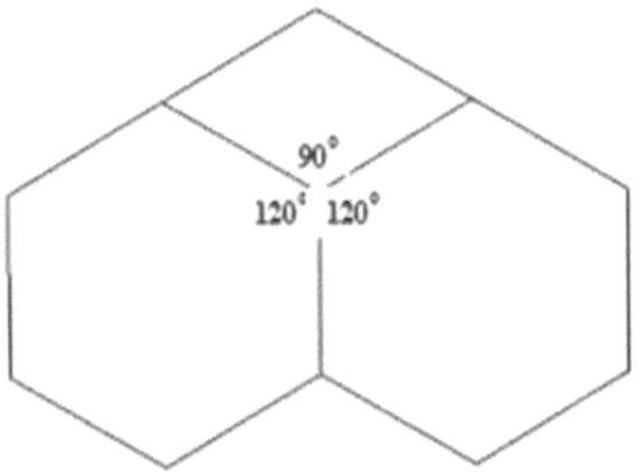

Fig. II-11. The Village of 24E'2

Solution:

$$120^{\circ} = 2\mathrm{Sin}^{-1}[\frac{\sqrt{3}}{2}\alpha] = 2\hat{\theta}_1 \qquad \text{where } \alpha = \frac{R}{\sqrt{R^2 - 1}}$$

$$90^{\circ} = 2\mathrm{Sin}^{-1}[\frac{\sqrt{2}}{2}\alpha] = 2\hat{\theta}_2$$

$$4\hat{\theta}_1 + 2\hat{\theta}_2 = 2\pi$$

$$\mathrm{Sin}(2\hat{\theta}_1) = \mathrm{Sin}(\pi - \hat{\theta}_2)$$

$$\vdots$$

$$2\left(\frac{\sqrt{3}}{2}\alpha\right)\sqrt{1 - \frac{3}{4}\alpha^2} = \frac{\sqrt{2}}{2}\alpha$$

Which yields: $R = \sqrt{10}$

* The value of **R** calculated by using **TSA** method (manual calculation) agrees with the result calculated by using the general method.

* All basic Es, E's, and TE's can be solved by the **TSA** method. In fact, using the TSA method to solve the basic configurations is even easier than using the general method.

The Name of an E-pack

There are three parts of the name of an E-pack. The first part is the family name. The second part shows how it is constructed. The third part tells what the arrangement is involved in the configuration. For example, an **E"** has a name as

$$\mathbf{48E' \ [6(8\ M)e +8(6\ A)]\ [21.\ (8\,\tilde{}\,8\,\tilde{}\,8)_e + 33.\ (\underline{12\ 6}\,{}^{-}\overset{2}{6})].}$$

The first part: ***48E'*** identifies that this **E"** is a member in the family of **48E'**.

The second part: [6(8 M)e +8(6 A)] indicates that the six 8-gons are serving as M-status with all 8 edges equally enlarged, and, the set of eight 6-gons are serving as A-status.

The third part: [21. $(8\,\tilde{}\,8\,\tilde{}\,8)_e$ **+ 33.** $(\underline{12\ 6}\,{}^{-}\overset{2}{6})$**]** tells that all 8-gons are arranged as the stack #**21** in the existing 8-gon stacks, which is $(8\,\tilde{}\,8\,\tilde{}\,8)_e$. And all 6-gons are arranged as the stack **#33** in the existing 6-gon list, which is $(\underline{12\ 6}\,{}^{-}\overset{2}{6})$.

It seems that the name is too long, but it gives very clear characteristic identification. One can look at it and draw the entire construction mentally (or on paper), and working on it can be easy. Sometimes, the second part can be eliminated, or even the # of the third part can be eliminated as well, so that the above configuration can be simplified as:

$$\mathbf{48E'\ [(8\,\tilde{}\,8\,\tilde{}\,8)_e + (\underline{12\ 6}\,{}^{-}\overset{2}{6})].}$$

[II – 6] The Amicable Even Distribution (AE):

There is a special category of E" called Amicable Even Distribution, which is constituted with two even distributions n_1E and n_2E having a mutual property that n_1E contains n_1 number of points and n_2 number of all regular k_1 - gon faces, and vice versa for the n_2E. These two configurations can be combined as

one configuration. (See the figures of five Es below.) The way of combining them is to add n_1 balls on n_2E with one ball on every face of n_2E and press them inward to create a new E". This new E" can be created in the same way as adding n_2 balls on n_1E. The new configuration is called the amicable E of n_1E and n_2E and is denoted as A (n_1+n_2) E. The component pair, n_1E and n_2E, are called Amicable pair. There are only three amicable pairs: 4E and 4E combined as A8E; 6E and 8E combined as A14E; and 12E and 20E combined as A32E. The A8E is in fact of the same constitution as 8E; so when we say AnE, we usually mean A14E or A32E. Note that the two component Es each has the same number of edges.

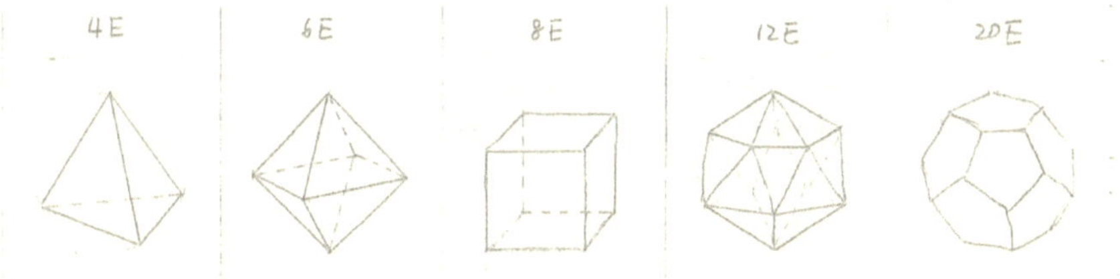

Fig. II-12. The five basic nEs

When adding n_1 balls on n_2E, both edges of and n_2E are enlarged to be $2e_1$ and $2e_2$, respectively. If $n_1 < n_2$, then $e_1 > e_2$. The two edges from the two components are at perpendicular orientation, but the longer one is under the shorter one, so on the surface, only the shorter ones are revealed. The number of faces of the new $A(n_1+n_2)E$ is n_1k_2, and the three edge-lengths are 2, 2, and $2e_2$, (while $e_1 > e_2$). Calculate R for $A(n_1+n_2)E$ through the following:

Let R_{n2} be the radius of the component n_2E, and S_2 = Sin(Pi / k_2). From the stack formula, we have $e_2 = \dfrac{2S_2\sqrt{R_{n2}^2 - 1}}{R_{n2}}$. And since

all edges of n_2E are equally enlarged, so $e_2 = \frac{R}{R_{n2}}$, which yields the radius R of the AnE as

$$R = \sqrt{2\left[S_2^2R_{n2}^2 + \sqrt{S_2^2R_{n2}^2\left(S_2^2R_{n2}^2 - 1\right)}\right]} \qquad \left[\text{or } R = \sqrt{2\left[S_1^2R_{n1}^2 + \sqrt{S_1^2R_{n1}^2\left(S_1^2R_{n1}^2 - 1\right)}\right]}\,\right]$$

***For A8E*: The R is the same as the R of 8E, which is**

$$\boxed{R = \sqrt{3} \doteq 1.7320508}$$

***For A14E:* It has 24 non-regular triangle faces of edges 2, 2, and 2 e_2, and R is:**

$$\boxed{R_{A14E} = \sqrt{3+\sqrt{3}} \doteq 2.1753277...} \quad \textbf{with } e_2 = 1.255926$$

***For A32E:* It has sixty nonregular triangle faces of edges 2, 2, and 2^{e_2}, and R is**

$$\boxed{R_{A32E} = \frac{1}{2}\sqrt{3\left(5+\sqrt{5}\right)+\sqrt{6\left(25+11\sqrt{5}\right)}} = 3.12084625...} \quad \textbf{with } e_2 = 1.113587$$

[II – 7] The Individual Families of E's

The calculation (using the general method) is shown below. The data is in DATAFILE: [DATA 2B] The E' and Families.

(The TSA method is not shown because the general method has to be performed while solving E"s.)

(1) The 12E'1 and Family

The values of x and y are determined by the stack formulae.

$$k = 2 \quad S = \mathrm{Sin}(\frac{\pi}{k}) = 1 \qquad C = \mathrm{Cos}(\frac{\pi}{k}) = 0$$

$$r_1 = \sqrt{x^2+y^2} \quad r_2 = \sqrt{2(R^2-r_1^2)}$$

$$h_1 = R - \sqrt{R^2-r_1^2} \quad h_2^2 = 4y^2 - r_1^2 - r_2^2 + 2r_1r_2$$

$$h_3 = \sqrt{R^2-r_2^2} \quad R = h_1 + h_2 + h_3$$

R ➔ : LT = RT

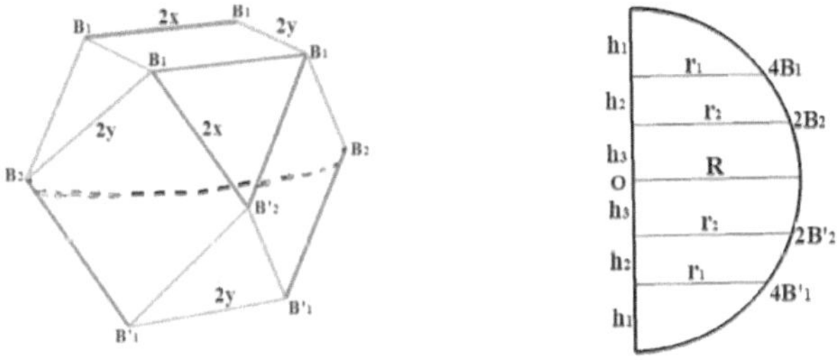

The setting and the layered diagram of 12E'1

For x = y = 1, we have the solution for the original 12E'1 as:

R = 2, Dv = 44.44444%, Ds = 75%

The 12E'1 family groups are:

(I)	8(3S)	(V)	8(3M)+6(4A)
(II)	4(3S)	(VI)	4(3M)+6(4A)
(III)	6(4S)e	(VII)	4(3S)+4(3S)
(IV)	6(4S)e,1	(VIII)	4(3M)+(4A)

The data of 12E'1 and its family are in the DATAFILE: [DATA 2B-1] The 12E'1 and Family.

(2) The 12E'2 and Family

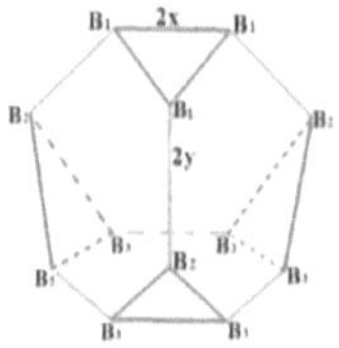

Front view of the setting

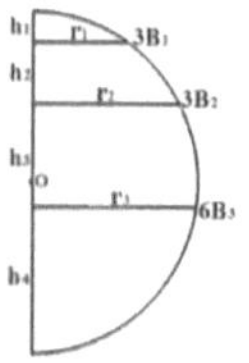

Layered diagram

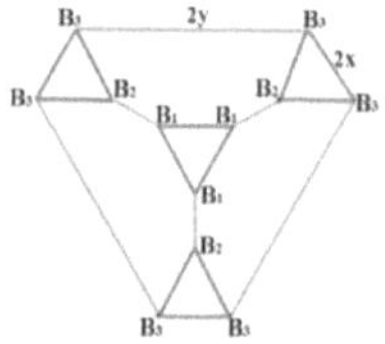

Top view

Calculation: The values of x and y are determined from the stack formulae upon the arrangement:

(1) $k = 3, \quad C = \text{Cos}\frac{\pi}{k}, \quad S = \text{Sin}\frac{\pi}{k}$

(2) $r_1 = \frac{x}{S} \qquad r_2 = \frac{x+y}{S} \qquad r_3 = \frac{1}{S}\sqrt{x^2+y^2+2Cxy}$

(3) $h_1 = R - \sqrt{R^2 - r_1^2} \qquad h_2^2 = 4y^2 - (r_2 - r_1)^2 \qquad h_3^2 = 4x^2 - r_2^2 - r_3^2 + 2r_2\sqrt{r_3^2 - x^2} \qquad h_4 = R - \sqrt{R^2 - r_3^2}$

(4) $R = \sum_{i=1}^{4} h_i$

Solve R by R → (4): LT = RT

For x = y = 1, we have the solution of the basic 12E'2 :

$$R = \sqrt{\frac{11}{2}} \doteq 2.34520788 \text{ Ds} = 54.54545\%$$

The family members of 12E'2 are grouped as:

(I) 4(3S) (II) 4(6S)e (III) 4(6S)3-e (IV) 4(6S)6-e (V) (3M)+(6A)

The data of 12E'2 and its family are in the DATAFILE: [DATA 2B-2] The 12E'2 and Family.

(3) The 24E'1 and Family

The mother 24E'1 contains six squares (distributed as 6E), eight triangles (distributed as 8E), and twelve squares (distributed as 12E'1). The eight triangles could be considered as two sets, each contains four 3-gons distributed as 4E. It is very complicated while adding balls on four different sets of faces of a configuration. There are three different family groups of E"s: Family A, Family B, and Family C.

***Family A*: In this family group, the six 4-gons are all squares of edge-lengths 2x, and the eight 3-gons are considered as one set with all edges of equal lengths, 2z. Hence, every 4-gon in the set of twelve 4-gons becomes rectangles of edge-lengths 2x and 2z. The Family A actually is a special case of Family B while x = y. So the calculation part is included in that for Family B.**

There are eight groups of E" members in Family A (red edge-length = 2x = 2y, green edge-length = 2z):

(I) 8(3S)e	[z = E3M, x = y = 1]
(II) 6(4S)e	[x = y = E4M, z = 1]
(III) 12(4S)e	[x = y = z = E4M]]
(IV) 12 $(4S)_{3 at e}$	[z =E4M, x = y = 1]
(V) 12 $(4S)_{4 at e}$	[x = y = E4M, z = 1]
(VI) 8(3M)e + 12(4A)	[z = e, x = y = E4A]
(VII) 6(4M)e + 12(4A)	[x = y = e, z = E4A]
(VIII) 8 $(3S)_z$ + 6 $(4S)_{x, x}$	[x = y = E4M, z = E3M]

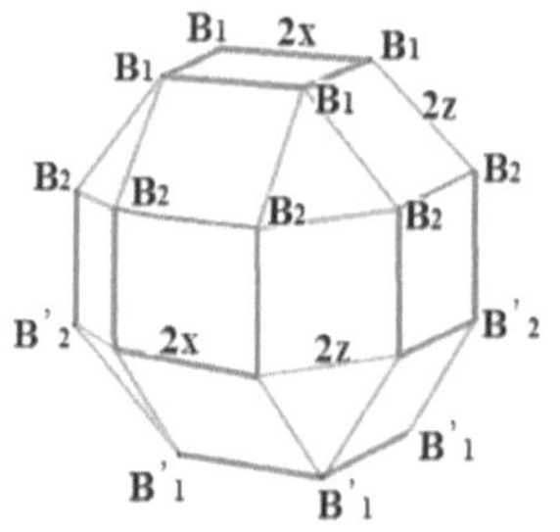

The 24E'1 Family A
(red edge = 2x = 2y, green edge = 2z)

Family B: **The six squares after certain arrangement of added balls become rectangles of edges 2x and 2y. The eight 3-gons are considered all in one set and each has edge-length 2z. The twelve squares become rhombuses with two parallel edges of lengths 2x and 2y; the two waist edges are of lengths 2z. (See Fig. 24E'1 Family B at right.) Actually, Family B is the general case, and Family A is a special case of family B while x = y**

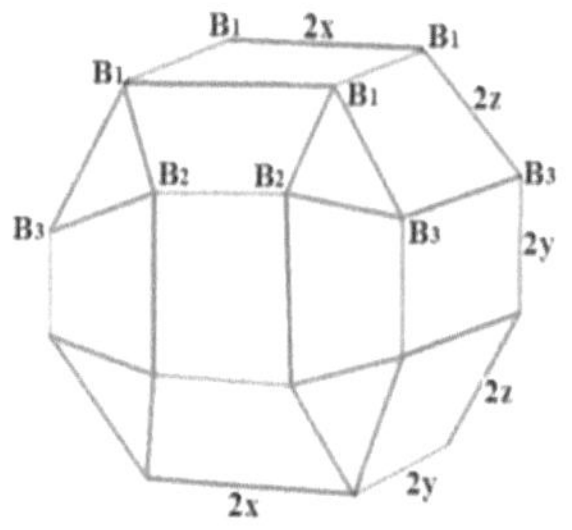

The 24E'1 Family B

(red edge = 2x, green edge = 2y, purple edge = 2z)

Calculation:

k = 2 C = 0, S = 1 **(1)**

$$r_1 = \sqrt{x^2 + y^2}, \quad r_2 = \sqrt{R^2 - x^2}, \quad r_3 = \sqrt{R^2 - y^2} \qquad (2, 3, 4)$$

$$h_1 = R - \sqrt{R^2 + r_1^2} \quad (5)$$

$$h_2^2 = 4z^2 - r_1^2 - r_2^2 + 2r_1 r_2 \mathrm{Cos}\theta_{12} \quad (6)$$

where $\mathrm{Cos}\theta_{12} = \frac{1}{r_1 r_2}[\sqrt{r_1^2 - x^2}\sqrt{r_2^2 - y^2} + xy]$

$$h_3 + h_3 = x \quad h_4 = y \quad (7, 8)$$

$$h_1 + h_2 + h_3 + h_4 = R \quad (9)$$

Solve by R ➔ (9): LT = RT

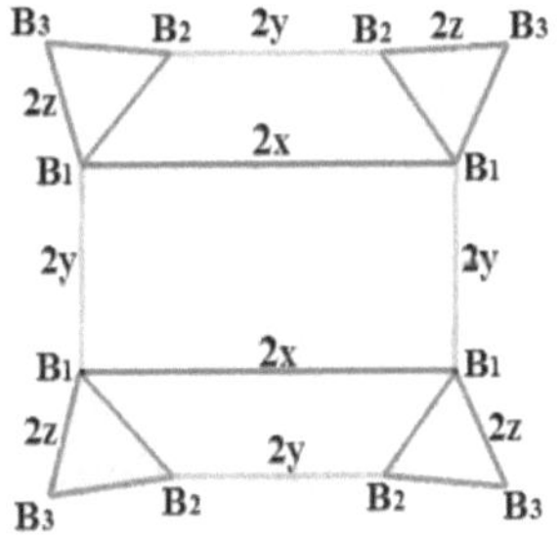

The top-view of 24E'1 Family B

There are two groups of E" members in the Family B:

(I) 6 (4 S)x, y [x > y, z = 1]
(II) 6 (4 S)x, y + 8 (3 S)z

While x = y = 1 yields the R and Ds of 24E'1, which is

$$R = \sqrt{5+2\sqrt{2}} \doteq 2.79793265 \qquad Ds = 76.644\%$$

***To compare the packing-efficiency of the Family-A and Family-B*: (See Data in DATA File : "DATA1-(24E'1)")**

By comparing the data of Family-A and Family-B, the packing of Family-B is much efficient than that of Family-A, (even the members of Family-A looks more normal than that of Family-B).

Example 1:

Family A-1 [6(4-S)e]: [1.(1 4)e] R = 3.359631
Family B-1 [6(4-S)e,1]: [2.(1 4)e,1] R = 3.230079

Example 2:

Family A-8 [8(3-S1) + 6(4-S2)e]: [2.(3 3) + 7.(4 4)e] R = 6.04119
Family B-2 [8(3-S1) + 6(4-S2)e,1]: [2.(3 3) + 8.(4 4)e,1] R = 5.16127

The data (both Family-A and Family-B) is in DATAFILE: [DATA 2B-3] The 24E'1 and Family.

***24E'1 Family-C:* The Family C is while the eight (3-gon)s are considered as two sets of four (3-gon)s.**

Suppose the six 4-gons remain as squares, then the total surface angle of point B1 is different than that of B2. Suppose the diagonal of each of the six 4-gons which connects the two larger triangles is shorter than the other one, as shown in figures below, so it could make the two total surface angles equal.

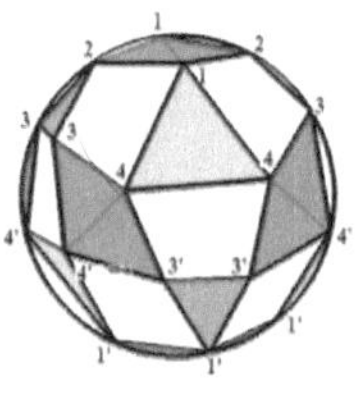

The Configuration of 24E'1-C

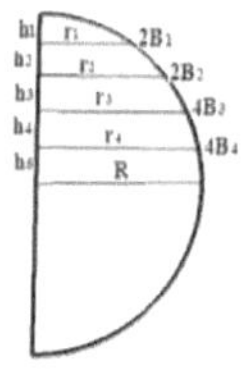

The Layered Diagram

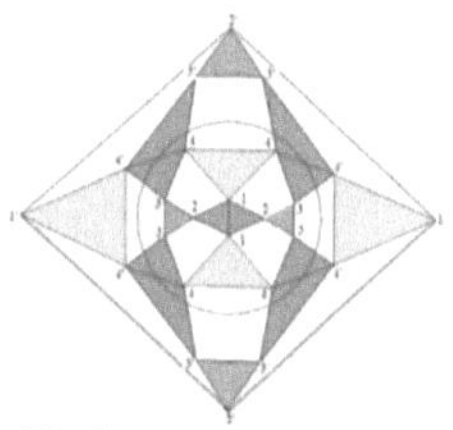

The Top-view Diagram

The Configuration of 24E'1-C The Layered Diagram The Top-view Diagram

Let the half-edge-length of the six 4-gons be 1 (it has to be 1 due to criterion-b) and the half length of the shorter diagonal be d, where $\sqrt{2} \geq d \geq 1$. The half-length x = 1, and y and z are from the stack formulae.

*** Take a close look at the top-view diagram, this figure is exactly the figure of 12E'2 with the four 3-gons the green triangles, and the four 6-gons are as that each one is loaded with a stack $(3\overset{\sim 2}{}6)$ (the purple triangles), so each of the 6-gon has three edges of length 2y and the other three edges of length 2d.**

Therefore the Family-C of 24E'1 is equivalent to 12E'2, and its data is already in the DATAFILE: [DATA 2B-2] 12E'2 and Family.

(4) The 24E'2 and Family

The 24E'2 has six 4-gons and eight 6-gons (where x = y = red, z = green). The calculation:

$k = 4,\quad C = \mathrm{Cos}(\frac{\pi}{k}) = \frac{\sqrt{2}}{2} \qquad S = \mathrm{Sin}(\frac{\pi}{k}) = \frac{\sqrt{2}}{2}$

$r_1 = \sqrt{2}\,x$

$r_2 = \sqrt{2}\,(x + y)$

$h_1 = R - \sqrt{R^2 - r_1^2}$

$h_2 = \sqrt{4z^2 - (r_2^2 - r_1^2)} = \sqrt{2}z$

$h_3 = \sqrt{R^2 - r_2^2}$

$R = h_1 + h_2 + h_3$

The setting of 24E'2

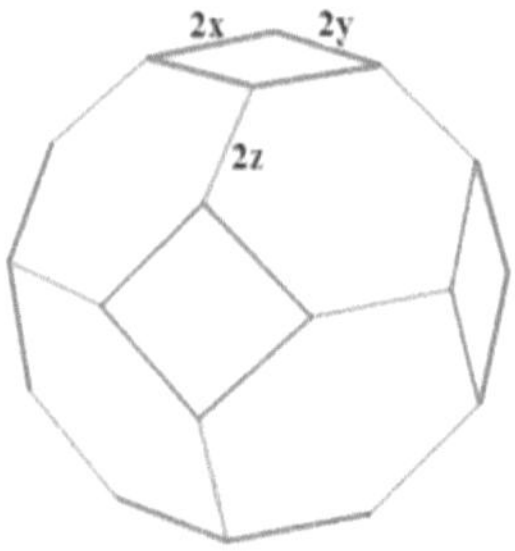

The setting of 24E'2

While x = y = z = 1, the solution of basic 24E'2 is:

$R = \sqrt{10} \doteq 3.16227766$ Dv = 33.28266 Ds = 60%

And $r_1 = \sqrt{2}$ $r_2 = \sqrt{8}$ $r_3 = R$ $h_1 = \sqrt{10} - \sqrt{8}$ $h_2 = \sqrt{2}$ $h_3 = \sqrt{2}$ $h_4 = 0$

While adding balls to create E"s, the 4-gons could be squares or rectangles, and the eight 6-gons can be performed as two sets of four 6-gons (of even or uneven 6-gons). The figures below are the settings of 24E'2. The layered diagram and the top-view diagram with three different edge-length 2x, 2y, and 2z are shown with different colors. The equations can be developed according to the diagrams:

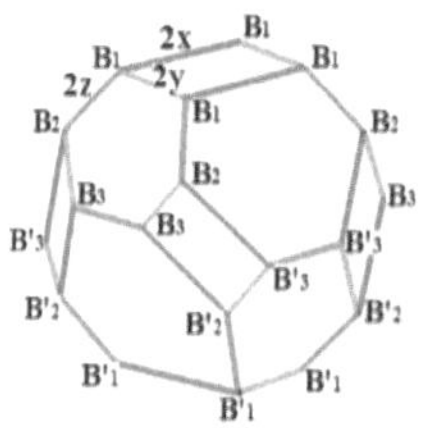

The 24E'2 while x < > y

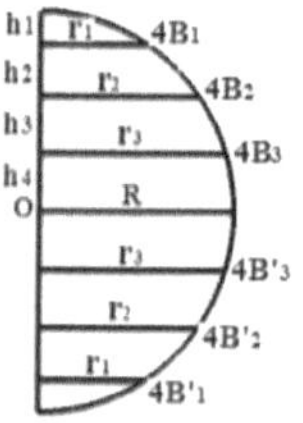

The layered diagram

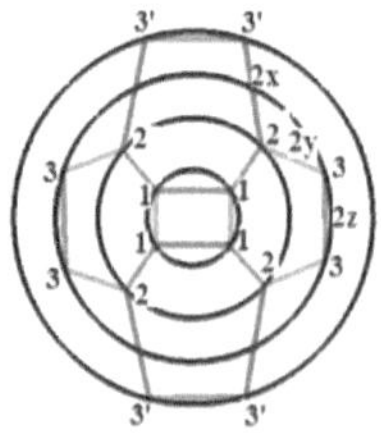

The top-view diagram

The value of x, y, and z are determined from the stack formulae based upon how the added balls are arranged.

$k = 2, S = 1, C = 0$ **(1)**

$r_1 = \sqrt{x^2+y^2} \quad r_2 = \sqrt{(x+z)^2+(y+z)^2}$ **(2), (3)**

$4h_4^2 = 4(x^2 + y^2) - 2r_3^2 + 2r_3^2 \mathrm{Cos}(90^\circ - 2\theta_{o3}) = 4(R^2 - r_3^2)$ **which yields**

$r_3 = \sqrt{[-z+\sqrt{2(R^2-r_1^2)}]^2+z^2}$ **(4)**

$h_1 = R - \sqrt{R^2-r_1^2}$ **(5)**

$h_2 = \sqrt{R^2-r_1^2} - \sqrt{R^2-r_2^2}$ **(6)**

$h_3^2 = 4y^2 - r_2^2 - r_3^2 + 2\left[\sqrt{r_2^2-(y+z)^2}\sqrt{r_3^2-z^2} + (y + z)z\right]$ **(7)**

$h_4 = \sqrt{R^2-r_3^2}$ **(8)**

$R = h_1 + h_2 + h_3 + h_4$ **(9)**

Find R by R → (9): LT = RT

While x = y = z = 1, yields the R of 24E'2 which is

$$R = \sqrt{10} \doteq 3.16227766 \text{ Ds} = 60\%$$

The family groups of 24E'2 are:

(1) six (4S)e	(7) four (6S)4 at e	(13) four (6M)e,1 + six (4A)
(2) six (4S)e,1	(8) four (6S)6 at e	(14) four (6M)e + four(6A)
(3) eight (6S)e	(9) six (4M)e + eight (6A)	(15) four (6M)e,1 +four (6A)
(4) eight (6S)4 at e	(10) six (4M)e + four (6A)	(16) six(4S1) + four (6S2)
(5) eight (6S)6 at e	(11) six(4M)e,1 + four (6A	(17) four (6S1) + four (6S2)
(6) four (6S)e	(12) four (6M)e + six (4A)	

The data of 24E'2 and its family are in the DATAFILE: [DATA 2B-4] The 24E'2 and Family.

(5) The 24E'3 and Family

$$k = 4,\ C = S = \frac{\sqrt{2}}{2}$$

$$r_1 = \frac{1}{S}\sqrt{x^2+y^2+2Cxy} \quad r_2 = \frac{1}{S}(y + \sqrt{2}x)$$

$$h_1 = R - \sqrt{R^2-r_1^2} \quad h_2 = 4x^2 - r_1^2 - r_2^2 + 2r_1r_2\frac{\sqrt{r_1^2-x^2}}{r_1^2} \quad h_3 = y$$

$$R = h_1 + h_2 + h_3$$

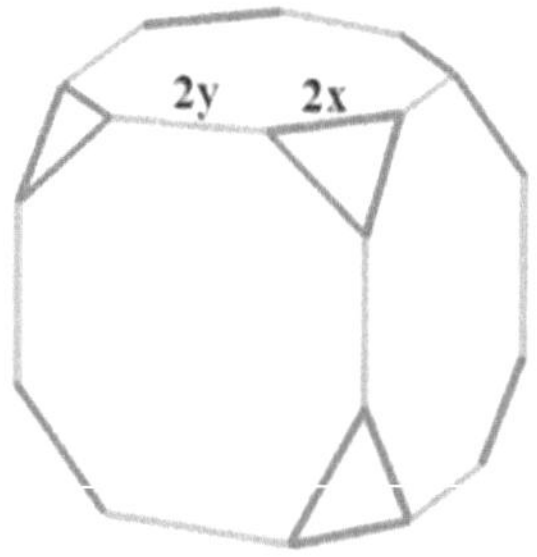

The 24E'3 and Family

While x = y = 1 yields the R of 24E'3 is

$$R = \sqrt{7+4\sqrt{2}} \doteq 3.55764729 \quad Ds = 47.405\%$$

There are five family groups of 24E'3:

(I)	(3S)e	[x = e, y = 1]
(II)	(8S)e	[x = e, y = e]
(III)	(8S)3 at e	[x = e, y =1]
(IV)	(8S) 8 at e	[x = 1, y = e]
(V)	(3M)e + (8A)	[x = e, y = y(A)]

The data of 24E'3 and its family are in the DATAFILE: [DATA 2B-5] The 24E'3 and Family.

(6) The 30E' and Family

The 30E' has twelve 5-gons and twenty 3-gons. The equations are:

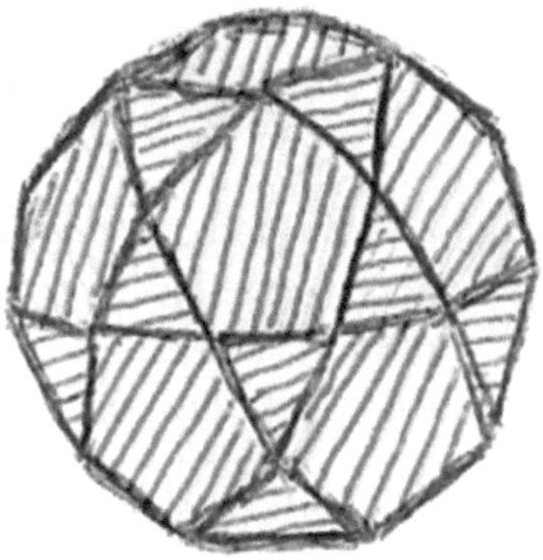

The setting of 30E'

$$k = 5 \quad C = \frac{1+\sqrt{5}}{4} \quad S = \sqrt{\frac{5-\sqrt{5}}{8}}$$

$$r_1 = \frac{e}{S} \quad r_2 = \frac{2Ce}{S}$$

$$h_1 = R - \sqrt{R^2 - r_1^2} \quad h_2^2 = 4e^2 - r_1^2 - r_2^2 + 2Cr_1r_2 \quad h_3 = \sqrt{R^2 - r_2^2}$$

$$R = h_1 + h_2 + h_3 \left(= \frac{e}{\text{Sin}(\frac{\pi}{10})} \right) = (1 + \sqrt{5})e$$

While e = 1, we have the R and Ds of 30E':

$R = 1 + \sqrt{5} \doteq 3.2360679775\ldots$ **Ds = 71.619%**

The family of 30E' is the smallest in all eleven E' families.

The data of 30E' and its family are in the DATAFILE: [DATA 2B-6] The 30E' and Family.

(7) The 48E' and Family

The analysis of the 48E' has already been illustrated as an example in **chapter II -- 4.**

This family is a huge family which contains **24** groups. Each group has from a little bit less than ten to about eight or nine hundred members, and totals to more than 3400.

This family is the second largest, next only to the family of **120E'.**

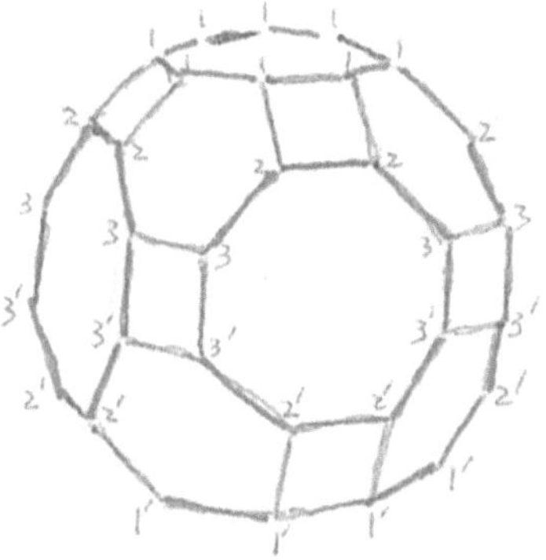

The setting of 48E'

The R and Ds of 48E' is

$R=\sqrt{13+6\sqrt{2}} \doteq 4.6352218$ **Ds = 56.686% The setting of 48E'**

The data of 48E' and its family are in the DATAFILE: [DATA 2B-7] The 48E' and Family.

(8) The 60E'1 and Family

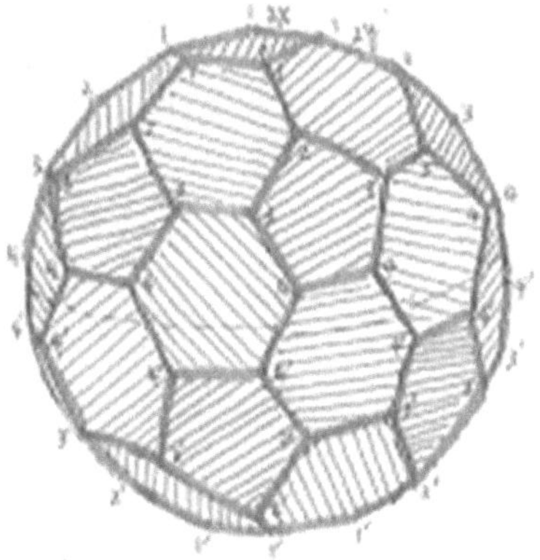

The setting of 60E'1

Calculation: The value of (x, y) are from stack formulae

$k = 5 \quad C = \cos\frac{\pi}{k} \qquad S = \sin\frac{\pi}{k} \qquad\qquad r_1 = \frac{x}{S}$

$$r_2 = \frac{1}{S}(x + y)$$

$$r_3 = \frac{1}{S}\sqrt{(2Cx)^2+y^2+2C(2Cx)y}$$

$$r_4 = \frac{1}{S}\sqrt{x^2+(x+y)^2+2Cx(x+y)}$$

$$r_0 = 0 \quad r_5 = R$$

$$h_i = \sqrt{R^2-r_{i-1}^2} - \sqrt{R^2-r_i^2}$$

$$h_2^2 = 4y^2 - (r_2 - r_1)^2$$

$$R = \sum_{i=1}^{5} h_i$$

While x = y = 1, we have the radius of 60E'1 as

$$\boxed{R = \sqrt{\tfrac{1}{2}(29+9\sqrt{5})} \doteq 4.95603732 \text{ Ds} = 51.069\%}$$

The data of 60E'1 and its family are in the DATAFILE: [DATA 2B-8] The 60E'1 and Family.

(9) The 60E'2 and Family

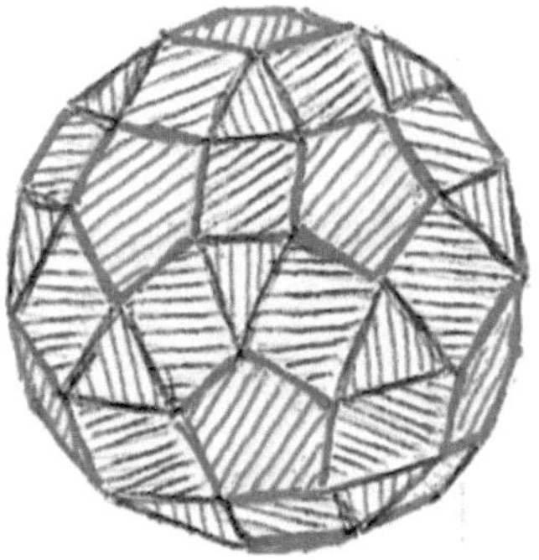

The setting of 60E'2

Calculation: The value of (x, y) is from the stack formulae (x = red, y = green):

$$k = 5 \quad C = \mathrm{Cos}\frac{\pi}{k} \qquad S = \mathrm{Sin}\frac{\pi}{k}$$

$$r_1 = \frac{x}{S}$$

$$r_2 = \frac{1}{S}\sqrt{x^2+y^2+2Cxy}$$

$$r_3 = \frac{1}{S}\sqrt{(2Cx)^2+y^2+2C(2Cx)y}$$

$$r_4 = \sqrt{\frac{2}{1+C}[R^2-(x^2+y^2)]}$$

$$r_0 = 0, \quad r_5 = R \quad h_i = \sqrt{R^2-r_{i-1}^2} - \sqrt{R^2-r_i^2}$$

$$h_4^2 = 4x^2 - r_3^2 - r_4^2 + 2r_4\sqrt{r_3^2-4C^2x^2} \quad \text{if } \frac{x}{y} < 2c$$

$$h_4^2 = 4y^2 - r_3^2 - r_4^2 + 2r_4\sqrt{r_3^2 - y^2} \text{ if } \frac{x}{y} > 2c$$

$$R = \sum_{i=1}^{5} h_i$$

** The value of 2c is called the "Equator index."

While $\frac{x}{y} < 2c$, then all five B_4 balls are at the north of the equator and all B'_4 are at the south of the equator.

While $\frac{x}{y} = 2c$, then all five B_4 balls and all five B'_4 are on the equator.

While $\frac{x}{y} > 2c$, then all five B_4 balls are at the south of the equator and all B'_4 are at the north of the equator.

The proof can be obtained from the above equations.

While x = y = 1, yields the R of 60E'2 which is

$$\boxed{R = \sqrt{11+4\sqrt{5}} = 4.465901\ldots \quad Ds = 75.21\%}$$

The data of 60E'2 and its family are in the DATAFILE: [DATA 2B-9] The 60E'2 and Family.

(10) The 60E'3 and Family

Calculation: The value of (x = red, y = green) is from the stack formulae:

$k = 5 \quad C = \cos\frac{\pi}{k} \qquad S = \sin\frac{\pi}{k} \qquad C_1 = \cos\frac{\pi}{10} \qquad S_1 = \sin\frac{\pi}{10}$

$$r_1 = \frac{1}{S}\sqrt{x^2+y^2+2Cxy}$$

$$r_2 = \frac{1}{S}(x + 2Cy)$$

$$r_3 = \frac{1}{S}(x + 2Cy + 2S_1x)$$

$$r_0 = 0\,,\quad r_5 = R$$

$$h_i = \sqrt{R^2-r_{i-1}^2} - \sqrt{R^2-r_i^2}$$

$$h_4^2 = 4y^2 - r_3^2 - r_4^2 + 2r_3\sqrt{r_4^2-y^2}$$

$$R = \sum_{i=1}^{5} h_i$$

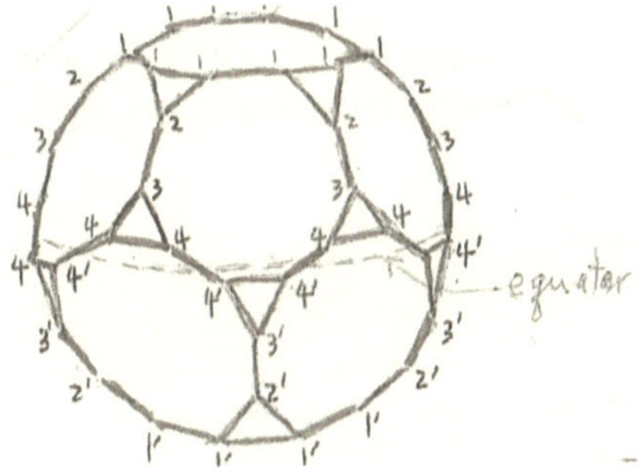

The setting of $60E'_3$

While x = y = 1, the R and Ds of $60E'_3$ are

$$\boxed{R = \sqrt{\tfrac{1}{2}(37+15\sqrt{5})} \doteq 5.938898\ldots \quad Ds = 42.528\%}$$

The data of 60E'3 and its family are in the DATAFILE: [DATA 2B-10] The 60E'3 and Family.

(11) The 120E' and Family

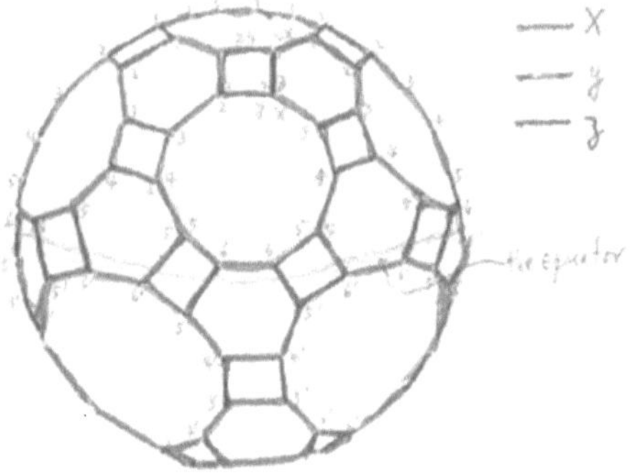

The setting of 120E'

$C = \mathrm{Cos}(\frac{\pi}{5})$ $S = \mathrm{Sin}(\frac{\pi}{5})$ $C_1 = \mathrm{Cos}(\frac{\pi}{10})$ $S_1 = \mathrm{Sin}(\frac{\pi}{10})$

The values of x, y, z are from the stack formulae:

$$r_1 = \frac{1}{S}\sqrt{x^2+y^2+2Cxy}$$

$$r_2 = \frac{1}{S}\sqrt{(x+z)^2+y^2+2C(x+z)y}$$

$$r_3 = \frac{1}{S}\sqrt{z^2+(y+2Cx)^2+2Cz(y+2Cx)}$$

$$r_4 = \frac{1}{S}\sqrt{z^2+(y+2Cx+2S_1y)^2+2Cz(y+2Cx+2S_1y)}$$

$$r_5 = \frac{1}{S}\sqrt{(x+2Cy)^2+(x+z)^2+2C(x+2Cy)(x+z)}$$

$$r_6 = \sqrt{\frac{1}{1+C}[-x\sqrt{1-C}+\sqrt{2(R^2-y^2-z^2)}]^2+x^2}$$

$$r_7 = R \quad r_0 = 0$$

$$h_i = \sqrt{R^2-r_{i-1}^2} - \sqrt{R^2-r_i^2} \quad i = 1, 2, \ldots 7$$

$$h_6 = \sqrt{4y^2-r_5^2-r_6^2+2[\sqrt{r_5^2-(x+2Cy)^2}\sqrt{r_6^2-x^2}+x(x+2Cy)]} \text{ if } \frac{y}{z} \le 2C$$

$$h_6 = \sqrt{4z^2-r_5^2-r_6^2+2[\sqrt{r_5^2-(x+z)^2}\sqrt{r_6^2-x^2}+x(x+z)]} \text{ if } \frac{y}{z} > 2C$$

$$h_7 = \sqrt{R^2-r_6^2}$$

$$R = \sum_1^7 h_i$$

While x = y = z = 1, it yields the values of R and Ds of 120E':

$$\boxed{R = \sqrt{31+12\sqrt{5}} \doteq 7.60478 \quad Ds = 51.874\%}$$

The data of 120E' and its family are in the DATAFILE: [DATA 2B-11] The 120E' and Family.

The DATAFILE The data of all E''s (up to four layers of stacks) of Es, E's, and AEs are very patiently done and have been double- and triple-checked. The data from manual calculations are 100% accurate. The data by computer are about 99% accurate. All data of this section are stored in **DATAFILE: [DATA 2B] The E's and Families.**

CHAPTER III

TWISTED EVEN DISTRIBUTION

[III – 1] The Twisted Even Distribution (TE)

General description: The twisted even-distribution (**TE**) is extended by twisting a certain E'. The candidate E' must have three sets of faces with one set of all squares called the third set or **T-set (of T-faces)**, one set of all triangles which is the second set colored green and are called **G-set**, and the set of Principal faces, **P-set of P-faces**. The P-faces could be all ⬠s, □s, or Δ s. There are only three E's, the $12E'_1$, $24E'_1$, and $60E'_2$, qualified to be twisted to form **TEs**. The three TEs are **T12E, T24E,** and **T60E**. The three mother **E's** and three daughter **TEs** are shown in the following figures, **Fig. III-1a** and **Fig. III-1b.**

Fig. III-1a. The three mother nE's:

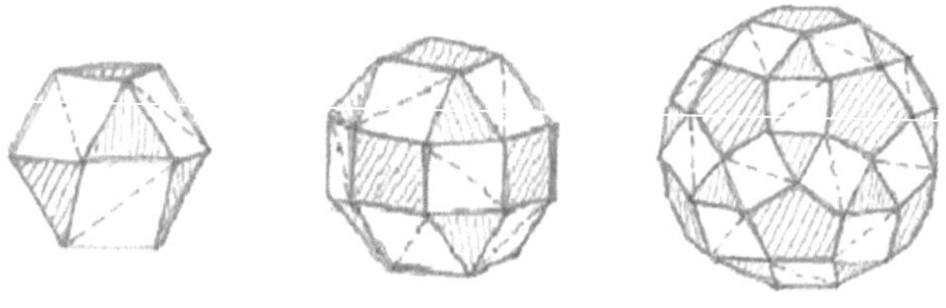

$12E'_1$ $24E'_1$ $60E'_2$

Fig. III-1b. The three daughter TEs:

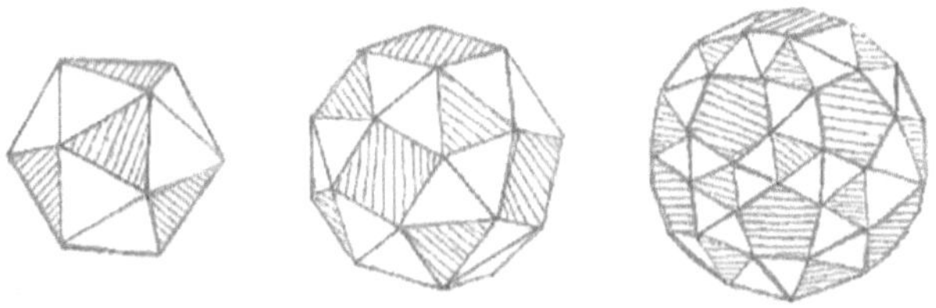

T12E **T24E** **T60E**

The process of twisting: The candidate E' is set with a P-face at the top as shown in **Fig. III-1a**. The twist axis is the axis OP of the system, where O is the center of the system and P is the center of the top P-face. In the process of twisting, the equator plane stays still, and the P-face will be right-hand (or left-hand) twisted about the twist axis OP at a certain angle called **twist angle (θ_T)**. In the mean time, all P-faces are automatically right-hand twisted at a same amount of angle about its own twist axis, and all the G-triangles are left-hand twisted at the same amount of angle about its OG axis (G is the center of the G-face). During the twisting, all □ s of the third set are changing their shapes toward a diamond shape. The process stops at the length of the short diagonal reaches **2** (it was $2\sqrt{2}$). While the twisting process is done, every diamond becomes a pair of unit twin-triangles, called T-triangles (**T- Δ s)**. The two component triangles of the twin T- Δ s are not on one plane since the twisting started.

The P-faces are usually marked with red strips, the second set of G- Δ s with green strips, and the □ s, or later the twin T- Δ s, with purple strips, or left white only with their common edge between the two component sister-triangles purpled. The edges of the P-faces are called **Edge-X**. The edges of the **G-** Δ s are called **Edge-Y**. And the common purple edges between the two component sister-triangles are called **Edge-Z**. Their lengths are **2x, 2y,** and **2z**, respectively.

The twist angle $\boldsymbol{\theta_T}$**:** In the process of creating a TE, while the square face of the mother configuration is twisted, one of the diagonal shrinks from $2\sqrt{2}$ **to 2.** The other diagonal is lengthened from $2\sqrt{2}$ **to** $2\sqrt{3}$. The radius of the mother E' shrinks from Rn to R.

The twist angle is the central angle of a corner ball on the square along the lengthened diagonal to the corner ball at the tip of the paired triangle of the new TE. **(See Fig. III-2a and b.)** The calculation of the θ_T is as following:

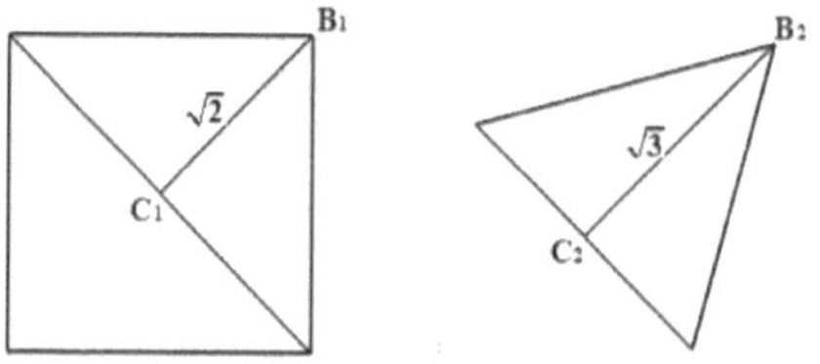

Fig. III-2a. The square becomes a paired triangle.

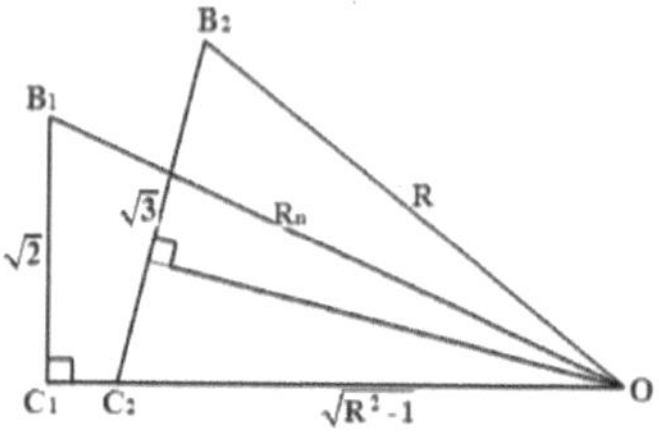

Fig. III-2b.

$$\theta_T = \angle B_1OB_2\,,\ \theta_1 = \angle C_1OB_1\,,$$

The half-diagonal $\sqrt{2}$ becomes $\sqrt{3}$

$$\theta_2 = \angle C_2OB_2\,,\ \theta_T = \theta_2 - \theta_1$$

Calculation

$$\sin\theta_1 = \frac{\sqrt{2}}{R_n} \qquad \cos\theta_1 = \frac{\sqrt{R_n^2-2}}{R_n}$$

$$\cos\theta_2 = \frac{R^2+(R^2+1)-3}{2R_nR\sqrt{R^2-1}} = \frac{R^2-2}{R\sqrt{R^2-1}} \qquad \sin\theta_2 = \frac{\sqrt{3R^2-4}}{R\sqrt{R^2-1}}$$

$$\theta_T = \theta_2 - \theta_1$$

$$\boxed{\sin(\theta_T) = \frac{1}{R_nR\sqrt{R^2-1}}\left[\sqrt{(R_n^2-2)(3R^2-4)} - \sqrt{2}(R^2-2)\right]}$$ --------The twist angle formula

The transition group: In creating a TE, when the mother E' started to be twisted, the diagonal (**Edge-Z**) of the square shrinks from $2\sqrt{2}$ **to 2**. During the process of twisting, all the temporary transition T-Δ s (with $\sqrt{2}$ **> z > 1**) are under WB situation. They all can survive independently, and so they are all E''s. But they don't belong to the family of mother **E'** or the new family of **TE**. There is an infinite number of such transition configurations existed. They are classified as a group called the "***Transition Group.***" Every individual one in this group has **x = 1, y = 1**, and $\sqrt{2} \geq z \geq 1$.

[III – 2] Solving the three TEs

The TEs have a general character that their village contains a P-face of a certain k-gon which intersects with four triangles. One of the four is the green triangle which does not adjacent with the P-face, and the other three are one and a half of the paired T-triangles (see **Fig. III-1b**). This character provides a promising way to use the total surface angle method to solve the TEs. By the TSA method, there only one diagram of the village is needed, and the calculation can bc done by hand (even it is a very time-consuming task and needs substantial patience). To use the general method, there are too many diagrams, too many equations with too long solving procedure, and finally, it still have to rely on computer to draw the approximate value of R (even if such value is very close to the proper virtual value). For comparison (or reference), the general method, especially with the top-view diagram, is attached before using the TSA method to solve the T24E and T60E.

(1) T12E

The **T12E (see Fig. III-1b)** has all **20** faces of unit triangles. *It is exactly the configuration 12E.* But the **12E** is considered to contain only one set of **20** triangles, and **T12E** has three sets of triangle faces, four red P-triangles (distributed as **4E**), four G-triangles (distributed as another **4E**), and six paired T-triangles (distributed as **6E**). While adding balls on **12E** to create its family members, the arrangements on every face of **20** Δ faces are all same. While creating the family of **T12E**, since it has three different sets of faces, there would be many options of different arrangements on such multi-sets of faces. Thus, even the **12E** and **T12E** are identical but their families are different, and so they are considered as two independent configurations.

The radius R and Ds of T12E is certainly the same as that of 12E, which is

$$R = \sqrt{\frac{1}{2}(5+\sqrt{5})} \doteq 1.90211303259.... \quad \text{and} \quad Ds = \frac{100(12)}{4R^2}\% = 82.91769...\%$$

The twist angle θ_T of T12E, using the twist angle formula:

$$\mathrm{Sin}(\theta_T) = \frac{1}{R_n R\sqrt{R^2-1}}\left[\sqrt{(R_n^2-2)(3R^2-4)} - \sqrt{2}(R^2-2)\right], \quad \theta_T = \underline{16.992708}^{\,0}$$

where $R_n = 2$ $\quad R = \sqrt{\frac{1}{2}(5+\sqrt{5})}$

(2) T24E

The virtual value of R of T24E is found by using the total surface angle (TSA) method. It is not found by using the general method. The following, using the general method, is attached just for reference.

Using the general method:

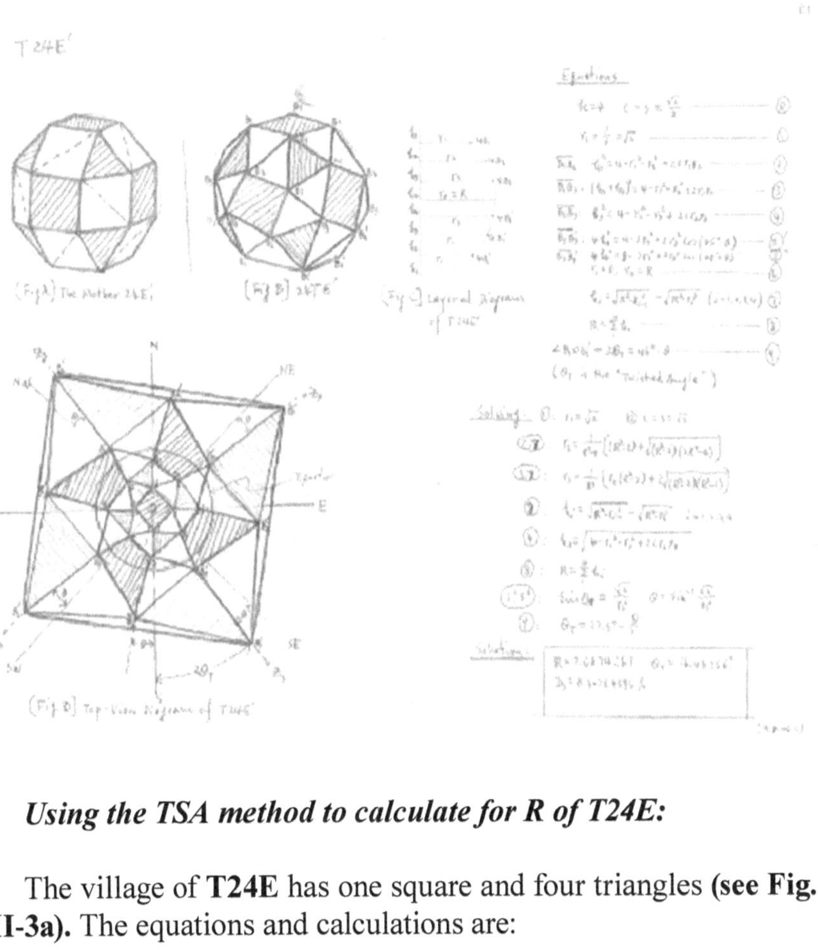

Using the TSA method to calculate for R of T24E:

The village of **T24E** has one square and four triangles **(see Fig. III-3a).** The equations and calculations are:

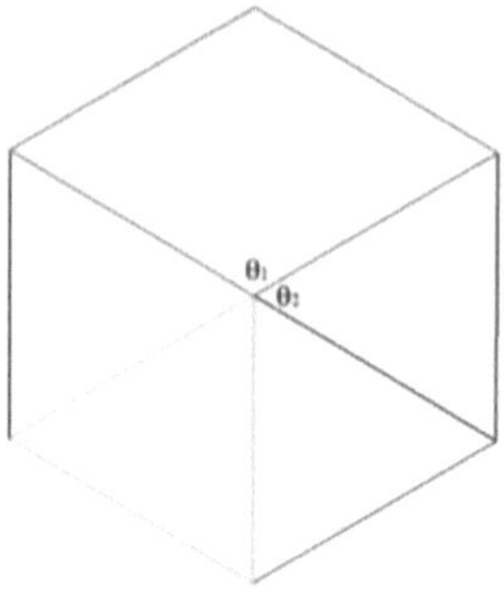

Fig. III-3a. The Village of T24E

$$\bar{\theta}_1 = 90^0 ,\ \hat{\theta}_1 = 2\mathrm{Sin}^{-1}(\frac{\sqrt{2}}{2}\alpha) \equiv 2u \text{ where } \alpha = \frac{R}{\sqrt{R^2-1}}$$

$$\bar{\theta}_2 = 60^0 ,\ \hat{\theta}_2 = 2\mathrm{Sin}^{-1}(\frac{1}{2}\alpha) \equiv 2v$$

$$2u + 8v = 2\pi \qquad \therefore \quad \mathrm{Sin}(4v) = \mathrm{Sin}(\pi - u)$$

$$\vdots$$

$$R^6 - 10R^4 + 22R^2 - 14 = 0$$

By manual calculation to solve the above cubic equation, it yields:

$$R = \sqrt{\frac{1}{3}\left[\left(199+\sqrt{297}\right)^{\frac{1}{3}} + \left(199-\sqrt{297}\right)^{\frac{1}{3}} + 10\right]} \doteq 2.6874267474892.... \qquad Ds = \frac{2400}{4R^2} = 83.0764595\%$$

The Twist angle: $\mathrm{Sin}(\theta_T) = \frac{1}{R_n R\sqrt{R^2-1}}[\sqrt{(R_n^2-2)(3R^2-4)} - \sqrt{2}(R^2-2)]$,

$$\theta_T = 8.467553^0$$

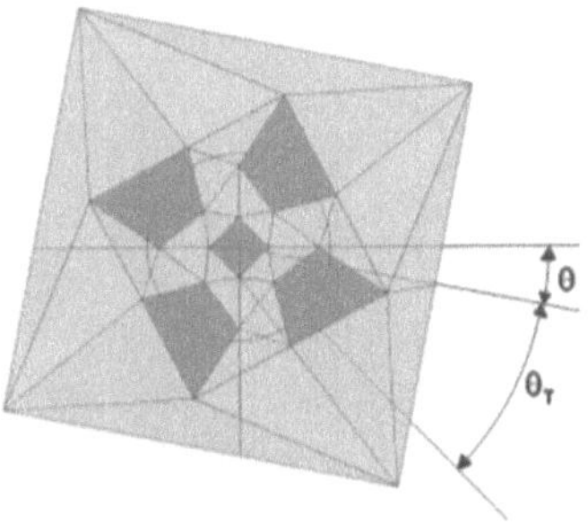

Fig. III-3b. The twist angle θ_T

(3) T60E

***Using the general method to solve T60E*:**

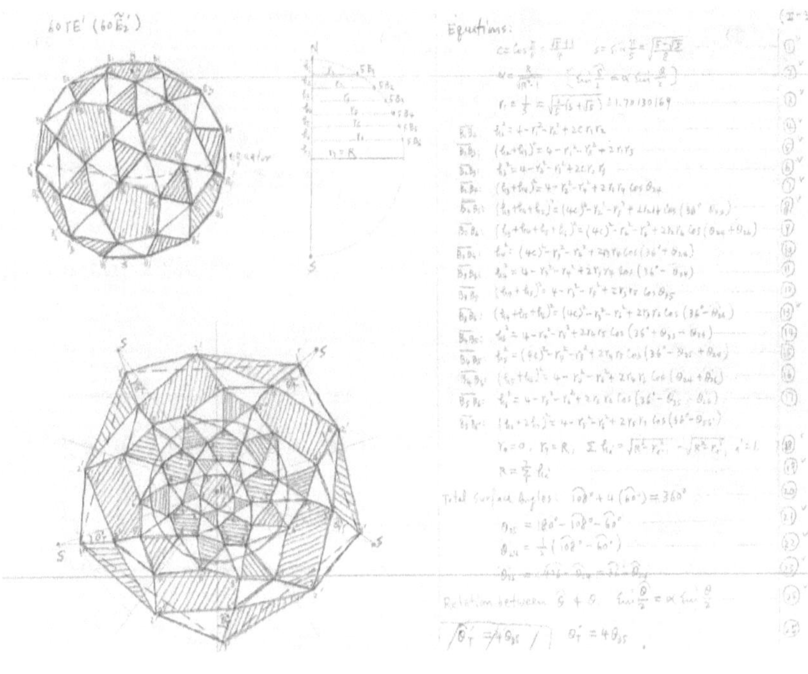

***Using the TSA method to solve T60E*:** The **T60E** has 12 pentagon P-faces, 20 green Δ s, and 30 paired twin triangles. The village contains a pentagon intersecting with four triangles. **(See Fig. III-4.)** The equations and calculations are:

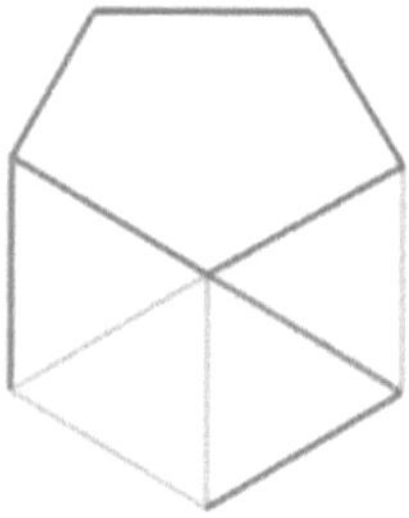

Fig. III-4. The Village of T60E

$$\bar{\theta}_1 = 108^\circ,\ \hat{\theta}_1 = 2\mathrm{Sin}^{-1}\left[\frac{1+\sqrt{5}}{4}\alpha\right] \equiv 2u$$

$$\bar{\theta}_2 = 60^\circ,\ \hat{\theta}_2 = 2\mathrm{Sin}^{-1}\left[\frac{1}{2}\alpha\right] \equiv 2v$$

$$\hat{\theta}_1 + 4\hat{\theta}_2 = 2\pi \quad \text{or} \quad \mathrm{Sin}(4v) = \mathrm{Sin}(\pi - u)$$

$$\vdots$$

$$2R^6 - \left(27+7\sqrt{5}\right)R^4 + \left(63+19\sqrt{5}\right)R^2 - \left(41+13\sqrt{5}\right) = 0$$

By manual calculation to solve the above cubic equation, it yields:

$$R = \sqrt{A + B + \frac{1}{6}(27+7\sqrt{5})} \doteq 4.311674750227544\ldots \text{ where}$$
$$\frac{A}{B} = \frac{1}{3}\sqrt[3]{\frac{1}{2}\left[\left(5112+2285\sqrt{5}\right) \pm \sqrt{64233+28728\sqrt{5}}\right]}$$

$$Ds = \frac{6000}{4R^2} \doteq 80.6862\%$$

The twist angle of T60E':

$$\mathrm{Sin}(\theta_T) = \frac{1}{R_n R\sqrt{R^2-1}}\left[\sqrt{(R_n^2-2)(3R^2-4)} - \sqrt{2}(R^2-2)\right], \qquad \theta_T = 4.9845665^{0}$$

[III – 3] The TE Families

The P-faces and the G-faces can serve as **S- or M-**status only, because they are odd-numbered polygons. Every T-triangle has three edges, Edge-X, -Y, and -Z, and they could be of different lengths while the P-face or G-face or both are loaded. The T-triangle can be loaded (for detailed analysis, see **"To load balls on a triangle face" with Fig III-6a, -6b on next few pages.)**

The **Edge-Z** has a special character. While the T-triangles are not loaded, the length **z** is flexible which plays a very important role in adjusting the well-bound situation or makes a difference for the pack between its existence or not. For a new member E", if one of the three angles $\bar{\phi}_1$, $\bar{\phi}_2$, $\bar{\phi}_3$ is <60 (see **Fig. III-5),** it is possible to lengthen the edge-Z to make $\widehat{\phi}_{21}$ $(= \widehat{\phi}_2 + \widehat{\phi}_1)$ smaller and smaller until it is $<\pi$, or to make $\bar{\phi}_3$ larger and larger until it equals 60^{0}. (Note that **z** was 1, but once it is lengthened, the angles $\bar{\phi}_1$ and $\bar{\phi}_2$ lose their common edge of length 2 to bind the two end balls, so these two angles together will need to be < than a π). (The $2x_2$ and $2y_2$ are the edge lengths of added balls on P-face and G-face, respectively, as shown in **Fig. III-5.)**

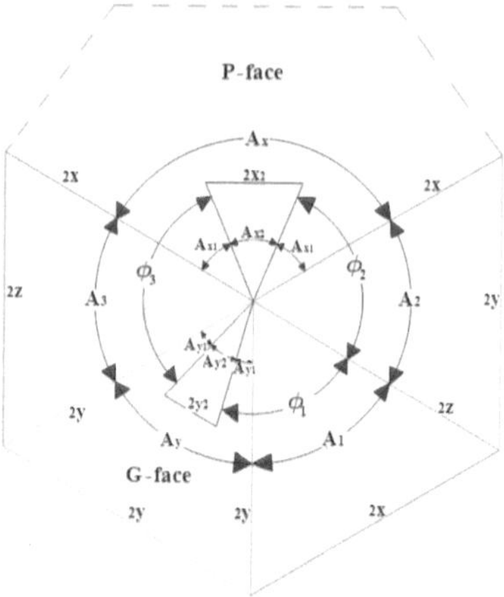

Fig. III-5. The general village of a TE
And many angles and edges

It is a fact that the shorter the Edge-Z is, the smaller the size of the pack is. So ***while the length of Edge-Z is at its minimum, the pack is at its best efficiency state.*** To find the most efficient pack of a certain arrangement, the following calculations are performed:

$$\widehat{A}_x = 1 - \frac{2c_4^2R^2}{R^2-x^2},$$

$$\widehat{A}_y = 1 - \frac{2c_3^2R^2}{R^2-y^2}$$

$$A = \frac{(y^2 \quad z^2-x^2)R^2-2y^2z^2}{2yz\sqrt{R^2-y^2}\sqrt{R^2-z^2}}$$

$$\widehat{A}_2 = \frac{(z^2+x^2-y^2)R^2-2z^2x^2}{2zx\sqrt{R^2-z^2}\sqrt{R^2-x^2}}$$

$$\widehat{A}_3 = \frac{(x^2+y^2-z^2)R^2-2x^2y^2}{2xy\sqrt{R^2-x^2}\sqrt{R^2-y^2}}$$

$$\widehat{A}_{x2} = \frac{(2-x_2^2)R^2-2}{2(R^2-1)}$$

$\hat{A}_{x1} = \frac{1}{2}(\hat{A}_x - \hat{A}_{x2})$ **[If P-face is unloaded, then** $\hat{A}_{x1} = \hat{A}_x$**]**

$$\hat{A}_{y2} = \frac{(2-y_2^2)R^2-2}{2(R^2-1)}$$

$\hat{A}_{y1} = \frac{1}{2}(\hat{A}_y - \hat{A}_{y2})$ **[If G-face is unloaded, then** $\hat{A}_{y1} = \hat{A}_y$**]**

$$\hat{A}_x + \hat{A}_y + \hat{A}_1 + \hat{A}_2 + \hat{A}_3 = 2\pi \qquad \textbf{(8)}$$

Find the value of R by R ➔ (8): LT = RT

***The range of the length of z*: Let $z_{\phi 3}$ be the shortest value of z while $\bar{\phi}_3 = 60^\circ$, and $z_{\phi 12}$ be the longest value of z while $\hat{\phi}_{12}$ (= $\hat{\phi}_1 + \hat{\phi}_2$) is *barely* $< \pi$. So, for the most efficient pack, the value of z should be the maximum value of ($z_{\phi 3}$, $z_{\phi 12}$, 1) while all $\bar{\phi}_1$, $\bar{\phi}_2$, $\bar{\phi}_3 \geq 60$. And obviously z should be in the range of $\sqrt{x^2+y^2} \geq z \geq x - y$ (if x > y).**

Therefore, for an existing family member E", the value of z should be in the range as:

$$\boxed{\sqrt{x^2+y^2} \geq z \geq \text{Max}[\, z_{\phi 3},\, z_{\phi 12},\, |x-y|_+ ,\, 1\,]} \qquad \textbf{(9)}$$

The equation (9) is called the **formula of range of z**. In constructing an E" for TE families, the formula of range of z plays a crucial roll. For a certain arrangement, if the new E" failed to meet criterion-a or -d while z = 1, it is possible to increase the value of z *slowly* until the first value appeared which yields the angle $\hat{\phi}_{12}$ (= $\hat{\phi}_1 + \hat{\phi}_2$) $< \pi$, or until $\bar{\phi}_3 = 60^\circ$, then such value of **z** yields the most efficient pack of the certain arrangement. If the value of **z** is increased up to $\sqrt{x^2+y^2}$ and it still failed in criterion-a or -d, then it indicates that such an arrangement does not exist. Also in some cases, **z** has to be **1** for criterion-b, and thus, it cannot be lengthened.

The lengthen-Z group: It is obvious that for a certain arrangement of balls added on the P-face or G-face, if there exists a value of **z,** then there is an infinite number of values of **z** existed in its range, and so there are an infinite number of E"s existed for just such a certain arrangement. And all these E"s (fit the formula of range of z) are in a group called the ***lengthen-Z group***. They are all the members of the family of the certain **TE.** but the most efficient of them is ***only*** while **z** is in the range **(pay attention for the "=" sign,** which means **"only").**

$$\sqrt{x^2+y^2} \geq z = \mathrm{Max}[\, z_{\phi 3},\ z_{\phi 12},\ |x-y|_+ ,\ 1\,]$$

The data of three TEs and their families are in the DATAFILE: [DATA 3] The TEs and Families.

Construction of TE families

There are five family groups in a TE family:

(1) P-S: The P-faces serve as single status. All edges of P-face are equally enlarged. (x, y, z) = (x, 1, z)
(2) G-S: The G-triangle faces serve as S-Status. (x, y, z) = (1, y, z)
(3) T-S: The T-triangles serve as S-Status. (x, y, z) = (z, z, z)
(4) (G-S1) + (T-S2): Since all G-faces and T-faces are all triangles, the cases (3) and (4) have the same radius R but contain different total number of balls, and thus, different Ds and Dv as well. (x, y, z) = (z, z, z)
(5) (P-S1) + (T-S2): The P-face and T-face are independently serve as S-Status. (x, y, z) = (z, z, z)

The value of z should be in the range of

$$\sqrt{x^2+y^2} \geq z \geq \text{Max}[\, z_{\phi 3},\ z_{\phi 12},\ |x-y|_+ ,\ 1\,]$$

To load balls on a triangle face: **While a triangle face serves as M- or S-status, all its three edges should be of equal length. One exception is in S-status with three edges of different lengths when only one ball is added on the triangle face (see Fig. III-6a and b). The three edges of the T-face should all be < $\sqrt{3_-}$ for the criterion-d across all edges. While x and y are known, the value of z is calculated as:**

$$r^2 = \frac{4x^2y^2z^2}{2(x^2y^2+y^2z^2+z^2x^2)-(x^4+y^4+z^4)} \qquad (1)$$

$$\text{Since } 4 = r^2 + (R - \sqrt{R^2-r^2}) \text{ yields } r^2 = \frac{4(R^2-1)}{R^2} \qquad (2)$$

Combining (1) and (2) yields:

$$z^2 = \frac{1}{R^2-1}[(x^2+y^2)r^2 - 2x^2y^2 + 2xy\sqrt{r^4-(x^2+y^2)r^2+x^2y^2}] \quad (3)$$

For the construction of (P-M) + (T-A), the G-edge's half-length y could be 1 or longer. If it is 1, then the central ball would have six or seven bindings, so y got to be > 1. Let y = z, equally lengthened, so that

$$y = z = \sqrt{\frac{1}{2}[r^2+\sqrt{r^2(r^2+x^2)}]} \text{ where } r^2 = \frac{4(R^2-1)}{R^2} \qquad (4)$$

For such case, the only possible arrangement on P-face would be (1 k), and y = z that both y and z should be < $\sqrt{3_-}$.

The (P-M) + (T-A) construction:

(1) For T12E: It is the same as (G-S1) + (T-S2).

(2) For T24E: 6(P4-M)+12(T3-A) [(1 4)+(1 3)]:

$$x = \frac{\sqrt{2(R^2-1)}}{R},\ y = z = [\sqrt{1+\sqrt{1.5}}\,]x$$

(3) For T60E: 12(P5-M)+60(T3-A) –(1 5)+(1 3):

$$x = \sqrt{\frac{1}{2}(5-\sqrt{5})}\,\frac{\sqrt{R^2-1}}{R},\ y = z = \sqrt{2+\sqrt{\frac{1}{2}(13-\sqrt{5})}}\,\frac{\sqrt{R^2-1}}{R} = 2.078\ldots > \sqrt{3}$$

Therefore, the #(2) T24E:[(P4-M)+(T3-A)][(1 4) + (1 3)] is the only possible case and the only one in the family of T24E.

[III – 4] The Angle of Adjacent Faces

To find the angle of adjacent faces of the basic five Es, eleven E's and three TEs (their faces are all unit regular polygons), the steps are as the following:

Let $A_{k_1\text{-}k_2}$ be the angle between two adjacent P_1 and P_2 faces of k_1 - and k_2 -gon;

D be the midpoint on the adjacent edge of P_1 -face and P_2 -face;

a_1 and a_2 be the lengths of P_1D and P_2D;

R_n be the radius of the mother configuration

and θ_1 and θ_2 be the central angle of a_1 and a_2

$$S_i = \operatorname{Sin}\frac{\pi}{k_i} \qquad C_i = \operatorname{Cos}\frac{\pi}{k_i}$$

$$a_{ki} = \frac{c_i}{s_i} \quad (i = 1 \text{ or } 2)$$

$$\theta_i = \mathrm{Sin}^{-1}\left[\frac{a_{ki}}{\sqrt{R_n^2 - 1}}\right]$$

$$\underline{A_{k_1\text{-}k_2} = \pi - (\theta_1 + \theta_2)}$$

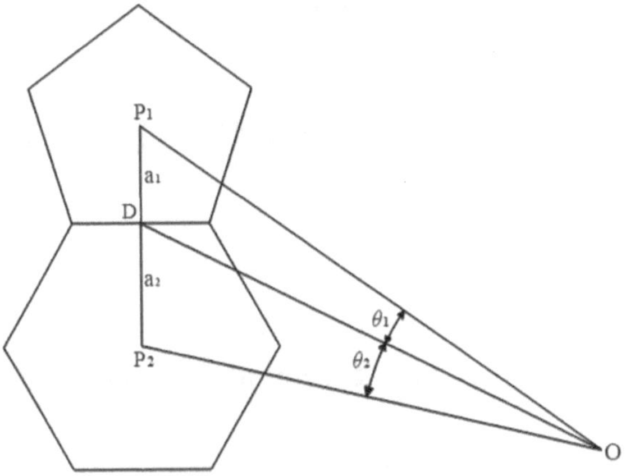

Fig. III-7. The angle of adjacent faces

For the five Es: $k_1 = k_2$ **and** $\theta_1 = \theta_2$**, so**

$$\underline{A_{k_1\text{-}k_2} = \pi - 2\theta}$$

The data of angle of adjacent faces of basic Es, E's, and TEs are in DATAFILE: [DATA 1-2].

CHAPTER IV

STRING EVEN DISTRIBUTION (SE)

[IV – 1] Definitions and General Descriptions

The sq-string and tr-string:

A string of consecutive unit squares (a square of edge-length 2 units) is called a **sq-string**. Every individual square of a sq-string is called a **sq-pearl**. A string of consecutive paired unit triangles is called a **tr-string.** Every individual paired triangle is called a **tr-pearl (see Fig. IV-1a and 1b).** The four corner balls of a tr-pearl are not on one plane. It is impossible for a string containing pearls of k-gon faces for k > 4, because otherwise the balls at the pearl corner will not be WB under criterion-a. There is a very special 6-gon pearl string case, which will be discussed in section [IV – 9].

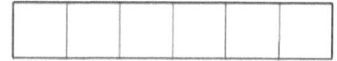

Fig. IV-1a. A sq-string containing 6 sq-pearls

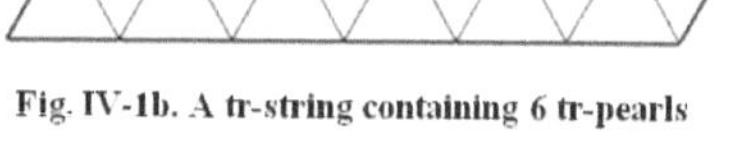

Fig. IV-1b. A tr-string containing 6 tr-pearls

The string even distribution (SE) (or string configuration): **(See Fig. IV-2a, -2b, -2c, and -2d)**

The SE is constituted by inserting strings between two adjacent faces of a mother E or E'. These two adjacent faces, called the ***principal faces***, (more precisely, the P_1-face and P_2-face), could be from the same set or two different sets of faces of the mother configuration. Every string inserted between a P_1-face and a P_2-face must contain the same number of same pearls. An SE with all sq-strings or all tr-strings is denoted as **Sq-SnE(k_1 - k_2, m)** or **Tr-SnE (k_1 - k_2, m)**, where **n** is from the mother nE or nE', k_1 and k_2 are the number of edges of P_1- and P_2-faces, and **m** is the number of pearls in every string. A string can contain an infinite number of pearls, so the SE could be an infinite ball packing.

In an SE pack, while k = 3, 4, or 5, the k-gon P-face should have every edge connected with a string. For k = 6, 8 or 10, the P-face can have only every next edge connected with a string, and the other half number of edges left empty (called the open edges). The reason **is** that because all open straight angles $\bar{\phi}_o$ should be equal or greater than 60°. (For all definitions mentioned above and in the next few paragraphs, refer to the **Fig. IV-2a, b, c, and d**.)

The open face: While a string is inserted between two P-faces, there are some open areas formed on the surface called the **open faces**. The open face is surrounded by strings, or by strings and the open edges of P-faces. The open face usually is not a polygon and all its corner points may not be on one plane, (because the strings are not straight lines). While every string contains only one pearl, the open-face could be a polygon and so with all corner balls on one plane. ***Every edge of an open face must be of length 2.***

The edges: The edge of the P-face which connects a string is called the **Edge-X** and is of length **2x**. The edge of the P-face which does not connect a string is called the **open edge** or **Edge-Y** and is of length **2y (Note that y = 0, while k = 3, 4, or 5; and y = 1, while k = 6, 8, or 10).** The edge between two pearls is called the **Edge-Z** and is of length **2z. (Note that the Edge-Z of a Tr-SE is different than the Edge-Z of TE.** The TE has only one paired triangle, and the Edge-Z of a TE is the common edge between the two components of the paired triangle. The Tr-SE could have more than one paired triangle, and the Edge-Z of a Tr-SE is the edge between two tr-pearls.) **The edge between two component of paired triangles of every tr-pearl is always of length 2.**

The basic SE**:** An SE with both P-faces unloaded (all edges are of lengths 2), and every string contains only one pearl (m = 1), is called a basic SE. ***The distance*** between two balls across Edge-X is $\mathbf{2d_x}$ and across Edge-Z is $\mathbf{2d_z}$.

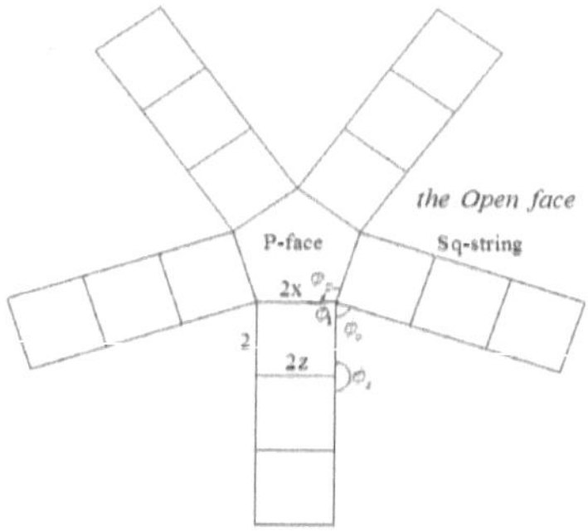

Fig. IV-2a

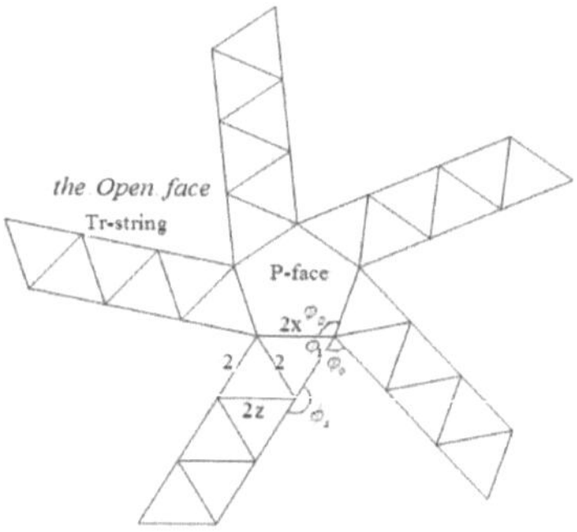

Fig. IV-2b

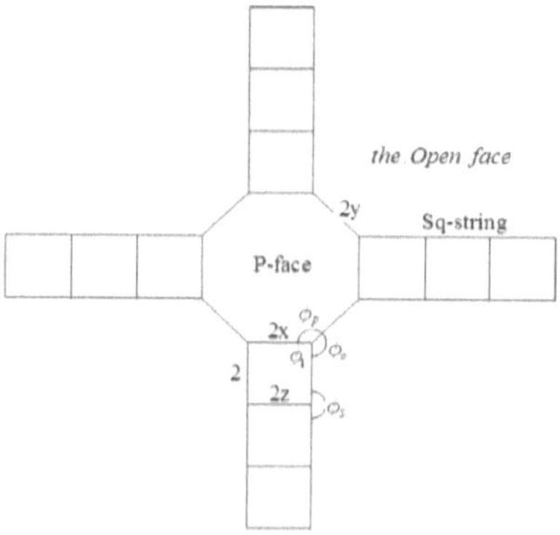

Fig. IV-2c]

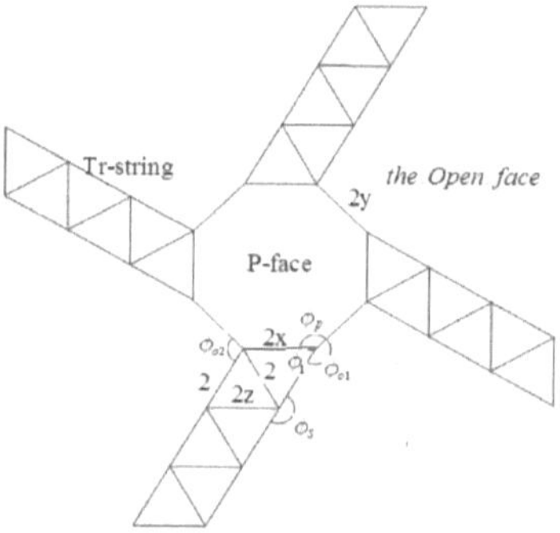

Fig. IV-2d

The angles:

** The inner angle of the P-face is called the **P-angle** (ϕ_p).
** The angle between two strings or between a string and an open edge of the P-face is called the **open angle** (ϕ_o).
For a Tr-SE, while k = 6, 8, or 10, there are two different open-angles, ϕ_{o1} and ϕ_{o2}, **(See Fig. IV-2d.)**
** The angle between two adjacent pearls at the open face side is called the **string angle** (ϕ_s).

It is obvious from the above figures that all $\hat{\phi}_s$ are $< \pi$, which indicates that ***all balls on the string are WB.***

[IV –2] The Sq-SnE(k1-k2, m)

While adding balls to create SE families, the edges of X and Z could be enlarged, and so the Sq-pearls could become rectangles, and the two end pearls **(Ep)** could become rhombuses. **The pearl edge at the open-face side is always of length 2. (See Fig. IV-3.)**

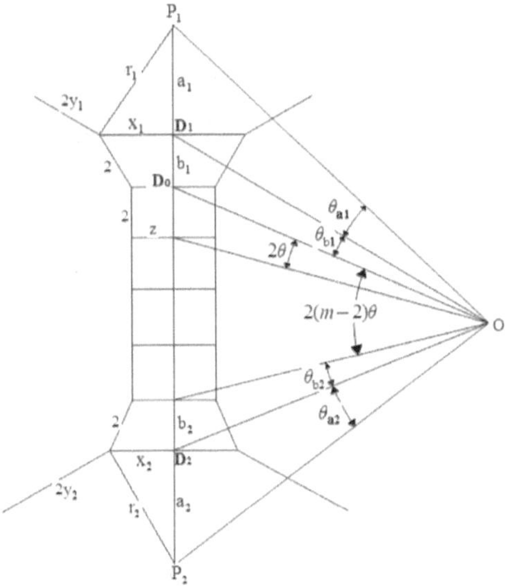

Fig. IV-3. The Sq-SnE (k1-k2, m)

* The point P_i is the center of P_i-face, **i = 1 or 2.**
* The point P_{oi} is the center of P_i-face of the mother configuration.
* The r_i is the radius of P_i-face (r_{oi} is that of the mother configuration).
* The D_1 **and** D_2 are the midpoints on Edge-**X1** and Edge-**X2.**
* The D_0 is the midpoint on the common edge of two adjacent pearls.
* $a_i = P_iD_i$ and $b_i = P_iD_0$ are the lengths defined as shown in the figure.
* The a_{oi} and b_{oi} are the corresponding lengths of the mother configuration.
* The θ_{ai} and θ_{bi} are the central angles of a_i and b_i.
* The 2θ is the central angle of every single pearl.
* The ψ_i is the central angle of a_{oi} of the mother configuration. (The mother configuration has no string between two P-faces, so the two points D_1 and D_2 are merged as one point D_{12}.)

Calculation:

While $k_i = 3, 4,$ or 5:

$$C_i = \mathrm{Cos}\left(\frac{\pi}{k_i}\right),\ S_i = \mathrm{Sin}\left(\frac{\pi}{k_i}\right),\ y_i = 0 \tag{1}$$

While $k_i = 6, 8,$ or 10:

$$C_i = \mathrm{Cos}\left(\frac{2\pi}{k_i}\right),\ S_i = \mathrm{Sin}\left(\frac{2\pi}{k_i}\right),\ y_i = 1 \tag{1}$$

The x_i and z are determined by stack formula.
While $m = 1, z = x_2$.

$$r_i = \frac{1}{S_i}\sqrt{x_i^2 + y_i^2 + 2C_i x_i y_i} \tag{2}$$

$$\mathrm{Sin}\left(\theta_{ai}\right) = \frac{\sqrt{r_i^2 - x_i^2}}{\sqrt{R^2 - x_i^2}} \tag{3}$$

$$\mathrm{Cos}\left(\theta_{bi}\right)=\frac{\left(R^2-x_i^2\right)+\left(R^2-z^2\right)-b_i^2}{2\sqrt{\left(R^2-x_i^2\right)\left(R^2-z^2\right)}} \quad (4)$$

where $b_i=\sqrt{4-\left(x_i-z\right)^2}$ **if m = 1 then** $b_2=0$

$$\mathrm{Sin}\left(\theta\right)=\frac{1}{\sqrt{R^2-z^2}} \quad \text{if m = 1 then } \theta=0 \quad (5)$$

For the mother configuration:

$$r_{oi}=\frac{1}{S_{oi}} \qquad a_{oi}=\frac{C_i}{S_i} \qquad \mathrm{Sin}(\psi_i)=\frac{a_{oi}}{\sqrt{R_n^2-1}} \quad (6)$$

For m > 1:

$$\left(\theta_{a1}+\theta_{a2}\right)+\left(\theta_{b1}+\theta_{b2}\right)+2\left(m-2\right)\theta=\psi_1+\psi_2 \quad (7)$$

For m = 1,

$$z=x_2,\ b_2=0,\ \text{and}\ \theta=0: \qquad \theta_{a1}+\theta_{a2}+\theta_{b1}=\psi_1+\psi_2 \quad (8)$$

Find R by R → Eq. (7) or (8): LT = RT

Solving a basic Sq-SE(k1-k2, 1): For to find the virtual value R of a basic Sq-SE, the *central angle method* is used. [The TSA method does not work, due to unknown angle measurement of the open angles].

***Examples:* to find the virtual value of R of basic Sq-S12E'1(4-3, 1)**

For the mother 12E'1: Ro = 2 N = 48

$$k_1=3 \quad k_2=4 \quad S_i=\mathrm{Sin}\left(\frac{\pi}{k_i}\right) \quad C_i=\mathrm{Cos}\left(\frac{\pi}{k_i}\right) \quad i=1,2$$

$$r_i = \frac{1}{S_i} \quad a_i = \sqrt{r_i^2 - 1}$$

$$\mathrm{Sin}\theta_i = \frac{a_i}{\sqrt{R^2-1}} \quad \mathrm{Sin}\theta_{oi} = \frac{a_i}{\sqrt{R_o^2-1}}$$

$$\theta = \theta_2 \qquad 3\theta = (\theta_{o1} + \theta_{o2}) - \theta_1$$

.

.

$$(R^2 - 2)(R^4 - 18R^2 + 41) = 0$$

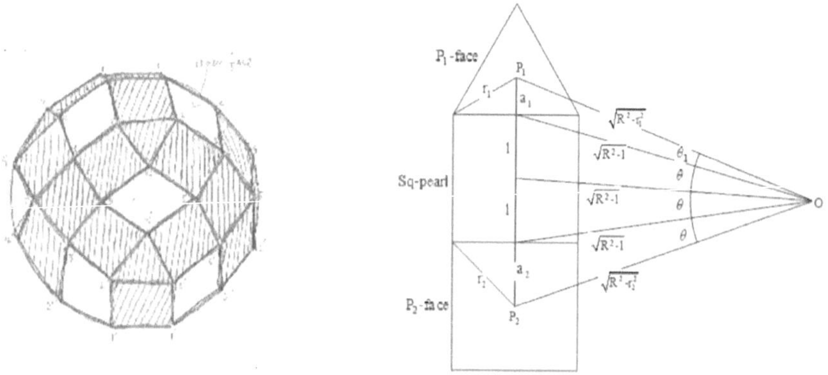

Fig. IV-4. The setting (left) and the central angle diagram of Sq-S12E'1(4-3, 1) (right)

$$\boxed{R = \sqrt{9+\sqrt{40}} \doteq 3.9146589\ldots}$$

Using the same method, the virtual value R of basic Sq-S24E'2(4-6, 1) is found as:

$$\boxed{R = \sqrt{\left[525 + \sqrt{1000}\right]^{\frac{1}{3}} + \left[525 - \sqrt{1000}\right]^{\frac{1}{3}} + 10} \doteq 5.11153338396559\ldots}$$

Most of the time, using the above way to find the virtual value of R would involve solving third ordered equation with astronomically large numbers, square roots, cubic roots, etc.,

and none of them can be generated by computer; all should be performed patiently by manual calculation. Still, some of them are almost impossible to solve, such as these two cases:

(1) If $k_1 \neq k_2$
(2) If one of the k_1 and k_2 is not 4

(For to find the virtual value R of other basic SEs besides these two cases, the author has tried very hard to use many different approaches, but has no success.)

The equivalent configurations of Sq-SnE(k-k, 1) while m = 1:
An Sq-SE transformed from a mother E with every string contains only one sq-pearl is equivalent to a certain E'. For example, the Sq-S4E(3, 1), which is transformed from 4E, has 4 triangle P-faces, 6 strings each containing 1 sq-pearl, and 4 open faces of unit squares. The open face is a square with all points on one plane, such a configuration is exactly the $12E'_1$. The five Sq-SnE(k-k, 1) from their mother nEs, each has a configuration equivalent to an E' as listed below.

(1) ***Sq-S4E(3-3, 1)*** = $12E'_1$ **with 4 △ P-faces, 6 one-sq-pearl strings, and 4 open △ - faces. R = 2**
(2) ***Sq-S6E(3-3, 1)*** = ***24E'1*****, 8 △ P-faces, 12 one-sq-pearl strings, and 6 open □ - faces. R =** $\sqrt{5+2\sqrt{2}}$
(3) ***Sq-S8E(4-4, 1)*** = ***24E'1*****, 6 □ P-faces, 12 one-sq-pearl strings, and 8 open △ - faces. R =** $\sqrt{5+2\sqrt{2}}$
(4) ***Sq-12E(3-3, 1)*** = ***60E'2*****, 20 △ P-faces, 30 one-sq-pearl strings, and 12 open 5-gon - faces. R =** $\sqrt{11+4\sqrt{5}}$
(5) ***Sq-20E(5-5, 1)*** = ***60E'2*****, 12 5-gon P-faces, 30 one-sq-pearl strings, and 20 open △ - faces. R =** $\sqrt{11+4\sqrt{5}}$

The data of the basic Sq-SnE(k1-k2, 1) and their equivalent configurations are in the DATAFILE: [DATA 5-1].

A string containing m sq-pearls is equivalent to a stack of

While m is even, the equivalent stack is $(2^{\sim}\overset{2}{4}{}^{-}\overset{2}{4}{}^{-}\overset{2}{4}\ldots{}^{-}\overset{2}{4})$**, which has m - 1 number of "**$\overset{2}{4}$**" after the first "2".**

While m is odd, the equivalent stack is $(4^{-}\overset{2}{4}{}^{-}\overset{2}{4}{}^{-}\overset{2}{4}\ldots{}^{-}\overset{2}{4})$**, which has m number of "**$\overset{2}{4}$**" in the series.**

Actually, the stack of the form $(k^{\sim}\overset{2}{2k}{}^{-}\overset{2}{2k}{}^{-}\overset{2}{2k}\ldots{}^{-}\overset{2}{2k})$ **is a sq-string containing m sq-pearls connected on a k-gon P-face where k is an odd number; the stack of the form** $(\overset{2}{2k}{}^{-}\overset{2}{2k}{}^{-}\overset{2}{2k}{}^{-}\overset{2}{2k}\ldots{}^{-}\overset{2}{2k})$ **is a sq-string containing m sq-pearls connected on a 2k-gon P-face.**

[IV–3] Construction of Families of Sq-SE and Criteria

Construction status:

P-S: One or both P-faces serve as single (S) status.
T-S: The string pearl faces (T-faces) serve as single (S) status.
(T-S1) + (P-S2): The string pearl faces (T-faces) and one or both P-faces are serving as S-status independently.

The criteria

While k = 3, 4, or 5: $S = \mathrm{Sin}(\frac{\pi}{k})$ $C = \mathrm{Cos}(\frac{\pi}{k})$

While k = 6, 8, or 10: $S = \mathrm{Sin}(\frac{2\pi}{k})$ $C = \mathrm{Cos}(\frac{2\pi}{k})$

(1) $\widehat{\phi}_0 < \pi$

$$\mathrm{Sin}\left(\frac{\widehat{\phi}_p}{2}\right) = \frac{R}{\sqrt{R^2\text{-}x^2}} \quad \mathrm{Sin}\left(\frac{\bar{\phi}_p}{2}\right) = \frac{CR}{\sqrt{R^2\text{-}x^2}}$$

$$\text{Cos}\left(\widehat{\phi}_1\right) = \frac{\left(R^2 - 2\right)x - R^2 z}{2\sqrt{R^2 - x^2}\sqrt{R^2 - 1}}$$

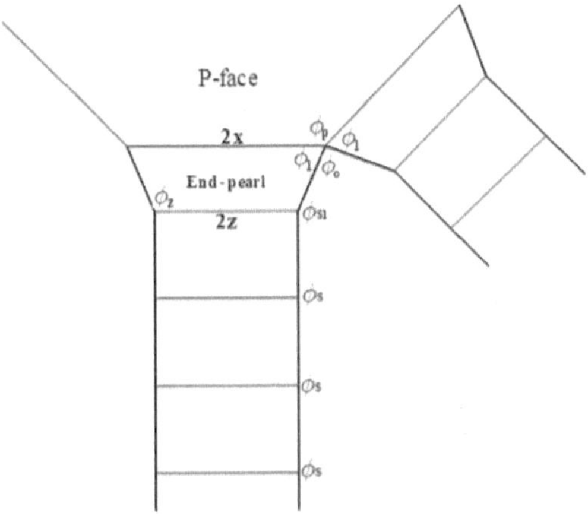

Fig. IV-5a. k = 3, 4, or 5

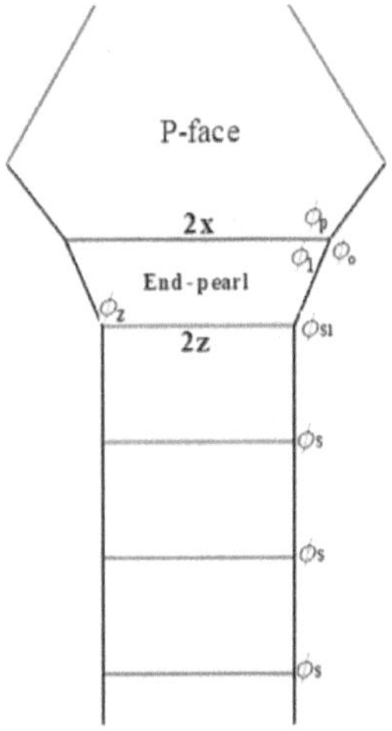

Fig. IV-5b. k = 6, 8, or 10

While k = 3, 4, or 5:

For $\widehat{\phi}_0 < \pi$, it needs $\widehat{\phi}_p + 2\widehat{\phi}_1 > \pi$, or $\frac{1}{2}\widehat{\phi}_p + \widehat{\phi}_1 > \frac{\pi}{2}$

Which yields $\boxed{x < \frac{1}{R^2-2}[2cR\sqrt{R^2-1} + R^2 z]}$ **(roughly** $x < z + 2c$**)**

While k = 6, 8, or 10:

For $\widehat{\phi}_0 < \pi$, it needs $\frac{1}{2}\widehat{\phi}_p > \frac{\pi}{2} - \frac{1}{2}\widehat{\phi}_1$.

Which yields $x < \frac{R^2(z+2c)}{R^2-4}$ **(roughly** $x < z + 2c$**)**

(2) $\bar{\phi}_0 \geq 60^{\circ}$ or dd $\geq$ 1

While m = 1 and k = 3, 4, or 5, if $x_1 > x_2$ (see Fig. IV-6), the straight angle $\bar{\phi}_0$ could be $< 60^{\circ}$. So it is necessary to find what x_1 should be so that $\bar{\phi}_0$ of P_2 -face would be $\geq 60^{\circ}$ or dd $\geq$ 1.

$$\widehat{\phi}_p + 2\widehat{\phi}_1 + \widehat{\phi}_0 = 2\pi$$

$$Cos\widehat{\phi}_1 = \frac{(R^2-2)x_2-R^2x_1}{2\sqrt{R^2-1}\sqrt{R^2-x_2^2}}$$

$$\text{Sin}\frac{\widehat{\phi}_p}{2} = \frac{C_2R}{\sqrt{R^2-x_2^2}} \qquad \text{Sin}\frac{\widehat{60}^{0}}{2} = \frac{R}{2\sqrt{R^2-1}}$$

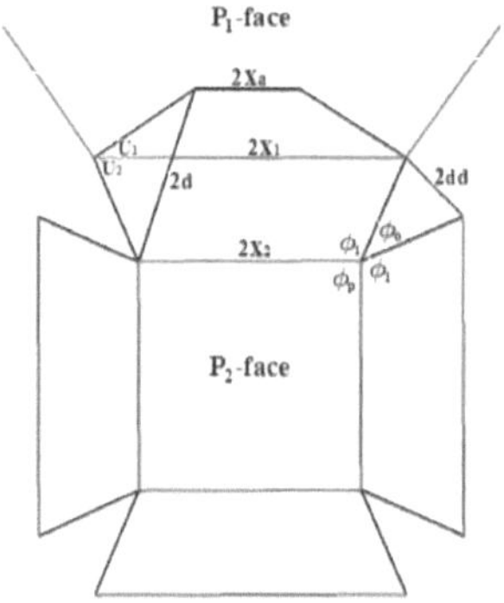

Fig. IV-6

Which yields

$$\boxed{x_1 \leq \frac{1}{R^2}\left[(R^2-2)x_2 + \sqrt{(S_2^2R^2 - x_2^2)(3R^2-4)}\right] - C_2 \quad \text{while } x_1 > x_2}$$

(vice versa for x1 < x2)

(3) $\widehat{\phi}_z + \widehat{\phi}_{za} < \pi$ (Fig. IV-9) (If there is no ball on the string pearls, then this criterion does not exist.)

Let z_a be the half-distance between two added balls parallel to Edge-Z on the string pearls. While the stacks are:

$$(1\,\overset{2}{4}):\ z_a = 0,\ z = \frac{\sqrt{3R^2-4}}{R}$$

$$(2^{-}\overset{2}{4}):\ z_a = 1,\ z = \frac{1}{R^2}\left[\left(R^2-2\right)+\sqrt{\left(R^2-1\right)\left(3R^2-4\right)}\right]$$

$$(2^{\sim}\overset{2}{4}):\ z_a = 0,\ z = \frac{2\sqrt{R^2-2}}{\sqrt{R^2-1}}$$

*** The stack $(2^{\sim}\overset{2}{4})$ can only be applied on a one-pearl string configuration due to the criterion-d across Edge-Z.**

Since $\mathrm{Cos}(\phi_{za}) = \dfrac{(R^2-2)z-R^2z_a}{2\sqrt{R^2-1}\sqrt{R^2-z^2}}$ $\quad \mathrm{Cos}\widehat{\phi_z} = \dfrac{(R^2-2)z-R^2x}{2\sqrt{R^2-1}\sqrt{R^2-z^2}}$

Which yields

$$\boxed{x < \frac{2(R^2-2)z}{R^2} - z_a}$$

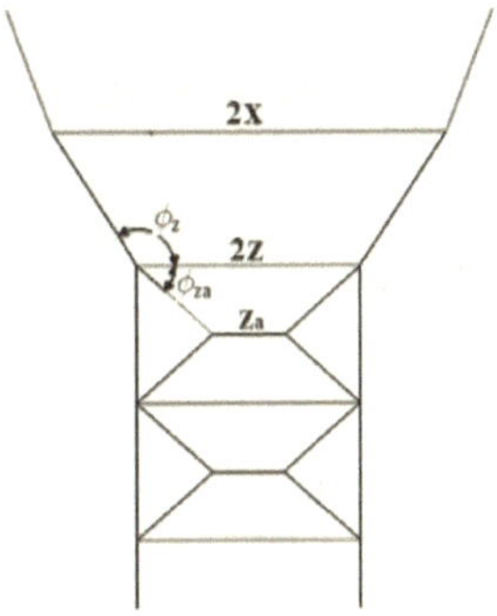

Fig. IV-7

(4) $d \geq 1$: (See Fig. IV-7) Using the d-formula to find the cross-edge distance d, which is

$$\boxed{d = \frac{\sqrt{2(R^2-1)}}{R}\sqrt{1-\mathrm{Cos}(\widehat{U}_1+\widehat{U}_2)}}$$

where

$$\mathrm{Cos}\widehat{U}_1 = \frac{(R^2-2)x_1-R^2x_a}{2\sqrt{R^2-1}\sqrt{R^2-x_1^2}} \quad \mathrm{Cos}\widehat{U}_2 = \frac{(R^2-2)x_1-R^2x_2}{2\sqrt{R^2-1}\sqrt{R^2-x_1^2}}$$

*** This criterion could be eliminated since for all k, the minimum $\widehat{U}_2$ is near 18^0, and $\mathrm{Cos}(18^0) \doteq .31$, and *therefore,* $d \geq 1$ *is always true.***

The range of x while m > 1:

While k=3, 4, or 5, (y=0) $\dfrac{2cR\sqrt{R^2-1}+R^2z}{R^2-2} > x \geq \max[1, \dfrac{R^2-4}{R^2}z]$

While k = 6, 8, or 10, (y = 1) $\dfrac{R^2(z+2c)}{R^2-4} > x \geq \max[1, \dfrac{R^2-4}{R^2}z]$

**** *Approximately, the range of x is:* $z+2C > x \geq z$**

***A question* ---- Can y be greater than 1?**

Answer: *Yes, y can be greater than 1.* But it is limited for y = x, and k = 6, 8, or 10 only, and the possible arrangements on P-faces are:

(1): $(10^{\sim}10)$ for z = 1, and R is limited to be < 7+, and k = 10 only.

(2): $(k^{\sim}k^{\sim}k)$ for z = 0 ($z = \dfrac{\sqrt{3R^2-4}}{R}$), and k = 6, 8, 10.

(3): $(k \bar{\ } k \tilde{\ } k)$ **for z = 0 (** $z = \frac{\sqrt{3R^2-4}}{R}$ **), and k = 8, 10.**

Actually, while y > 1, there must be a layer above the base layer with y = 0 or 1. *Therefore, the case of y > 1 can be eliminated.*

***The data*:**

The data of basic Sq-SEs and their equivalent configurations are in DATAFILE: [DATA 4-1].
The data of basic Sq-SE while m = 1, 2, …, 100 are in DATAFILE: [DATA 4-2].
The data of Sq-SnE(k-k, m) and families are in DATAFILE: [DATA 4A-1, 2, 3, 4, 5].
The data of Sq-SnE'(k1-k2, m) and families are in DATAFILE: [DATA 4B-1, 2] (only the family of Sq- S12E'1 and the family of Sq-S30E' are completed).

[IV– 4] The End-pearl (Ep) and Special One-pearl Sq-SE

Since z is always ≤ x, so the Ep could be a square, a rhombus, or a rectangle. (Fig. IV-8a, 8b, 8c).

A rhombus Ep cannot be loaded with any stacks (except the special one-pearl cases as shown in Fig. IV-10a, b).

Fig. IV-8a: While P-face is unloaded (x = 1), then the Ep is a square

Fig. IV-8b: If m = 1 and $x_1 \neq x_2$, or m > 1 and x > z, then the Ep is a rhombus.

Fig. IV-8c: If m = 1 and x1 = x2 (both P-faces are loaded), then the Ep is a rectangle. Fig. IV-8a Fig. IV-8b Fig. IV-8c

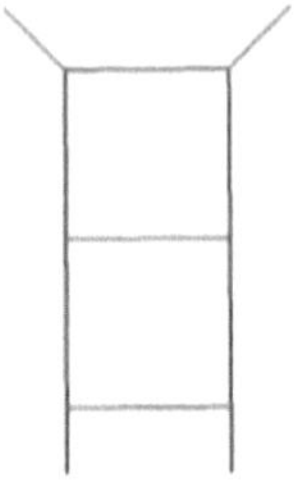

Fig. IV-8a

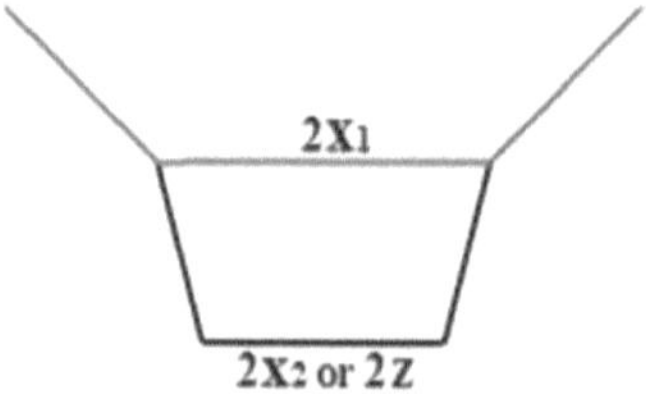

Fig. IV-8b

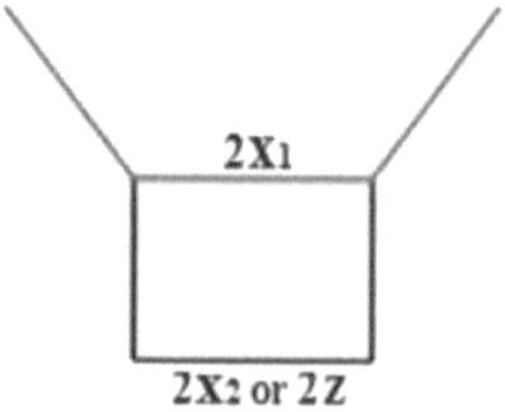

Fig. IV-8c

There are only three possible stacks, $(1\ \overset{2}{4})$, $(2\ \overset{-2}{4})$ and $(2\ \overset{\sim 2}{4})$, available to be loaded on the sq-string pearls. (See Fig. IV-9a, b, c.)

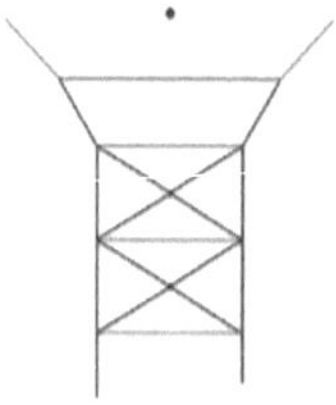

Fig. IV-9a. $(1\ \overset{2}{4})$

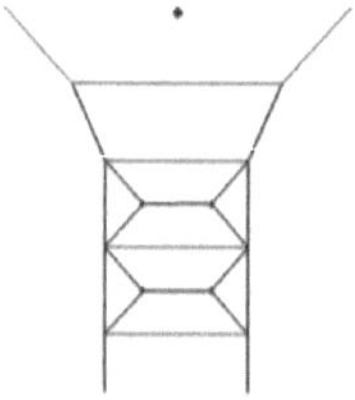

Fig. IV-9b. $(2\ \overset{-2}{4})$

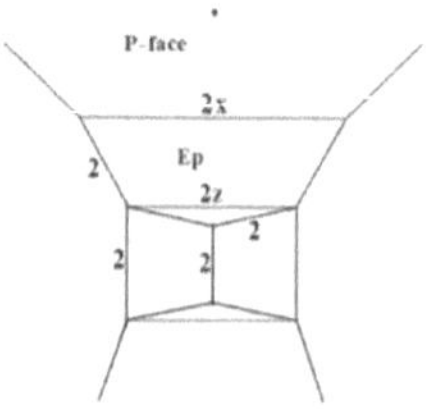

Fig. IV-9c. $(2\ \overset{\sim 2}{4})$

The special one-pearl Ep configuration **($z = 0$, $z_s > 1$)**

A one-pearl Ep cannot be loaded unless it is a rectangle. But as the *special one-pearl Ep,* it can as in the following:

Between two P-faces, there one or two balls can be inserted as shown in Fig. IV-10a, b. Both cases look like they are rhombus Ep. Actually, they are not, since all Ep and all pearls have Edge-Zs of length 2, but in these cases, Zs > 2.

During the calculation, they are considered as:

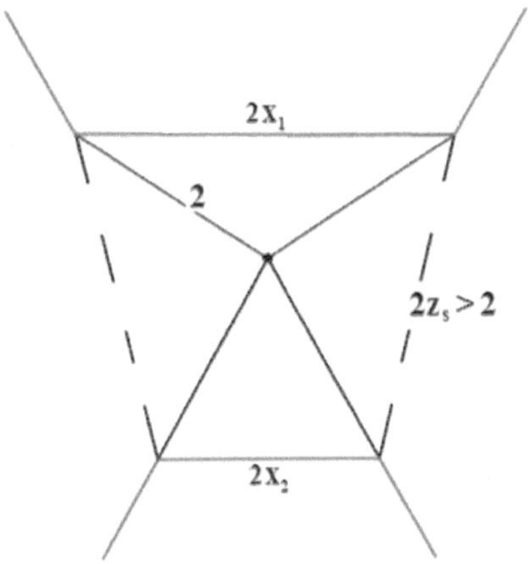

Fig. IV-10a

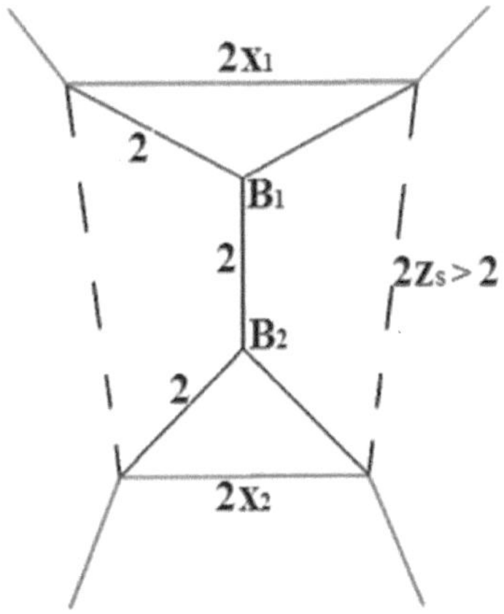

Fig. IV-10b

For case A (Fig. IV-10a), it is considered as m = 2, z = 0.
For case B (Fig. IV-10b), it is considered as m = 3, z = 0.

The limitation of half-length x of Edge-X is

$$\text{While } k = 3, 4, \text{ or } 5: \; x < \frac{2CR\sqrt{R^2 - 1}}{R^2 - 2}$$

$$\text{While } k = 6, 8, \text{ or } 10: \; x < \frac{2CR^2}{R^2 - 4}$$

** While loading on P-faces, due to the limitation of the length of x, the only possible stacks on 4-gon P-face is the stack (1 4), and on 5-gon P-face, it is the stack (1 5).

The data of one-pearl and special one-pearl-Ep of Sq-SE are in DATAFILE: [DATA 4-3].

[IV – 5] The Tr-SnE (k1-k2, m)

The Tr-SnE(k_1-k_2, m) is formed by inserting a string of **m** number of paired triangles (so there are **2m** number of triangles in every string) between two k_1- and k_2**-gon** P-faces of a mother **nE** or **nE'. [See Fig. IV-11a, the standard Tr-SnE(k_1-k_2, m).]** The inner adjacent edge of the tr-pearl (**B_0B_{11}, B_1B_{22}, ...**) is always of length **2**. The common edge between two pearls (**B_1B_{11}, B_2B_{22}, ...**) is called the **Edge-Z** and is of length **2z,** which is flexible, i.e., **z** can be $\geq$**1.** The open side edge of every tr-pearl (**B_0B_1, B_1B_2, ...**) is of length **2.** The two end triangles of a string are called the end-pearl **(Ep).** ***It is impossible to add balls on the tr-pearls.***

Calculation:

** A character with a sign ' means it is the point on the surface above the point of the character.

** A straight line with a sign ' means both end-points of the line are on the surface above the straight line.

The **2θ** is the central angle of a tr-pearl which is not the Ep of the string.

The Rn, ki, ksi, and Pi are defined the same as before.

The Pi is the center of the Pi-face; **i = 1, 2.**

The Di is the midpoint on Edge-Xi.

The Bj and Bjj are the j-th ball on the string from P_1- **face.**

The Tj is the midpoint of Edge-Z (BjBjj) of jth pearl.

The $\mathbf{a_1 = P_1B_{11}}$ $\mathbf{d_1 = P_1T_1}$ $\mathbf{d_2 = P_2T_m}$.

The $\mathbf{LL = P_1T_m}$ $\mathbf{TT = T_1T_m}$ $\mathbf{PP = P_1P_2}$.

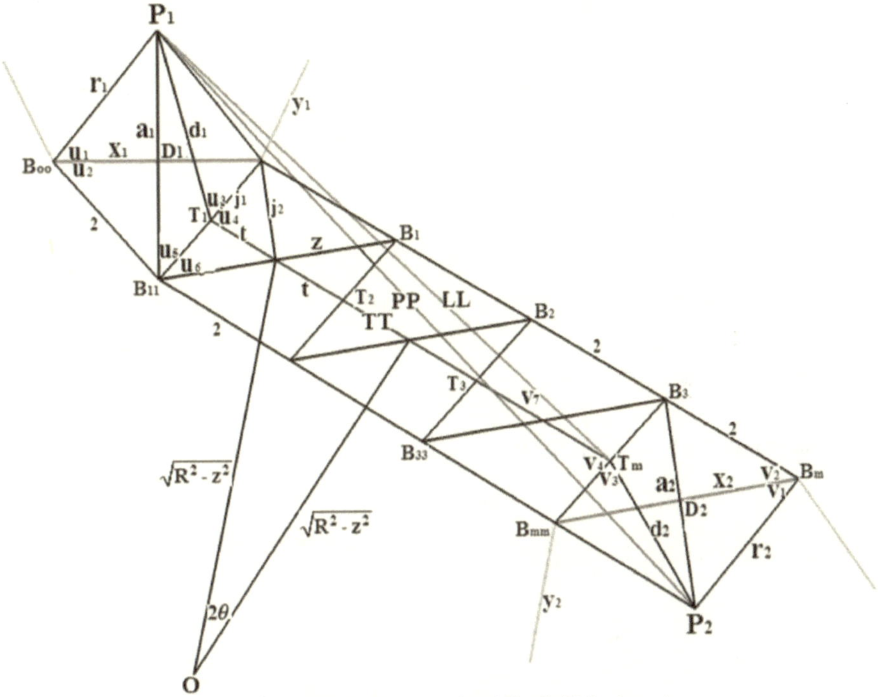

[Fig. IV-11a] The standard Tr-SnE(k1-k2, m)

Solution steps:

(1) Find x_1 and x_2 from the stack formulae, and

Set z as z = 1 or $z = \min[\, z_d = \sqrt{\dfrac{3R^2 - 4}{R^2 - 1}},\ z_x = \dfrac{R^2 x}{\sqrt{(R^2 - 2)^2 - R^2 x^2}}\,]$,

where $x = \min[\, x_1, x_2\,]$.

(2) $C_i = \mathrm{Cos}(\dfrac{\pi}{k_{si}})$ $S_i = \mathrm{Sin}(\dfrac{\pi}{k_{si}})$ **where** k_{si} **is the number of strings connected on Pi-face.**

$$r_i = \frac{1}{S_i}\sqrt{x_i^2 + y_i^2 + 2C_i x_i y_i} \quad r_i' = \sqrt{2R[R - \sqrt{R^2 - r_i^2}\,]}$$

$$x_i' = \sqrt{2R[R - \sqrt{R^2 - x_i^2}\,]} \quad z' = \sqrt{2R[R - \sqrt{R^2 - z^2}\,]}$$

$j_1 = 1 \quad j_1' = \sqrt{2R[R-\sqrt{R^2-1}]}$

$j_2 = \sqrt{4-z^2} \quad j_2' = \sqrt{2R[R-\sqrt{R^2-j_2^2}]}$

$\hat{u}_1 = [r_1', 2x_1, r_1'] \quad \hat{v}_1 = [r_2', 2x_2, r_2']$

$\hat{u}_2 = [2, 2x_1, 2] \quad \hat{v}_2 = [2, 2x_2, 2]$

$d_1' = [r_1', j_1', (\hat{u}_1 + \hat{u}_2)] \quad d_2' = [r_2', j_1', (\hat{v}_1 + \hat{v}_2)]$

$a_1' = [r_1', 2, (\hat{u}_1 + \hat{u}_2)]$

$\hat{u}_5 = [a_1', 2, r_1'] \quad \hat{u}_6 = [2, 2z, 2]$

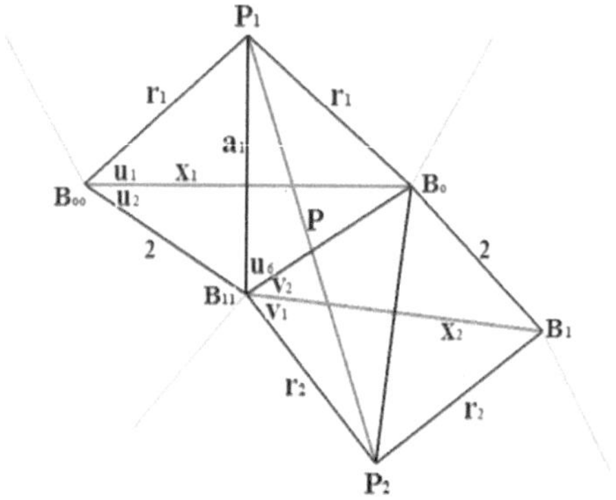

Fig. IV-11b. Tr-SnE(k1-k2, 1)

(3) If m = 1 (Fig. IV-11b) then $P' = [a_1', r_2', (\hat{v}_1 + \hat{v}_2 + \hat{u}_6)]$ **, go to (5).**

(4) $\sin\theta = \dfrac{1}{\sqrt{R^2 - 1}} \quad t' = R\sin\theta$

$\hat{u}_3 = [d_1', j_1', r_1'] \quad \hat{v}_3 = [d_2', j_1', r_2']$

$\hat{u}_4 = [t', j_1', j_2'] \quad \hat{v}_4 = \hat{u}_4$

$T' = 2R\sin[(m-1)\theta] \quad L' = [d_1', T', (\hat{u}_3 + \hat{u}_4)]$

$\hat{v}_7 = [L', T', d_1'] \quad P' = [L', d_2', (\hat{v}_3 + \hat{v}_4 + \hat{v}_7)]$

(5) To find the mother LLo (= $P'_{01}P'_{02}$): (Fig. IV-11c)

$$S_{oi} = \mathrm{Sin}\left(\frac{\pi}{k_i}\right) \qquad C_{oi} = \mathrm{Cos}\left(\frac{\pi}{k_i}\right)$$

$$a_{oi} = \frac{C_{oi}}{S_{oi}}$$

$$\mathrm{Sin}(\theta_i) = \frac{a_{oi}}{\sqrt{Rn^2 - 1}} \qquad P_o' = 2R_n \,\mathrm{Sin}\left[\frac{\theta_1+\theta_2}{2}\right]$$

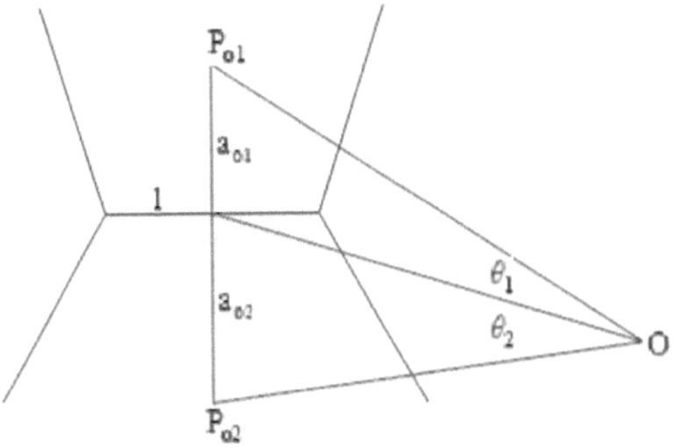

Fig. IV-11c. Two adjacent P-faces of the mother configuration

(6) To find the R, Dv and Ds:

$$R = \frac{P'}{Po'} Rn \quad Dv = \frac{100N}{(R-1)^3}\% \quad Ds = \frac{100N}{4R^2}\%$$

**** The notation of $\hat{u} = [e_1, e_2, e_0]$ is that the cosine of the surface angle $\hat{u}$ which is calculated by using the formula**

$$\mathrm{Cos}(\hat{u}) = \frac{(e_1^2 + e_2^2 - e_o^2)R^2 - 2e_1^2e_2^2}{2e_1e_2\sqrt{R^2 - e_1^2}\sqrt{R^2 - e_2^2}}$$

**** The notation $e_0 = [e_1, e_2, \hat{u}]$ is that the length e_0 which is calculated by using the formula**

$$e_0 = \frac{1}{R}\sqrt{\left(e_1^2 + e_2^2\right)R^2 - 2e_1e_2\left(e_1e_2 + \sqrt{R^2 - e_1^2}\sqrt{R^2 - e_2^2}\,\mathrm{Cos}(\hat{u})\right)}\,.$$

The following figures are the setting, the layered diagram, and the top-view diagram of Tr-S12E'1(4-3, 1) as an illustration of solving a Tr-SnE'(k1-k2, m).

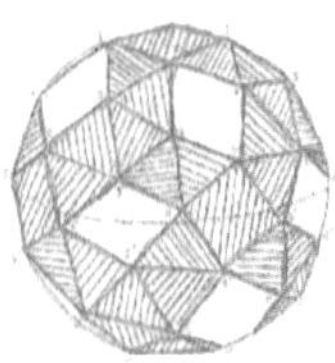

Fig. IV-12a. The setting

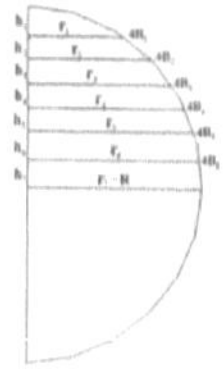

Fig. IV-12b. The layered diagram

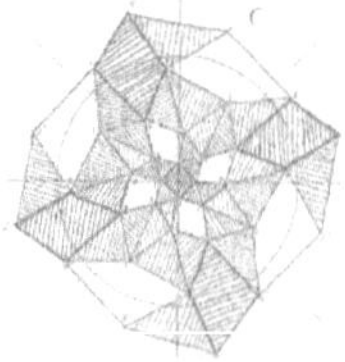

Fig. IV-12c. The top-view diagram

The solution: The R of Tr-S12E'1(4-3, m) is

$$R = 3.7701264\ Ds = 84.42468$$

**** The value of R determined through this procedure is from *R à: LT = RT, which does not yield the virtual value.* So far, no virtual value of R of a Tr-SnE'(k1-k2, 1) has been successfully found except while the mothers of the Tr- SEs are nEs, and they are equivalent to the three TEs as in the following:**

The Tr-SnE configuration from a mother nE while m = 1 is a TE

(1) *Tr-S4E (3, 1) = T12E* with 4 △ - P-faces, 4 △ - Open-faces and 6 Tr-strings of one tr-pearl.

$$R = \sqrt{\frac{1}{2}(5+\sqrt{5})} \doteq 1.90211303259....$$

(2) ***Tr-S6E (3, 1) = T24E*** **with 6 □ - P-faces, 8 △ - Open-faces, and 12 Tr-strings of one tr-pearl.**

$$R = \sqrt{\frac{1}{3}\left[\left(199 + \sqrt{297}\right)^{\frac{1}{3}} + \left(199 - \sqrt{297}\right)^{\frac{1}{3}} + 10\right]} \doteq 2.6874267474892....$$

(3) ***Tr-S8E (4, 1) = T24E*** **with 8 △ - P-faces, 6 □ - Open-faces, and 12 Tr-strings of one tr-pearl.**

R = the same as that of Tr-S6E(3,1)

(4) ***Tr-S12E (5, 1) = T60E*** **with 12 - P-faces, 20 △ - Open faces, and 30 Tr-strings of one tr-pearl.**

$$R = \sqrt{A + B + \frac{1}{6}\left(27 + 7\sqrt{5}\right)} \doteq 4.311674750227544....$$

where $\frac{A}{B} = \frac{1}{3}\sqrt[3]{\frac{1}{2}\left[\left(5112 + 2285\sqrt{5}\right) \pm \sqrt{64233 + 28728\sqrt{5}}\right]}$

(5) ***Tr-S20E (3, 1) = T60E*** **with 20 △ - P-faces, 12 5-gon-Open-faces, and 30 Tr-strings.**

R = the same as that of Tr-S12E(5, 1)

A situation while inserting tr-strings:

If the mother configuration has a third set of all squares, it is impossible to insert a tr-string between the square of this set with any P-faces due to criterion-a. (See **Fig. IV-13** at right, the two corner angles $\phi > \pi$.)

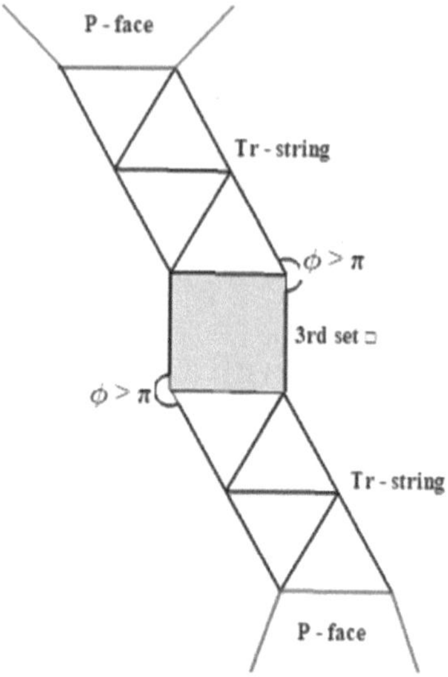

Fig. IV-13

The mother configurations $\mathbf{24E_1'}$, $\mathbf{48E'}$, $\mathbf{60E_2'}$ and **120E'** have a third set of squares; these squares cannot be connected with a tr-strings.

Therefore, it is impossible to transform $\mathbf{24E_1'}$ and $\mathbf{60E_2'}$ to **Tr-SE.**

In **48E'**, it is possible to insert tr-strings between every **6-gon** and **8-gon** P-faces.

In **120E'**, it is possible to insert tr-strings between every **6-gon** and **10-gon** P-faces.

[Note that there is no problem for inserting sq-strings in any E or E'.]

[IV – 6] Equations and Criteria for Constructing Tr-SE Families

Equations:

$C = \mathrm{Cos}(\frac{\pi}{k_s})$ $\quad S = \mathrm{Sin}(\frac{\pi}{k_s})$ **While k = 3, 4, 5 ks = k**

While k = 6, 8, 10 ks = k/2

For k = 3, 4, or 5: (See Fig. IV-14a.)

$$\mathrm{Sin}\left(\frac{\widehat{\phi}_p}{2}\right) = \frac{CR}{\sqrt{R^2 - x^2}}$$

$$\widehat{\phi}_o = 2\pi - \left(\widehat{\phi}_p + 2\widehat{\phi}_1 + \widehat{\phi}_z\right)$$

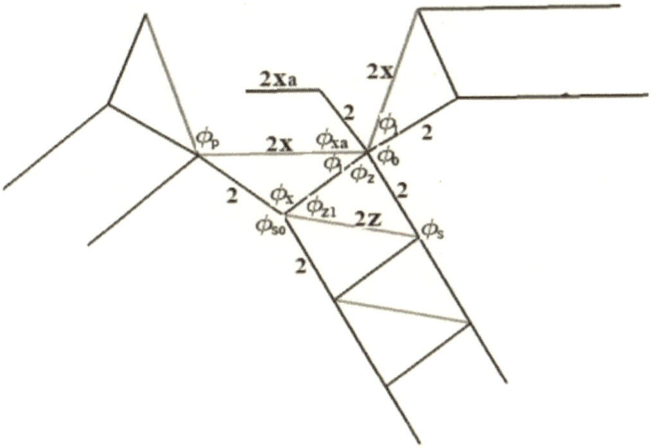

Fig. IV-14a. While k = 3, 4, 5

For k = 6, 8, or 10: (See Fig. IV-14b.)

$$\mathrm{Cos}\left(\widehat{\phi}_p\right) = -\frac{CR^2 + x}{\sqrt{R^2 - x^2}\sqrt{R^2 - 1}}$$

$$\widehat{\phi}_o = 2\pi - \left(\widehat{\phi}_p + \widehat{\phi}_1 + \widehat{\phi}_z\right)$$

$$\widehat{\phi}_{oo} = 2\pi - \left(\widehat{\phi}_p + \widehat{\phi}_1\right)$$

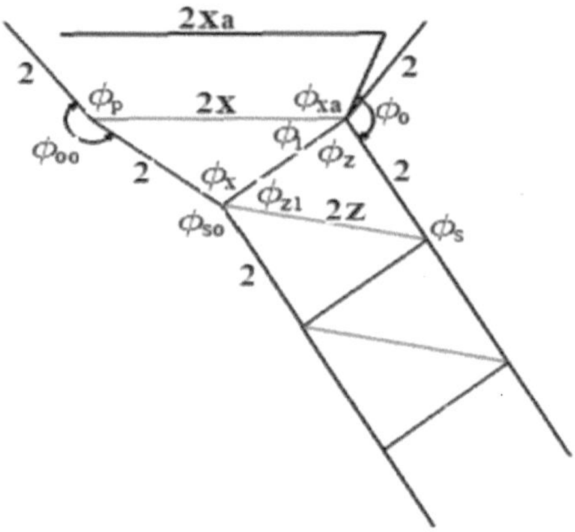

Fig. IV-14b. While k = 6, 8, 10

For all k: $\operatorname{Sin}\left(\frac{\phi_x}{2}\right) = \frac{xR}{2\sqrt{R^2-1}}$

$$\operatorname{Sin}\left(\frac{\hat{\phi}_z}{2}\right) = \frac{zR}{2\sqrt{R^2-1}}$$

$$\operatorname{Cos}\left(\hat{\phi}_{xa}\right) = \frac{\left(R^2-2\right)x - R^2 x_a}{2\sqrt{R^2-x^2}\sqrt{R^2-1}}$$

$$\operatorname{Cos}\left(\hat{\phi}_1\right) = \frac{\left(R^2-2\right)x}{2\sqrt{R^2-x^2}\sqrt{R^2-1}}$$

$$\operatorname{Cos}\left(\hat{\phi}_{z1}\right) = \frac{\left(R^2-2\right)z}{2\sqrt{R^2-z^2}\sqrt{R^2-1}}$$

$$\hat{\phi}_{so} = 2\pi - \left(\hat{\phi}_x + 2\hat{\phi}_{z1}\right)$$

The criteria for constructing Tr-SE:

The above equations can provide the requirements of the criteria for constructing Tr-SEs:

Criterion – 1: **For** $\widehat{\phi}_x < \pi$ **, which requires** $\frac{2\sqrt{R^2-1}}{R} > x \geq 1$

Criterion – 2: **For** $\widehat{\phi}_{so} < \pi$ **, which requires** $z < \frac{R^2 x}{\sqrt{(R^2-2)^2 + R^2 x^2}}$

Criterion – 3: **For** $d_z \geq 1$**, which requires** $z \leq \sqrt{\frac{3R^2-4}{R^2-1}}$

Criterion – 4: **For** $d_x \geq 1$**, which requires** $x \leq \sqrt{\frac{3R^2-4}{R^2-1}}$ **while** $x_a = 0$

$d_x \geq 1$ **is guaranteed, while** $x_a \geq 1$

Criterion – 5: **For** $\widehat{\phi}_{oo} < \pi$ **, which requires** $x < \frac{2CR^2}{R^2-4}$

Criterion – 6: **Can y > 1 while k = 6, 8, or 10?**

Answer**: Yes, y can be > 1. But this case is ruled out for the same reason as that for Sq-SE.**

Criterion – 7: **While x is in its range, what should** x_a **be? Or what kind of stacks can be stacked on P-faces?**

Answer:

For k = 3, 4, or 5: $x_a = 0$ **only, --------- only (1 k) is possible.**

For k = 6, 8, or 10: $x_a = 0$ **---------------- only** $(4\ \overset{\sim 2}{8})$ **and** $(5\ \overset{\sim 2}{10})$ **are possible.**

$x_a = 1$ **-------------- only** $(5\ \overset{-\ 2}{10})$ **is possible.**

$x_a = \sqrt{\frac{5-\sqrt{5}}{2}}$ **------ only** $(1\ 5\ \overset{\sim 2}{10})$ **and** $(1\ 5\ \overset{-\ 2}{10})$ **are possible**

Criterion – 8: **Is $\hat{\phi}_0 < \pi$?**

Answer:

Yes, it is true for all k and all m.

Criterion – 9: $\bar{\phi}_0 \geq 60^\circ$ **?**

$\bar{\phi}_0 \geq 60^\circ$ ***is true for all k while m > 1, but not always true for m = 1.***

While m = 1, **the $\hat{\phi}_{01}$ or $\hat{\phi}_{02}$ could be $< 60^\circ$.**

The following is a list for one-tr-pearl which can be inserted between two P-faces. (The P-faces could be loaded or not).

(The sign "–" means a one-tr-pearl string can be inserted between.)

A 3-gon P-face – all other P-faces except (1 3).	**(1 3) – (1 3) only.**
A 4-gon P-face – All, except (1 5) and (4 8).	**(1 4) – all, except an unloaded 5-gon, (1 3), and (1 5).**
A 5-gon P-face – All, except (1 3) and (1 4).	**(1 5) – all, except a 5-gon, (1 3), and (1 4).**

(** ***The 4-gons mentioned above are not in the third set of 4-gons of a mother configuration.)***

A 6-gon P-face -- All, except (1 3)	
An 8-gon P-face – All, except (1 3).	$(4\overset{2}{\sim}8)$ **–All, except 4-gon, and (1 3).**

A 10-gon P-face – All, but (1 3).

$(5\overset{\sim 2}{10})$ **– All but (1 3).**

$(5\overset{-2}{10})$ **– all, but (1 3).**

$(1\ 5\overset{\sim 2}{10})$ **– all, but (1 3)**

$(1\ 5\overset{-2}{10})$ **– all, but (1 3)**

Note: *1. Between two the same k-gon P-faces, it can always have a string inserted. The loaded arrangement on these two P-faces should be identical.

***2. There are some k-gon-faces that are never adjacent with some other k-gon-faces in all nE or nE', such as a 5-gon is never adjacent with an 8-gon or a 10-gon; or an 8-gon is never adjacent with a 10-gon.**

So in the above list, it does not need to mention the possibilities of inserting strings between these never-neighbored k-gon faces.

Conclusion of the criteria for constructing Tr-SE:

It is impossible to add balls on a tr-pearl.

For k = 3, 4, or 5: The only possible stack to add on a 3, 4, or 5-gon P-face is (1 k), $x_a = 0$.

For this case, the range of x is: $\sqrt{\frac{3R^2 - 4}{R^2 - 1}} \geq x \geq 1$

For k = 6, 8, or 10: y = 1, the only possible stacks to add on are: $(4\overset{\sim 2}{8})$, $(5\overset{\sim 2}{10})$, $(1\ 5\overset{\sim 2}{10})$, $(5\overset{-2}{10})$, $(1\ 5\overset{-2}{10})$

For all k, the range of x is: min $\left[\sqrt{\frac{3R^2-4}{R^2-1}},\ \left(\frac{2CR^2}{R^2-4}\right)_{-}\right] \geq x \geq 1$

And the range of z is: min $\left(\sqrt{\frac{3R^2-4}{R^2-1}}, \left(\frac{R^2x}{\sqrt{(R^2-2)^2+R^2x^2}}\right)^-\right) \geq z \geq 1$

While m = 1, the possible arrangements on two P-faces are the same as listed in *Criterion 9.*

The data:

The data of the basic Tr-SnE(k1-k2, 1) are in the DATAFILE: [DATA 5-1] The Basic Tr-SE and Equivalent Configurations.

The data for comparison of R of Sq-SE and Tr-SE are in DATAFILE: [DATA 5-2].

The data of Tr-SE and their families are in the DATAFILE: [DATA 5-3].

The data of Tr-SE' and their families are in DATAFILE: [DATA 5-4].

[IV – 7] Long-string Even Distributions

The shape of an SE with very long strings (no matter it is a Sq-SnE or Tr-SnE or whether the P-faces are loaded or not) resembles to a certain **nE or AnE**. The very long strings of the new configuration look like the lines of edges on the new surface. The P-face, regardless of what k-gon is, or loaded or not, would be condensed as a point on the surface. The open faces become the faces of the new configuration. **[See Fig. IV-15a, 15b, 15c]**

Let **Pi** (i = 1 and 2) be the centers of the P-faces and **Pi'** be the point on the surface above the point **Pi.**

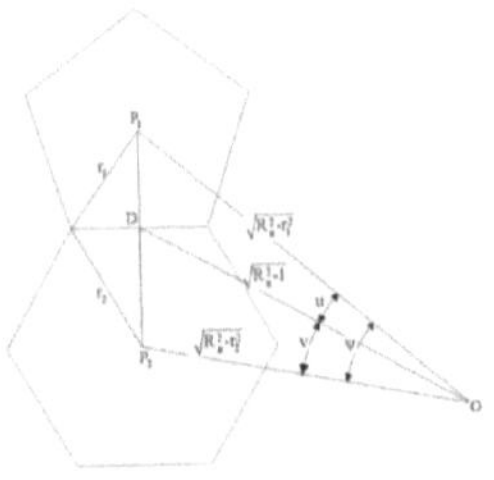

Fig. IV-15a

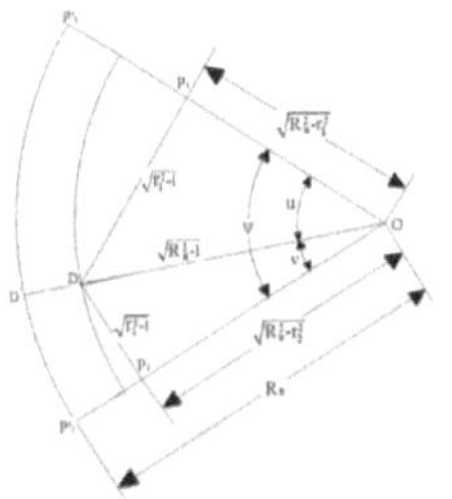

Fig. IV-15b

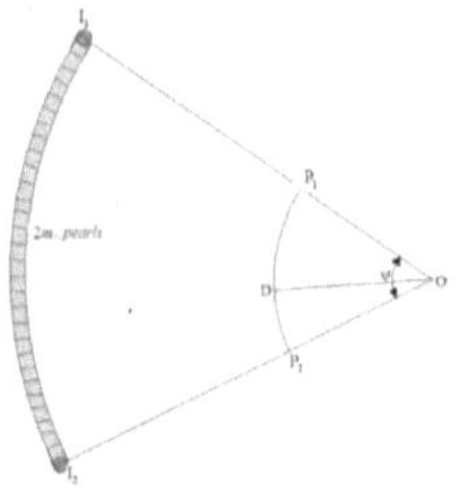

Fig. IV-15c

The [Fig. IV-15a] shows the two P-faces of a mother configuration.

The [Fig. IV-15b] is the plane which contains Pi, Pi' and the center O of the mother configuration. The angles U and V are defined as shown in the figure.

The [Fig. IV-15c] is an illustration while the string is very long and so the two P-faces are condensed as two points **I1** and **I2** on the surface.

The angles $\boldsymbol{\theta_i}$ and Ψ are defined as shown in the figures.

Calculation:

The **Rn**, k_1, k_2, and **m** are known, and

$$S_i = \mathrm{Sin}\left(\frac{\pi}{k_i}\right) \quad \text{and} \quad C_i = \mathrm{Cos}\left(\frac{\pi}{k_i}\right)$$

$$r_i = \frac{1}{S_i} \qquad (1)$$

$$\mathrm{Sin}\left(U_i\right) = \frac{\sqrt{r_i^2 - 1}}{\sqrt{R_n^2 - 1}} \qquad (2)$$

$$\psi = U_1 + U_2 \qquad (3)$$

$$P_1'P_2' = R_n\psi \tag{4}$$

$$I_1I_2 = 2m = R\psi \tag{5}$$

$$R = \frac{2m}{\psi} = \frac{2m}{U_1 + U_2} \tag{6}$$

The Eq. (6) is called the "long string formula" .

Note: If m is not a very large number, the R of Sq-SE or Tr-SE should not be calculated this way.

The data

The data of the R and Ds of basic Sq-SE and Tr-SE (for comparison), and their long-string resemblance configurations are in the DATAFILE: [DATA 5-2].

The data of the Tr-SnE(k-k, m), for m = 1, 2, 3, 10, 100 and families are in the DATAFILE: [DATA 5-3].

The data of the Tr-SnE'(k1-k2, m), for m = 1, 2, 3, 10, 100 and families are in the DATAFILE: [DATA 5-4].

**** Most of the high surface density configurations are among the TEs and Tr-SEs.**

The data of high Ds configurations are in DATAFILE:

[DATA 5-5].

[IV -- 8] A Special String Configuration

There is a very special class of string configuration, which has string pearl 6-gons, denoted as 6-gon-SE. The 6-gon-string can only exist between two P-faces of 5-gons or two 10-gons. In other words, only the mother E or E' with twelve 5-gons or twelve 10-gons can be transformed to 6-gon-SEs.

This special string configuration is initiated from a new high technology in science concerns fullerene, nano-carbon-tube, etc., all with the construction of 6-gon of carbon molecules. The category of fullerene involves buckminsterfullerene (a 60-carbon molecule constructed as the **60E'1**) with some string of hexagons inserted between two hexagons. Each of these two hexagons is one of the 5 hexagons surrounding a pentagon. The 6-gon-SE has no open-face, because every open-face is filled with many other hexagons. Every hexagon in the 6-gon-SE has all edges of lengths **2**, but most of them are non-regular 6-gons [see **Fig. IV-16**].

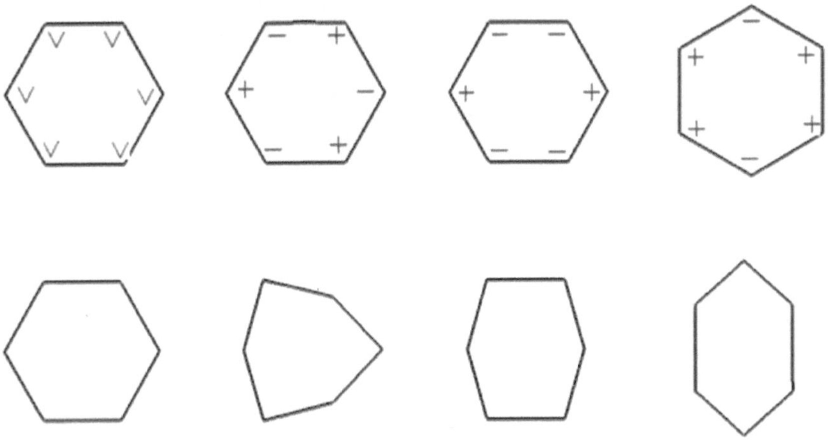

Fig. IV-16

In Fig. IV-16. The sign "v" means the straight angle is exactly 120 deg. (The corresponding surface angle is > 120 deg.)

The sign "+" means the straight angle is > 120 deg. (The corresponding surface angle is certainly > 120 deg.)

The sign "–" means the straight angle is < 120 deg. (The corresponding surface angle could be <, =, or > 120 deg.)

The top four hexagons are while all inner angles are close to 120 deg.

The bottom four hexagons are while the inner angles are obviously =, or >, or < than 120 deg.

There is no three regular hexagon which can be intersected at one point on a surface, because otherwise, the total surface angle of these three will exceed 360 deg. While inserting a 6-gon string between two P-faces, there could have many three-6-gon joint at one point cases. The point that three 6-gon jointed is called the **"3-corner,"** or just **3C**. So, ***if there is a 3C existing on a surface, then at least one of these three 6-gons is not a regular 6-gon, or at least one of the three 3C straight angles is less than 120 deg.***

Some examples:

(1) To insert one 6-gon string between every 5-gon of a 20E, it becomes **20E**$_{(5^{-}5)}$**.** [See **Fig. IV-17**]. The notation is

6-gon-S20E(5-5, 1) = 20E$_{(5^{-}5)}$.

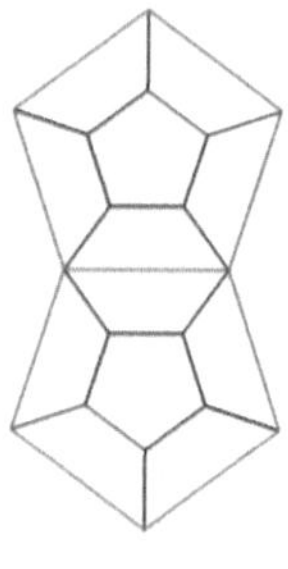

Fig. IV-17

For this case there are three one-pearl-6-gon strings joint at one point forms a 3C. All these three 6-gons are similar irregular 6-gons, that the two angles adjacent with the pentagon are > 120 deg. And the 3C angles are < 120 deg.

We already know the R, N and Ds of **20E** $(5^{-}5)$ are

R = 5.72943574
N = 80
Ds = 60.93
e = 2.0443907 [the enlarged edge of 20E $(5^{-}5)$ **]**

From the length of the enlarged edge **e**, we can calculate the 3C angles.

(2) To insert two 6-gons between every two 5-gons of a 20E, which is equivalent to that adding a stack $(5^{-}5^{\sim}\overset{2}{10})$ on every 10-gon of **60E'3** with 3-gon at the enlarged edge, **[see Fig. IV-18].** The notation of the configuration is

6-gon-S20E(5-5, 2) = 60E'3 $(5^{-}5^{\sim}\overset{2}{10})$ **3 at e**

R = 8.58517
N = 180 Fig. IV-18
Ds = 61.054
e = 1.825888

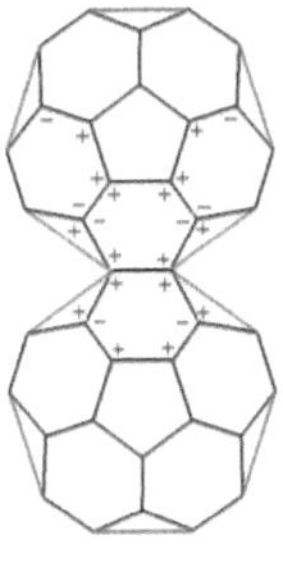

Fig. IV-18

(3-1) To insert three 6-gons between every two 5-gons of a 20E, which is equivalent as to insert one 6-gon between every 10-gon of a **60E'3 [see Fig. IV-19, the point on the 3-gon face is not an added ball. It is just the center of the 3-gon face]**, with every 10-gon of the 60E'3 loaded with a stack $(5^{-}5^{\sim}\overset{2}{10}^{\sim}\overset{2}{10})$, and the string is at the enlarged edge of the 10-gon. The notation of the configuration is

6-gon-S20E(5-5, 3) = 60E'3[(10S)10 at e, 59. $(5^{-}5^{\sim}\overset{2}{10}^{\sim}\overset{2}{10})$ **]**

R = 10.512126
N = 300
Ds = 67.87%
e = 2.649571 Fig. IV-19

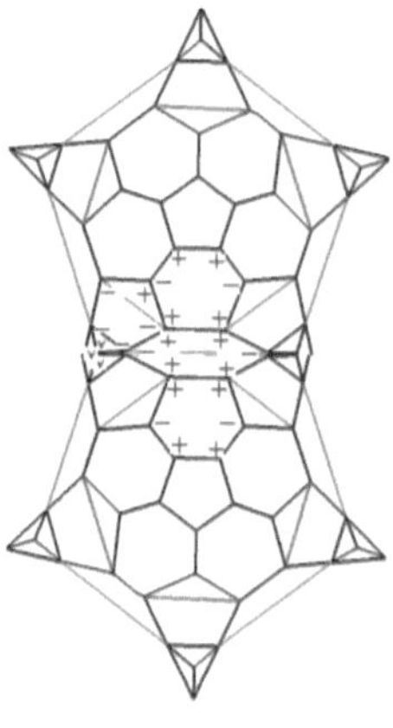

Fig. IV-19

(3-2) To insert three 6-gons between every two 5-gons of a 20E, which is equivalent as to insert one 6-gon between every two 10-gons of a **60E'3 [see Fig. IV-19, the point on the 3-gon face is an added ball, and this configuration is different than that of the above (3-1) case]**. This time, the 3-gons serve as M-faces as loaded with one ball. And the 10-gon faces serve as A-faces with the loaded stack $(5^{-}5^{\sim}\overset{2}{10}^{\sim}\overset{2}{10})$, and the string is at the enlarged edge of the 10-gon. The notation is

6-gon-S20E(5-5, 3) = 60E'3[(3M)+(10A)] [1.(1 3) + 59. $(5^{-}5^{\sim}\overset{2}{10}{}^{\sim}\overset{2}{10})$ **]**

R = 11.611788
N = 320
Ds = 59.33%
e = 2.220481 (the enlarged edge of 10-gon)
e = 1.862631 (the enlarged edge of 3-gon)

**** The 6-gon-pearl SE is a new division in the class of SE. The study of it is in a very primary stage. The calculation for solving a general case of a 6-gon-SE(k1-k2, m) is to be developed. The question of whether it can be an infinite packing is still in waiting.**

POSTSCRIPTS

A Tribute to the "Buckminsterfulerine"

The author was amazed by the chemical character and structure of the 60-carbon-molecule, C60, nicknamed Buckminsterfulerene (or Bucky ball). She cut papers to make 12-pentagon and 20-hexigon and paste them to imitate the C60. She calculated its radius. Then she started to cut and paste papers to make the five basic Es, eleven E's and so on, and do their calculations. All this study was initiated from the **Buckminsterfulerene.** The structure of Buckminsterfullerene is same as that of **60E'1.**

The head signs

The head signs, such as $^{\sim}$, $^{-}$, $^{\vee}$, and ^{2}, as well as the head signs of the surface angle $\widehat{\phi}$ and the straight angle $\bar{\phi}$, developed in this study have been used almost everywhere. They play very important roles in this study. Without these signs, it is impossible to see the construction of arrangements of a system and the structure with many different distances, locations, and angles among balls and layers, and so the entire system would be very confused. The consequence would be no equations can be developed, no analysis can be processed, and simply this study would be jeopardized. Hence, it is appreciated for the invention of these signs, even though they look like very small things.

So far the highest surface density **of all packs, including all Es, E's, AEs, TE's, SEs, and all E"s, is the**

T60E' [2.(5 $\tilde{}$ 5) + 2.(3 $\tilde{}$ 3)]. Its R and Ds are approximately

R ≑ 7.2896
Ds ≑ 84.68473 %

The second highest Ds is that of Tr-S30E'(5-3, 1):

R ≑ 5.9581525
Ds ≑ 84.50804 %

The third highest Ds is that of Tr-S30E'(5- $\dot{3}$, 1) (similar to T60E: (P5-S) 2. (5 $\tilde{}$ 5) or (G3-S) 2. (3 $\tilde{}$ 3)). Its R and Ds are

R 5.9585
Ds ≑ 84.49843 %

The author has been struggling **and working very hard in attempting to find the virtual values of R of basic Sq-SnE'(k1-k2, 1) and Tr- SnE'(k1-k2, 1), but failed in most of them. She hopes someday, someone will find them.**

Anyone who have found the virtual value of R of any Sq-SnE(k1-k2, 1) while $k_1 \neq k_2$ or one of the k_1 and k_2 is not 4 or any Tr-SnE(k_1 - k_2, 1) while the mother configuration is not an nE, please release your method to the author at ***ycachang@gmail.com*** **or send it by mail to: Ying-chien Chang, 7 Coventry Lane, Athens, OH 45701.**

It will be very highly appreciated!

Some open questions and deep thoughts**:**

** ***Question 1:*** **How to pack 5 balls in a spherical container under the packing criteria?**
** ***Question 2:*** **How to explain the central-symmetric situation of 4E?**

The data:

* **The data by hand calculation of all basic Es, E's, AEs, TEs, Tr-SEs and some of the Sq-SEs are 100% accurate.**
* **The data generated by computer are mostly about 99.5% accurate.**
* **The data in the DATAFILE were very patiently done and have been double- and triple-checked.**
* **All E"s are constructed with at most four layers of balls, including the base layer.**

***A scanned-in datafile*, which covers all detailed data (such as all enlarged edges and the inner angles, even the reason while an E" failed in any criterion) of Es, E's, AEs, TEs, and SEs (Sq-SEs and Tr-SEs), and their family member E"s is not included in this printed transcripts. Such data file is available by request for research purposes.**

***A thought* aftermath on the study of Even Distribution:**
It is realized that no infinite ball stack can be added on any kind of faces. The only case of infinite number of ball packing is the SE packing with long, long strings containing an infinite number of pearls. Every infinite long string packing looks like one of the five original Es or one of the two AEs (A14E and A32E). This fact makes sense in that everything started from the original five Es, and so many varieties of E-extensions are generated from there; finally, everything returns to the origin.

Further study

In this study, all configurations of Es and E-Extensions have all their points floating on a spherical surface and so are called the **floating E-packing.** The spherical ball packing actually covers more than just the floating E-packing. There is a parallel way of packing to the floating packing; it's called **solid packing**. And there is an independent class of spherical ball packing called **intruding packing**, which is an asymmetric packing. All these will be included in the author's succeeding research efforts.

APPENDIX – The Existing Stacks

3-gon	S M A	R=50, Em ≐
$1\ (1\ 3)$	× ∨ ×	1.732
$2\ (3\ {}^{\sim}3)$	∨ ∨ ×	2
$3\ (3\ {}^{-}3)$	× ∨ ×	2.732
$4\ (\underline{6\ 3})$	∨ ∨ ×	3
$8\ (3\ {}^{\sim}3\ {}^{-}3)$	× ∨ ×	3.732
$16\ (\underline{6\ 3}\ {}^{-}3)$	× ∨ ×	4.732

4-gon	S M A	R=50, Em ≐	Stacks	S M A	R=50, Em ≐	Stacks	S M A	R=50, Em ≐
$1\ (1\ \overset{\vee}{4})\mathrm{e}$	× ∨ ×	1.414	$19\ (2\ {}^{\sim}\overset{2}{4}\ {}^{-}\overset{2}{4})$	∨ ∨ ∨	4	$54\ (1\ 4\ {}^{\sim}\overset{2}{8}\ {}^{\sim}4)\mathrm{e}$	∨ ∨ × Rc<6.191	2.956 (R=6.191)
$2\ (1\ \overset{1}{4})$	∨ ∨ ∨	1.732	$20\ (2\ {}^{-}\overset{2}{4}\ {}^{\sim}\overset{2}{4})$	× × ∨		$68\ (2\ {}^{\sim}\overset{2}{4}\ {}^{-}\overset{2}{4}\ {}^{\sim}\overset{2}{4})$	× × ∨	
$3\ (2\ {}^{\sim}\overset{\vee}{4})\mathrm{e}$	∨ ∨ ×	1.822	$21\ (2\ {}^{-}\overset{2}{4}\ {}^{-}\overset{2}{4})$	× × ∨		$69\ (2\ {}^{\sim}\overset{2}{4}\ {}^{-}\overset{2}{4}\ {}^{-}\overset{2}{4})$	∨ ∨ ∨	5.986
$4\ (2\ {}^{\sim}\overset{1}{4})$	∨ ∨ ∨	2	$23\ (4\ {}^{\sim}4\ {}^{-}4)\mathrm{e}$	× ∨ ×	3.346	$96\ (4\ {}^{-}\overset{2}{4}\ {}^{-}\overset{2}{4}\ {}^{\sim}\overset{2}{4})$	× × ∨	
$5\ (2\ {}^{-}\overset{1}{4})$	∨ ∨ ∨	2.732	$26\ (4\ {}^{\sim}4\ {}^{-}\overset{2}{4})$	× × ∨		$97\ (4\ {}^{-}\overset{2}{4}\ {}^{-}\overset{2}{4}\ {}^{-}\overset{2}{4})$	∨ ∨ ∨	6.977
$6\ (4\ {}^{\sim}4)\mathrm{e}$	∨ ∨ ×	1.932	$28\ (4\ {}^{-}\overset{2}{4}\ {}^{\sim}\overset{2}{4})$	× × ∨		$103\ (4\ {}^{-}4\ {}^{\sim}\overset{2}{8}\ {}^{-}4)\mathrm{e}$	∨ ∨ ×	3.926
$7\ (4\ {}^{-}4)\mathrm{e}$	× ∨ ×	2.414	$29\ (4\ {}^{-}\overset{2}{4}\ {}^{-}\overset{2}{4})$	∨ ∨ ∨	4.99	$107\ (4\ {}^{\sim}\overset{2}{8}\ {}^{\sim}4\ {}^{-}4)\mathrm{e}$	× ∨ ×	4.405
$8\ (4\ {}^{-}\overset{2}{4})$	∨ ∨ ∨	3	$30\ (4\ {}^{\sim}\overset{2}{8}\ {}^{\sim}4)\mathrm{e}$	∨ ∨ ×	3	$111\ (4\ {}^{-}\overset{2}{8}\ {}^{\sim}4\ {}^{-}4)\mathrm{e}$	× ∨ ×	4.4
$9\ (\underline{8\ 4})\mathrm{e}$	∨ ∨ ×	2.932	$32\ (4\ {}^{-}\overset{2}{8}\ {}^{\sim}4)\mathrm{e}$	∨ ∨ × Rc<9.805	(R=9.805) 2.958	$114\ (4\ {}^{\sim}\overset{2}{8}\ {}^{\sim}4\ {}^{-}\overset{2}{4})$	× × ∨	
$10\ (1\ 4\ {}^{\sim}4)\mathrm{e}$	∨ ∨ ×	2	$35\ (\underline{8\ 4}\ {}^{-}4)\mathrm{e}$	× ∨ ×	4.336	$116\ (4\ {}^{-}\overset{2}{8}\ {}^{\sim}4\ {}^{-}\overset{2}{4})$	× × ∨	
$13\ (1\ \overset{2}{4}\ {}^{\sim}\overset{2}{4})$	× × ∨		$36\ (\underline{8\ 4}\ {}^{-}\overset{2}{4})$	× × ∨		$153\ (8\ {}^{-}\overset{2}{8}\ {}^{\sim}8\ {}^{-}\overset{2}{4})$	∨ ∨ ×	
$14\ (1\ \overset{2}{4}\ {}^{-}\overset{2}{4})$	× × ∨		$43\ (1\ 4\ {}^{\sim}4\ {}^{-}4)\mathrm{e}$	× ∨ ×	3.41			
$18\ (2\ {}^{\sim}\overset{2}{4}\ {}^{\sim}\overset{2}{4})$	× × ∨		$46\ (1\ 4\ {}^{\sim}4\ {}^{-}\overset{2}{4})$	× × ∨				

5-gon	S M A	R=50, Em ≐	Stacks	S M A	R=50, Em ≐	Stacks	S M A	R=50, Em ≐
$1\ (1\ 5)$	× ∨ ×	1.175	$11\ (5\ {}^{\sim}\overset{2}{10}\ {}^{\sim}5)$	∨ ∨ ×	2.899	$37\ (5\ {}^{\sim}5\ {}^{\sim}\overset{2}{10}\ {}^{\sim}5)$	∨ ∨ × R=10.01297	(R=10.01297) 2.96
$2\ (5\ {}^{\sim}5)$	∨ ∨ ×	1.826	$13\ (5\ {}^{-}\overset{2}{10}\ {}^{\sim}5)$	∨ ∨ ×	2.911	$42\ (5\ {}^{-}5\ {}^{\sim}\overset{2}{10}\ {}^{-}5)$	∨ ∨ ×	3.722
$3\ (5\ {}^{-}5)$	× ∨ ×	2.174	$16\ (\underline{10\ 5}\ {}^{-}5)$	× ∨ ×	3.992	$46\ (5\ {}^{\sim}\overset{2}{10}\ {}^{\sim}5\ {}^{-}5)$	× ∨ ×	4.066
$4\ (\underline{10\ 5})$	∨ ∨ ×	2.824	$22\ (1\ 5\ {}^{\sim}5\ {}^{-}5)$	× ∨ ×	3.072	$50\ (5\ {}^{-}\overset{2}{10}\ {}^{\sim}5\ {}^{-}5)$	× ∨ ×	4.079
$5\ (1\ 5\ {}^{\sim}5)$	∨ ∨ ×	1.901	$25\ (1\ 5\ {}^{\sim}\overset{2}{10}\ {}^{\sim}5)$	∨ ∨ ×	2.946	$86\ (10\ {}^{-}\overset{2}{10}\ {}^{\sim}10\ {}^{-}5)$	∨ ∨ ×	4.714
$8\ (5\ {}^{\sim}5\ {}^{-}5)$	× ∨ ×	2.998	$27\ (1\ 5\ {}^{-}\overset{2}{10}\ {}^{\sim}5)$	∨ ∨ ×	2.966			

6-gon	S M A	R=50, Em ≐	Stacks	S M A	R=50, Em ≐	Stacks	S M A	R=50, Em ≐
$1\ (3\ {}^{\sim}\overset{2}{6})$	∨ ∨ ∨	1.731	$23\ (6\ {}^{\sim}6\ {}^{-}\overset{2}{6})$	× × ∨		$73\ (3\ {}^{\sim}\overset{2}{6}\ {}^{-}\overset{2}{6}\ {}^{\sim}\overset{2}{6})$	× × ∨	
$2\ (3\ {}^{-}\overset{2}{6})$	∨ ∨ ∨	1.998	$25\ (6\ {}^{-}\overset{2}{6}\ {}^{\sim}\overset{2}{6})$	× × ∨		$74\ (3\ {}^{\sim}\overset{2}{6}\ {}^{-}\overset{2}{6}\ {}^{-}\overset{2}{6})$	∨ ∨ ∨	5.18
$3\ (6\ {}^{\sim}6)\mathrm{e}$	∨ ∨ ×	1.731	$26\ (6\ {}^{-}\overset{2}{6}\ {}^{-}\overset{2}{6})$	∨ ∨ ∨	4.454	$75\ (3\ {}^{-}\overset{2}{6}\ {}^{\sim}\overset{2}{6}\ {}^{\sim}\overset{2}{6})$	× × ∨	

4 $(6{}^{-}6)$e	∨ ∨ ×	2.73	32 $(\underline{12\ 6}{}^{-}6)$e	× ∨ ×	3.72	76 $(3{}^{-}\overset{2}{6}{}^{\sim}\overset{2}{6}{}^{-}\overset{2}{6})$	∨ ∨ ∨	4.722
5 $(6{}^{-}\overset{2}{6})$	∨ ∨ ∨	2.73	33 $(\underline{12\ 6}{}^{-}\overset{2}{6})$	∨ ∨ ∨ Rc<6.462		81 $(\underline{6\ 3}{}^{-}3{}^{\sim}\overset{2}{6})$	× × ∨	
6 $(\underline{12\ 6})$e	∨ ∨ ×	2.73	36 $(12{}^{-}\overset{2}{12}{}^{\sim}6)$e	∨ ∨ × Rc<8.428	2.944 (R=8.428)	82 $(\underline{6\ 3}{}^{-}3{}^{-}\overset{2}{6})$	∨ ∨ ∨	5.71
7 $(1\ 3{}^{\sim}\overset{2}{6})$	∨ ∨ ∨	1.977	43 $(1\ 3{}^{\sim}\overset{2}{6}{}^{\sim}\overset{2}{6})$	∨ ∨ ∨	2.998	85 $(\underline{6\ 3}{}^{-}\overset{2}{6}{}^{\sim}\overset{2}{6})$	× × ∨	
8 $(1\ 3{}^{-}\overset{2}{6})$	∨ ∨ ∨	2.73	44 $(1\ 3{}^{\sim}\overset{2}{6}{}^{-}\overset{2}{6})$	∨ ∨ ∨	3.704	86 $(\underline{6\ 3}{}^{-}\overset{2}{6}{}^{-}\overset{2}{6})$	× × ∨	
9 $(3{}^{\sim}3{}^{\sim}\overset{2}{6})$	∨ ∨ ∨	2	45 $(1\ 3{}^{-}\overset{2}{6}{}^{\sim}\overset{2}{6})$	× × ∨		96 $(6{}^{\sim}6{}^{\sim}6{}^{-}6)$e	× ∨ ×	3
10 $(3{}^{\sim}3{}^{-}\overset{2}{6})$	∨ ∨ ∨	2.996	46 $(1\ 3{}^{-}\overset{2}{6}{}^{-}\overset{2}{6})$	× × ∨		103 $(6{}^{\sim}6{}^{\sim}6{}^{-}\overset{2}{6})$	∨ ∨ ∨ Rc<33.17677	3.99 (R=50)
11 $(3{}^{-}3{}^{\sim}\overset{2}{6})$	× × ∨		49 $(3{}^{\sim}3{}^{-}\overset{2}{3}{}^{\sim}\overset{2}{6})$	× × ∨		113 $(6{}^{-}\overset{2}{6}{}^{-}\overset{2}{6}{}^{\sim}\overset{2}{6})$	× × ∨	
12 $(3{}^{-}3{}^{-}\overset{2}{6})$	∨ ∨ ∨	3.725	50 $(3{}^{\sim}3{}^{-}\overset{2}{3}{}^{-}\overset{2}{6})$	∨ ∨ ∨	4.719	114 $(6{}^{-}\overset{2}{6}{}^{-}\overset{2}{6}{}^{-}\overset{2}{6})$	∨ ∨ ∨	6.17
13 $(3{}^{\sim}\overset{2}{6}{}^{\sim}\overset{2}{6})$	∨ ∨ ∨	2.976	55 $(3{}^{\sim}3{}^{\sim}\overset{2}{6}{}^{\sim}\overset{2}{6})$	∨ ∨ ∨	2.998	115 $(6{}^{\sim}6{}^{\sim}\overset{2}{12}{}^{-}6)$e	∨ ∨ ×	2.975
14 $(3{}^{\sim}\overset{2}{6}{}^{-}\overset{2}{6})$	∨ ∨ ∨	3.459	56 $(3{}^{\sim}3{}^{\sim}\overset{2}{6}{}^{-}\overset{2}{6})$	∨ ∨ ∨	3.726	117 $(6{}^{\sim}6{}^{-}\overset{2}{12}{}^{\sim}6)$e	∨ ∨ ×	2.998
15 $(3{}^{-}\overset{2}{6}{}^{\sim}\overset{2}{6})$	∨ ∨ ∨	2.998	57 $(3{}^{\sim}3{}^{-}\overset{2}{6}{}^{\sim}\overset{2}{6})$	× × ∨		119 $(6{}^{-}6{}^{\sim}\overset{2}{12}{}^{\sim}6)$e	∨ ∨ ×	2.998
16 $(3{}^{-}\overset{2}{6}{}^{-}\overset{2}{6})$	× × ∨		58 $(3{}^{\sim}3{}^{-}\overset{2}{6}{}^{-}\overset{2}{6})$	× × ∨		121 $(6{}^{-}6{}^{-}\overset{2}{12}{}^{\sim}6)$e	∨ ∨ × Rc<9.2826	2.954 (R=9.2826)
17 $(\underline{6\ 3}{}^{\sim}\overset{2}{6})$	× × ∨		61 $(3{}^{-}3{}^{-}\overset{2}{6}{}^{\sim}\overset{2}{6})$	× × ∨		151 $(\underline{12\ 6}{}^{\sim}\overset{2}{12}{}^{\sim}6)$e	∨ ∨ × R<6.897	2.916 (R=6.897)
18 $(\underline{6\ 3}{}^{-}\overset{2}{6})$	∨ ∨ ∨	3.992	62 $(3{}^{-}3{}^{-}\overset{2}{6}{}^{-}\overset{2}{6})$	× × ∨		152 $(\underline{12\ 6}{}^{\sim}\overset{2}{12}{}^{-}6)$e	∨ ∨ ×	3.725
19 $(6{}^{\sim}6{}^{\sim}6)$e	∨ ∨ ×	2	71 $(3{}^{\sim}\overset{2}{6}{}^{\sim}\overset{2}{6}{}^{\sim}\overset{2}{6})$	× × ∨				
20 $(6{}^{\sim}6{}^{-}6)$e	∨ ∨ ×	2.727	72 $(3{}^{\sim}\overset{2}{6}{}^{\sim}\overset{2}{6}{}^{-}\overset{2}{6})$	∨ ∨ ∨	4.7			

8-gon	S M A	R=50, Em ≐	Stacks	S M A	R=50, Em ≐	Stacks	S M A	R=50, Em ≐
1 $(4{}^{\sim}\overset{2}{8})$	∨ ∨ ∨	1.413	34 $(\underline{16\ 8}{}^{-}8)$e	× ∨ ×	3.339	78 $(4{}^{-}\overset{2}{8}{}^{\sim}4{}^{-}\overset{2}{8})$	∨ ∨ ∨ Rc3<9.805	3.89 (R=9.805)
2 $(4{}^{-}\overset{2}{8})$	∨ ∨ ∨	1.516	35 $(\underline{16\ 8}{}^{-}\overset{2}{8})$	× ∨ × Rc<13.7755		81 $(4{}^{\sim}\overset{2}{8}{}^{\sim}\overset{2}{8}{}^{\sim}\overset{2}{8})$	∨ ∨ ∨ Rc<8.9173 9	2.95 (R=8.9173 9)
3 $(8{}^{\sim}8)$e	∨ ∨ ×	1.585	38 $(16{}^{-}\overset{2}{16}{}^{\sim}8)$e	∨ ∨ ×	2.987	82 $(4{}^{\sim}\overset{2}{8}{}^{\sim}\overset{2}{8}{}^{-}\overset{2}{8})$	∨ ∨ ∨	4.077
4 $(8{}^{-}8)$e	× ∨ ×	1.763	41 $(1\ 4{}^{\sim}4{}^{\sim}\overset{2}{8})$	∨ ∨ ∨	1.931	83 $(4{}^{\sim}\overset{2}{8}{}^{-}\overset{2}{8}{}^{\sim}\overset{2}{8})$	∨ ∨ ∨ Rc<7.697	2.932 (R=7.697)
5 $(8{}^{-}\overset{2}{8})$	∨ ∨ ∨	2.411	42 $(1\ 4{}^{\sim}4{}^{-}\overset{2}{8})$	∨ ∨ ∨	2.513	84 $(4{}^{\sim}\overset{2}{8}{}^{-}\overset{2}{8}{}^{-}\overset{2}{8})$	∨ ∨ ∨	4.227
6 $(\underline{16\ 8})$e	∨ ∨ ∨	2.583	45 $(1\ 4{}^{\sim}\overset{2}{8}{}^{\sim}\overset{2}{8})$	∨ ∨ ∨	2.805	85 $(4{}^{-}\overset{2}{8}{}^{\sim}\overset{2}{8}{}^{\sim}\overset{2}{8})$	∨ ∨ ∨ Rc<8.236	2.949 (R=8.236)
7 $(1\ 4{}^{\sim}\overset{2}{8})$	∨ ∨ ∨	1.675	46 $(1\ 4{}^{\sim}\overset{2}{8}{}^{-}\overset{2}{8})$	∨ ∨ ∨	3.084	86 $(4{}^{-}\overset{2}{8}{}^{\sim}\overset{2}{8}{}^{-}\overset{2}{8})$	∨ ∨ ∨	4.132
8 $(1\ 4{}^{-}\overset{2}{8})$	∨ ∨ ∨	1.929	47 $(1\ 4{}^{-}\overset{2}{8}{}^{\sim}\overset{2}{8})$	∨ ∨ ∨	2.906	91 $(\underline{8\ 4}{}^{-}4{}^{\sim}\overset{2}{8})$	× × ∨	
9 $(2\ 4{}^{\sim}\overset{2}{8})$	∨ ∨ ∨	1.873	48 $(1\ 4{}^{-}\overset{2}{8}{}^{-}\overset{2}{8})$	× × ∨		92 $(\underline{8\ 4}{}^{-}4{}^{-}\overset{2}{8})$	∨ ∨ ∨	4.841
10 $(2\ 4{}^{-}\overset{2}{8})$	∨ ∨ ∨	2.336	53 $(2\ 4{}^{\sim}\overset{2}{8}{}^{\sim}\overset{2}{8})$	∨ ∨ ∨	2.885	95 $(\underline{8\ 4}{}^{-}\overset{2}{8}{}^{\sim}\overset{2}{8})$	× × ∨	
11 $(4{}^{\sim}4{}^{\sim}\overset{2}{8})$	∨ ∨ ∨	1.909	54 $(2\ 4{}^{\sim}\overset{2}{8}{}^{-}\overset{2}{8})$	∨ ∨ ∨	3.278	96 $(\underline{8\ 4}{}^{-}\overset{2}{8}{}^{-}\overset{2}{8})$	× × ∨	
12 $(4{}^{\sim}4{}^{-}\overset{2}{8})$	∨ ∨ ∨	2.445	55 $(2\ 4{}^{-}\overset{2}{8}{}^{\sim}\overset{2}{8})$	∨ ∨ ∨	2.995	106 $(8{}^{\sim}8{}^{\sim}8{}^{-}8)$e	× ∨ ×	2.69

13 $(4\,{}^{-}4\,{}^{\sim 2}8)$	∨ ∨ ∨	2	56 $(2\,4\,{}^{-2}8\,{}^{-2}8)$	× × ∨		110 $(8\,{}^{-}8\,{}^{\sim}8\,{}^{-}8)$e	× ∨ ×	2.749
14 $(4\,{}^{-}4\,{}^{-2}8)$	∨ ∨ ∨	2.925	59 $(4\,{}^{\sim}4\,{}^{-}4\,{}^{\sim 2}8)$	× × ∨		113 $(8\,{}^{\sim}8\,{}^{\sim}8\,{}^{-2}8)$	∨ ∨ ∨	3.834
15 $(4\,{}^{\sim 2}8\,{}^{\sim 2}8)$	∨ ∨ ∨	2.672	60 $(4\,{}^{\sim}4\,{}^{-}4\,{}^{-2}8)$	∨ ∨ ∨	3.85	115 $(8\,{}^{-}8\,{}^{\sim}8\,{}^{-2}8)$	∨ ∨ ∨	2.913
16 $(4\,{}^{\sim 2}8\,{}^{-2}8)$	∨ ∨ ∨	2.823	65 $(4\,{}^{\sim}4\,{}^{\sim 2}8\,{}^{\sim 2}8)$	∨ ∨ ∨	2.9	121 $(8\,{}^{-2}8\,{}^{\sim 2}8\,{}^{\sim 2}8)$	× × ∨	
17 $(4\,{}^{-2}8\,{}^{\sim 2}8)$	∨ ∨ ∨	2.728	66 $(4\,{}^{\sim}4\,{}^{\sim 2}8\,{}^{-2}8)$	∨ ∨ ∨	3.317	122 $(8\,{}^{-2}8\,{}^{\sim 2}8\,{}^{-2}8)$	∨ ∨ ∨	4.402
18 $(4\,{}^{-2}8\,{}^{-2}8)$	× × ∨		67 $(4\,{}^{\sim}4\,{}^{-2}8\,{}^{\sim 2}8)$	∨ ∨ ∨ Rc<24.106073	2.993 (R=24.106073)	123 $(8\,{}^{-2}8\,{}^{-2}8\,{}^{\sim 2}8)$	× × ∨	
19 $(\underline{8\ 4}\,{}^{\sim 2}8)$	× × ∨		68 $(4\,{}^{\sim}4\,{}^{-2}8\,{}^{-2}8)$	× × ∨		124 $(8\,{}^{-2}8\,{}^{-2}8\,{}^{-2}8)$	∨ ∨ ∨	5.216
20 $(\underline{8\ 4}\,{}^{-2}8)$	∨ ∨ ∨	3.44	69 $(4\,{}^{-}4\,{}^{\sim 2}8\,{}^{\sim 2}8)$	∨ ∨ ∨	2.929	125 $(8\,{}^{\sim}8\,{}^{\sim 2}16\,{}^{\sim}8)$e	∨ ∨ ×	2.76
21 $(8\,{}^{\sim}8\,{}^{\sim}8)$e	∨ ∨ ×	1.93	70 $(4\,{}^{-}4\,{}^{\sim 2}8\,{}^{-2}8)$	∨ ∨ ∨	3.407	127 $(8\,{}^{\sim}8\,{}^{-2}16\,{}^{\sim}8)$e	∨ ∨ ×	2.791
22 $(8\,{}^{\sim}8\,{}^{-}8)$e	∨ ∨ ×	2.347	71 $(4\,{}^{-}4\,{}^{-2}8\,{}^{\sim 2}8)$	∨ ∨ ∨ Rc<7.22931	2.923 (R=7.22931)	129 $(8\,{}^{-}8\,{}^{\sim 2}16\,{}^{\sim}8)$e	∨ ∨ ×	2.841
23 $(8\,{}^{-}8\,{}^{\sim}8)$e	∨ ∨ ∨	1.99	72 $(4\,{}^{-}4\,{}^{-2}8\,{}^{-2}8)$	× × ∨		131 $(8\,{}^{-}8\,{}^{-2}16\,{}^{\sim}8)$e	∨ ∨ ×	2.8877
25 $(8\,{}^{\sim}8\,{}^{-2}8)$	∨ × × $R_\phi < 6.13625$	2.848 (R=6.13625)	73 $(4\,{}^{\sim 2}8\,{}^{\sim}4\,{}^{\sim 2}8)$	× × ∨		170 $(16\,{}^{-2}16\,{}^{\sim}8\,{}^{-}8)$e	× ∨ ×	3.741
27 $(8\,{}^{-2}8\,{}^{\sim 2}8)$	∨ ∨ ∨	2.998	74 $(4\,{}^{\sim 2}8\,{}^{\sim}4\,{}^{-2}8)$	∨ ∨ ∨	3.509	173 $(16\,{}^{-2}16\,{}^{\sim}8\,{}^{-2}8)$	× × ∨	
28 $(8\,{}^{-2}8\,{}^{-2}8)$	∨ ∨ ∨	3.817	77 $(4\,{}^{-2}8\,{}^{\sim}4\,{}^{\sim 2}8)$	× × ∨				

10-gon	S M A	R=50, Em ≑	Stacks	S M A	R=50, Em ≑	Stacks	S M A	R=50, Em ≑
1 $(5\,{}^{\sim 2}10)$	∨ ∨ ∨	1.174	33 $(\underline{20\ 10}\,{}^{-2}10)$	× × ∨		74 $(5\,{}^{\sim 2}10\,{}^{-2}10\,{}^{-2}10)$	∨ ∨ ∨	3.511
2 $(5\,{}^{-2}10)$	∨ ∨ ∨	1.207	36 $(20\,{}^{-2}20\,{}^{\sim}10)$e	∨ ∨ ×	2.894	75 $(5\,{}^{-2}10\,{}^{\sim 2}10\,{}^{\sim 2}10)$	∨ ∨ ∨	2.952
3 $(10\,{}^{\sim}10)$e	∨ ∨ ×	1.485	39 $(1\,5\,{}^{\sim}5\,{}^{\sim 2}10)$	∨ ∨ ∨	1.777	76 $(5\,{}^{-2}10\,{}^{\sim 2}10\,{}^{-2}10)$	∨ ∨ ∨	3.499
4 $(10\,{}^{-}10)$e	× ∨ ×	1.616	40 $(1\,5\,{}^{\sim}5\,{}^{-2}10)$	∨ ∨ ∨	2.106	81 $(\underline{10\ 5}\,{}^{-}5\,{}^{\sim 2}10)$	× × ∨	
5 $(10\,{}^{-2}10)$	∨ ∨ ∨	2.172	43 $(1\,5\,{}^{\sim 2}10\,{}^{\sim 2}10)$	∨ ∨ ∨	2.409	82 $(\underline{10\ 5}\,{}^{-}5\,{}^{-2}10)$	∨ ∨ ∨	4.188
6 $(\underline{20\ 10})$e	∨ ∨ ×	2.481	44 $(1\,5\,{}^{\sim 2}10\,{}^{-2}10)$	∨ ∨ ∨	2.482	85 $(\underline{10\ 5}\,{}^{-2}10\,{}^{\sim 2}10)$	× × ∨ Rc<9.279	
7 $(1\,5\,{}^{\sim 2}10)$	∨ ∨ ∨	1.311	45 $(1\,5\,{}^{-2}10\,{}^{\sim 2}10)$	∨ ∨ ∨	2.459	86 $(\underline{10\ 5}\,{}^{-2}10\,{}^{-2}10)$	× × ∨	
8 $(1\,5\,{}^{-2}10)$	∨ ∨ ∨	1.382	46 $(1\,5\,{}^{-2}10\,{}^{-2}10)$	× × ∨		96 $(10\,{}^{\sim}10\,{}^{\sim}10\,{}^{-}10)$e	× ∨ ×	2.437
9 $(5\,{}^{\sim}5\,{}^{\sim 2}10)$	∨ ∨ ∨	1.737	49 $(5\,{}^{\sim}5\,{}^{-}5\,{}^{\sim 2}10)$	× × ∨		100 $(10\,{}^{-}10\,{}^{\sim}10\,{}^{-}10)$e	× ∨ ×	2.511
10 $(5\,{}^{\sim}5\,{}^{-2}10)$	∨ ∨ ∨	2.032	50 $(5\,{}^{\sim}5\,{}^{-}5\,{}^{-2}10)$	∨ . ∨ ∨	3.199	103 $(10\,{}^{\sim}10\,{}^{\sim}10\,{}^{-2}10)$	∨ ∨ ∨	3.555

$11\ (5^{-}5^{\sim}\overset{2}{10})$	∨ ∨ ∨	1.9	$55\ (5^{\sim}5^{\sim}\overset{2}{10}^{\sim}\overset{2}{10})$	∨ ∨ ∨	2.685	$105\ (10^{-}10^{\sim}10^{-}\overset{2}{10})$	∨ ∨ ∨	3.67
$12\ (5^{-}5^{-}\overset{2}{10})$	∨ ∨ ∨	2.378	$56\ (5^{\sim}5^{\sim}\overset{2}{10}^{-}\overset{2}{10})$	∨ ∨ ∨	2.906	$107\ (10^{\sim}10^{-}\overset{2}{10}^{\sim}\overset{2}{10})$	× × ∨ Rc<9.1211	
$13\ (5^{\sim}\overset{2}{10}^{\sim}\overset{2}{10})$	∨ ∨ ∨	2.308	$57\ (5^{\sim}5^{-}\overset{2}{10}^{\sim}\overset{2}{10})$	∨ ∨ ∨	2.838	$108\ (10^{\sim}10^{-}\overset{2}{10}^{-}\overset{2}{10})$	∨ ∨ ∨ $R_{\phi 2}$ <13 64	3.9 (R=13.64)
$14\ (5^{\sim}\overset{2}{10}^{-}\overset{2}{10})$	∨ ∨ ∨	2.345	$58\ (5^{\sim}5^{-}\overset{2}{10}^{-}\overset{2}{10})$	× × ∨		$111\ (10^{-}\overset{2}{10}^{\sim}\overset{2}{10}^{\sim}\overset{2}{10})$	× × ∨ Rd<8.835	
$16\ (5^{-}\overset{2}{10}^{-}\overset{2}{10})$	× × ∨		$59\ (5^{-}5^{\sim}\overset{2}{10}^{\sim}\overset{2}{10})$	∨ ∨ ∨	2.774	$112\ (10^{-}\overset{2}{10}^{\sim}\overset{2}{10}^{-}\overset{2}{10})$	∨ ∨ ∨	4.06
$17\ (\underline{10\ 5}^{\sim}\overset{2}{10})$	× × ∨ Rc<7.48		$60\ (5^{-}5^{\sim}\overset{2}{10}^{-}\overset{2}{10})$	∨ ∨ ∨	3.068	$113\ (10^{-}\overset{2}{10}^{-}\overset{2}{10}^{\sim}\overset{2}{10})$	× × ∨	
$18\ (\underline{10\ 5}^{-}\overset{2}{10})$	∨ ∨ ∨	3.026	$61\ (5^{-}5^{-}\overset{2}{10}^{\sim}\overset{2}{10})$	∨ ∨ ∨	2.964	$114\ (10^{-}\overset{2}{10}^{-}\overset{2}{10}^{-}\overset{2}{10})$	∨ ∨ ∨	4.499
19 $(10^{\sim}10^{\sim}10)e$	∨ ∨ ×	1.825	$62\ (5^{-}5^{-}\overset{2}{10}^{-}\overset{2}{10})$	× × ∨		115 $(10^{\sim}10^{\sim}\overset{2}{20}^{\sim}10)e$	∨ ∨ × R_{f3} >13.645	2.414 (R=13.645)
20 $(10^{\sim}10^{-}10)e$	∨ ∨ ×	2.099	$63\ (5^{\sim}\overset{2}{10}^{\sim}5^{\sim}\overset{2}{10})$	× × ∨		117 $(10^{\sim}10^{-}\overset{2}{20}^{\sim}10)e$	∨ ∨ × R_{f3} >13.645	2.414 (R=13.645)
21 $(10^{-}10^{\sim}10)e$	∨ ∨ ×	1.9	$64\ (5^{\sim}\overset{2}{10}^{\sim}5^{-}\overset{2}{10})$	∨ ∨ ∨	3.1	119 $(10^{-}10^{\sim}\overset{2}{20}^{\sim}10)e$	∨ ∨ ×	2.609
$23\ (10^{\sim}10^{-}\overset{2}{10})$	∨ ∨ ∨ $R_{\phi 2}$ <13.6 4	2.902 (R=13.64)	$67\ (5^{-}\overset{2}{10}^{\sim}5^{\sim}\overset{2}{10})$	× × ∨		121 $(10^{-}10^{-}\overset{2}{20}^{\sim}10)e$	∨ ∨ × R_{f3} >8.9983	2.627 (R=50)
$25\ (10^{-}\overset{2}{10}^{\sim}\overset{2}{10})$	∨ ∨ ∨	2.897	$68\ (5^{-}\overset{2}{10}^{\sim}5^{-}\overset{2}{10})$	∨ ∨ ∨	3.113	$147\ (\underline{20\ 10}^{\sim}\overset{2}{20}^{\sim}10)$	∨ ∨ ∨	2.985
$26\ (10^{-}\overset{2}{10}^{-}\overset{2}{10})$	∨ ∨ ∨	3.339	71 $(5^{\sim}\overset{2}{10}^{\sim}\overset{2}{10}^{\sim}\overset{2}{10})$	∨ ∨ ∨	2.944	148 $(\underline{20\ 10}^{\sim}\overset{2}{20}^{-}10)e$	∨ ∨ ×	3.297
32 $(\underline{20\ 10}^{-}10)e$	× ∨ ×	3.089	72 $(5^{\sim}\overset{2}{10}^{\sim}\overset{2}{10}^{-}\overset{2}{10})$	∨ ∨ ∨	3.474	149 $(\underline{20\ 10}^{-}\overset{2}{20}^{\sim}10)e$	∨ ∨ × Rc<16.14130 8	2.985 (R=16.141308)
$33\ (\underline{20\ 10}^{-}\overset{2}{10})$	× × ∨		73 $(5^{\sim}\overset{2}{10}^{-}\overset{2}{10}^{\sim}\overset{2}{10})$	∨ ∨ ∨	2.955	$161\ (20^{-}\overset{2}{20}^{\sim}10^{-}\overset{2}{10})$	× × ∨	

End of [APPENDIX] ---- Existing Stacks

DATAFILE

[DATA 1-1] The Basic Es, E's, and AEs, (and TEs Excluded)

**** fi is the number of faces of ki-gon. L is the number of edges of the configuration. v is the number of faces content in the village. The boxed Ds is while Ds > 80 %.**

Configuration	L	f_1 k_1	f_2 k_2	f_3 k_3	v	R	Dv	Ds
4E	6	4 3			3	$\sqrt{\frac{3}{2}}$ = 1.2247448	36.32615	66.666667
6E	12	8 3			4	$\sqrt{2}$ = 1.41421356	42.640686	75
8E	12	6 4			3	$\sqrt{3}$ = 1.7320508	39.230482	66.666667
12E	30	20 3			5	$\sqrt{\frac{1}{2}(5+\sqrt{5})}$ = 1.902113	49.095114	82.917961 (boxed)
20E	30	12 5			3	$\frac{\sqrt{3}}{2}(1+\sqrt{5})$ = 2.802517	36.376132	63.661002
$12E'_1$	24	6 4	8 3		4	2 = 2	44.444444	75
$12E'_2$	18	4 3	4 6		3	$\sqrt{\frac{11}{2}}$ = 2.34520788	32.056188	54.545455
$24E'_1$	48	6 4	8 3	12 4	4	$\sqrt{5+2\sqrt{2}}$ = 2.79793265	43.809618	76.643749
$24E'_2$	36	6 4	8 6		3	$\sqrt{10}$ = 3.16227766	33.282662	60
$24E'_3$	36	8 3	6 8		3	$\sqrt{7+4\sqrt{2}}$ = 3.55764729	25.350648	47.405144
30E'	60	12 5	20 3		4	$1+\sqrt{5}$ = 3.236068	39.466849	71.618627
48E'	96	6 8	8 6	12 4	3	$\sqrt{13+6\sqrt{2}}$ = 4.6352218	26.82305	55.852189
$60E'_1$	90	12 5	20 6		3	$\sqrt{\frac{1}{2}(29+9\sqrt{5})}$ = 4.95603732	28.397425	61.069185
$60E'_2$	120	12 5	20 3	30 4	4	$\sqrt{11+4\sqrt{5}}$ = 4.469590	36.667973	75.209594
$60E'_3$	90	20 3	12 10		3	$\sqrt{\frac{1}{2}(37+15\sqrt{5}}$ = 5.938898	17.9589	42.528447
120E'	180	12 10	20 6	30 4	3	$\sqrt{31+12\sqrt{5}}$ = 7.604789	18.834779	51.873663
A14E		24 3	24 3		3 4	$\sqrt{3+\sqrt{3}}$ = 2.1753277	43.728277	73.963702
A32E		60 3	60 3		12 20	$\frac{1}{2}\sqrt{3(5+\sqrt{5})+\sqrt{6(25+11\sqrt{5})}}$ = 3.1208482	45.728833	82.138105 (boxed)
T12E	30	20 3			5	$\sqrt{\frac{1}{2}(5+\sqrt{5})}$ = 1.902113	49.095114	82.917961 (boxed)
T24E		6 4	8 3	24 3	5	(see below)* = 2.68742675	47.867532	83.07646 (boxed)
T60E		12 5	20 3	60 3	5	(see below)* = 4.31167475	40.036564	80.6862 (boxed)

* The R of T24E is $R=\sqrt{\frac{1}{3}\left[\sqrt[3]{199+\sqrt{297}}+\sqrt[3]{199-\sqrt{297}}+10\right]} \doteq 2.6874267474892...$

* The R of T60E is $R=\sqrt{A+B+\frac{1}{6}\left(27+7\sqrt{5}\right)} \doteq 4.311674750...$

where $\frac{A}{B}=\frac{1}{3}\sqrt[3]{\frac{1}{2}\left[\left(5112+2285\sqrt{5}\right)\pm\sqrt{64233+28728\sqrt{5}}\right]}$

End of [DATA 1-1] The basic Es, E's, AEs, and TEs

[DATA 1-2] The Angle of Adjacent Faces

Configurations	Rn	$A_{k_1\text{-}k_2} = \pi - (\theta_1 + \theta_2)$	$A_{k_1\text{-}k_2}$
4E	$\sqrt{\frac{3}{2}}$	$A_{3\text{-}3} = \pi - 2\mathrm{Sin}^{-1}(\sqrt{\frac{2}{3}})$	1.23095942 rad 70.5287754 deg
6E	$\sqrt{2}$	$A_{3\text{-}3} = \pi - 2\mathrm{Sin}^{-1}(\sqrt{\frac{1}{3}})$	1.01063324 rad 109.471221 deg
8E	$\sqrt{3}$	$A_{4\text{-}4} = \frac{\pi}{2}$	$\pi/2$ rad 90^{0} deg
12E	$\sqrt{\frac{5+\sqrt{5}}{2}}$	$A_{3\text{-}3} = \pi - 2\mathrm{Sin}^{-1}(\sqrt{\frac{3-\sqrt{5}}{6}})$	2.4118645 rad 138.189685 deg
20E	$\sqrt{\frac{3(3+\sqrt{5})}{2}}$	$A_{5\text{-}5} = \pi - 2\mathrm{Sin}^{-1}(\sqrt{\frac{5-\sqrt{5}}{10}})$	2.0344439 rad 116.56505 deg
$12E'_1$	2	$A_{3\text{-}4} = \pi - [\mathrm{Sin}^{-1}(\frac{1}{3}) + \mathrm{Sin}^{-1}(\sqrt{\frac{1}{3}})]$	2.1862760 rad 125.26439 deg
$12E'_2$	$\sqrt{\frac{11}{2}}$	$A_{3\text{-}6} = \pi - [\mathrm{Sin}^{-1}(\sqrt{\frac{2}{27}}) + \mathrm{Sin}^{-1}(\sqrt{\frac{2}{3}})]$ $A_{6\text{-}6} = \pi - 2\mathrm{Sin}^{-1}(\sqrt{\frac{2}{3}})$	1.9106332 rad 109.47122 deg 1.2309594 rad 70.528779 deg
$24E'_1$	$\sqrt{5+2\sqrt{2}}$	$A_{3\text{-}4} = \pi - [\mathrm{Sin}^{-1}(\sqrt{\frac{2-\sqrt{2}}{12}}) + \mathrm{Sin}^{-1}(\sqrt{\frac{2-\sqrt{2}}{4}})]$ $A_{4\text{-}4} = \frac{3\pi}{4}$	2.5261129 rad. 144.73561 deg. 2.3561945 rad. 135 deg.
$24E'_2$	$\sqrt{10}$	$A_{4\text{-}6} = \pi - [\mathrm{Sin}^{-1}(\frac{1}{3}) + \mathrm{Sin}^{-1}(\sqrt{\frac{1}{3}})]$ $A_{6\text{-}6} = \pi - 2\mathrm{Sin}^{-1}(\sqrt{\frac{1}{3}})$	2.1862760 rad 125.26439 deg 1.9106332 rad 109.47122 deg
$24E'_3$	$\sqrt{7+4\sqrt{2}}$	$A_{3\text{-}8} = \frac{3\pi}{4} - \mathrm{Sin}^{-1}(\frac{\sqrt{2}-1}{\sqrt{6}})$ $A_{8\text{-}8} = \frac{\pi}{2}$	2.1862760 rad 125.26439 deg $\pi/2$ rad 90 deg
30E'	$1+\sqrt{5}$	$A_{3\text{-}5} = \pi - [\mathrm{Sin}^{-1}(\sqrt{\frac{5-2\sqrt{5}}{15}}) + \mathrm{Sin}^{-1}(\frac{1}{5})]$	2.4892345 rad 142.62264 deg
48E'	$\sqrt{13+6\sqrt{2}}$	$A_{4\text{-}6} = \frac{7\pi}{8} - \mathrm{Sin}^{-1}(\frac{\sqrt{2-\sqrt{2}}}{12})$ $A_{6\text{-}8} = \frac{7\pi}{8} - \mathrm{Sin}^{-1}(\frac{\sqrt{2+\sqrt{2}}}{12})$ $A_{4\text{-}8} = \frac{7\pi}{8} - [\mathrm{Sin}^{-1}(\frac{\sqrt{2+\sqrt{2}}}{12} + \mathrm{Sin}^{-1}(\frac{\sqrt{2-\sqrt{2}}}{12})]$	2.5261129 rad 144.73561 deg 2.1862760 rad 125.26439 deg 2.3598558 rad 135.20978 deg
$60E'_1$	$\sqrt{\frac{29+9\sqrt{5}}{2}}$	$A_{5\text{-}6} = \pi - [\mathrm{Sin}^{-1}(\sqrt{\frac{5+\sqrt{5}}{90}}) + \mathrm{Sin}^{-1}(\sqrt{\frac{3-\sqrt{5}}{6}})]$ $A_{6\text{-}6} = \pi - 2\mathrm{Sin}^{-1}(\sqrt{\frac{3-\sqrt{5}}{6}})$	2.4892345 rad 142.62263 deg 2.411865 rad 138.189685 deg
$60E'_2$	$\sqrt{11+4\sqrt{5}}$	$A_{3\text{-}4} = \pi - [\mathrm{Sin}^{-1}(\sqrt{\frac{5-2\sqrt{5}}{30}}) + \mathrm{Sin}^{-1}(\sqrt{\frac{5-2\sqrt{5}}{10}})]$	2.7808521 rad 159.33109 deg 2.5880183 rad 148.28253 deg

		$A_{5\text{-}4} = \pi - [\mathrm{Sin}^{-1}(\sqrt{\frac{1}{10}}) + \mathrm{Sin}^{-1}(\sqrt{\frac{5-2\sqrt{5}}{10}})]$	
$60E'_3$	$\sqrt{\frac{37+15\sqrt{5}}{2}}$	$A_{3\text{-}10} = \pi - [\mathrm{Sin}^{-1}(\sqrt{\frac{7-3\sqrt{5}}{30}}) + \mathrm{Sin}^{-1}(\sqrt{\frac{5-\sqrt{5}}{10}})]$	2.4902345 rad 142.62263 deg
		$A_{10\text{-}10} = \pi - 2\mathrm{Sin}^{-1}(\sqrt{\frac{5-\sqrt{5}}{10}})$	2.0344439 rad 116.56505 deg
120E'	$\sqrt{31+12\sqrt{5}}$	$A_{4\text{-}6} = \pi - [\mathrm{Sin}^{-1}(\sqrt{\frac{5-2\sqrt{5}}{30}}) + \mathrm{Sin}^{-1}(\sqrt{\frac{5-2\sqrt{5}}{10}})]$	2.7767288 rad 159.09484 deg
		$A_{6\text{-}10} = \pi - [\mathrm{Sin}^{-1}(\sqrt{\frac{5-2\sqrt{5}}{10}}) + \mathrm{Sin}^{-1}(\sqrt{\frac{1}{6}})]$	2.4892345 rad 142.62263 deg
		$A_{4\text{-}10} = \pi - [\mathrm{Sin}^{-1}(\sqrt{\frac{1}{6}}) + \mathrm{Sin}^{-1}(\sqrt{\frac{5-2\sqrt{5}}{30}})]$	2.5880183 rad 148.282525 deg
T24E'	2.68742675...	$A_{4\text{-}3} = \pi - [\mathrm{Sin}^{-1}(\sqrt{\frac{1}{3(R_n^2-1)}}) + \mathrm{Sin}^{-1}(\sqrt{\frac{1}{R_n^2-1}})]$	2.5252867 rad 144.68827 deg
		$A_{3\text{-}3} = \pi - 2\mathrm{Sin}^{-1}(\sqrt{\frac{1}{3(R_n^2-1)}})$	2.6744481 rad 153.23459 deg
T60E'	4.31167475...	$A_{5\text{-}3} = \pi - [\mathrm{Sin}^{-1}(\sqrt{\frac{5+2\sqrt{5}}{5(R_n^2-1)}}) + \mathrm{Sin}^{-1}(\sqrt{\frac{1}{3(R_n^2-1)}})]$	2.7145477 rad 155.53213 deg
		$A_{3\text{-}3} = \pi - 2\mathrm{Sin}^{-1}(\sqrt{\frac{1}{3(R_n^2-1)}})$	2.8654007 rad 164.175366 deg

$a_3 = \sqrt{\frac{1}{3}}$ $a_4 = 1$ $a_5 = \sqrt{\frac{5+2\sqrt{5}}{5}}$ $a_6 = \sqrt{3}$ $a_8 = 1+\sqrt{2}$ $a_{10} = \sqrt{5+2\sqrt{5}}$

End of [DATA 1-2] The Angles of Adjacent Faces of basic Es, E's, and TEs (text: III-4)

[DATA 2A] The Five Es and Families

	4E	6E	8E	12E	20E
Rn	$\sqrt{\frac{3}{2}} \doteq 1.2247448$	$\sqrt{2} \doteq 1.41421356$	$\sqrt{3} \doteq 1.7320508$	$\sqrt{\frac{1}{2}(5+\sqrt{5})} \doteq 1.902113$	$\frac{\sqrt{3}(1+\sqrt{5})}{2} \doteq 2.802517$
	1. (1 3) = [8E]	1. (1 3) =[A14E]	1. (1 4) = [A14E]	1. (1 3) = [A32E]	1. (1 5) = [A32E]
R e Dv Ds	1.7320508 1.41421356 39.2305 % 66.6667 %	2.17532778 1.53819 43.7283 % 73.9637 %	2.17532778 1.255926 43.7283 % 73.9637 %	3.12084625 1.64072597 45.7289 % 82.1382 %	3 12084625 1.1135869 45.7289 % 82.1382 %
	3. $(3\ \bar{3})$	3. $(3\ \bar{3})$	3. $(2\ \check{\tilde{4}})_e$ = [20E]	3. $(3\ \bar{3})$	3. $(5\ \bar{5})$
R e Dv Ds	2.6373988 2.153446 33.2466 57.5062	3.35963 2.3756175 36.2053 66.44745	2.802517 1.618034 36.376 63.661	4.87474728 2.56282613 35.5112 75.7475	5.72943574 2.04439068 26.25145 60.9264
	8. $(3\ \tilde{3}\ \bar{3})$	8. $(3\ \tilde{3}\ \bar{3})$	7. $(4\ \bar{4})$	8. $(3\ \tilde{3}\ \bar{3})$	8. $(5\ \tilde{5}\ \bar{5})$
R e Dv Ds	3.5832135 2.925682 2.908357 % 54.5196	4.5746615 3.23477 31.1701 64.5083	3.64116025 2.10222483 32.0089 60.3407	6.64925146 3.49571848 29.4929 74.63945	7.8817544 2.8123839 19.9817 56.3407
	16. $(\underline{6\ 3}\ \bar{3})$	16. $(\underline{6\ 3}\ \bar{3})$	23. $(4\ \tilde{4}\ \bar{4})$	16. $(\underline{6\ 3}\ \bar{3})$	16. $(\underline{10\ 5}\ \bar{5})$
R e Dv Ds	4.550374 3.715365 23.3934 48.2954	5.805387 4.105028 24.7478 57.8592	5.02106428 2.898913 25.6548 55.5311	8.4340292 4.434027 22.867 67.4794	10.54332542 3.76209116 13.0028 44.9795
			35. $(\underline{8\ 4}\ \bar{4})$		22. $(1\ 5\ \tilde{5}\ \bar{5})$
R e Dv Ds			6.54967967 3.78142476 18.5911 46.622		8.08225632 2.88392735 20.2891 58.1726
			43. $(1\ 4\ \tilde{4}\ \bar{4})$		
R e Dv Ds			5.1629992 2.98086 26.986 58.147		

End of [DATA 2A] The five Es and Families

[DATA 2B] The Eleven E's and Families

The sign × is for nonexisting case which failed in any WB criteria. The sign • is for a good pack.

[DATA 2B – 1] The 12E'1 and Family

Ro = 2

Family groups:

(1) 8(3S)
(2) 4(3S)
(3) 6(4S)e
(4) 6(4S)e,1
(5) 4(3S)x + 4(3S)y
(6) 4(3M)x + 6(4A)x,y

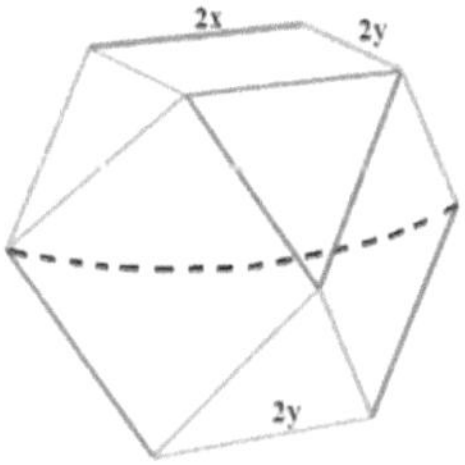

The 12E'1

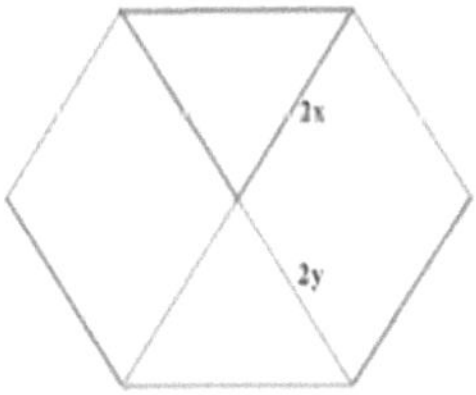

The village of 12E'1

12E'1 Group (1): 8 (3 S)		
	2. $(3\ \tilde{3})$	4. $(\underline{6\ 3})$
R =	3.855773	5.747475

12E'1 Group (2): 4 (3 S) [x = e, y = 1]						
	1. (1 3)	2. $(3\ \tilde{3})$	3. $(3\ \bar{3})$	4. $(\underline{6\ 3})$	8. $(3\ \tilde{3}\ \bar{3})$	16. $(\underline{6\ 3}\ \bar{3})$
R =	2.63738	2.933522	3.557648	3.9040127	4.500636	5.459738

12E'1 Group (3): 6(4 S)e								
	6. $(4\ \tilde{4})_e$	9. $(\underline{8\ 4})_e$	10. $(1\ 4\ \tilde{4})_e$	30. $(4\ \overset{\sim 2 \sim}{8}\ 4)_e$	32. $(4\ \overset{-2 \sim}{8}\ 4)_e$	54. $(1\ 4\ \overset{\sim 2 \sim}{8}\ 4)_e$	103. $(4\ \bar{4}\ \overset{\sim 2 -}{8}\ 4)e$	153. $(8\ \overset{-2}{8}\ \overset{\sim 2 -}{8}\ 4)_e$
R =	3.609352	5.472721	3.8188	5.681422	5.7121715	5.756305	7.3345892	92086148

12E'1 Group (4): 6(4-S) [x = e, y = 1]								
	2. $(1\ \overset{2}{4})$	4. $(2\ \overset{\sim 2}{4})$	5. $(2\ \overset{-2}{4})$	8. $(4\ \overset{-2}{4})$	19. $(2\ \overset{\sim 2}{4}\ \overset{-2}{4})$	29. $(2\ \overset{-2}{4}\ \overset{-2}{4})$	69. $(2\ \overset{\sim 2}{4}\ \overset{-2}{4}\ \overset{-2}{4})$	97. $(4\ \overset{-2}{4}\ \overset{-2}{4}\ \overset{-2}{4})$
R =	2.5779355	2.927577	3.5576475	3.914659	4.92606	5.9495438	6.97985196	8.014389

12E'1 Group (5): $4(3\,S)_x + 4(3\,S)_y$		1. (1 3)	3. $(3\ \bar{3})$	8. $(3\ \tilde{3}\ \bar{3})$	16. $(\underline{6\ 3}\ \bar{3})$
2. $(3\ \tilde{3})$	R =	3.5832139	4.5006359	5.4369409	6.387737
4. $(\underline{6\ 3})$	R =	4.5503752	5.4597371	6.3877347	7.3300402

12E'1 (6): 4(3M)x + 6(4A)x,y		1. (1 3)	2. $(3\ \tilde{3})$	3. $(3\ \bar{3})$	4. $(\underline{6\ 3})$	8. $(3\ \tilde{3}\ \bar{3})$	16. $(\underline{6\ 3}\ \bar{3})$
2. $(1\ \overset{2}{4})$	R =	×	×	×	×	×	×
4. $(2\ \overset{\sim 2}{4})$	R =	×	×	×	×	×	×
5. $(2\ \overset{-2}{4})$	R =	×	3.46988	×	×	×	×
8. $(4\ \overset{-2}{4})$	R =	4.289074	4.3704365	×	×	×	×
13. $(1\ \overset{2}{4}\ \overset{\sim 2}{4})$	R =	×	5.05069	×	×	×	×
14. $(1\ \overset{2}{4}\ \overset{-2}{4})$	R =	4.525179	4.7081993	5.01171	5.065017	×	×
18. $(2\ \overset{\sim 2}{4}\ \overset{\sim 2}{4})$	R =	5.274046	5.3362267	×	×	×	×
19. $(2\ \overset{\sim 2}{4}\ \overset{-2}{4})$	R =	4.5621244	4.7806741	5.1820232	5.291944	×	×
20. $(2\ \overset{-2}{4}\ \overset{\sim 2}{4})$	R =	×	6.0341933	×	×	×	×
21. $(2\ \overset{-2}{4}\ \overset{-2}{4})$	R =	4.508536	×	5.4210255	5.6305185	5.954146	×
26. $(4\ \tilde{4}\ \overset{-2}{4})$	R =	×	×	×	×	×	×
28. $(4\ \overset{-2}{4}\ \overset{\sim 2}{4})$	R =	6.282396	6.333761	×	×	×	×
29. $(4\ \overset{-2}{4}\ \overset{-2}{4})$	R =	×	×	5.4613212	5.707865	6.122294	×
36. $(\underline{8\ 4}\ \overset{-2}{4})$	R =	×	×	×	×	7.843716	×

46. $(1\ 4\ \overset{\sim}{4}\ \overset{-2}{4})$	R =	×	5.756906	6.127718	6.21128	×	×
68. $(2\ \overset{\sim 2}{4}\ \overset{-2}{4}\ \overset{\sim 2}{4})$	R =	7.3033627	7.348143	×	×	×	×
69. $(2\ \overset{\sim 2}{4}\ \overset{-2}{4}\ \overset{-2}{4})$	R =	×	×	×	×	6.387772	7.0861505
96. $(4\ \overset{-2}{4}\ \overset{-2}{4}\ \overset{\sim 2}{4})$	R =	8.3317504	8.37226	×	×	×	×
97. $(4\ \overset{-2}{4}\ \overset{-2}{4}\ \overset{-2}{4})$	R =	×	×	×	×	×	7.33066
114. $(4\ \overset{\sim 2 \sim}{8}\ 4\ \overset{-2}{4})$	R =	×	×	×	7.62381	7.98691	×
116. $(4\ \overset{-2 \sim}{8}\ 4\ \overset{-2}{4})$	R =	×	×	7.422083	7.634793	7.999231	×

---------------------- End of 12E'1 Family ----------------------

[DATA 2B – 2] The 12E'2 and Family

$$\mathbf{Ro} = \sqrt{\frac{11}{2}} \doteq 2.34520788$$

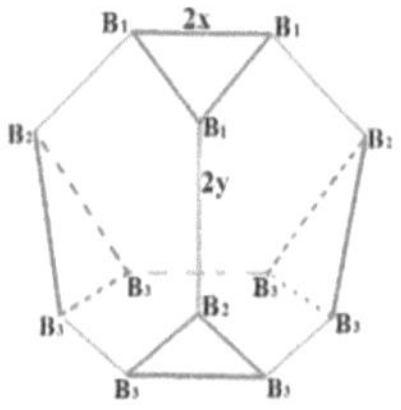

The setting

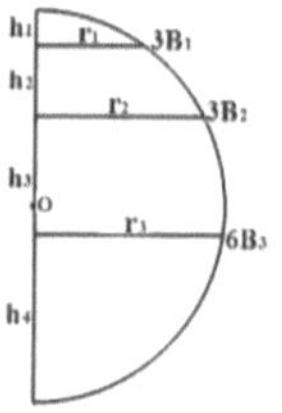

Layered diagram

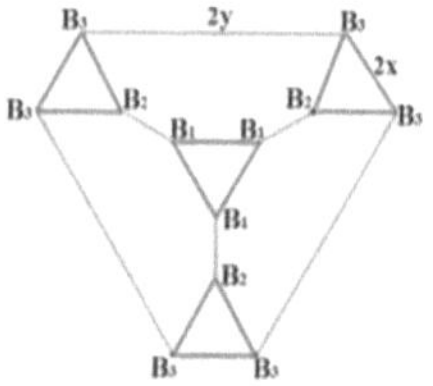

Top view

Family groups:

(1) 4(3S)e (2) 4(6S)e (3) 4(6S)3-e (4) 4(6S)6-e (5) 4(3M)e + 4(6A)

12E'₂ Group (1): 4(3 S)e		
	2. (3 $\tilde{3}$)	4. ($\underline{6\ 3}$)
R	R = $\sqrt{7+4\sqrt{2}}$ = 3.557647	R = 4.7936058

12E'₂ Group (2): 4(6 S)e			
	3. (6 $\tilde{6}$)e	6. ($\underline{12}$ 6)e	152. ($\underline{12}$ 6 $\overset{\sim 2}{12}$ 6)e
R	3.592147	5.7624393	7.781275

12E'₂ Group (3): 4(6 S)3 at e									
	1. (3 $\overset{\sim 2}{6}$)	5. (6 $\overset{-2}{6}$)	7. (1 3 $\overset{\sim 2}{6}$)	9. ($\tilde{3}$ 3 $\overset{\sim 2}{6}$)	13. (3 $\overset{\sim 2}{6}$ $\overset{\sim 2}{6}$)	14. (3 $\overset{\sim 2}{6}$ $\overset{-2}{6}$)	15. (3 $\overset{-2}{6}$ $\overset{\sim 2}{6}$)	26. (6 $\overset{-2}{6}$ $\overset{-2}{6}$)	43. (1 3 $\overset{\sim 2}{6}$ $\overset{\sim 2}{6}$)
R	2.927576	4.114857	3.3520613	3.4698796	4.541716	4.92606	4.641806	6.127416	4.673917
	44. (1 3 $\overset{\sim 2}{6}$ $\overset{-2}{6}$)	55. ($\tilde{3}$ 3 $\overset{\sim 2}{6}$ $\overset{\sim 2}{6}$)	56. ($\tilde{3}$ 3 $\overset{\sim 2}{6}$ $\overset{-2}{6}$)	72. (3 $\overset{\sim 2}{6}$ $\overset{\sim 2}{6}$ $\overset{-2}{6}$)	74. (3 $\overset{\sim 2}{6}$ $\overset{-2}{6}$ $\overset{-2}{6}$)	76. (3 $\overset{-2}{6}$ $\overset{\sim 2}{6}$ $\overset{-2}{6}$)	103. (6 $\tilde{6}$ $\tilde{6}$ $\overset{-2}{6}$)	114. (6 $\overset{-2}{6}$ $\overset{-2}{6}$ $\overset{-2}{6}$)	
R	5.274046	4.696413	5.336228	6.49306	6.9798501	6.562865	5.918914	8.1850171	

12E'₂ Group(4): 4(6 S)6 at e								
	1. (3 $\overset{\sim 2}{6}$)	2. (3 $\overset{-2}{6}$)	5. (6 $\overset{-2}{6}$)	8. (1 3 $\overset{-2}{6}$)	10. ($\tilde{3}$ 3 $\overset{-2}{6}$)	12. ($\bar{3}$ 3 $\overset{-2}{6}$)	13. (3 $\overset{\sim 2}{6}$ $\overset{\sim 2}{6}$)	14. (3 $\overset{\sim 2}{6}$ $\overset{-2}{6}$)
R	2.7979326	2.933522	3.7229424	3.5832135	3.8557725	4.5006358	4.0764564	4.3384965
	15 (3 $\overset{-2}{6}$ $\overset{\sim 2}{6}$)	18. ($\underline{6\ 3}$ $\overset{-2}{6}$)	26. (6 $\overset{-2}{6}$ $\overset{-2}{6}$)	43. (1 3 $\overset{\sim 2}{6}$ $\overset{\sim 2}{6}$)	44. (1 3 $\overset{\sim 2}{6}$ $\overset{-2}{6}$)	50. ($\tilde{3}$ 3 $\overset{-2}{3}$ $\overset{-2}{6}$)	55. ($\tilde{3}$ 3 $\overset{\sim 2}{6}$ $\overset{\sim 2}{6}$)	56. ($\tilde{3}$ 3 $\overset{\sim 2}{6}$ $\overset{-2}{6}$)
R	4.1638877	4.8099033		4.204331	4.621103	5.436941	d < 1 ×	4.6788755
	72. (3 $\overset{\sim 2}{6}$ $\overset{\sim 2}{6}$ $\overset{-2}{6}$)	74. (3 $\overset{\sim 2}{6}$ $\overset{-2}{6}$ $\overset{-2}{6}$)	76. (3 $\overset{-2}{6}$ $\overset{\sim 2}{6}$ $\overset{-2}{6}$)	82. ($\underline{6\ 3}$ $\tilde{3}$ $\overset{-2}{6}$)	103. (6 $\tilde{6}$ $\tilde{6}$ $\overset{-2}{6}$)	114 (6 $\overset{-2}{6}$ $\overset{-2}{6}$ $\overset{-2}{6}$)		
R	5.572744	5.923016	5.637128	6.387734	5.197666	6.857786		

12E'2 Group (5): 4(3M) + 4(6A) •== existing packs ×== nonexisting

	1	2	5	7	8	9	10	11	12	13	14	15	16	17	18	23	25	26	33	43	44	45	46	49	50
1			•		•		•			•	•		•			•	•			•	•		•		
2			•		•		•			•	•	•	•			•	•	•		•	•	•	•		
3									•		•				•	•		•	•		•		•		
4									•		•				•			•	•		•		•		
8																		•	•						•
16																									
							35	36	37	38	61	62	71	72	73	74	75	76	81	82	85	86	103	113	114
1							•	•					•		•		•						•	•	
2							•	•	•	•	•		•		•		•				•		•	•	
3								•		•		•		•		•		•					•		
4								•		•		•		•		•		•				•	•		
8										•		•		•		•		•				•			•
16																•				•					•

12E'2 Group (5): 4(3M) + 4(6A) •==existing packs ×== nonexisting									
	5 (6 $\overset{-2}{6}$)	8 (1 3 $\overset{-2}{6}$)	10 ($\tilde{3}$ 3 $\overset{-2}{6}$)	12 ($\bar{3}$ 3 $\overset{-2}{6}$)	13 (3 $\overset{\sim 2}{6}$ $\overset{\sim 2}{6}$)	14 (3 $\overset{\sim 2}{6}$ $\overset{-2}{6}$)	15. (3 $\overset{-2}{6}$ $\overset{\sim 2}{6}$)	16 (3 $\overset{-2}{6}$ $\overset{-2}{6}$)	18 ($\underline{6\ 3}$ $\overset{-2}{6}$)
1	4.015496	3.64116	3.806599	×	4.621103	4.5261985	×	4.664796	×
2	4.0849458	3.8066	3.992627	×	4.678876	4.663389	4.879707	4.821858	×

3	×	×	×	4.831668	×	4.890578	×	×	5.014501
4	×	×	×	5.014501	×	4.92763	×	×	
	23 $(6\,\overset{\sim}{6}\,\overset{-2}{6})$	25 $(6\,\overset{-2}{6}\,\overset{\sim 2}{6})$	26 $(6\,\overset{-2}{6}\,\overset{-2}{6})$	33 $(\underline{12\ 6}\,\overset{-2}{6})$	43 $(1\,3\,\overset{\sim 2}{6}\,\overset{\sim 2}{6})$	44 $(1\,3\,\overset{\sim 2}{6}\,\overset{-2}{6})$	45 $(1\,3\,\overset{-2}{6}\,\overset{\sim 2}{6})$	46 $(1\,3\,\overset{-2}{6}\,\overset{-2}{6})$	50 $(3\,\overset{\sim}{3}\,\overset{-2}{3}\,\overset{-2}{6})$
1	5.17476	5.535139	×	×	4.891326	4.7122262	×	5.021065	×
2	5.304516	5.58079	5.266753	×	4.943738	4.871142	5.524191	5.233437	×
3	5.510482	×	5.7100735	7.3009285	×	5.16454	×	5.67021	×
4	×	×	5.859566	7.426613	×	5.242729	×	5.8170515	×
8	×	×	6.086448	7.605308	×	×	×	×	6.041189
	55 $(3\,\overset{\sim}{3}\,\overset{\sim 2}{6}\,\overset{\sim 2}{6})$	56 $(3\,\overset{\sim}{3}\,\overset{\sim 2}{6}\,\overset{-2}{6})$	57 $(3\,\overset{\sim}{3}\,\overset{-2}{6}\,\overset{\sim 2}{6})$	58 $(3\,\overset{\sim}{3}\,\overset{-2}{6}\,\overset{-2}{6})$	61 $(3\,\overset{-}{3}\,\overset{-2}{6}\,\overset{\sim 2}{6})$	62 $(3\,\overset{-}{3}\,\overset{-2}{6}\,\overset{-2}{6})$	71 $(3\,\overset{\sim 2}{6}\,\overset{\sim 2}{6}\,\overset{\sim 2}{6})$	72 $(3\,\overset{\sim 2}{6}\,\overset{\sim 2}{6}\,\overset{-2}{6})$	73 $(3\,\overset{\sim 2}{6}\,\overset{-2}{6}\,\overset{\sim 2}{6})$
1	4.943739	4.747124	×	×	×	×	5.827481	×	6.1823993
2	4.994695	4.908974	5.776412	5.3577645	6.4271585	×	5.869607	×	6.2231276
3	×	5.2113875	×	5.848065	×	6.2128594	×	5.907726	×
4	×	5.294957	×	6.017037	×	6.4374475	×	6.08001	×
8	×	×	×	6.298637	×	6.862748	×	6.372757	×
	74 $(3\,\overset{\sim 2}{6}\,\overset{-2}{6}\,\overset{-2}{6})$	75 $(3\,\overset{-2}{6}\,\overset{\sim 2}{6}\,\overset{\sim 2}{6})$	76 $(3\,\overset{-2}{6}\,\overset{\sim 2}{6}\,\overset{-2}{6})$	82 $(\underline{6\ 3}\,\overset{-}{3}\,\overset{-2}{6})$	85 $(\underline{6\ 3}\,\overset{-2}{6}\,\overset{\sim 2}{6})$	86 $(\underline{6\ 3}\,\overset{-2}{6}\,\overset{-2}{6})$	103 $(6\,\overset{\sim}{6}\,\overset{\sim}{6}\,\overset{-2}{6})$	113 $(6\,\overset{-2}{6}\,\overset{-2}{6}\,\overset{\sim 2}{6})$	114 $(6\,\overset{-2}{6}\,\overset{-2}{6}\,\overset{-2}{6})$
1	×	5.8881	×	×	×	×	5.700343	7.1092466	×
2	×	5.9292776	×	×	6.701623	×	5.849997	7.1448124	×
3	6.113885	×	5.945498	×	×	×	6.123341	×	×
4	6.319608	×	6.1214935	×	×	6.5747485	6.184771	×	×
8	6.698713	×	6.424627	×	×	7.053932	×	×	7.3174395
16	6.977853	×	×	7.263434	×	×	×	×	7.9047895

------------------------ End of 12E'2 Family ------------------------

[DATA 2B – 3A] The 24E'1 Family A [x = y = e, z = 1]

Ro = $\sqrt{5+2\sqrt{2}} \doteq 2.79793265$

Family A groups:

(1) 6(4 S)e	[z = E3M, x = y = 1]
(2) 8(3 S)e	[x = y = E4M, z = 1]
(3) 12(4 S)e	[x = y = z = E4M]]
(4) 12 $(4\,S)_{3\,at\,e}$	[z =E4M, x = y = 1]
(5) 12 $(4\,S)_{4\,at\,e}$	[x = y = E4M, z = 1]
(6) 8(3M)e + 12(4A)	[z = e, x = y = E4A]
(7) 6(4M)e + 12(4A)	[x = y = e, z = E4A]
(8) 8 $(3\,S)_{z}$ + 6 $(4\,S)_{x,\,x}$	[x = y = E4M, z = E3M]

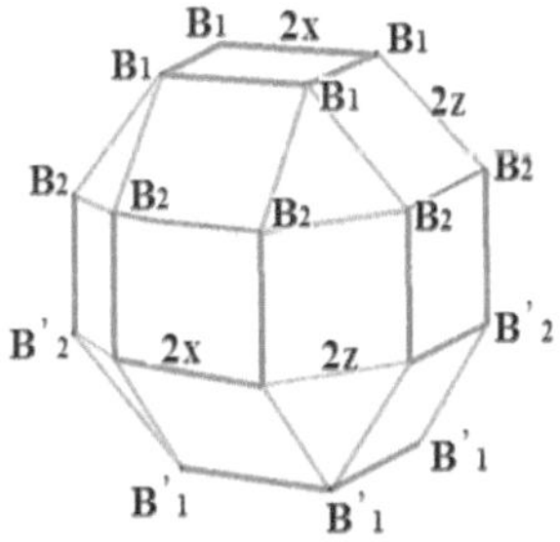

The 24E'1 Family A

24E'1 Family A-Group (1): 6(4 S)e (x = y = e, z = 1)								
	1. $(1\ \check{4})e$	3. $(2\ \overset{\sim\vee}{4})e$	6. $(4\ \tilde{4})e$	7. $(4\ \bar{4})e$	9. $(\underline{8\ 4})e$	10. $(1\ 4\ \tilde{4})e$	23. $(4\ \tilde{4}\ \bar{4})e$	30. $(4\ \overset{\sim 2\sim}{8}\ 4)e$
R	3.359631	3.974709	4.163887	4.831668	5.702636	4.322215	6.212856	95.872809
	32. $(4\ \overset{-2\sim}{8}\ 4)e$ [Rc<9.805]	35. $(\underline{8\ 4}\ \bar{4})e$	43. $(1\ 4\ \tilde{4}\ \bar{4})e$	54. $(1\ 4\ \overset{\sim 2\sim}{8}\ 4)e$ [Rc<6.191]	107. $(4\ \overset{\sim 2\sim}{8}\ 4\ \bar{4})e$	111. $(4\ \overset{-2\sim}{8}\ 4\ \bar{4})e$	153. $(8\ \overset{-2\sim}{8}\ \overset{2-}{8}\ 4)e$	
R	5.897159	7.728704	6.341144	5.928411	7.8730405	7.886068	8.8083128	

24E'1 Family A-Group (2): 8(3 S)e (x = y = 1, z = e)						
	1. $(1\ 3)$	2. $(3\ \tilde{3})$	3. $(3\ \bar{3})$	4. $(\underline{6\ 3})$	8. $(3\ \tilde{3}\ \bar{3})$	16. $(\underline{6\ 3}\ \bar{3})$
R =	3.64116	3.992627	4.831668	5.219395	6.04119	7.2634337

24E'1 Family A-Group (3): 12(4 S)e (x = y = z)							
	6. $(4\ \tilde{4})e$	9. $(\underline{8\ 4})e$	10. $(1\ 4\ \tilde{4})e$	30. $(4\ \overset{\sim 2\sim}{8}\ 4)e$	32. $(4\ \overset{-2\sim}{8}\ 4)e$ [Rc<9.805]	54. $(1\ 4\ \overset{\sim 2\sim}{8}\ 4)e$ [Rc<6.191]	153. $(8\ \overset{-2\sim}{8}\ \overset{2-}{8}\ 4)e$
R	5.244047	7.953303	5.489277	8.2050404	8.2255416	×	13.39068

24E'1 FamilyA-Group(4): 12 (4 S)3 at e (x = y = 1, z = e)								
	2. $(1\ \overset{2}{4})$	4. $(2\sim\overset{1}{4})$	5. $(2-\overset{1}{4})$	8. $(4-\overset{1}{4})$	19. $(2\sim\overset{2}{4}-\overset{1}{4})$	29. $(2-\overset{2}{4}-\overset{2}{4})$	69. $(2\sim\overset{2}{4}-\overset{1}{4}-\overset{2}{4})$	97. $(4-\overset{1}{4}-\overset{1}{4}-\overset{1}{4})$
R	3.609352	3.991015	4.831668	5.222466	6.4701969	7.726284	8.9872098	10.25119

24E'1 Family A-Group (5): 12 (4 S)4 at e (x = y = e, z = 1)								
	2. $(1\ \overset{2}{4})$	4. $(2\sim\overset{1}{4})$	5. $(2-\overset{1}{4})$	8. $(4-\overset{1}{4})$	19. $(2\sim\overset{1}{4}-\overset{1}{4})$	29. $(2-\overset{2}{4}-\overset{2}{4})$	69. $(2\sim\overset{2}{4}-\overset{1}{4}-\overset{2}{4})$	97. $(4-\overset{1}{4}-\overset{1}{4}-\overset{1}{4})$
R	3.855773	4.338497	5.436941	5.23015	7.524367	9.1338505	10.74787	12.36465

24E'1 Family A-Group (6): 8(3M)e + 12(4A)							
		1. $(1\ 3)$	2. $(3\ \tilde{3})$	3. $(3\ \bar{3})$	4. $(\underline{6\ 3})$	8. $(3\ \tilde{3}\ \bar{3})$	16. $(\underline{6\ 3}\ \bar{3})$
5. $(2-\overset{1}{4})$	R	4.891366	4.994696	×	×	×	×
8. $(4-\overset{1}{4})$	R	6.436909	6.513945	×	×	×	×
19. $(2\sim\overset{1}{4}-\overset{1}{4})$	R	×	7.041899	7.578854	7.6748498	×	×
29. $(2-\overset{2}{4}-\overset{2}{4})$	R	×	×	×	8.2127	8.763612	×
69. $(2\sim\overset{1}{4}-\overset{1}{4}-\overset{1}{4})$	R	×	×	×	×	×	9.97204
97. $(4-\overset{1}{4}-\overset{1}{4}-\overset{1}{4})$	R	×	×	×	×	×	×

24E'1 Family A-Group (7): 6(4M)e + 12(4A) (x = y = E4M, z = E4A)									
		2 $(1\ \overset{1}{4})$	4 $(2\ \overset{\sim 1}{4})$	5 $(2\ \overset{-1}{4})$	8 $(4\ \overset{-2}{4})$	19 $(2\ \overset{\sim 2-2}{4\ 4})$	29 $(4\ \overset{-2-2}{4\ 4})$	69 $(2\ \overset{\sim 2-2-2}{4\ 4\ 4})$	97 $(4\ \overset{-2-2-2}{4\ 4\ 4})$
1. $(1\ \check{4})e$	R	3.8188	5.006776	4.4928345	5.714074	×	×	×	×
3 $(2\ \overset{\sim\vee}{4})e$	R	×	×	4.922732	6.116966	6.40922	×	×	×
6 $(4\ \tilde{4})e$	R	×	×	5.0303151	6.2119324	6.5748665	×	×	×

7 (4 4)e	R	×	×	5.3413885	6.4798934	7.1717994	×	×	×
9 (8 4)e	R	×	×	×	×	7.7194082	8.072301	×	×
10 (1 4 4)e	R	×	×	5.1088445	6.27607	6.69104	×	×	×
23 (4 4 4)e	R	×	×	×	×	7.97443	8.5750652	×	×
30 (4 8 4)e	R	×	×	×	×	7.7965355	8.2050404	×	×
32 (4 8 4)e [Rc<9.805]	R	×	×	×	×	7.803347	8.215764	×	×
35 (8 4 4)e	R	×	×	×	×	×	9.521559	10.08264	×
43 (1 4 4 4)e	R	×	×	×	×	8.012397	8.6688791	×	×
107 (4 8 4 4)e	R	×	×	×	×	×	95.702042	10.189405	×
111 (4 8 4 4)e	R	×	×	×	×	×	9.57259	10.193764	×
153 (8 8 8 4)e	R	×	×	×	×	×	9.6068091	10.8179	11.130755

24E'1 Family A-Group (8): 8(3S)e1+6(4S)e2

		1. (1 3)	2. (3 3)	3. (3 3)	4. (6 3)	8. (3 3 3)	16. (6 3 3)
1. (1 4)e	R	×	4.5746615	×	5.805386	×	×
3 (2 4)e	R	×	5.180886	×	6.404669	×	×
6 (4 4)e	R	5.021064	5.357764	6.2128595	6.5747485	7.418579	8.635007
7 (4 4)e	R	×	6.04119	×	7.263434	×	×
9 (8 4)e	R	6.5496195	6.874918	7.7287045	8.07352	8.923685	10.130463
10 (1 4 4)e	R	5.162998	5.494072	6.341147	16.6986455	7.538688	8.7497783
23 (4 4 4)e	R	×	7.418579	×	8.635009	×	×
30 (4 8 4)e	R	6.707363	7.028026	7.8730405	8.213803	9.05825	10.25785
32 (4 8 4)e [Rc<9.805]	R	6.726398	7.045185	7.886068	8.225321	9.066758	10.263 ×
35 (8 4 4)e	R	×	8.923685	×	10.130465	×	×
43 (1 4 4 4)e	R	×	7.538688	×	8.7497782	×	×
107 (4 8 4 4)e	R	×	9.05825	×	10.25785	×	×
111 (4 8 4 4)e	R	×	9.066757	×	10.26299	×	×
153. (8 8 8 4)e	R	9.6439596	9.95819	10.80383	11.13075	11.9795	13.1677565

End of [DATA 2B-3A] 24E'1 Family-A

[DATA 2B – 3B] The 24E'1 Family B (x = e, y = 1, z = 1 or e)

Family-B groups:

(1) 6(4 S)x,1 (x = e, y = z = 1) (2) 6(4 S)x,1 + 8(3 S)z (x = e, y = 1, z = e)

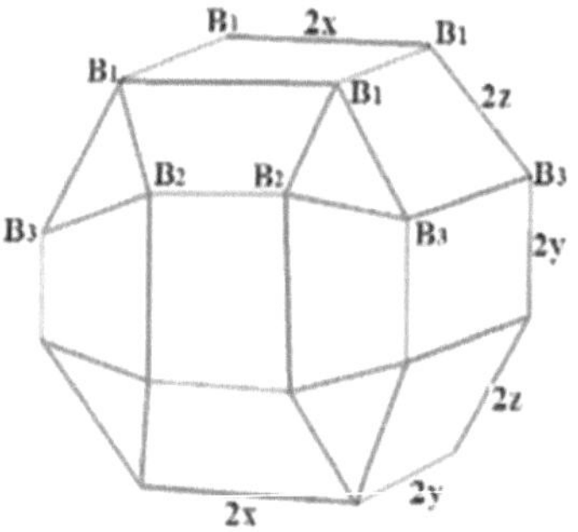

24E'1 Family-B, Group (1): 6 (4 S)x,1 (x = x, y = 1, z = 1)								
	2. $(1\ \overset{2}{4})$	4. $(2^{\sim}\overset{1}{4})$	5. $(2^{-}\overset{2}{4})$	8. $(4^{-}\overset{2}{4})$	19. $(2^{\sim}\overset{1}{4}{}^{-}\overset{1}{4})$	29. $(2^{-}\overset{2}{4}{}^{-}\overset{2}{4})$	69. $(2^{\sim}\overset{1}{4}{}^{-}\overset{1}{4}{}^{-}\overset{1}{4})$	97. $(4^{-}\overset{1}{4}{}^{-}\overset{1}{4}{}^{-}\overset{1}{4})$
R	3.2360619	3.3965332	3.5576472	×	×	×	×	×

24E'1 Family-B, Group (2): 6 (4S)x,1 + 8(3 S)z [x = x, y = 1, z = z]									
		2. $(1\ \overset{2}{4})$	4. $(2^{\sim}\overset{1}{4})$	5. $(2^{-}\overset{1}{4})$	8. $(4^{-}\overset{1}{4})$	19. $(2^{\sim}\overset{1}{4}{}^{-}\overset{1}{4})$	29. $(2^{-}\overset{2}{4}{}^{-}\overset{2}{4})$	69. $(2^{\sim}\overset{1}{4}{}^{-}\overset{1}{4}{}^{-}\overset{1}{4})$	97. $(4^{-}\overset{1}{4}{}^{-}\overset{1}{4}{}^{-}\overset{1}{4})$
1. (1 3)	R =	4.122486	4.304896	4.645739	4.767091	×	×	×	×
2. $(3\ \tilde{}\ 3)$	R =	4.4752547	4.6580965	5.024576	5.16127	5.428818	×	×	×
3. $(3\ \bar{}\ 3)$	R =	5.335061	5.5232635	5.938897	6.097245	6.530447	6.765235	×	×
4. $(\underline{6\ 3})$	R =	5.7150948	5.9004171	6.3218482	6.4825365	6.9462	7.2534535	×	×
8. $(3\ \tilde{}\ 3\ \bar{}\ 3)$	R =	6.55364	6.75321	7.192198	7.362106	7.888657	8.306167	8.582531	×
16. $(\underline{6\ 3}\ \bar{}\ 3)$	R =	77797157	7.969536	8.436747	8.611962	9.185455	9.68216	10.08976	10.38761

End of [DATA2B—3B] 24E'1 Family B

[DATA 2B – 4] The 24E'2 and Family:

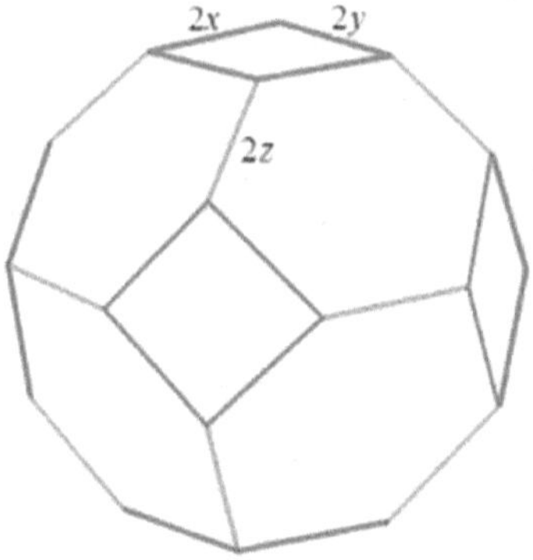

$$Ro = \sqrt{10} \doteq 3.16227766$$

Family groups:

(1) 6(4S)e **(2) 6(4S)e,1** **(3) 8(6S)e**
(4) 8(6S)4 at e **(5) 8(6S)6 at e** **(6) 4(6S)e**
(7) 4(6S)4 at e **(8) 4(6S)6 at e** **(9) 6(4M)e + 8(6A)**
(10) 6(4M)e + 4(6A) **(11) 6(4M)e,1 + 4(6A)** **(12) 4(6M)e + 6(4A)**
(13) 4(6M)e,1 + 6(4A) **(14) 4(6M)e + 4(6A)** **(15) 4(6M)e,1 +4(6A)**
(16) 6(4S1) + 4(6S2) **(17) 4(6S1) + 4(6S2)**

24E'2 Group (1): 6(4S)e (x = y, z = 1)							
	6 (4 $\overset{\sim}{4}$)e	9 ($\underline{8\ 4}$)e	10 (1 4 $\overset{\sim}{4}$)e	30 (4 $\overset{\sim 2}{8}$ $\overset{\sim}{4}$)e	32 (4 $\overset{-2}{8}$ $\overset{\sim}{4}$)e [Rc<9.805]	103 (4 $\overset{-}{4}$ $\overset{\sim 2}{8}$ $\overset{-}{4}$)e	153 (8 $\overset{-2}{8}$ $\overset{\sim 2}{8}$ $\overset{-}{4}$)e
R	4.8316699	6.6826625	5.006776	6.870816	6.89261	8.76323	10.41168

24E'2 Group (2): 6(4S)e,1 (x = e, y = z = 1)								
	2. (1 $\overset{2}{4}$)	4. (2 $\overset{\sim 2}{4}$)	5. (2 $\overset{-2}{4}$)	8. (4 $\overset{-2}{4}$)	19. (2 $\overset{\sim 2}{4}$ $\overset{-2}{4}$)	29. (2 $\overset{-2}{4}$ $\overset{-2}{4}$)	69. (2 $\overset{\sim 2}{4}$ $\overset{-2}{4}$ $\overset{-2}{4}$)	97. (4 $\overset{-2}{4}$ $\overset{-2}{4}$ $\overset{-2}{4}$)
R	3.8130699	4.114857	4.7936051	5.111534	6.127415	7.1532934	8.1850172	9.220447

24E'2 Group (3): 8(6S)e (x = y = z)								
3 (6 $\overset{\sim}{6}$)e	6 ($\underline{12\ 6}$)e	19 (6 $\overset{\sim}{6}$ $\overset{\sim}{6}$)e	36 (12 $\overset{-2}{12}$ $\overset{\sim}{6}$)e [Rc<8.428]	115	117	119	121	152 ($\underline{12\ 6}$ $\overset{\sim 2}{12}$ $\overset{-}{6}$)e
5.1920131	8.243906	× (d < 1)	9.327995 ×	× (d < 1)	× (d < 1)	× (d < 1)	× (d < 1)	11.29193

24E'2 Group (4): 8 (6S)4 at e (x = y, z = 1)									
	1. (3 $\overset{\sim 2}{6}$) .	5. (6 $\overset{-2}{6}$)	7. (1 3 $\overset{\sim 2}{6}$)	9. (3 $\overset{\sim}{3}$ $\overset{\sim 2}{6}$)	13. (3 $\overset{\sim 2}{6}$ $\overset{\sim 2}{6}$)	14. (3 $\overset{\sim 2}{6}$ $\overset{-2}{6}$)	15. (3 $\overset{-2}{6}$ $\overset{\sim 2}{6}$)	26. (6 $\overset{-2}{6}$ $\overset{-2}{6}$)	43. (1 3 $\overset{\sim 2}{6}$ $\overset{\sim 2}{6}$)
R	4.338498	6.198517	4.891326	4.994696	6.756405	7.52437	6.858471	9.395365	6.865955
	44 (1 3 $\overset{\sim 2}{6}$ $\overset{-2}{6}$)	55 (3 $\overset{\sim}{3}$ $\overset{\sim 2}{6}$ $\overset{\sim 2}{6}$)	56 (3 $\overset{\sim}{3}$ $\overset{\sim 2}{6}$ $\overset{-2}{6}$)	72 (3 $\overset{\sim 2}{6}$ $\overset{\sim 2}{6}$ $\overset{-2}{6}$)	74 (3 $\overset{\sim 2}{6}$ $\overset{-2}{6}$ $\overset{-2}{6}$)	76 (3 $\overset{-2}{6}$ $\overset{\sim 2}{6}$ $\overset{-2}{6}$)	103 (6 $\overset{\sim}{6}$ $\overset{\sim}{6}$ $\overset{-2}{6}$)	114 (6 $\overset{-2}{6}$ $\overset{-2}{6}$ $\overset{-2}{6}$)	
R	8.01944	6.876401	8.0837	9.903224	10.74788	9.97555	8.76323	12.621595	

24E'2 Group (5): 8 (6S)6 at e (x = y = 1, z = e)								
	1. (3 $\overset{\sim 2}{6}$)	2. (3 $\overset{-2}{6}$)	5. (6 $\overset{-2}{6}$)	7. (1 3 $\overset{\sim 2}{6}$)	8. (1 3 $\overset{-2}{6}$)	9. (3 $\overset{\sim}{3}$ $\overset{\sim 2}{6}$)	10. (3 $\overset{\sim}{3}$ $\overset{-2}{6}$)	12. (3 $\overset{-}{3}$ $\overset{-2}{6}$)
R	3.914659	4.163888	5.1115338	× (d < 1)	5.0210654	× (d < 1)	5.3577659	6.2128595
	13. (3 $\overset{\sim 2}{6}$ $\overset{\sim 2}{6}$)	14. (3 $\overset{\sim 2}{6}$ $\overset{-2}{6}$)	15. (3 $\overset{-2}{6}$ $\overset{\sim 2}{6}$)	18. ($\underline{6\ 3}$ $\overset{-2}{6}$)	26. (6 $\overset{-2}{6}$ $\overset{-2}{6}$)	43. (1 3 $\overset{\sim 2}{6}$ $\overset{\sim 2}{6}$)	44 (1 3 $\overset{\sim 2}{6}$ $\overset{-2}{6}$)	55 (3 $\overset{\sim}{3}$ $\overset{\sim 2}{6}$ $\overset{\sim 2}{6}$)
R	5.501725	5.949544	× (d < 1)	6.57475	7.1532905	× (d < 1)	6.282395	× (d < 1)
	56 (3 $\overset{\sim}{3}$ $\overset{\sim 2}{6}$ $\overset{-2}{6}$)	72 (3 $\overset{\sim 2}{6}$ $\overset{\sim 2}{6}$ $\overset{-2}{6}$)	74 (3 $\overset{\sim 2}{6}$ $\overset{-2}{6}$ $\overset{-2}{6}$)	76 (3 $\overset{-2}{6}$ $\overset{\sim 2}{6}$ $\overset{-2}{6}$)	103 (6 $\overset{\sim}{6}$ $\overset{\sim}{6}$ $\overset{-2}{6}$)	114 (6 $\overset{-2}{6}$ $\overset{-2}{6}$ $\overset{-2}{6}$)		
R	6.333764	7.50248	8.01439	7.56276	× (d < 1)	9.220445		

24E'2 Group (6): 4 (6S)e (x = z, y = 1)									
	3. (6 $\overset{\sim}{6}$)e	6 ($\underline{12\ 6}$)e	19 (6 $\overset{\sim}{6}$ $\overset{\sim}{6}$)e	36 (12 $\overset{-2}{12}$ $\overset{\sim}{6}$)e [Rc<8.428]	115 (6 $\overset{\sim}{6}$ $\overset{\sim 2}{12}$ $\overset{-}{6}$)e	117 (6 $\overset{\sim}{6}$ $\overset{-2}{12}$ $\overset{\sim}{6}$)e	119 (6 $\overset{-}{6}$ $\overset{\sim 2}{12}$ $\overset{\sim}{6}$)e	121 (6 $\overset{-}{6}$ $\overset{-2}{12}$ $\overset{\sim}{6}$)e	152 ($\underline{12\ 6}$ $\overset{\sim 2}{12}$ $\overset{-}{6}$)e
R	4.5423438	6.6647933	5.2671364	7.5081105	7.2331908	7.3424962	7.399056	7.5275939	8.778547

24E'₂ Group (7): 4 (6S)₄ at e (x = e, y = z = 1)

	1. $(3\,\overset{\sim2}{6})$	2. $(3\,\overset{-2}{6})$	5. $(6\,\overset{-2}{6})$	7. $(1\,3\,\overset{\sim2}{6})$	8. $(1\,3\,\overset{-2}{6})$	9. $(3\,\tilde{3}\,\overset{\sim2}{6})$	10. $(3\,\tilde{3}\,\overset{-2}{6})$	12. $(3\,\bar{3}\,\overset{-2}{6})$
R	3.7229425	3.904013	4.6352213	4.0154964	4.5503752	4.084945	4.809903	5.4597375
	13. $(3\,\overset{\sim2}{6}\,\overset{\sim2}{6})$	14. $(3\,\overset{\sim2}{6}\,\overset{-2}{6})$	15. $(3\,\overset{-2}{6}\,\overset{\sim2}{6})$	18. $(\underline{6\,3}\,\overset{-2}{6})$	26. $(6\,\overset{-2}{6}\,\overset{-2}{6})$	43. $(1\,3\,\overset{\sim2}{6}\,\overset{\sim2}{6})$	44. $(1\,3\,\overset{\sim2}{6}\,\overset{-2}{6})$	55. $(3\,\tilde{3}\,\overset{\sim2}{6}\,\overset{\sim2}{6})$
R	4.950893	5.270575	5.023113	5.7475225	6.1985167	5.0410612	5.5351384	5.0549577
	36 $(12\,\overset{-2}{12}\,\tilde{6})$e [Rc<8.428]	56 $(3\,\tilde{3}\,\overset{\sim2}{6}\,\overset{-2}{6})$	72 $(3\,\overset{\sim2}{6}\,\overset{\sim2}{6}\,\overset{-2}{6})$	74 $(3\,\overset{\sim2}{6}\,\overset{-2}{6}\,\overset{-2}{6})$	76 $(3\,\overset{-2}{6}\,\overset{\sim2}{6}\,\overset{-2}{6})$	103 $(6\,\tilde{6}\,\tilde{6}\,\overset{-2}{6})$		
R		5.5807902	6.481769	6.8577893	6.53627	6.0268558		

24E'₂ Group (8): 4 (6S)₆ at e (x = y = 1, z = e)

	1. $(3\,\overset{\sim2}{6})$	5. $(6\,\overset{-2}{6})$	7 $(1\,3\,\overset{\sim2}{6})$	9 $(3\,\tilde{3}\,\overset{\sim2}{6})$	13 $(3\,\overset{\sim2}{6}\,\overset{\sim2}{6})$	14 $(3\,\overset{\sim2}{6}\,\overset{-2}{6})$	15 $(3\,\overset{-2}{6}\,\overset{\sim2}{6})$	26 $(6\,\overset{-2}{6}\,\overset{-2}{6})$	43 $(1\,3\,\overset{\sim2}{6}\,\overset{\sim2}{6})$
R	3.914659	5.1115334	4.2890745	4.370437	5.501724	5.949544	5.584664	7.1532952	5.5987242
	44 $(1\,3\,\overset{\sim2}{6}\,\overset{-2}{6})$	55 $(3\,\tilde{3}\,\overset{\sim2}{6}\,\overset{\sim2}{6})$	56 $(3\,\tilde{3}\,\overset{\sim2}{6}\,\overset{-2}{6})$	72 $(3\,\overset{\sim2}{6}\,\overset{\sim2}{6}\,\overset{-2}{6})$	74 $(3\,\overset{\sim2}{6}\,\overset{-2}{6}\,\overset{-2}{6})$	76 $(3\,\overset{-2}{6}\,\overset{\sim2}{6}\,\overset{-2}{6})$	103 $(6\,\tilde{6}\,\tilde{6}\,\overset{-2}{6})$	114 $(6\,\overset{-2}{6}\,\overset{-2}{6}\,\overset{-2}{6})$	
R	6.2823947	5.611375	6.333763	7.502478	8.0143877	7.5627605	6.850024	9.2204499	

24E'2 Group (9): 6 (4M)e + 8(6A) (Outline)

	1	2	5	7	8	9	10	11	12	13	14	15	16	17	18	23	25	26	43	44	45	46
1	•	•	•	•		•		•		•	•			•			•		•			
3		•	•		•		•			•	•		•			•	•		•	•		
6			•		•		•			•	•		•			•	•		•	•		
7			•		•		•			•	•	•	•			•	•	•	•	•	•	•
9									•		•				•	•		•		•		•
10					•		•			•	•	•	•			•	•		•	•	•	
23									•		•				•	•				•		•
30									•		•				•	•		•		•		•
32									•		•				•	•		•		•		•
35																		•				
43									•		•				•	•		•		•		•
103																		•				
107																		•				
111																		•				
153																						

	49	50	55	56	57	58	61	62	71	72	73	74	75	76	81	82	85	86	103	113	114	
1	•		•						•		•		•		•					•		
3			•	•					•		•		•						•	•		
6			•	•					•		•		•						•	•		
7			•	•	•		•		•		•		•				•		•	•		
9				•		•				•				•					•			
10			•	•	•		•		•		•		•				•		•	•		
23				•		•				•		•		•					•			
30				•		•				•				•					•			
32				•		•				•				•					•			
35		•						•		•		•		•				•			•	
43				•		•		•		•		•		•					•			
103		•						•		•		•		•				•			•	
107		•						•		•		•		•				•			•	
111		•						•		•		•		•				•			•	
153												•				•					•	

24E'₂ Group (9): 6(4M)e +8(6A) [x = y = E4M, z = E6A]									
1. (1 4)e		1 (3 ${}^{\sim}\overset{2}{6}$)	2 (3 ${}^{-}\overset{2}{6}$)	5 (6 ${}^{-}\overset{2}{6}$)	7 (1 3 ${}^{\sim}\overset{2}{6}$)	9 (3 ${}^{\sim}3$ ${}^{\sim}\overset{2}{6}$)	11 (3 ${}^{-}3$ ${}^{\sim}\overset{2}{6}$)	13 (3 ${}^{\sim}\overset{2}{6}$ ${}^{\sim}\overset{2}{6}$)	14 (3 ${}^{\sim}\overset{2}{6}$ ${}^{-}\overset{2}{6}$)
	R	4.322215	4.330097	5.513092	5.162998	5.494071	6.3411439	6.3512435	6.0323385
		17 ($\underline{6\ 3}$ ${}^{\sim}\overset{2}{6}$)	25 (6 ${}^{-}\overset{2}{6}$ ${}^{\sim}\overset{2}{6}$)	43 (1 3 ${}^{\sim}\overset{2}{6}$ ${}^{\sim}\overset{2}{6}$)	49 (3 ${}^{\sim}3$ ${}^{-}\overset{2}{3}$ ${}^{\sim}\overset{2}{6}$)	55 (3 ${}^{\sim}3$ ${}^{\sim}\overset{2}{6}$ ${}^{\sim}\overset{2}{6}$)	71 (3 ${}^{\sim}\overset{2}{6}$ ${}^{\sim}\overset{2}{6}$ ${}^{\sim}\overset{2}{6}$)	73 (3 ${}^{\sim}\overset{2}{6}$ ${}^{-}\overset{2}{6}$ ${}^{\sim}\overset{2}{6}$)	75 (3 ${}^{-}\overset{2}{6}$ ${}^{\sim}\overset{2}{6}$ ${}^{\sim}\overset{2}{6}$)
	R	6.6986455	7.5497333	6.67804	7.5386856	6.726227	7.892861	8.410451	7.9500524
		81 ($\underline{6\ 3}$ ${}^{-}3$ ${}^{\sim}\overset{2}{6}$)	113 (6 ${}^{-}\overset{2}{6}$ ${}^{-}\overset{2}{6}$ ${}^{\sim}\overset{2}{6}$)						
	R	8.749775	9.6140437						
3 (2 ${}^{\sim}$4)e		2 (3 ${}^{-}\overset{2}{6}$)	5 (6 ${}^{-}\overset{2}{6}$)	8 (1 3 ${}^{-}\overset{2}{6}$)	10 (3 ${}^{\sim}3$ ${}^{-}\overset{2}{6}$)	13 (3 ${}^{\sim}\overset{2}{6}$ ${}^{\sim}\overset{2}{6}$)	14 (3 ${}^{\sim}\overset{2}{6}$ ${}^{-}\overset{2}{6}$)	16 (3 ${}^{-}\overset{2}{6}$ ${}^{-}\overset{2}{6}$)	23 (6 ${}^{\sim}6$ ${}^{-}\overset{2}{6}$)
	R	4.691993	5.846523	5.32576	5.509722	6.6764033	6.499544	6.673619	7.3566401
		25 (6 ${}^{-}\overset{2}{6}$ ${}^{\sim}\overset{2}{6}$)	43 (1 3 ${}^{\sim}\overset{2}{6}$ ${}^{\sim}\overset{2}{6}$)	44 (1 3 ${}^{\sim}\overset{2}{6}$ ${}^{-}\overset{2}{6}$)	55 (3 ${}^{\sim}3$ ${}^{\sim}\overset{2}{6}$ ${}^{\sim}\overset{2}{6}$)	56 (3 ${}^{\sim}3$ ${}^{\sim}\overset{2}{6}$ ${}^{-}\overset{2}{6}$)	71 (3 ${}^{\sim}\overset{2}{6}$ ${}^{\sim}\overset{2}{6}$ ${}^{\sim}\overset{2}{6}$)	73 (3 ${}^{\sim}\overset{2}{6}$ ${}^{-}\overset{2}{6}$ ${}^{\sim}\overset{2}{6}$)	75 (3 ${}^{-}\overset{2}{6}$ ${}^{\sim}\overset{2}{6}$ ${}^{\sim}\overset{2}{6}$)
	R	7.863625	6.996642	6.691939	7.0423567	6.717785	8.2009635	8.7203373	8.2557539
		103 (6 ${}^{\sim}6$ ${}^{\sim}6$ ${}^{-}\overset{2}{6}$)	113 (6 ${}^{-}\overset{2}{6}$ ${}^{-}\overset{2}{6}$ ${}^{\sim}\overset{2}{6}$)						
		7.940663	9.9185005						
6 (4 ${}^{\sim}$4)e		5 (6 ${}^{-}\overset{2}{6}$)	8 (1 3 ${}^{-}\overset{2}{6}$)	10 (3 ${}^{\sim}3$ ${}^{-}\overset{2}{6}$)	13 (3 ${}^{\sim}\overset{2}{6}$ ${}^{\sim}\overset{2}{6}$)	14 (3 ${}^{\sim}\overset{2}{6}$ ${}^{-}\overset{2}{6}$)	16 (3 ${}^{-}\overset{2}{6}$ ${}^{-}\overset{2}{6}$)	23 (6 ${}^{\sim}6$ ${}^{-}\overset{2}{6}$)	25 (6 ${}^{-}\overset{2}{6}$ ${}^{\sim}\overset{2}{6}$)
	R	5.926501	5.45569	5.652276	6.751782	6.6191145	6.805401	7.472923	7.934045
		43 (1 3 ${}^{\sim}\overset{2}{6}$ ${}^{\sim}\overset{2}{6}$)	44 (1 3 ${}^{\sim}\overset{2}{6}$ ${}^{-}\overset{2}{6}$)	55 (3 ${}^{\sim}3$ ${}^{\sim}\overset{2}{6}$ ${}^{\sim}\overset{2}{6}$)	56 (3 ${}^{\sim}3$ ${}^{\sim}\overset{2}{6}$ ${}^{-}\overset{2}{6}$)	71 (3 ${}^{\sim}\overset{2}{6}$ ${}^{\sim}\overset{2}{6}$ ${}^{\sim}\overset{2}{6}$)	73 (3 ${}^{\sim}\overset{2}{6}$ ${}^{-}\overset{2}{6}$ ${}^{\sim}\overset{2}{6}$)	75 (3 ${}^{-}\overset{2}{6}$ ${}^{\sim}\overset{2}{6}$ ${}^{\sim}\overset{2}{6}$)	103 (6 ${}^{\sim}6$ ${}^{\sim}6$ ${}^{-}\overset{2}{6}$)
	R	7.069816	6.823678	7.1149125	6.850919	8.2696453	8.7886768	8.3238492	8.068317
		113 (6 ${}^{-}\overset{2}{6}$ ${}^{-}\overset{2}{6}$ ${}^{\sim}\overset{2}{6}$)							
	R	9.98447							
7 (4 ${}^{-}$4)e		5 (6 ${}^{-}\overset{2}{6}$)	8 (1 3 ${}^{-}\overset{2}{6}$)	10 (3 ${}^{\sim}3$ ${}^{-}\overset{2}{6}$)	13 (3 ${}^{\sim}\overset{2}{6}$ ${}^{\sim}\overset{2}{6}$)	14 (3 ${}^{\sim}\overset{2}{6}$ ${}^{-}\overset{2}{6}$)	15 (3 ${}^{-}\overset{2}{6}$ ${}^{\sim}\overset{2}{6}$)	16 (3 ${}^{-}\overset{2}{6}$ ${}^{-}\overset{2}{6}$)	23 (6 ${}^{\sim}6$ ${}^{-}\overset{2}{6}$)
	R	6.15575	5.9038975	6.155291	6.9624475	7.043446	7.249025	7.2844	7.8899506
		25 (6 ${}^{-}\overset{2}{6}$ ${}^{\sim}\overset{2}{6}$)	26 (6 ${}^{-}\overset{2}{6}$ ${}^{-}\overset{2}{6}$)	43 (1 3 ${}^{\sim}\overset{2}{6}$ ${}^{\sim}\overset{2}{6}$)	44 (1 3 ${}^{\sim}\overset{2}{6}$ ${}^{-}\overset{2}{6}$)	45 (1 3 ${}^{-}\overset{2}{6}$ ${}^{\sim}\overset{2}{6}$)	46 (1 3 ${}^{-}\overset{2}{6}$ ${}^{-}\overset{2}{6}$)	55 (3 ${}^{\sim}3$ ${}^{\sim}\overset{2}{6}$ ${}^{\sim}\overset{2}{6}$)	56 (3 ${}^{\sim}3$ ${}^{\sim}\overset{2}{6}$ ${}^{-}\overset{2}{6}$)
	R	8.124318	7.8443996	7.272636	7.302448	8.0906624	7.8262155	7.315738	7.3357094
		57 (3 ${}^{\sim}3$ ${}^{-}\overset{2}{6}$ ${}^{\sim}\overset{2}{6}$)	61 (3 ${}^{-}3$ ${}^{-}\overset{2}{6}$ ${}^{\sim}\overset{2}{6}$)	71 (3 ${}^{\sim}\overset{2}{6}$ ${}^{\sim}\overset{2}{6}$ ${}^{\sim}\overset{2}{6}$)	73 (3 ${}^{\sim}\overset{2}{6}$ ${}^{-}\overset{2}{6}$ ${}^{\sim}\overset{2}{6}$)	75 (3 ${}^{-}\overset{2}{6}$ ${}^{\sim}\overset{2}{6}$ ${}^{\sim}\overset{2}{6}$)	85 ($\underline{6\ 3}$ ${}^{-}\overset{2}{6}$ ${}^{\sim}\overset{2}{6}$)	103 (6 ${}^{\sim}6$ ${}^{\sim}6$ ${}^{-}\overset{2}{6}$)	113 (6 ${}^{-}\overset{2}{6}$ ${}^{-}\overset{2}{6}$ ${}^{\sim}\overset{2}{6}$)
	R	8.409573	9.259956	8.4537216	8.9693608	8.506186	9.594235	8.540739	10.154649
9 ($\underline{8\ 4}$)e		12 (3 ${}^{-}3$ ${}^{-}\overset{2}{6}$)	14 (3 ${}^{\sim}\overset{2}{6}$ ${}^{-}\overset{2}{6}$)	18 ($\underline{6\ 3}$ ${}^{-}\overset{2}{6}$)	23 (6 ${}^{\sim}6$ ${}^{-}\overset{2}{6}$)	26 (6 ${}^{-}\overset{2}{6}$ ${}^{-}\overset{2}{6}$)	44 (1 3 ${}^{\sim}\overset{2}{6}$ ${}^{-}\overset{2}{6}$)	46 (1 3 ${}^{-}\overset{2}{6}$ ${}^{-}\overset{2}{6}$)	56 (3 ${}^{\sim}3$ ${}^{\sim}\overset{2}{6}$ ${}^{-}\overset{2}{6}$)
	R	7.305602	7.404095	7.508272	8.2272429	8.459447	7.7405415	8.438001	7.781829
		58 (3 ${}^{\sim}3$ ${}^{-}\overset{2}{6}$ ${}^{-}\overset{2}{6}$)	72 (3 ${}^{\sim}\overset{2}{6}$ ${}^{\sim}\overset{2}{6}$ ${}^{-}\overset{2}{6}$)	76 (3 ${}^{-}\overset{2}{6}$ ${}^{\sim}\overset{2}{6}$ ${}^{-}\overset{2}{6}$)	103 (6 ${}^{\sim}6$ ${}^{\sim}6$ ${}^{-}\overset{2}{6}$)				
	R	8.6438232	8.6678424	8.6976125	8.9566188				
10. (1 4 ${}^{\sim}$4)e		8 (1 3 ${}^{-}\overset{2}{6}$)	10 (3 ${}^{\sim}3$ ${}^{-}\overset{2}{6}$)	13 (3 ${}^{\sim}\overset{2}{6}$ ${}^{\sim}\overset{2}{6}$)	14 (3 ${}^{\sim}\overset{2}{6}$ ${}^{-}\overset{2}{6}$)	15 (3 ${}^{-}\overset{2}{6}$ ${}^{\sim}\overset{2}{6}$)	16 (3 ${}^{-}\overset{2}{6}$ ${}^{-}\overset{2}{6}$)	23 (6 ${}^{\sim}6$ ${}^{-}\overset{2}{6}$)	25 (6 ${}^{-}\overset{2}{6}$ ${}^{\sim}\overset{2}{6}$)
	R	5.551834	5.756906	6.801483	6.7027455	7.09134	6.89736	7.551856	7.978454
		43 (1 3 ${}^{\sim}\overset{2}{6}$ ${}^{\sim}\overset{2}{6}$)	44 (1 3 ${}^{\sim}\overset{2}{6}$ ${}^{-}\overset{2}{6}$)	45 (1 3 ${}^{-}\overset{2}{6}$ ${}^{\sim}\overset{2}{6}$)	55 (3 ${}^{\sim}3$ ${}^{\sim}\overset{2}{6}$ ${}^{\sim}\overset{2}{6}$)	56 (3 ${}^{\sim}3$ ${}^{\sim}\overset{2}{6}$ ${}^{-}\overset{2}{6}$)	57 (3 ${}^{\sim}3$ ${}^{-}\overset{2}{6}$ ${}^{\sim}\overset{2}{6}$)	61 (3 ${}^{-}3$ ${}^{-}\overset{2}{6}$ ${}^{\sim}\overset{2}{6}$)	71 (3 ${}^{\sim}\overset{2}{6}$ ${}^{\sim}\overset{2}{6}$ ${}^{\sim}\overset{2}{6}$)
	R	7.11742	6.91522	7.943157	7.1620345	6.943668	8.265532	9.122817	8.3125463
		73 (3 ${}^{\sim}\overset{2}{6}$ ${}^{-}\overset{2}{6}$ ${}^{\sim}\overset{2}{6}$)	75 (3 ${}^{-}\overset{2}{6}$ ${}^{\sim}\overset{2}{6}$ ${}^{\sim}\overset{2}{6}$)	85 ($\underline{6\ 3}$ ${}^{-}\overset{2}{6}$ ${}^{\sim}\overset{2}{6}$)	103 (6 ${}^{\sim}6$ ${}^{\sim}6$ ${}^{-}\overset{2}{6}$)	113 (6 ${}^{-}\overset{2}{6}$ ${}^{-}\overset{2}{6}$ ${}^{\sim}\overset{2}{6}$)			
	R	8.8307748	8.366337	9.460009	8.154099	10.024159			
23. (4 ${}^{\sim}$4 ${}^{-}$4)e		12 (3 ${}^{\sim}3$ ${}^{\sim}\overset{2}{6}$)	14 (3 ${}^{\sim}\overset{2}{6}$ ${}^{-}\overset{2}{6}$)	18 ($\underline{6\ 3}$ ${}^{-}\overset{2}{6}$)	23 (6 ${}^{\sim}6$ ${}^{-}\overset{2}{6}$)	44 (1 3 ${}^{\sim}\overset{2}{6}$ ${}^{-}\overset{2}{6}$)	46 (1 3 ${}^{-}\overset{2}{6}$ ${}^{-}\overset{2}{6}$)	56 (3 ${}^{\sim}3$ ${}^{\sim}\overset{2}{6}$ ${}^{-}\overset{2}{6}$)	58 (3 ${}^{\sim}3$ ${}^{-}\overset{2}{6}$ ${}^{-}\overset{2}{6}$)
	R	7.67626	7.520486	7.924658	8.3226323	7.9417047	8.79382	7.991517	9.0456995
		72 (3 ${}^{\sim}\overset{2}{6}$ ${}^{\sim}\overset{2}{6}$ ${}^{-}\overset{2}{6}$)	74 (3 ${}^{\sim}\overset{2}{6}$ ${}^{-}\overset{2}{6}$ ${}^{-}\overset{2}{6}$)	76 (3 ${}^{-}\overset{2}{6}$ ${}^{\sim}\overset{2}{6}$ ${}^{-}\overset{2}{6}$)	103 (6 ${}^{\sim}6$ ${}^{\sim}6$ ${}^{-}\overset{2}{6}$)				
	R	9.071241	9.408105	9.10677	9.147787				

30 (4 8 4)e		12 (3 3 6)	14 (3 6 6)	18 (6 3 6)	23 (6 6 6)	26 (6 6 6)	44 (1 3 6 6)	46 (1 3 6 6)	56 (3 3 6 6)
	R	7.4114631	7.448295	7.623807	8.264884	8.552577	7.8019605	8.5306935	7.84492
		58 (3 3 6 6)	72 (3 6 6 6)	76 (3 6 6 6)	103 (6 6 6 6)				
	R	8.745795	8.770035	8.800965	9.011005				
32 (4 8 4)e [Re<9.805]		12 (3 3 6)	14 (3 6 6)	18 (6 3 6)	23 (6 6 6)	26 (6 6 6)	44 (1 3 6 6)	46 (1 3 6 6)	56 (3 3 6 6)
	R	7.422083	7.4523513	7.647935	8.267609	8.5593753	7.80737	8.537494	7.850428 Ds = 77.885
		58 (3 3 6 6)	72 (3 6 6 6)	76 (3 6 6 6)	103 (6 6 6 6)				
	R	8.757825	8.777022	8.807955	9.014345				
35. (8 4 4)e		26 (6 6 6)	50 (3 3 3 6)	62 (3 3 6 6)	72 (3 6 6 6)	74 (3 6 6 6)	76 (3 6 6 6)	86 (6 3 6 6)	114 (6 6 6 6)
	R	9.386826	9.527735	10.63521	9.80867	10.37272	9.863428	10.89063	11.24499
43 (1 4 4 4)e]		12 (3 3 6)	14 (3 6 6)	18 (6 3 6)	23 (6 6 6)	26 (6 6 6)	44 (1 3 6 6)	46 (1 3 6 6)	56 (3 3 6 6)
	R	7.748687	7.524884	8.005882	8.3192595	8.88237	7970701	8.85787	8.022771
		58 (3 3 6 6)	62 (3 3 6 6)	72 (3 6 6 6)	74 (3 6 6 6)	76 (3 6 6 6)	103 (6 6 6 6)		
	R	9.118797	9.654593	9.144547	9.49548	9.18114	9.171757		
103 (4 4 8 4)e		26 (6 6 6)	50 (3 3 3 6)	62 (3 3 6 6)	72 (3 6 6 6)	74 (3 6 6 6)	76 (3 6 6 6)	86 (6 3 6 6)	114 (6 6 6 6)
	R	9.265327	9.1687	10.28924	9.60671	10.0721	9.6525	10.49974	10.78002
107 (4 8 4 4)e		26 (6 6 6)	50 (3 3 3 6)	62 (3 3 6 6)	72 (3 6 6 6)	74 (3 6 6 6)	76 (3 6 6 6)	86 (6 3 6 6)	114 (6 6 6 6)
	R	9.395287	9.60849	10.706226	9.84334	10.433263	9.900732	10.9714	11.34083
111 (4 8 4 4)e		26 (6 6 6)	50 (3 3 3 6)	62 (3 3 6 6)	72 (3 6 6 6)	74 (3 6 6 6)	76 (3 6 6 6)	86 (6 3 6 6)	114 (6 6 6 6)
	R	9.395423	9.612628	10.7085156	9.844841	10.43546	9.902333	10.973678	11.34303
153 (8 8 8 4)e		74 (3 6 6 6)	82 (6 3 3 6)	114 (6 6 6 6)					
	R	10.72492	11.04147	11.93575					

End of 24E'2 Group (9): 6(4M)e +8(6A) [x = y = E4M, z = E6A]

24E'2 Family Group (10): 6(4M)e + 4(6A) [x = y = F4M, z = F6A] – Outline																							
	2	5	8	10	12	14	16	18	23	26	44	46	50	56	58	62	72	74	76	82	86	103	114
6					•			•		•		•			•		•		•				
9													•			•		•			•		
10				•	•			•		•	•	•		•	•		•		•			•	
30								•				•	•		•	•	•	•	•		•		
32					•			•				•	•		•	•	•	•	•		•		
103																				•			•
153																							

Every dotted cell in the chart of Group (10) has the same solution as in the corresponding cell of Group (9), except the Ds in Group (10) has higher value than that in Group (9) due to more balls in Group (10).

End of 24E'2 Group (10): 6(4M)e + 4(6A)

24E'₂ Family Group (11): 6(4M)e + 4(6A) [x = y = F4M, z = F6A] – Outline

	2	5	8	10	12	14	16	18	23	26	44	46	50	56	58	62	72	74	76	82	86	103	114
2			•	•	•	•	•	•	•	•	•	•		•	•		•		•			•	
4			•	•	•		•	•		•	•	•		•	•		•	•	•			•	
5					•			•		•		•	•		•	•	•	•	•		•		
8					•			•					•		•	•	•	•	•		•		•
19													•							•	•		•
29																				•			
69																							
97																							

24E'₂ Family Group (11): 6(4M)e + 4(6A) [x = F4M, y = 1, z = F6A]

		$8(1\ 3\ \overset{-2}{6})$	$10(3\ \overset{\sim}{3}\ \overset{-2}{6})$	$12(3\ \overset{-}{3}\ \overset{-2}{6})$	$14(3\ \overset{\sim 2}{6}\ \overset{-2}{6})$	$16(3\ \overset{-2}{6}\ \overset{-2}{6})$	$18(\underline{6\ 3}\ \overset{-2}{6})$	$23(6\ \overset{\sim}{6}\ \overset{-2}{6})$
$2(1\ \overset{1}{4})$	R	4.525178	4.7081995	5.01171	5.684588	5.852315	5.0650165	6.5324442
$4(2\ \overset{\sim 1}{4})$	R	4.5621243	4.78065	5.182023	×	5.892121	5.291943	×
$5(2\ \overset{-1}{4})$	R	×	×	5.421025	×	×	5.6305185	×
$8(4\ \overset{-2}{4})$	R	×	×	5.461322	×	×	5.7078649	×
		$26(6\ \overset{-2}{6}\ \overset{-2}{6})$	44 $(1\ 3\ \overset{\sim 2}{6}\ \overset{-2}{6})$	$46(1\ 3\ \overset{-2}{6}\ \overset{-2}{6})$	50 $(3\ \overset{\sim}{3}\ \overset{-2}{3}\ \overset{-2}{6})$	56 $(3\ \overset{\sim}{3}\ \overset{\sim 2}{6}\ \overset{-2}{6})$	$58(3\ \overset{\sim}{3}\ \overset{-2}{6}\ \overset{-2}{6})$	$62(3\ \overset{-}{3}\ \overset{-2}{6}\ \overset{-2}{6})$
$2(1\ \overset{1}{4})$	R	6.2327649	5.880296	6.2160289	×	5.9094755	6.2935473	×
$4(2\ \overset{\sim 1}{4})$	R	6.373349	5.9248716	6.3505437	×	5.959573	6.469746	×
$5(2\ \overset{-1}{4})$	R	6.5548932	×	6.519331	5.9541457	×	6.7259503	7.1273449
$8(4\ \overset{-2}{4})$	R	×	×	×	6.122294	×	6.7650853	7.256272
$19(2\ \overset{\sim 2}{4}\ \overset{-2}{4})$	R	×	×	×	6.3877725	×	×	×
		72 $(3\ \overset{\sim 2}{6}\ \overset{\sim 2}{6}\ \overset{-2}{6})$	74 $(3\ \overset{\sim 2}{6}\ \overset{-2}{6}\ \overset{-2}{6})$	76 $(3\ \overset{-2}{6}\ \overset{\sim 2}{6}\ \overset{-2}{6})$	$82(\underline{6\ 3}\ \overset{-}{3}\ \overset{-2}{6})$	$86(\underline{6\ 3}\ \overset{-2}{6}\ \overset{-2}{6})$	103 $(6\ \overset{\sim}{6}\ \overset{\sim}{6}\ \overset{-2}{6})$	114 $(6\ \overset{-2}{6}\ \overset{-2}{6}\ \overset{-2}{6})$
$2(1\ \overset{1}{4})$	R	6.3093	×	6.318067	×	×	7.1280753	×
$4(2\ \overset{\sim 1}{4})$	R	6.497808	6.591085	6.5170862	×	×	7.1666834	×
$5(2\ \overset{-1}{4})$	R	6.7747348	7.0195285	6.8121869	×	7.229839	×	×
$8(4\ \overset{-2}{4})$	R	6.821618	7.117918	6.8655321	×	7.3996027	×	7.5526411
$19(2\ \overset{\sim 2}{4}\ \overset{-2}{4})$	R	×	×	×	7.0861505	7.6742094	×	8.041012
29 $(4\ \overset{-2}{4}\ \overset{-2}{4}\ \overset{-2}{4})$	R	×	×	×	7.330662	×	×	×

24E'₂ Group (12): 4(6M)e + 6(4A)

	5	8	14	19	21	26	29	36	46	69	97	114	116
3			•										
6					•								
19	•	•	•	•					•				
36													
115					•		•					•	•
117				•	•		•		•			•	•
119				•	•	•	•		•			•	•
121				•									
152													

24E'₂ Group (12): 4(6M)e + 6(4A) [x = z = F6M, y = F4A]

		$5\,(2\ \overset{-2}{4})$	$8\,(4\ \overset{-2}{4})$	$14\,(1\ \overset{2}{4}\ \overset{-2}{4})$	$19\,(2\ \overset{\sim 2}{4}\ \overset{-2}{4})$	$21\,(2\ \overset{-2}{4}\ \overset{-2}{4})$	$46\,(1\ 4\ \tilde{4}\ \overset{-2}{4})$	
$3\,(6\ \tilde{6})e$	R	×	×	6.532443	×	×	×	
$6\,(\underline{12\ 6})e$	R	×	×	×	×	8.590274	×	
$19\,(6\ \tilde{6}\ \tilde{6})e$	R	5.918914	6.850025	7.129076	7.166681	×	8.154098	
		$19\,(2\ \overset{\sim 2}{4}\ \overset{-2}{4})$	$21\,(2\ \overset{-2}{4}\ \overset{-2}{4})$	$26\,(4\ \tilde{4}\ \overset{-2}{4})$	$29\,(4\ \overset{-2}{4}\ \overset{-2}{4})$	$46\,(1\ 4\ \tilde{4}\ \overset{-2}{4})$	$114\,(4\ \overset{\sim 2\sim}{8}\ 4\ \overset{-2}{4})$	$116\,(4\ \overset{-2\sim}{8}\ 4\ \overset{-2}{4})$
$115\,(6\ \tilde{6}\ \overset{\sim 2}{12}\ \bar{6})e$	R	×	9.118926	×	9.145381	×	11.120131	11.122444
$117\,(6\ \tilde{6}\ \overset{-2}{12}\ \tilde{6})e$	R	8.952924	9.211408	×	9.24353	9.90015	11.204946	11.207175
$119\,(6\ \bar{6}\ \overset{\sim 2}{12}\ \tilde{6})e$	R	8.974713	9.23571	9.797162	9.269037	9.914997	11.217496	11.219699
$121\,(6\ \bar{6}\ \overset{-2}{12}\ \tilde{6})e$	R	9.034019	×	×	×	×	×	×

24E'₂ Group (13): 4(6M)e,₁ + 6(4A)

	P. 1	5	8	14	19	21	26	29	P. 2	36	46	69	97	114	116
1		•	•	•											
2		•	•	•	•						•				
5		•		•	•	•	•			•	•			•	•
7		•	•	•	•						•				
8		•	•	•	•	•	•				•			•	
9		•	•	•	•						•				
10				•	•	•	•	•			•			•	•
12						•		•		•				•	•
13				•	•	•	•	•			•			•	•
14					•	•	•	•		•	•			•	•
15				•	•	•	•	•			•			•	•
18						•		•		•		•		•	•
26								•		•		•		•	•
43				•	•	•	•	•			•			•	•
44					•	•		•		•				•	•
50								•				•			
55				•	•	•	•	•			•			•	•
56					•	•		•		•				•	•
72								•				•			
74								•				•	•		
76								•				•			
82												•	•		
103						•		•		•		•		•	•
113												•	•		

24E'2 Group (13): 4(6M)e,1 + 6(4A) [x = F6M, y = F4A, z = 1]

P. 1		5. $(2^{-}\overset{2}{4})$	8. $(4^{-}\overset{2}{4})$	14. $(1\ \overset{2}{4}^{-}\overset{2}{4})$	19. $(2^{\sim}\overset{2}{4}^{-}\overset{2}{4})$	21. $(2^{-}\overset{2}{4}^{-}\overset{2}{4})$	26. $(4^{\sim}4^{-}\overset{2}{4})$	29. $(2^{-}\overset{2}{4}^{-}\overset{2}{4})$
1. $(3^{\sim}\overset{2}{6})$	R	4.5417159	5.5017241	5.6845877	×	×	×	×
2. $(3^{-}\overset{2}{6})$	R	4.6918068	5.5846632	5.8523148	5.892121	×	×	×
5. $(6^{-}\overset{2}{6})$	R	4.810074	×	6.2327695	6.3733513	6.5548937	7.2155515	×
7. $(1\ 3^{\sim}\overset{2}{6})$	R	4.6739171	5.5987242	5.880296	5.9248716	×	×	×
8. $(1\ 3^{-}\overset{2}{6})$	R	4.8092665	56.366245	6.216029	6.3505438	6.5193316	7.2016285	×
9. $(3^{\sim}3^{\sim}\overset{2}{6})$	R	4.6964128	5.611376	5.9094747	5.9595726	×	×	×
10. $(3^{\sim}3^{-}\overset{2}{6})$	R	×	×	6.2935473	6.4697408	6.7259501	7.2894103	6.7650841
12. $(3^{-}3^{-}\overset{2}{6})$	R	×	×	×	×	7.1273449	×	7.256273
13. $(3^{\sim}\overset{2}{6}^{\sim}\overset{2}{6})$	R	×	×	6.3093011	6.4978072	6.7747337	7.302962	6.8216183
14. $(3^{\sim}\overset{2}{6}^{-}\overset{2}{6})$	R	×	×	×	6.5910862	7.0195271	7.3360178	7.1179187
15. $(3^{-}\overset{2}{6}^{\sim}\overset{2}{6})$	R	×	×	6.3180669	6.5170762	68.12187	7.3136404	6.8655322
18. $(\underline{6\ 3}^{-}\overset{2}{6})$	R	×	×	×	×	7.2298391	×	7.3996027
26. $(6^{-}\overset{2}{6}^{-}\overset{2}{6})$	R	×	×	×	×	×	×	7.55264
43. $(1\ 3^{\sim}\overset{2}{6}^{\sim}\overset{2}{6})$	R	×	×	6.3188607	6.5187905	6.8150093	7.3141782	6.8687303
44. $(1\ 3^{\sim}\overset{2}{6}^{-}\overset{2}{6})$	R	×	×	×	6.5943834	7.1432941	×	9.2762545
50. $(3^{\sim}3^{-}3^{-}\overset{2}{6})$	R	×	×	×	×	×	×	7.581186
55. $(3^{\sim}3^{\sim}\overset{2}{6}^{\sim}\overset{2}{6})$	R	×	×	6.3197381	6.5208842	6.8188813	7.3150818	6.8732125
56. $(3^{\sim}3^{\sim}\overset{2}{6}^{-}\overset{2}{6})$	R	×	×	×	6.5893547	7.158303	×	7.2961329
72. $(3^{\sim}\overset{2}{6}^{\sim}\overset{2}{6}^{-}\overset{2}{6})$	R	×	×	×	×	×	×	7.5830786
74. $(3^{\sim}\overset{2}{6}^{-}\overset{2}{6}^{-}\overset{2}{6})$	R	×	×	×	×	×	×	7.4910215
76. $(3^{-}\overset{2}{6}^{\sim}\overset{2}{6}^{-}\overset{2}{6})$	R	×	×	×	×	×	×	7.5821545
82. $(\underline{6\ 3}^{-}3^{-}\overset{2}{6})$	R	×	×	×	×	×	×	×
103. $(6^{\sim}6^{\sim}6^{-}\overset{2}{6})$	R	×	×	×	×	7.2751934	×	7.47353292

24E'2 Group (13): 4 (6M)e,1 + 6 (4A) [x = F6M, y = F4A, z = 1]
P. 2

P. 2		36. $(\underline{8\ 4}^{-}\overset{2}{4})$	46 $(1\ 4^{\sim}4^{-}\overset{2}{4})$	69 $(2^{\sim}\overset{2}{4}^{-}\overset{2}{4}^{-}\overset{2}{4})$	97 $(4^{-}\overset{2}{4}^{-}\overset{2}{4}^{-}\overset{2}{4})$	114 $(4^{\sim}\overset{2}{8}^{\sim}4^{-}\overset{2}{4})$	116 $(4^{-}\overset{2}{8}^{\sim}4^{-}\overset{2}{4})$	
2. $(3^{-}\overset{2}{6})$	R	×	6.89736	×	×	×	×	
5. $(6^{-}\overset{2}{6})$	R	8.4594454	7.3317797	×	×	8.5525766	8.5593752	

7. $(1\ 3^{\sim}\overset{2}{6})$	R	×	6.9155219	×	×	×	×	
8. $(1\ 3^{-}\overset{2}{6})$	R	×	7.3170354	×	×	8.5306935	×	
9. $(3^{\sim}3^{\sim}\overset{2}{6})$	R	×	6.9436681	×	×	×	×	
10. $(3^{\sim}3^{-}\overset{2}{6})$	R	×	7.4151109	×	×	8.745795	8.7528252	
12. $(3^{-}3^{-}\overset{2}{6})$	R	9.0377049	×	×	×	9.1651053	9.1727652	
13. $(3^{\sim}\overset{2}{6}{}^{\sim}\overset{2}{6})$	R	×	7.4303873	×	×	8.7700344	8.7770218	
14. $(3^{\sim}\overset{2}{6}{}^{-}\overset{2}{6})$	R	8.9291357	7.4865897	×	×	9.0471758	9.0545555	
15. $(3^{-}\overset{2}{6}{}^{\sim}\overset{2}{6})$	R	×	7.4432199	×	×	8.8009648	8.8079544	
18. $(\underline{6\ 3}^{-}\overset{2}{6})$	R	9.126554	×	7.6742094	×	9.2657198	9.2737899	
26. $(6^{-}\overset{2}{6}{}^{-}\overset{2}{6})$	R	9.1649088	×	8.0410122	×	9.3320356	9.3414388	
43. $(1\ 3^{\sim}\overset{2}{6}{}^{\sim}\overset{2}{6})$	R	×	7.4438217	×	×	8.8014407	8.8084331	
44. $(1\ 3^{\sim}\overset{2}{6}{}^{-}\overset{2}{6})$	R	9.0437455	×	×	×	9.1713195	9.1789709	
50. $(3^{\sim}3^{-}3^{-}\overset{2}{6})$	R	×	×	8.1874665	×	×	×	
55. $(3^{\sim}3^{\sim}\overset{2}{6}{}^{\sim}\overset{2}{6})$	R	×	7.4449032	×	×	8.8535074	8.810496	
56. $(3^{\sim}3^{\sim}\overset{2}{6}{}^{-}\overset{2}{6})$	R	9.0556035	×	×	×	9.1843296	9.1920161	
72. $(3^{\sim}\overset{2}{6}{}^{\sim}\overset{2}{6}{}^{-}\overset{2}{6})$	R	×	×	8.2190729	×	×	×	
74. $(3^{\sim}\overset{2}{6}{}^{-}\overset{2}{6}{}^{-}\overset{2}{6})$	R	×	×	8.4323917	8.7580772	×	×	
76. $(3^{-}\overset{2}{6}{}^{\sim}\overset{2}{6}{}^{-}\overset{2}{6})$	R	×	×	8.2453275	×	×	×	
82. $(\underline{6\ 3}^{-}3^{-}\overset{2}{6})$	R	×	×	8.57474	9.13446	×	×	
103. $(6^{\sim}6^{\sim}6^{-}\overset{2}{6})$	R	9.1515707	×	7.816892	×	9.2959295	9.3041921	
114. $(6^{-}\overset{2}{6}{}^{-}\overset{2}{6}{}^{-}\overset{2}{6})$	R	×	×	8.5468087	9.3978467	×	×	

End of 24E'2 Group (13): 4(6M)e,1 + 6(4A)

24E'₂ Group (14): 4(6M)e + 4(6A) [x = x = F6M, y = F6A]	2	5	8	10	12	14	16	18	23	26	44	46	50	56	58	62	72	74	76	82	86	103	114
3			•	•	•	•	•		•	•	•	•		•	•		•		•			•	
6					•			•		•		•	•		•	•	•	•	•		•		
19		•	•	•	•	•	•	•	•	•	•	•		•	•		•	•	•			•	
36					•	•		•	•	•	•	•	•	•	•	•	•	•	•		•	•	
115				•	•			•		•		•	•	•	•	•	•	•	•		•		
117					•			•		•	•	•	•	•	•	•	•	•	•		•	•	
119					•			•		•	•	•	•	•	•	•	•	•	•		•	•	
121					•	•			•	•	•	•	•		•	•	•	•	•		•	•	
152													•			•	•	•			•		•

24E'₂ Group (14): 4(6M)e + 4(6A) [x = z = F6M, y = F6A]		5 (6 6)	8 (1 3 6)	10 (3 3 6)	12 (3 3 6)	14 (3 6 6)	16 (3 6 6)	18 (6 3 6)
3 (6 6)e	R	×	5.1747602	5.304517	5.510481	6.0823566	6.207934	×
6 (12 6)e	R	×	×	×	7.3009307	×	×	7.4266127
19 (6 6 6)e	R	6.026854	5.7003433	5.8499954	6.1233417	6.560919	6.7069518	6.1845
36 (12 12 6)e [Rc<8.428]	R	×	×	×	7.9598368	7.990896	×	8.10681
115 (6 6 12 6)e	R	×	×	7.261495	7.7381803	×	×	7.884585
117 (6 6 12 6)e	R	×	×	×	7.81931	×	×	7.96921
119 (6 6 12 6)e	R	×	×	×	7.854013	×	×	8.003395
121 (6 6 12 6)e	R	×	×	×	7.941125	7.977985	×	×
		23 (6 6 6)	26 (6 6 6)	44 (1 3 6 6)	46 (1 3 6 6)	50 (3 3 3 6)	56 (3 3 6 6)	58 (3 3 6 6)
3 (6 6)e	R	6.74293	6.4821455	6.225123	6.4714964	×	6.245463	6.5225656
6 (12 6)e	R	×	8.1890103	×	8.1740652	7.6053063	×	8.308414
19 (6 6 6)e	R	7.2006232	7.042095	6.7222369	7.029465	×	6.7441934	7.108127
36 (12 12 6)e [Rc<8.428]	R	×	×	8.2371746	×	8.361827	8.2663397	×
115 (6 6 12 6)e	R	×	8.6061329	×	8.5906997	8.1285519	8.1033292	8.7444587
117 (6 6 12 6)e	R	×	8.681254	8.138644	8.665803	8.223879	8.168381	8.822664
119 (6 6 12 6)e	R	×	8.70484	8.163142	8.689502	8.258853	8.192786	8.84607
121 (6 6 12 6)e	R	8.563797	8.769405	8.227644	8.754356	8.351025	×	8.910189
152 (12 6 12 6)e	R	×	×	×	×	9.4369953	×	×

		62 $(3\,{}^{-}3\,{}^{-2}6\,{}^{-2}6)$	72 $(3\,{}^{\sim 2}6\,{}^{\sim 2}6\,{}^{-2}6)$	74 $(3\,{}^{\sim 2}6\,{}^{-2}6\,{}^{-2}6)$	76 $(3\,{}^{-2}6\,{}^{\sim 2}6\,{}^{-2}6)$	86 $(\underline{6\,3}\,{}^{-2}6\,{}^{-2}6)$	103 $(6\,{}^{\sim}6\,{}^{\sim}6\,{}^{-2}6)$	114 $(6\,{}^{-2}6\,{}^{-2}6\,{}^{-2}6)$
3 $(6\,{}^{\sim}6)$e	R	×	8.5308654	×	6.535113	×	7.2006232	×
6 $(\underline{12\,6})$e	R	8.5489183	8.3260331	8.4880003	8.3459468	8.5883718	×	×
19 $(6\,{}^{\sim}6\,{}^{\sim}6)$e	R	×	7.1221797	7.1726449	7.1332261	×	7.6675574	×
115 $(6\,{}^{\sim}6\,{}^{\sim 2}12\,{}^{-}6)$e	R	9.042564	8.7618613	8.959724	8.7840791	9.111017	×	×
117 $(6\,{}^{\sim}6\,{}^{-2}12\,{}^{\sim}6)$e	R	9.129748	8.839925	9.043479	8.862483	9.2026	9.040813	×
119 $(6\,{}^{-}6\,{}^{\sim 2}12\,{}^{\sim}6)$e	R	9.153316	8.86316	9.066777	8.885633	9.226741	9.057265	×
121 $(6\,{}^{-}6\,{}^{-2}12\,{}^{\sim}6)$e	R	9.216903	8.926797	9.13007	8.949036	9.291367	9.102283	×
152 $(\underline{12\,6}\,{}^{\sim 2}12\,{}^{-}6)$e	R	10.30012	9.8647	10.162196	×	10.43163	×	10.596723

End of 24E’2 (14): 4 (6M)e + 4 (6A) [x = z = F6M, y = F4A]

24E'2 (15): 4(6M)e,1 + 4(6A) – Outline																		
P. 1-1,2,3,4	2	5	8	10	12	13	14	15	16	18	23	25	26	43	44	45	46	
1	•	•	•	•	•	•	•		•	•	•	•	•	•	•		•	
2	•	•	•	•		•	•		•		•	•	•	•	•		•	
5		•		•	•		•		•	•	•		•		•		•	
7		•	•	•	•	•	•	•	•	•	•	•	•	•	•	•	•	
8		•	•	•	•		•		•	•	•		•		•		•	
9		•	•	•	•	•	•	•	•	•	•	•	•	•	•	•	•	
10				•	•		•			•	•		•		•		•	
	50	55	56	57	58	61	62	71		72	73	74	75	76	82	85	86	103
1		•	•		•			•		•	•		•	•				•
2		•	•					•			•		•					•
5	•		•		•		•			•		•		•			•	•
7		•	•	•	•	•		•		•	•	•	•	•		•		•
8			•		•					•				•				•
9		•	•	•	•	•	•	•		•	•	•	•	•		•		•
10			•		•					•		•		•				•
P. 2-1, 2	2	5	8	10	12	13	14	15	16	18	23	25	26	43	44	45	46	50
12					•					•			•		•			•
13					•		•			•	•		•		•		•	•
14					•		•			•			•		•		•	•
15					•		•			•	•		•		•		•	•
18										•			•					•
26													•					•
43					•		•			•	•		•		•		•	•
44										•			•		•			•
50																		
55					•		•			•	•		•		•		•	•
56										•			•					•
	55	56	57	58	61	62	71	72	73	74	75	76	82	85	86	103	113	114
12		•		•		•		•		•		•			•			
13		•		•		•		•		•		•			•	•		•
14		•		•		•		•		•		•	•		•	•		•
15		•		•		•		•		•		•			•	•		•
18						•		•		•		•			•			•
26								•		•		•	•		•			•
43		•		•		•		•		•		•			•	•		•
44		•		•		•		•		•		•	•		•			•
50								•		•		•	•		•			•
55		•		•		•		•		•		•			•	•		•
56		•				•		•		•		•	•		•	•		•
P. 3	2	5	8	10	12	13	14	15	16	18	23	25	26	43	44	45	46	50
72																		
74																		
76																		
82																		
103													•					•
114																		
	55	56	57	58	61	62	71	72	73	74	75	76	82	85	86	103	113	114
72										•		•	•					•
74													•					•
76										•			•					•
82													•					•
103						•		•		•		•	•		•			•
114																		

24E'2 (15): 4(6M)e,1 + 4(6A) [z = F6M, x = F6A, y = 1]

P. 1-1	$2\,(3\,^{-2}6)$	$5\,(6\,^{-2}6)$	$8\,(1\,3\,^{-2}6)$	$10\,(3\,^{\sim}3\,^{-2}6)$	$12\,(3\,^{-}3\,^{-2}6)$	$13\,(3\,^{\sim 2}6\,^{\sim 2}6)$	$14\,(3\,^{\sim 2}6\,^{-2}6)$	$15\,(3\,^{-2}6\,^{\sim 2}6)$	$16\,(3\,^{-2}6\,^{-2}6)$
$1\,(3\,^{\sim 2}6)$	4.0764564	4.9508927	4.5261986	4.6633894	4.8905785	5.572743	5.426976	×	5.5555818
$2\,(3\,^{-2}6)$	4.163888	5.0231131	4.6647956	4.821858	×	5.637128	5.5555814	×	5.7026365
$5\,(6\,^{-2}6)$	×	5.111538	×	5.2667529	5.7100737	×	5.8626044	×	6.069107
$7\,(1\,3\,^{\sim 2}6)$	×	5.0410612	4.7122262	4.8711427	5.1645401	5.647737	5.581168	5.85847	5.7297915
$8\,(1\,3\,^{-2}6)$	×	5.1170571	5.0210654	5.2334363	5.6762086	×	5.8466375	×	6.049718
$9\,(3\,^{\sim}3\,^{\sim 2}6)$	×	5.0549572	4.7471239	4.908974	5.211387	5.6574625	5.60549	5.867212	5.756929
$10\,(3\,^{\sim}3\,^{-2}6)$	×	×	×	5.3577659	5.8480647	×	5.9141275	×	×

P. 1-2	18. $(\underline{6\,3}\,^{-2}6)$	$23\,(6\,^{\sim}6\,^{-2}6)$	$25\,(6\,^{-2}6\,^{\sim 2}6)$	$26\,(6\,^{-2}6\,^{-2}6)$	$43\,(1\,3\,^{\sim 2}6\,^{\sim 2}6)$	$44\,(1\,3\,^{\sim 2}6\,^{-2}6)$	$45\,(1\,3\,^{-2}6\,^{\sim 2}6)$	$46\,(1\,3\,^{-2}6\,^{-2}6)$	
$1\,(3\,^{\sim 2}6)$	4.9276231	6.0823565	6.4817259	5.8626049	5.8274815	5.5811689	×	5.846638	
$2\,(3\,^{-2}6)$	×	6.207934	6.53627	6.0691077	5.888101	5.7297923	×	6.048718	
$5\,(6\,^{-2}6)$	5.8595664	6.4821455	×	6.6017181	×	6.1003329	×	6.575607	
$7\,(1\,3\,^{\sim 2}6)$	5.2427276	6.225122	6.541273	6.100333	5.896583	5.7568148	6.500937	6.080812	
$8\,(1\,3\,^{-2}6)$	5.8170521	6.4714965	×	6.575607	×	6.0808104	×	6.54962	
$9\,(3\,^{\sim}3\,^{\sim 2}6)$	5.2949565	6.2454615	6.547457	6.1356681	5.904983	5.7840611	6.507316	6.115774	
$10\,(3\,^{\sim}3\,^{-2}6)$	6.0170377	6.5225655	×	6.7299644	×	6.1751677	×	1.701955	

P. 1-3	$50\,(3\,^{\sim}3\,^{-2}3\,^{-2}6)$	$55\,(3\,^{\sim}3\,^{\sim 2}6\,^{\sim 2}6)$	$56\,(3\,^{\sim}3\,^{\sim 2}6\,^{-2}6)$	$57\,(3\,^{\sim}3\,^{-2}6\,^{\sim 2}6)$	$58\,(3\,^{\sim}3\,^{-2}6\,^{-2}6)$	$61\,(3\,^{-}3\,^{-2}6\,^{\sim 2}6)$	$62\,(3\,^{-}3\,^{-2}6\,^{-2}6)$	$71\,(3\,^{\sim 2}6\,^{\sim 2}6\,^{\sim 2}6)$	
$1\,(3\,^{\sim 2}6)$	×	5.8696065	5.60549	×	5.914127	×	×	6.756401	
$2\,(3\,^{-2}6)$	×	5.929278	5.7569287	×	×	×	×	6.807846	
$5\,(6\,^{-2}6)$	6.086448	×	6.135669	×	6.729965	×	7.0322997	×	
$7\,(1\,3\,^{\sim 2}6)$	×	5.937409	5.7840611	6.748477	6.175167	7.398585	×	6.811789	
$8\,(1\,3\,^{-2}6)$	×	×	6.115774	×	6.701956	×	×	×	
$9\,(3\,^{\sim}3\,^{\sim 2}6)$	×	5.945568	5.811664	6.754179	6.213516	7.40298	6.336633	6.817204	
$10\,(3\,^{\sim}3\,^{-2}6)$	×	×	6.2135155	×	6.874916	×	7.229217	×	

P. 1-4	$72\,(3\,^{\sim 2}6\,^{\sim 2}6\,^{-2}6)$	$73\,(3\,^{\sim 2}6\,^{-2}6\,^{\sim 2}6)$	$74\,(3\,^{\sim 2}6\,^{-2}6\,^{-2}6)$	$75\,(3\,^{-2}6\,^{\sim 2}6\,^{\sim 2}6)$	$76\,(3\,^{-2}6\,^{\sim 2}6\,^{-2}6)$	$85\,(\underline{6\,3}\,^{-2}6\,^{\sim 2}6)$	$86\,(\underline{6\,3}\,^{-2}6\,^{-2}6)$	$103\,(6\,^{\sim}6\,^{\sim}6\,^{-2}6)$	
$1\,(3\,^{\sim 2}6)$	5.931103	7.13327	×	6.807846	5.940089	×	×	6.565919	
$2\,(3\,^{-2}6)$	×	7.182814	×	6.858471	×	×	×	6.7069518	
$5\,(6\,^{-2}6)$	6.766688	×	6.951165	×	6.7950295	×	7.109405	7.042095	
$7\,(1\,3\,^{\sim 2}6)$	6.2003676	7.18576	6.2757175	6.852224	6.2165944	7.6641895	×	6.7222369	
$8\,(1\,3\,^{-2}6)$	6.7385335	×	×	×	6.7665618	×	×	7.0294605	
$9\,(3\,^{\sim}3\,^{\sim 2}6)$	6.2395551	7.1904934	6.3220202	6.8674811 Ds=57.249	6.2566	7.668154	×	6.7441934	
$10\,(3\,^{\sim}3\,^{-2}6)$	6.9142482	×	7.1296503	×	6.9457515	×	×	7.1081255	

End of 24E'2 (15): 4(6M)e,1 + 4(6A)
[z = F6M, x = F6A, y = 1] P. 1- 1, 2, 3, 4

(continue) 24E'₂ (15): 4(6M)e,1 + 4(6A) [z = F6M, x = F6A, y = 1] P. 2

P. 2-1	12 (3 3 6)	14 (3 6 6)	18 (6 3 6)	23 (6 6 6)	26 (6 6 6)	44 (1 3 6 6)	46 (1 3 6 6)	50 (3 3 3 6)	56 (3 3 6 6)
12 (3 3 6)	6.2128595	×	6.4374477	×	7.0322993	6.284931	×	6.8627495	6.3366339
13 (3 6 6)	5.9077247	5.9311021	6.0800097	6.530865	6.7666881	6.2003675	6.738533	6.372757	6.2395505
14 (3 6 6)	6.1138857	5.9495423	6.3196065	×	6.9511649	6.275718	6.9193514	6.698715	6.3220202
15 (3 6 6)	5.9454982	5.9400894	6.1214935	6.535111	6.7950295	6.2165944	6.766562	6.424625	6.2565997
18 (6 3 6)	×	×	6.5747487	×	7.1094052	×	×	7.053933	×
26 (6 6 6)	×	×	×	×	7.16	×	×	7.317442	×
43 (1 3 6 6)	5.9505375	5.9411571	6.1264308	6.535407	6.7971952	6.218354	6768745	6.429683	6.2584112
44 (1 3 6 6)	×	×	6.4634978	×	7.0448534	6.285	×	6.892563	6.3358356
55 (3 3 6 6)	5.955617	5.942201 108, 76.466	6.1317022	6.5357473	6.8001488	6.2202861	6.771687	6.435674	6.2604263 120, 76.544
56 (3 3 6 6)	×	×	6.4837249	×	7.0563033	×	×	6.919013	6.3337628

P. 2-2	58 (3 3 6 6)	62 (3 3 6 6)	72 (3 6 6 6)	74 (3 6 6 6)	76 (3 6 6 6)	82 (6 3 3 6)	86 (6 3 6 6)	103 (6 6 6 6)	114 (6 6 6 6)
12 (3 3 6)	7.229217	77287045	7.275575	7.5795506	7.3160135	×	7.890417	×	×
13 (3 6 6)	6.914248	7.2755749	6.9536241	7.1734188	6.985462	×	7.380758	7.1221798	7.488965
14 (3 6 6)	7.12965	7.5795507	7.173418	7.4469688	7.2107155	6.977853	7.722004	7.1726449	7.9036858
15 (3 6 6)	6 945752	7.316014	6.985462	7.2107155	7.0178148	×	7.425207	7.1332261	7.5422503
18 (6 3 6)	×	7.890417	7.3807585	7.7220041	7.4252068	×	8.073519	×	8.3277116
26 (6 6 6)	×	×	7.4889675	7.9036849	7.5422466	7.904789	8.327712	×	8.64197
43 (1 3 6 6)	6.94793	7.318186	6.9876059	7.2128717	7.0199472	×	7.427416	7.1338714	7.5446736
44 (1 3 6 6)	7.243717	7.7464158	7.2901904	7.5963542	7.330846	7.266594	7.9089802	×	8.1275525
50 (3 3 3 6)	×	×	7.503887	7.963659	7.563186	8.069795	8.428883	×	8.774855
55 (3 3 6 6)	6.51092	7.326887	6.9907687	7.2163756	7.0231335	×	7.431376	7.1348804	7.5492761
56 (3 3 6 6)	×	7.767333	7.3048303	7.6151545	7.3459255	7.3039996	7.932236	7.1423781	8.154973

End of 24E'2 (15): 4(6M)e,1 + 4(6A) [z = F6M, x = F6A, y = 1] P. 2- 1, 2

(continue) 24E'₂ (15): 4(6M)e,₁ + 4(6A) [z = F6M, x = F6A, y = 1] P. 3

P. 3	26 (6 6 6)	50 (3 3 3 6)	62 (3 3 6 6)	72 (3 6 6 6)	74 (3 6 6 6)	76 (3 6 6 6)	82 (6 3 3 6)	86 (6 3 6 6)	114 (6 6 6 6)
72 (3 6 6 6)	×	×	×	×	7.9758635	7.564061	8.108219	×	8.7982932
74 (3 6 6 6)	×	×	×	×	×	×	8.377752	×	9.0029388
76 (3 6 6 6)	×	×	×	×	7.9848029	×	8.139672	×	8.8208031
82 (6 3 3 6)	×	×	×	×	×	×	8.635	×	9.167117
103 (6 6 6 6)	7.143253	7.174963	7.973084	7.43479	7.796146	7.4816968	7.667229	8.164433	8.4317375

End of 24E'2 (15): 4(6M)e,1 + 4(6A) [z = F6M, x = F6A, y = 1]

24E'2 Group (16): 6(4S)e1,1 + 4(6S)e2,1 [x = e1, y = 1, z = e2]									
	1 (3 6)	5. (6 6)	7 (1 3 6)	9 (3 3 6)	13 (3 6 6)	14 (3 6 6)	15 (3 6 6)	26 (6 6 6)	43 (1 3 6 6)
2 (1 4)	× $\hat{\phi}_{gr} > \pi$	× $\hat{\phi}_{gr} > \pi$	× $\hat{\phi}_{gr} > \pi$	5.05069	× $\hat{\phi}_{gr} > \pi$	× $\hat{\phi}_{gr} > \pi$	× $\hat{\phi}_{gr} = \pi$	× $\hat{\phi}_{gr} > \pi$	6.291435
4 (2 4)	4.92606	6.1274107	5.2740457	5.3362268	6.4930566	6.9698534	6.562866	8.1850172	6.5691298
5 (2 4)	× $\hat{\phi}_{gr} > \pi$	× $\hat{\phi}_{gr} > \pi$	× $\hat{\phi}_{gr} > \pi$	6.0341933	× $\hat{\phi}_{gr} > \pi$	× $\hat{\phi}_{gr} > \pi$	× $\hat{\phi}_{gr} = \pi$	× $\hat{\phi}_{gr} > \pi$	7.276059
8 (4 4)	8.9495441	7.1535952	6.2823957	6.333762	7.5024778	8.0143877	7.5629605	9.2204433	7.565492
19 (2 4 4)	6.9798533	8.1850172	7.303363	7.3481442	8.5228487	9.0517049	8.5762314	10.2583708	8.5772834
29 (4 4 4)	8.143877	9.22045	8.3317608	8.3722587	9.5501848	10.0909628	9.5985093	11.2980475	9.5987267
69 (2 4 4 4)	9.0517087	10.2583708	9.3648457	9.4024244	10.592169	11.1316095	10.6267037	12.33901	10.6264959
97 (4 4 4 4)	10.0909628	11.2980475	10.4010597	10.4365573	11.6173806	12.17331	11.659	13.3809247	11.658585
	44 (1 3 6 6)	55 (3 3 6 6)	56 (3 3 6 6)	72 (3 6 6 6)	74 (3 6 6 6)	76 (3 6 6 6)	103 (6 6 6 6)	114 (6 6 6 6)	
2 (1 4)	× $\hat{\phi}_{gr} > \pi$	6.3003523	× $\hat{\phi}_{gr} > \pi$	× $\hat{\phi}_{gr} > \pi$	× $\hat{\phi}_{gr} > \pi$	× $\hat{\phi}_{gr} > \pi$	7.5464704	× $\hat{\phi}_{gr} > \pi$	
4 (2 4)	7.303363	6.5769475	7.348142	8.5228505	9.0517049	8.576235	7.818844	10.2583708	
5 (2 4)	× $\hat{\phi}_{gr} > \pi$	7.2819344	× $\hat{\phi}_{gr} > \pi$	× $\hat{\phi}_{gr} > \pi$	× $\hat{\phi}_{gr} > \pi$	× $\hat{\phi}_{gr} > \pi$	8.5286335	× $\hat{\phi}_{gr} > \pi$	
8 (4 4)	8.3317608	7.5707376	8.3722587	9.5501848	10.0909628	9.5985093	8.8111682	11.2980475	
19 (2 4 4)	9.3648458	8.510437	9.4024244	10.58217	11.1316094	10.6267037	9.818562	10.33901	
29 (4 4 4)	10.4010578	9.6015686	10.4365573	11.6173806	12.17331	11.659	10.8359436	13.3809247	
69 (2 4 4 4)	11.4394583	10.628733	11.473401	12.6548992	13.2157894	12.694264	11.8601803	14.4236102	
97 (4 4 4 4)	12.4794393	11.6604094	12.512237	13.6941113	14.2589211	13.7316862	12.8892437	15.466844	

End of [DATA 2B –4] 24E'2 Group (16): 6(4S)e1,1 + 4(6S)e2,1 [x = e1, y = 1, z = e2]

24E'$_2$ Group (17): 4(6S)$e_{1,1}$ + 4(6S)$e_{2,1}$ [x = e_1, y = e_2, z = 1] ---- Outline The blank cells are diagonally mirror-imaged packs; for example: 12-8 is the same as 8-12.																								
	1	2	5	7	8	9	10	12	13	14	15	18	26	43	44	50	55	56	72	74	76	82	103	114
1	•							P. 1																
2	×	×																						
5	•	×	•																					
7	•	•	•	•																				
8	×	×	×	×	×																			
9	•	•	•	•	•	•																		
10	×	×	×	×	×	•	×																	
12	×	×	×	×	×	•	×	×																
13	•	•	•	•	×	•	•	×	•							P. 2								
14	•	×	•	•	×	•	×	×	•	•														
15	•	•	•	•	×	•	×	×	•	•	•													
18	×	×	×	×	×	•	×	×	•	×	×	×												
26	•	×	•	•	×	•	×	×	•	•	•	×	•											
43	•	•	•	•	•	•	•	•	•	•	•	•	•	•										
44	•	×	•	•	×	•	×	×	•	•	•	×	•	•	•									
50	×	×	×	×	×	•	×	×	×	×	×	×	×	•	×	×								
55	•	×	•	•	•	•	•	•	•	•	•	•	•	•	•	•	•							P. 3
56	•	×	•	•	×	•	×	×	•	•	•	×	•	•	•	×	•	•						
72	•	×	•	•	×	•	×	×	•	•	•	×	•	•	•	×	•	•	•					
74	•	×	•	•	×	•	×	×	•	•	•	×	•	•	•	×	•	•	•	•				
76	•	×	•	•	×	•	×	×	•	•	•	×	•	•	•	×	•	•	•	•	•			
82	×	×	×	×	×	•	×	×	×	×	×	×	×	•	×	×	•	×	×	×	×	×		
103	•	•	•	•	•	•	•	•	•	•	•	•	•	•	•	•	•	•	•	•	•	•	•	
114	•	×	•	•	×	•	×	×	•	•	•	×	•	•	•	×	•	•	•	•	•	×	•	•

24E'$_2$ Group (17): 4(6S)$e_{1,1}$ + 4(6S)$e_{2,1}$ [x = e_1, y = e_2, z = 1] The blank cells are diagonally mirror-imaged packs, that is, for example: 12-8 is same as 8-12.								
P. 1-1	1 (3 $\tilde{6}^{2}$)	2 (3 $\bar{6}^{2}$)	5 (6 $\bar{6}^{2}$)	7 (1 3 $\tilde{6}^{2}$)	8 (1 3 $\bar{6}^{2}$)	9 ($\tilde{3}$ 3 $\tilde{6}^{2}$)	10 ($\tilde{3}$ 3 $\bar{6}^{2}$)	12 ($\bar{3}$ 3 $\bar{6}^{2}$)
1 (3 $\tilde{6}^{2}$)	R=4.338498							
2 (3 $\bar{6}^{2}$)	×	×						
5 (6 $\bar{6}^{2}$)	5.270575	×	6.198516					
7 (1 3 $\tilde{6}^{2}$)	4.6211036	×	5.535139	4.891326				
8 (1 3 $\bar{6}^{2}$)	×	×	×	×	×			
9 ($\tilde{3}$ 3 $\tilde{6}^{2}$)	4.6788766	4.879707	5.5807912	4.9437382	5.524192	4.9946952		
10 ($\tilde{3}$ 3 $\bar{6}^{2}$)	×	×	×	×	×	5.7764113	×	
12 ($\bar{3}$ 3 $\bar{6}^{2}$)	×	×	×	×	×	6.4271585	×	×
13 (3 $\tilde{6}^{2}$ $\tilde{6}^{2}$)	5.5727432	×	6.481726	5.8274811	×	5.8696058	6.690934	×

$14\,(3{}^{\sim}\overset{2}{6}{}^{-}\overset{2}{6})$	5.9230153	×	6.8577893	6.1823991	×	6.2231277	×	×
$15\,(3{}^{-}\overset{2}{6}{}^{\sim}\overset{2}{6})$	5.6371276	×	6.5362722	5.888109	×	5.9292777	6.7442181	×
$18\,(\underline{6\,3}{}^{-}\overset{2}{6})$	×	×	×	×	×	6.7016218	×	×
P. 1-2	$1\,(3{}^{\sim}\overset{2}{6})$	$2\,(3{}^{-}\overset{2}{6})$	$5\,(6{}^{-}\overset{2}{6})$	$7\,(1\,3{}^{\sim}\overset{2}{6})$	$8\,(1\,3{}^{-}\overset{2}{6})$	$9\,(3{}^{\sim}3{}^{\sim}\overset{2}{6})$	$10\,(3{}^{\sim}3{}^{-}\overset{2}{6})$	$12\,(3{}^{-}3{}^{-}\overset{2}{6})$
$26\,(6{}^{-}\overset{2}{6}{}^{-}\overset{2}{6})$	6.8577897	×	7.7906754	7.1092461	×	7.1448128	×	×
$43\,(1\,3{}^{\sim}\overset{2}{6}{}^{\sim}\overset{2}{6})$	5.6477372	5.8584698	6.5412729	5.8965833	6.5009339	5.9374091	6.7484767	7.3985846
$44\,(1\,3{}^{\sim}\overset{2}{6}{}^{-}\overset{2}{6})$	6.182399	×	7.1092461	6.4369103	×	6.4756539	×	×
$50\,(3{}^{\sim}3{}^{-}\overset{2}{3}{}^{-}\overset{2}{6})$	×	×	×	×	×	7.347794	×	×
$55\,(3{}^{\sim}3{}^{\sim}\overset{2}{6}{}^{\sim}\overset{2}{6})$	5.6574624	5.8672116	6.547457	5.9049829	6.5073164	5.9455678	6.7541783	7.402982
$56\,(3{}^{\sim}3{}^{\sim}\overset{2}{6}{}^{-}\overset{2}{6})$	6.2231277	×	7.1448124	6.4756549	×	6.513943	×	×
$72\,(3{}^{\sim}\overset{2}{6}{}^{\sim}\overset{2}{6}{}^{-}\overset{2}{6})$	7.1332715	×	8.0561499	7.3800648	×	7.4141242	×	×
$74\,(3{}^{\sim}\overset{2}{6}{}^{-}\overset{2}{6}{}^{-}\overset{2}{6})$	7.524367	×	8.4603782	7.7730916	×	7.8062489	×	×
$76\,(3{}^{-}\overset{2}{6}{}^{\sim}\overset{2}{6}{}^{-}\overset{2}{6})$	7.1828154	×	8.0999418	7.4276273	×	7.4612806	×	×
$82\,(\underline{6\,3}{}^{-}3{}^{-}\overset{2}{6})$	×	×	×	×	×	8.282394	×	×
$103\,(6{}^{\sim}6{}^{\sim}6{}^{-}\overset{2}{6})$	6.636173	6.851589	9.5199135	6.8739326	7.4899238	6.908532	7.7335135	8.3808798
$114\,(6{}^{-}\overset{2}{6}{}^{-}\overset{2}{6}{}^{-}\overset{2}{6})$	8.4603782	×	9.3953643	8.7049803	×	8.7355093	×	×
P. 2	$13\,(3{}^{\sim}\overset{2}{6}{}^{\sim}\overset{2}{6})$	$14\,(3{}^{\sim}\overset{2}{6}{}^{-}\overset{2}{6})$	$15\,(3{}^{-}\overset{2}{6}{}^{\sim}\overset{2}{6})$	$18\,(\underline{6\,3}{}^{-}\overset{2}{6})$	$26\,(6{}^{-}\overset{2}{6}{}^{-}\overset{2}{6})$	$43\,(1\,3{}^{\sim}\overset{2}{6}{}^{\sim}\overset{2}{6})$	$44\,(1\,3{}^{\sim}\overset{2}{6}{}^{-}\overset{2}{6})$	$50\,(3{}^{\sim}3{}^{-}\overset{2}{3}{}^{-}\overset{2}{6})$
$13\,(3{}^{\sim}\overset{2}{6}{}^{\sim}\overset{2}{6})$	6.756401							
$14\,(3{}^{\sim}\overset{2}{6}{}^{-}\overset{2}{6})$	7.1332715	7.52437						
$15\,(3{}^{-}\overset{2}{6}{}^{\sim}\overset{2}{6})$	6.8078459	7.1828154	6.8584706					
$18\,(\underline{6\,3}{}^{-}\overset{2}{6})$	×	×	×	×				
$26\,(6{}^{-}\overset{2}{6}{}^{-}\overset{2}{6})$	8.0561499	8.4603782	8.0999417	×	9.3953643			
$43\,(1\,3{}^{\sim}\overset{2}{6}{}^{\sim}\overset{2}{6})$	6.8117897	7.1857598	6.8622238	7.6641887	8.101255	6.865955		
$44\,(1\,3{}^{\sim}\overset{2}{6}{}^{-}\overset{2}{6})$	3.7800652	7.7730921	7.4276278	×	8.7049803	9.4300091	0.01944	
$50\,(3{}^{\sim}3{}^{-}\overset{2}{3}{}^{-}\overset{2}{6})$	×	×	×	×	×	8.312958	×	×
$55\,(3{}^{\sim}3{}^{\sim}\overset{2}{6}{}^{\sim}\overset{2}{6})$	6.8172042	7.1904933	6.867482	7.668153	8.1046557	6.8711856	7.4342921	8.316159
$56\,(3{}^{\sim}3{}^{\sim}\overset{2}{6}{}^{-}\overset{2}{6})$	7.4141242	7.806248	7.4612806	×	8.7355037	7.4635871	8.0516638	×
$72\,(3{}^{\sim}\overset{2}{6}{}^{\sim}\overset{2}{6}{}^{-}\overset{2}{6})$	8.3171954	8.7213693	8.3593268	×	9.6505685	8.3603435	8.9634451	×
$74\,(3{}^{\sim}\overset{2}{6}{}^{-}\overset{2}{6}{}^{-}\overset{2}{6})$	8.7213693	9.1338505	8.7621121	×	10.07052	8.7627721	9.3768669	×
$76\,(3{}^{-}\overset{2}{6}{}^{\sim}\overset{2}{6}{}^{-}\overset{2}{6})$	8.3593269	8.7621065	8.401058	×	9.6878151	8.4020257	9.0030493	×
$82\,(\underline{6\,3}{}^{-}3{}^{-}\overset{2}{6})$	×	×	×	×	×	9.24088	×	×
$103\,(6{}^{\sim}6{}^{\sim}6{}^{-}\overset{2}{6})$	7.7769262	8.1629672	7.8205287	8.6396986	9.0711446	7.8221666	8.4007525	9.288323
$114\,(6{}^{-}\overset{2}{6}{}^{-}\overset{2}{6}{}^{-}\overset{2}{6})$	9.6505684	10.07052	9.6878133	×	11.0065713	9.6879393	10.3111395	×
P. 3	$55\,(3{}^{\sim}3{}^{\sim}\overset{2}{6}{}^{\sim}\overset{2}{6})$	$56\,(3{}^{\sim}3{}^{\sim}\overset{2}{6}{}^{-}\overset{2}{6})$	$72\,(3{}^{\sim}\overset{2}{6}{}^{\sim}\overset{2}{6}{}^{-}\overset{2}{6})$	$74\,(3{}^{\sim}\overset{2}{6}{}^{-}\overset{2}{6}{}^{-}\overset{2}{6})$	$76\,(3{}^{-}\overset{2}{6}{}^{\sim}\overset{2}{6}{}^{-}\overset{2}{6})$	$82\,(\underline{6\,3}{}^{-}3{}^{-}\overset{2}{6})$	$103\,(6{}^{\sim}6{}^{\sim}6{}^{-}\overset{2}{6})$	$114\,(6{}^{-}\overset{2}{6}{}^{-}\overset{2}{6}{}^{-}\overset{2}{6})$

55 $(3^{\sim}3^{\sim}\overset{2}{6}^{\sim}\overset{2}{6})$	6.876397							
56 $(3^{\sim}3^{\sim}\overset{2}{6}^{-}\overset{2}{6})$	7.4678003	8.0837						
72 $(3^{\sim}\overset{2}{6}^{\sim}\overset{2}{6}^{-}\overset{2}{6})$	8.3634458	8.9931998	9.9032206					
74 $(3^{\sim}\overset{2}{6}^{-}\overset{2}{6}^{-}\overset{2}{6})$	8.7655483	9.4060645	10.3229604	10.747873				
76 $(3^{-}\overset{2}{6}^{\sim}\overset{2}{6}^{-}\overset{2}{6})$	8.4050766	9.0326046	9.9394908	10.3582644	9.9755444			
82 $(\underline{6\ 3}^{-}3^{-}\overset{2}{6})$	9.24331	×	×	×	×	×		
103 $(6^{\sim}6^{\sim}6^{-}\overset{2}{6})$	7.8257601	8.4310747	9.322144	9.7308602	9.3596568	10.209263	8.763227	
114 $(6^{-}\overset{2}{6}^{-}\overset{2}{6}^{-}\overset{2}{6})$	9.6901059	10.3388104	11.2553945	11.6849303	11.2884603	×	10.6509299	12.62159
The blank cells are diagonally mirror-imaged packs, that is, for example: 12-8 is same as 8-12.								

End of 24E'2 Group (17): 4(6S)e1,1 + 4(6S)e2,1 [x = e1, y = e2, z = 1]

End of 24E'2 and Family

[DATA 2B – 5] The 24E'3 and Family

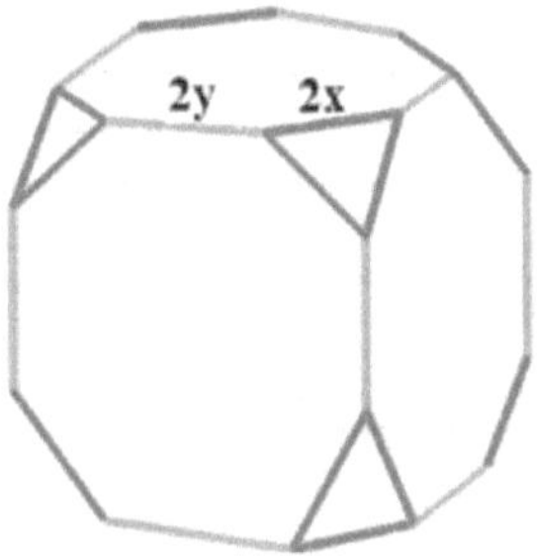

Ro = $\sqrt{7+4\sqrt{2}} \doteq 3.55764729$

Five family groups:	**(1) (3S)e**	**[x = e, y = 1]**
	(2) (8S)e	**[x = e, y = e]**
	(3) (8S)3 at e	**[x = e, y =1]**
	(4) (8S) 8 at e	**[x = 1, y = e]**
	(5) (3M)e + (8A)	**[x = e, y = y(A)]**

24E'3 Group (1): (3S) [x = e, y = 1]		2. (3 ~ 3)	4. (6 3)
	R	5.436941	7.33004

24E'3 Group (2) (8S)e [x = y = e]		3. (8 ~ 8)e	6. (16 8)e	21. (8 ~ 8 ~ 8)e	23. (8 ‾ 8 ~ 8)e	38. (16 ‾2 16 ~ 8)e
	R	5.229966	8.661318	× (d < 1)	× (d < 1)	× (d < 1)

24E'3 Group (3): (8S)3 at e [x = e, y = 1]

	1. (4 ~2 8)	5. (8 ‾2 8)	7. (1 4 ~2 8)	9. (2 4 ~2 8)	11. (4 ~ 4 ~2 8)	13. (4 ‾ 4 ~2 8)	15. (4 ~2 8 ~2 8)	16. (4 ~2 8 ‾2 8)
R	3.991015	5.81901	4.492834	4.922732	5.030315	5.3413885	6.2876503	6.4702
	17. (4 ‾2 8 ~2 8)	25. (8 ~ 8 ‾2 8) [R φ <6.13625]	27. (8 ‾2 8 ~2 8)	28. (8 ‾2 8 ‾2 8)	35. (16 8 ‾2 8) [Rc<13.7755]	41. (1 4 ~ 4 ~2 8)	45. (1 4 ~2 8 ~2 8)	46. (1 4 ~2 8 ‾2 8)
R	6.386057	7.53… ×	7.2010295	8.285964	10.159325	5.108844	6.592768	6.955871
	47. (1 4 ‾2 8 ~2 8)	53. (2 4 ~2 8 ~2 8)	54. (2 4 ~2 8 ‾2 8)	55. (2 4 ‾2 8 ~2 8)	65. (4 ~ 4 ~2 8 ~2 8)	66. (4 ~ 4 ~2 8 ‾2 8)	67. (4 ~ 4 ‾2 8 ~2 8)	69. (4 ‾ 4 ~2 8 ~2 8)
R	6.8056591	6.809588	7.3430355	7.1275325	6.8558645	7.430762	7.1959435	6.977465
	70 (4 ‾ 4 ~2 8 ‾2 8)	71. (4 ‾ 4 ‾2 8 ~2 8) [Rc<7.22931]	81. (4 ~2 8 ~2 8 ~2 8)	82 (4 ~2 8 ~2 8 ‾2 8)	83. (4 ~2 8 ‾2 8 ~2 8)	84. (4 ~2 8 ‾2 8 ‾2 8)	85 (4 ‾2 8 ~2 8 ~2 8) [Rc<8.236]	86. (4 ‾2 8 ~2 8 ‾2 8)
R	7.670866	7.33363 ×	7.310695	8.7621932	7.3312405	8.9872098	7.3232066	8.8631625
	113. (8 ~ 8 ~ 8 ‾2 8)	122. (8 ‾2 8 ~2 8 ‾2 8)	124. (8 ‾2 8 ‾2 8 ‾2 8)					
R	8.551781	9.537988	10.79722					

24E'3 Group (4): 6(8S)8 at e [x = 1, y = e]

	1. (4 ~2 8)	2. (4 ‾2 8)	5. (8 2 8)	7. (1 4 ~2 8)	8. (1 4 ‾2 8)	9. (2 v 4 ~2 8)	10. (2 v 4 ‾2 8)	11. (4 ~ 4 ~2 8)
R =	3.914659	3.992627	5.426976	4.330097	4.574662	4.691993	5.180886	× (d < 1)
	12. (4 ~ 4 ‾2 8)	13. (4 ‾ 4 ~2 8)	14. (4 ‾ 4 ‾2 8)	15. (4 ~2 8 ~2 8)	16. (4 ~2 8 ‾2 8)	17. (4 ‾2 8 ~2 8)	20. (8 4 ‾2 8)	25. (8 ~ 8 ‾2 8) [R φ <6.13625]
R	5.357764	5.056526	6.04119	5.810981	5.949544	5.891069	6.87492	R=6.8… ×
	27. (8 ‾2 8 ~2 8)	28. (8 ‾2 8 ‾2 8)	41. (1 4 ~ 4 ~2 8)	42. (1 4 ~ 4 ‾2 8)	45. (1 4 ~2 8 ~2 8)	46. (1 4 ~2 8 ‾2 8)	47. (1 4 ‾2 8 ~2 8)	53. (2 4 ~2 8 ~2 8)
R	6.613904	7.446967	× (d < 1)	5.4940705	6.072365	6.351243	6.250389	6.263501
	54. (2 4 ~2 8 ‾2 8)	55. (2 4 ‾2 8 ~2 8)	60. (4 ~ 4 ‾ 4 ‾2 8)	65. (4 ~ 4 ~2 8 ~2 8)	66. (4 ~ 4 ~2 8 ‾2 8)	67. (4 ~ 4 ‾2 8 ~2 8)	69. (4 ‾ 4 ~2 8 ~2 8)	70. (4 ‾ 4 ~2 8 ‾2 8)
R	6.6764033	× (d < 1)	7.418579	6.305624	6.751782	× (d < 1)	6.419428	6.962446
	71. (4 ‾ 4 ‾2 8 ~2 8)	74. (4 ~2 8 ~ 4 ‾2 8)	78. (4 ‾2 8 ~ 4 ‾2 8)	81. (4 ~2 8 ~2 8 ~2 8)	82. (4 ~2 8 ~2 8 ‾2 8)	83. (4 ~2 8 ‾2 8 ~2 8)	84. (4 ~2 8 ‾2 8 ‾2 8)	85. (4 ‾2 8 ~2 8 ~2 8)
R	6.305624	7.028023	7.0451835	6.305624	7.838871	6.305624	8.014387	6.305624
	86. (4 ‾2 8 ~2 8 ‾2 8)	92. (8 4 ‾ 4 ‾2 8)	113. (8 ~ 8 ~ 8 ‾2 8)	121. (8 ‾2 8 ~2 8 ~2 8)	122. (8 ‾2 8 ~2 8 ‾2 8)	124. (8 ‾2 8 ‾2 8 ‾2 8)		
R	7.921772	8.92368	7.711766	6.305624	8.517516	9.505343		

24E'₃ Group (5) (3M)e + (8A) [x = E3M, y = E8A]

	2	5	8	10	12	14	15	16	17	18	20	27	28	35	42	45	46	47	48	53	54	55	56	60	65	66
1		•		•	•	•	•	•	•	•		•	•	•	•	•	•		•	•	•		•		•	•
2		•		•	•	•	•	•	•	•		•	•	•	•	•	•	•	•	•	•	•	•		•	•
3											•		•	•			•				•			•		•
4											•		•	•							•			•		•
8																										
16																										

	67	68	69	70	71	72	73	74	77	78	81	82	83	84	85	86	92	95	96	113	115	121	122	123	124	173
1		•	•	•							•	•	•	•	•	•				•	•	•		•		
2	•	•	•	•	•	•					•	•	•	•	•	•		•		•	•	•		•		•
3				•		•		•		•		•		•		•			•	•	•		•		•	•
4				•		•		•		•		•		•		•			•	•	•		•		•	•
8												•		•		•	•						•		•	•
16																									•	

24E'₃ Group (5) (3M)e + (8A) [x = E3M, y = E8A]

		5. $(8\ \overset{-2}{8})$	10. $(2\ \overset{\vee}{4}\ \overset{-2}{8})$	12. $(4^{\sim}4\ \overset{-2}{8})$	14. $(4^{-}4\ \overset{-2}{8})$	15. $(4\ \overset{\sim2}{8}\ \overset{\sim2}{8})$	16. $(4\ \overset{\sim2}{8}\ \overset{-2}{8})$	17. $(4\ \overset{-2}{8}\ \overset{\sim2}{8})$	18. $(4\ \overset{-2}{8}\ \overset{-2}{8})$
1. (1 3)	R	5.756815	5.325759	5.455689	5.903897	6.282396	6.26974	6.4046853	6.379906
2. $(3^{\sim}3)$	R	5.811664	5.509722	5.652275	6.15529	R6.333762	6.3806685	6.4555052	6.502949
	R	20. $(\underline{8\ 4}\ \overset{-2}{8})$	27. $(8\ \overset{-2}{8}\ \overset{\sim2}{8})$	28. $(8\ \overset{-2}{8}\ \overset{-2}{8})$	35. $(\underline{16\ 8}\ \overset{-2}{8})$	42. $(1\ 4^{\sim}4\ \overset{-2}{8})$	45. $(1\ 4\ \overset{\sim2}{8}\ \overset{\sim2}{8})$	46. $(1\ 4\ \overset{\sim2}{8}\ \overset{-2}{8})$	47. $(1\ 4\ \overset{-2}{8}\ \overset{\sim2}{8})$
1. (1 3)	R	×	7.764082	7.40991	9.998442	5.551834	6.678039	6.15818	×
2. $(3^{\sim}3)$	R	×	7.806323	7.640725	10.245207	5.756907	6.726226	6.7608244	7.042409
3. $(3^{-}3)$	R	7.305602	×	8.090053	10.736016	×	×	6.955105	×
4. $(\underline{6\ 3})$	R	7.508272	×	8.2029845	10.86074	×	×	×	×
		48. $(1\ 4\ \overset{-2}{8}\ \overset{-2}{8})$	53. $(2\ 4\ \overset{\sim2}{8}\ \overset{\sim2}{8})$	54. $(2\ 4\ \overset{\sim2}{8}\ \overset{-2}{8})$	55. $(2\ 4\ \overset{-2}{8}\ \overset{\sim2}{8})$	56 $(2\ 4\ \overset{-2}{8}\ \overset{-2}{8})$	60 $(4^{\sim}4^{-}4\ \overset{-2}{8})$	65. $(4^{\sim}4\ \overset{\sim2}{8}\ \overset{\sim2}{8})$	66. $(4^{\sim}4\ \overset{\sim2}{8}\ \overset{-2}{8})$
1. (1 3)	R	6.8693905	6.99664	6.8728395	7.593173	7.296305	×	7.0698145	6.929521
2. $(3^{\sim}3)$	R	7.042359	7.042358	7.042318	7.636896	7.516148	×	7.1149114	7.1041415
3. $(3^{-}3)$	R	×	×	7.317061	×	×	7.676259	×	7.395243
4. $(\underline{6\ 3})$	R	×	×	×	×	×	7.924658	×	×
		67 $(4^{\sim}4\ \overset{-2}{8}\ \overset{\sim2}{8})$	68 $(4^{\sim}4\ \overset{-2}{8}\ \overset{-2}{8})$	69. $(4^{-}4\ \overset{\sim2}{8}\ \overset{\sim2}{8})$	70. $(4^{-}4\ \overset{\sim2}{8}\ \overset{-2}{8})$	71. $(4^{-}4\ \overset{-2}{8}\ \overset{\sim2}{8})$	72 $(4^{-}4\ \overset{-2}{8}\ \overset{-2}{8})$	74. $(4\ \overset{\sim2}{8}{}^{\sim}4\ \overset{-2}{8})$	78. $(4\ \overset{-2}{8}{}^{\sim}4\ \overset{-2}{8})$
1. (1 3)	R	×	7.4037535	7.2725365	7.082753	8.452776	×	×	×
2. $(3^{\sim}3)$	R	7.802855	7.636484	7.31574	7.27054	8.493175	8.078428	×	×
3. $(3^{-}3)$	R	×	×	×	7.602156	×	8.678987	7.411462	7.422083
4. $(\underline{6\ 3})$	R	×	×	×	7.661652	×	8.853257	7.62381	7.634792
		81. $(4\ \overset{\sim2}{8}\ \overset{\sim2}{8}\ \overset{\sim2}{8})$	82. $(4\ \overset{\sim2}{8}\ \overset{\sim2}{8}\ \overset{-2}{8})$	83. $(4\ \overset{\sim2}{8}\ \overset{-2}{8}\ \overset{\sim2}{8})$	84. $(4\ \overset{\sim2}{8}\ \overset{-2}{8}\ \overset{-2}{8})$	85. $(4\ \overset{-2}{8}\ \overset{\sim2}{8}\ \overset{\sim2}{8})$	86. $(4\ \overset{-2}{8}\ \overset{\sim2}{8}\ \overset{-2}{8})$	92. $(\underline{8\ 4}^{-}4\ \overset{-2}{8})$	95 $(\underline{8\ 4}\ \overset{-2}{8}\ \overset{\sim2}{8})$
1. (1 3)	R	8.153697	7.63902	8.33176	7.729547	8.236236	7.683397	×	×
2. $(3^{\sim}3)$	R	8.19441	7.90193	8.37226	8.0094485	8.276665	R7.953481	×	9.287254
3. $(3^{-}3)$	R	×	8.43266	×	8.5815424	×	8.501521	×	×
4. $(\underline{6\ 3})$	R	×	8.579865	×	8.74495	×	8.655606	×	×
8. $(3^{\sim}3^{-}3)$	R	×	8.76197	×	8.978084	×	8.860331	×	×
		96 $(\underline{8\ 4}\ \overset{-2}{8}\ \overset{-2}{8})$	113. $(8^{\sim}8^{\sim}8\ \overset{-2}{8})$	115 $(8^{-}8^{\sim}8\ \overset{-2}{8})$	121. $(8\ \overset{-2}{8}\ \overset{\sim2}{8}\ \overset{\sim2}{8})$	122 $(8\ \overset{-2}{8}\ \overset{\sim2}{8}\ \overset{-2}{8})$	123 $(8\ \overset{-2}{8}\ \overset{-2}{8}\ \overset{\sim2}{8})$	124. $(8\ \overset{-2}{8}\ \overset{-2}{8}\ \overset{-2}{8})$	173 $(16\ \overset{-2}{16}{}^{\sim}8\ \overset{-2}{8})$
1. (1 3)	R	×	8.33931	8.54184	8.815903	×	9.815522	×	×

2. $(3^{\sim}3)$	R	×	8.5105357	8.720153	8.85295	×	9.85195	×	11.32011
3. $(3^{-}3)$	R	9.243711	8.792504	9.02536	×	8.945437	×	9.57236	11.947673
4. $(\underline{6\ 3})$	R	9.479168	×	9.067288	×	9.1391206	×	9.85457	12.122178
8. $(3^{\sim}3^{-}3)$	R	×	×	×	×	9.468839	×	10.410235	12.409351
16. $(\underline{6\ 3}^{-}3)$	R	×	×	×	×	×	×	10.79156	×

End of 24E'3 and Family

[DATA 2B–6] The 30E’ and Family

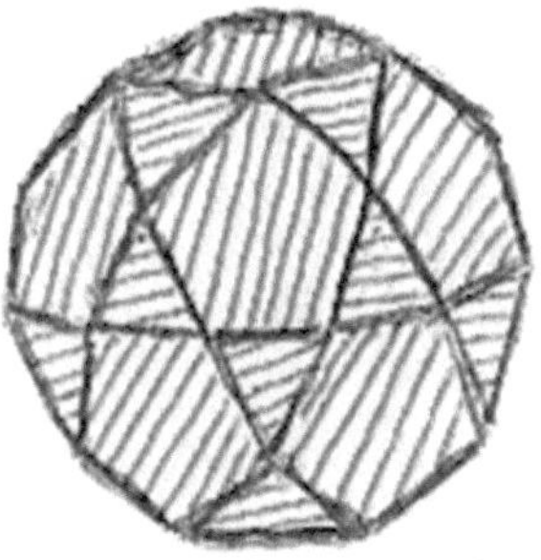

$Ro = 1 + \sqrt{5} \doteq 3.236068$

Two family groups: (1): 12(5 S) (2): 20(3 S)

30E’ Group (1): 12(5 S)		2. (5 ~ 5)		4. (<u>10 5</u>)	5 (1 5 ~ 5)	11 (5 ~ $\overset{2}{10}$ ~ 5)	13 (5 – $\overset{2}{10}$ ~ 5)	25. (1 5 ~ $\overset{2}{10}$ ~ 5)
	R	5.7127974		9.4513788	5.95081	9.0681124	9.1031733	9.24463
		27 (1 5 – $\overset{2}{10}$ ~ 5)		42. (5 – 5 ~ $\overset{2}{10}$ – 5)	(5 ~ $\overset{2}{10}$ ~ 5 – 5)	50. (5 – $\overset{2}{10}$ ~ 5 – 5)	86. (10 – $\overset{2}{10}$ ~ $\overset{2}{10}$ – 5)	
	R	R = 9.31249		11.60644	12.612995	12.65035	14.73487	

30E’ Group (2): 20(3 S)		2. (3 ~ 3)	4. (<u>6 3</u>)

The 30E’ has the smallest family in all eleven nE’s.

[DATA 2B – 7] The 48E' and Family

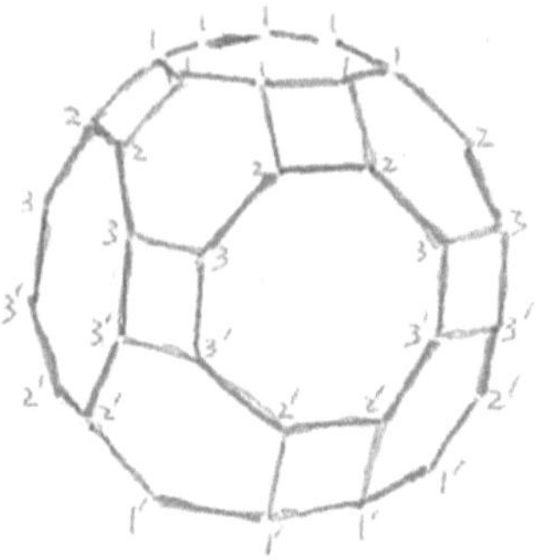

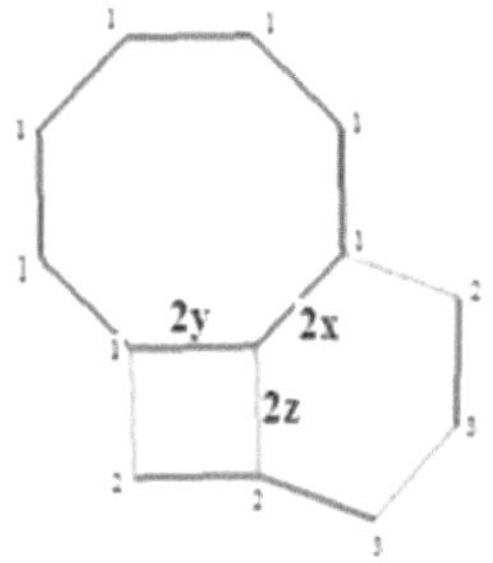

Ro = $\sqrt{13+6\sqrt{2}} \doteq 4.6352218$

Family groups:

(1) 6(8S)e	(4) 8(6S)e	(7) 12(4S)e
(2) 6(8S)6-e	(5) 8(6S)4-e	(8) 12(4S)8-e
(3) 6(8S)4-e	(6) 8(6S)8-e	(9) 12(4S)6-e
(10) 6(8M)e + 8(6A)	(14)) 8(6M)e + 12(4A)	(18) 12(4M)e + 6(8A)
(11) 6(8M)e +12(4A)	(15) 8(6M)e + 6(8A)	(19) 12(4M)e + 8(6A)
(12) 6(8M)e,1 + 8(6A)	(16) 8(6M)e,1 + 12(4A)	(20) 12(4M)e,1 + 6(8A)
(13) 6(8M)e,1 + 12(4A)	(17) 8(6M)e,1 + 6(8A)	(21) 12(4M)e,1 + 8(6A)
(22) 6(8S1) + 8(6S2)	(23) 8(6S1) + 12(4S2)	(24) 12(4S1) + 6(8S2)

48E' Group (1): 6(8S)e [x = y = E8M, z = 1]

3	6	21	23	38	125	127	129	131
$(8\,\overset{\sim}{8})_e$	$(\underline{16\ 8})_e$	$(8\,\overset{\sim}{8}\,\overset{\sim}{8})_e$	$(8\,\overset{-}{8}\,\overset{\sim}{8})_e$	$(16\,\overset{-2}{16}\,\overset{\sim}{8})_e$	$(8\,\overset{\sim}{8}\,\overset{\sim 2}{16}\,\overset{\sim}{8})_e$	$(8\,\overset{\sim}{8}\,\overset{-2}{16}\,\overset{\sim}{8})_e$	$(8\,\overset{-}{8}\,\overset{\sim 2}{16}\,\overset{\sim}{8})_e$	$(8\,\overset{-}{8}\,\overset{-2}{16}\,\overset{\sim}{8})_e$
6.42094 4	9.80457	7.620565	7.882801	11.282045	10.252761	10.335046	10.56849	10.698417

48E' Group (2): 6(8S)6 at e [x = E8M, y = 1, z = 1]

1. $(4\,\overset{\sim 2}{8})$	5. $(8\,\overset{-2}{8})$	7. $(1\,4\,\overset{\sim 2}{8})$	9. $(2\,4\,\overset{\sim 2}{8})$	11. $(4\,\overset{\sim}{4}\,\overset{\sim 2}{8})$	13 $(4\,\overset{-}{4}\,\overset{\sim 2}{8})$	15. $(4\,\overset{\sim 2}{8}\,\overset{\sim 2}{8})$	16. $(4\,\overset{\sim 2}{8}\,\overset{-2}{8})$	17 $(4\,\overset{-2}{8}\,\overset{\sim 2}{8})$
5.222465	7.0431683	5.714074	6.116966	6.211932	6.479892	7.5178302	7.726284	7.6181586
25. $(8\,\overset{\sim}{8}\,\overset{-2}{8})$ [R φ <6.136 25]	27. $(8\,\overset{-2}{8}\,\overset{\sim 2}{8})$	28. $(8\,\overset{-2}{8}\,\overset{-2}{8})$	41. $(1\,4\,\overset{\sim}{4}\,\overset{\sim 2}{8})$	45 $(1\,4\,\overset{\sim 2}{8}\,\overset{\sim 2}{8})$	46 $(1\,4\,\overset{\sim 2}{8}\,\overset{-2}{8})$	47 $(1\,4\,\overset{-2}{8}\,\overset{\sim 2}{8})$	53 $(2\,4\,\overset{\sim 2}{8}\,\overset{\sim 2}{8})$	54 $(2\,4\,\overset{\sim 2}{8}\,\overset{-2}{8})$
×	8.350176	9.538672	6.27607	7.8050395	8.208314	8.0135379	8.0017132	8.585726
55 $(2\,4\,\overset{-2}{8}\,\overset{\sim 2}{8})$	65 $(4\,\overset{\sim}{4}\,\overset{\sim 2}{8}\,\overset{\sim 2}{8})$	66. $(4\,\overset{\sim}{4}\,\overset{\sim 2}{8}\,\overset{-2}{8})$	67 $(4\,\overset{\sim}{4}\,\overset{-2}{8}\,\overset{\sim 2}{8})$	69 $(4\,\overset{-}{4}\,\overset{\sim 2}{8}\,\overset{\sim 2}{8})$	70 $(4\,\overset{-}{4}\,\overset{\sim 2}{8}\,\overset{-2}{8})$	71. $(4\,\overset{-}{4}\,\overset{-2}{8}\,\overset{\sim 2}{8})$ [Rc<7.22931]	81 $(4\,\overset{\sim 2}{8}\,\overset{\sim 2}{8}\,\overset{\sim 2}{8})$	82 $(4\,\overset{\sim 2}{8}\,\overset{\sim 2}{8}\,\overset{-2}{8})$
8.29751	8.041889	8.669004	8.351253	8.1430556	8.889229	×	8.419603	10.015079
83 $(4\,\overset{\sim 2}{8}\,\overset{-2}{8}\,\overset{\sim 2}{8})$ [Rc<7.697]	84 $(4\,\overset{\sim 2}{8}\,\overset{-2}{8}\,\overset{-2}{8})$	85 $(4\,\overset{-2}{8}\,\overset{\sim 2}{8}\,\overset{\sim 2}{8})$ Rc<8.236]	86 $(4\,\overset{-2}{8}\,\overset{\sim 2}{8}\,\overset{-2}{8})$	113 $(8\,\overset{\sim}{8}\,\overset{\sim}{8}\,\overset{-2}{8})$	115 $(8\,\overset{-}{8}\,\overset{\sim}{8}\,\overset{-2}{8})$	122 $(8\,\overset{-2}{8}\,\overset{\sim 2}{8}\,\overset{-2}{8})$	124 $(8\,\overset{-2}{8}\,\overset{-2}{8}\,\overset{-2}{8})$	
×	10.25119	×	10.116223	9.748704	9.941089	10.750671	12.05945	

48E' Group (3): 6(8S)4 at e [x = 1, y = E8M, z = 1]

	1. $(4\,\overset{\sim 2}{8})$	2. $(4\,\overset{-2}{8})$	5. $(8\,\overset{-2}{8})$	7. $(1\,4\,\overset{\sim 2}{8})$	8. $(1\,4\,\overset{-2}{8})$	9. $(2\,4\,\overset{\sim 2}{8})$	10. $(2\,4\,\overset{-2}{8})$	11. $(4\,\overset{\sim}{4}\,\overset{\sim 2}{8})$
R	5.111334	5.219395	R6.601718	5.513092	R5.805385	5.846522	6.404669	5.926502
	12. $(4\,\overset{\sim}{4}\,\overset{-2}{8})$	13. $(4\,\overset{-}{4}\,\overset{\sim 2}{8})$	14. $(4\,\overset{-}{4}\,\overset{-2}{8})$	15. $(4\,\overset{\sim 2}{8}\,\overset{\sim 2}{8})$	16. $(4\,\overset{\sim 2}{8}\,\overset{-2}{8})$	17. $(4\,\overset{-2}{8}\,\overset{\sim 2}{8})$	20. $(\underline{8\,4}\,\overset{-2}{8})$	25. $(8\,\overset{\sim}{8}\,\overset{-2}{8})$ [R φ <6.13625]
R	6.574749	6.15575	7.263434	6.99043	7.1532935	7.072388	8.073518	×
	27. $(8\,\overset{-2}{8}\,\overset{\sim 2}{8})$	28. $(8\,\overset{-2}{8}\,\overset{-2}{8})$	41. $(1\,4\,\overset{\sim}{4}\,\overset{\sim 2}{8})$	42. $(1\,4\,\overset{\sim}{4}\,\overset{-2}{8})$	45 $(1\,4\,\overset{\sim 2}{8}\,\overset{\sim 2}{8})$	46. $(1\,4\,\overset{\sim 2}{8}\,\overset{-2}{8})$	47. $(1\,4\,\overset{-2}{8}\,\overset{\sim 2}{8})$	53. $(2\,4\,\overset{1}{8}\,\overset{1}{8})$
R	7.70889	8.641968	5.981599	6.6986465	7.232635	7.549731	7.40646	7.402175
	54. $(2\,4\,\overset{\sim 2}{8}\,\overset{-2}{8})$	55. $(2\,4\,\overset{-2}{8}\,\overset{\sim 2}{8})$	60. $(4\,\overset{\sim}{4}\,\overset{-}{4}\,\overset{-2}{8})$	65. $(4\,\overset{\sim}{4}\,\overset{\sim 2}{8}\,\overset{\sim 2}{8})$	66. $(4\,\overset{\sim}{4}\,\overset{\sim 2}{8}\,\overset{-2}{8})$	67. $(4\,\overset{\sim}{4}\,\overset{-2}{8}\,\overset{\sim 2}{8})$	69. $(4\,\overset{-}{4}\,\overset{\sim 2}{8}\,\overset{\sim 2}{8})$	70. $(4\,\overset{-}{4}\,\overset{\sim 2}{8}\,\overset{-2}{8})$
R	7.863624	7.6566355	8.635012	7.437655	7.934044	7.707527	7.5292456	8.124318
	71. $(4\,\overset{-}{4}\,\overset{-2}{8}\,\overset{\sim 2}{8})$	74. $(4\,\overset{\sim 2}{8}\,\overset{\sim}{4}\,\overset{-2}{8})$	78. $(4\,\overset{-2}{8}\,\overset{\sim}{4}\,\overset{-2}{8})$	81. $(4\,\overset{\sim 2}{8}\,\overset{\sim 2}{8}\,\overset{\sim 2}{8})$	82. $(4\,\overset{\sim 2}{8}\,\overset{\sim 2}{8}\,\overset{-2}{8})$	83. $(4\,\overset{\sim 2}{8}\,\overset{-2}{8}\,\overset{\sim 2}{8})$ [Rc< 7.697]	84. $(4\,\overset{\sim 2}{8}\,\overset{-2}{8}\,\overset{-2}{8})$	85. $(4\,\overset{-2}{8}\,\overset{\sim 2}{8}\,\overset{\sim 2}{8})$
R	7.78559	8.213804	8.225317	7.784688	9.03349	×	9.220449	7.7908146
	86. $(4\,\overset{-2}{8}\,\overset{\sim 2}{8}\,\overset{-2}{8})$	92. $(\underline{8\,4}\,\overset{-}{4}\,\overset{-2}{8})$	113. $(8\,\overset{\sim}{8}\,\overset{\sim}{8}\,\overset{-2}{8})$	115.	122. $(8\,\overset{-2}{8}\,\overset{\sim 2}{8}\,\overset{-2}{8})$	124. $(8\,\overset{-2}{8}\,\overset{-2}{8}\,\overset{-2}{8})$		
R	9.116544	10.130465	8.847374	9.014974	9.667357	10.707078		

48E' Group (4): 8(6S)e [x = e, y = 1, z = e]

	3. $(6\ {}^{\sim}6)e$	6 $(\underline{12\ 6})e$	19 $(6\ {}^{\sim}6\ {}^{\sim}6)e$	115 $(6\ {}^{\sim}6\ {}^{\sim 2}12\ {}^{\sim}6)e$	117 $(6\ {}^{\sim}6\ {}^{-2}12\ {}^{\sim}6)e$	119 $(6\ {}^{-}6\ {}^{\sim 2}12\ {}^{\sim}6)e$	121 $(6\ {}^{-}6\ {}^{-2}12\ {}^{\sim}6)e$	152 $(\underline{12\ 6}\ {}^{\sim 2}12\ {}^{-}6)e$
R	6.742931 4	9.765609 7	7.6675573	10.564302	10.698353	10.72108	[Rc<9.2826] ×	12.804175

48E' Group (5): 8(6S)8 at e [x = e, y = z = 1]

	1. $(3\ {}^{\sim 2}6)$	5. $(6\ {}^{-2}6)$	7. $(1\ 3\ {}^{\sim 2}6)$	9. $(3\ {}^{\sim}3\ {}^{\sim 2}6)$	13. $(3\ {}^{\sim 2}6\ {}^{\sim 2}6)$	14. $(3\ {}^{\sim 2}6\ {}^{-2}6)$	15. $(3\ {}^{-2}6\ {}^{\sim 2}6)$
R	5.9230146	7.790672	6.436909	6.513945	8.3171963	9.1338505	8.4010568
	26. $(6\ {}^{-2}6\ {}^{-2}6)$	33. $(\underline{12\ 6}\ {}^{-2}6)$	43. $(1\ 3\ {}^{\sim 2}6\ {}^{\sim 2}6)$	44. $(1\ 3\ {}^{\sim 2}6\ {}^{-2}6)$	55. $(3\ {}^{\sim}3\ {}^{\sim 2}6\ {}^{\sim 2}6)$	56. $(3\ {}^{\sim}3\ {}^{\sim 2}6\ {}^{-2}6)$	72. $(3\ {}^{\sim 2}6\ {}^{\sim 2}6\ {}^{-2}6)$
R	11.006575	×	8.402993	9.618565	8.409077	9.6758266	11.502622
	74. $(3\ {}^{\sim 2}6\ {}^{-2}6\ {}^{-2}6)$	76. $(3\ {}^{-2}6\ {}^{\sim 2}6\ {}^{-2}6)$	103. $(6\ {}^{\sim}6\ {}^{\sim}6\ {}^{-2}6)$	114. $(6\ {}^{-2}6\ {}^{-2}6\ {}^{-2}6)$			
R	12.36465	11.567395	10.306283	R14.238996			

48E' Group (6): 8(6S)4 at e [x = y = 1, z = e]

	1. $(3\ {}^{\sim 2}6)$	2. $(3\ {}^{-2}6)$	5. $(6\ {}^{-2}6)$	7. $(1\ 3\ {}^{\sim 2}6)$	8. $(1\ 3\ {}^{-2}6)$	9. $(3\ {}^{\sim}3\ {}^{\sim 2}6)$	10. $(3\ {}^{\sim}3\ {}^{-2}6)$
R =	5.4269762	5.702636	6.601718	5.756815	6.54962	5.811664	6874917
	12. $(3\ {}^{-}3\ {}^{-2}6)$	13. $(3\ {}^{\sim 2}6\ {}^{\sim 2}6)$	14. $(3\ {}^{\sim 2}6\ {}^{-2}6)$	15. $(3\ {}^{-2}6\ {}^{\sim 2}6)$	18. $(\underline{6\ 3}\ {}^{-2}6)$	26. $(6\ {}^{-2}6\ {}^{-2}6)$	33. $(\underline{12\ 6}\ {}^{-2}6)$
R =	7.728904	6.953624	7.446969	7.17813	8.073519	8.64197	×
	43. $(1\ 3\ {}^{\sim 2}6\ {}^{\sim 2}6)$	44. $(1\ 3\ {}^{\sim 2}6\ {}^{-2}6)$	50. $(3\ {}^{\sim}3\ {}^{-2}3\ {}^{-2}6)$	55. $(3\ {}^{\sim}3\ {}^{\sim 2}6\ {}^{\sim 2}6)$	56. $(3\ {}^{\sim}3\ {}^{\sim 2}6\ {}^{-2}6)$	72. $(3\ {}^{\sim 2}6\ {}^{\sim 2}6\ {}^{-2}6)$	74. $(3\ {}^{\sim 2}6\ {}^{-2}6\ {}^{-2}6)$
R =	7.022074	7.764082	8.923687	7.028444	7.806323	8.973748	9.505343
	76. $(3\ {}^{-2}6\ {}^{\sim 2}6\ {}^{-2}6)$	82. $(\underline{6\ 3}\ {}^{-}3\ {}^{-2}6)$	103. $(6\ {}^{\sim}6\ {}^{\sim}6\ {}^{-2}6)$	114. $(6\ {}^{-2}6\ {}^{-2}6\ {}^{-2}6)$			
R =	9.024416		8.257584	10.70707			

48E' (7) 12(4S)e [x = 1, y = e, z = e]

	6 $(4\ {}^{\sim}4)e$	9 $(\underline{8\ 4})e$	10 $(1\ 4\ {}^{\sim}4)e$	30 $(4\ {}^{\sim 2}8\ {}^{\sim}4)e$	32 $(4\ {}^{-2}8\ {}^{\sim}4)e$	103. $(4\ {}^{-}4\ {}^{\sim 2}8\ {}^{-}4)e$	153. $(8\ {}^{-2}8\ {}^{\sim 2}8\ {}^{-}4)e$
R =	5.71164	9.818828	6.542351	10.049706	(Rc<9.805) ×	12.52992	15.247266

48E' (8) 12(4S)8 at e [x = 1, y = e, z = 1]

	2. $(1\ {}^{2}4)$	4. $(2\ {}^{\sim 2}4)$	5. $(2\ {}^{-2}4)$	8. $(4\ {}^{-2}4)$	19. $(2\ {}^{\sim 2}4\ {}^{-2}4)$	29. $(4\ {}^{-2}4\ {}^{-2}4)$	69. $(2\ {}^{\sim 2}4\ {}^{-2}4\ {}^{-2}4)$	97. $(4\ {}^{-2}4\ {}^{-2}4\ {}^{-2}4)$
R =	5.7475225	6.198516	7.330038	7.7906745	9.395365	11.006571	12.6215702	14.23898

48E' (9) 12(4S)6 at e [x = 1, y = 1, z = e]

	2. $(1\ {}^{2}4)$	4. $(2\ {}^{\sim 2}4)$	5. $(2\ {}^{-2}4)$	8. $(4\ {}^{-2}4)$	19. $(2\ {}^{\sim 2}4\ {}^{-2}4)$	29. $(4\ {}^{-2}4\ {}^{-2}4)$	69. $(2\ {}^{\sim 2}4\ {}^{-2}4\ {}^{-2}4)$	97. $(4\ {}^{-2}4\ {}^{-2}4\ {}^{-2}4)$
R =	5.4727210	5.819008	6.682662	7.0431679	8.2859644	9.538671	10.797212	12.0594497

48E' Group (10): 6(8M)e + 8(6A) [x = y = E8M, z = E6A]

	2	5	8	10	12	13	14	15	16	18	23	25	26	43	44	45	46	50
3			•	•	•		•		•	•			•		•		•	
6					•					•			•				•	
21	•	•	•	•	•	•	•		•		•	•	•	•	•		•	
23	•	•	•	•	•	•	•		•	•	•	•	•	•	•		•	
38				•	•		•		•	•	•		•		•		•	•
125				•	•					•			•		•		•	•
127				•	•				•	•			•		•		•	•
129				•	•				•	•			•		•		•	•
131				•	•				•	•			•		•		•	•
170																		

	55	56	57	58	61	62	71	72	73	74	75	76	82	85	86	103	113	114
3		•		•				•				•				•		
6				•		•		•		•		•			•			
21	•	•		•			•	•	•	•	•	•				•	•	
23	•	•		•			•	•	•	•	•	•				•	•	
38		•		•		•		•		•		•			•	•		
125		•		•		•		•		•		•						
127		•		•		•		•		•		•						
129		•		•		•		•		•		•				•		
131		•		•		•		•		•		•				•		
170																		

48E' Group (10): 6(8M)e + 8(6A) [x = y = E8M, z = E6A]

P. 1	2. $(3^{-}\overset{2}{6})$	5. $(6^{-}\overset{2}{6})$	8. $(1\ 3^{-}\overset{2}{6})$	10. $(3^{\sim}3^{-}\overset{2}{6})$	12. $(3^{-}3^{-}\overset{2}{6})$	13. $(3^{\sim}\overset{2}{6}^{\sim}\overset{2}{6})$	14. $(3^{\sim}\overset{2}{6}^{-}\overset{2}{6})$	16 $(3^{-}\overset{2}{6}^{-}\overset{2}{6})$
3. $(8^{\sim}8)_e$ R =	×	×	7.3640816	7.4952404	×	×	8.5523897	8.6855131
6. $(\underline{16\ 8})_e$	×	×	9.99844	×	10.7360158	×	×	×
21. $(8^{\sim}8^{\sim}8)_e$	7.7117656	8.8473735	8.3393101	8.5105357	8.7925047	9.6804175	9.4990249	9.6711935
23. $(8^{-}8^{\sim}8)_e$	7.8932608	9.0149769	8.5484	8.7201524	9.025351	9.8406491	9.688871	9.868176
38. $(16^{-}\overset{2}{16}^{\sim}8)_e$	×	×	×	11.3201069	11.9476743	×	12.0897232	12.4041706
125. $(8^{\sim}8^{\sim}\overset{2}{16}^{\sim}8)_e$	×	×	×	10.59625	11.14392	×	×	×
127. $(8^{\sim}8^{-}\overset{2}{16}^{\sim}8)_e$	×	×	×	10.650604	11.208045	×	×	11.786533
129. $(8^{-}8^{\sim}\overset{2}{16}^{\sim}8)_e$	×	×	×	10.817376	11.392093	×	×	11.940967
131. $(8^{-}8^{-}\overset{2}{16}^{\sim}8)_e$	×	×	×	10.911772	11.50209	×	×	12.036968
P. 2	18 $(\underline{6\ 3}^{-}\overset{2}{6})$	23 $(6^{\sim}6^{-}\overset{2}{6})$	25 $(6^{-}\overset{2}{6}^{\sim}\overset{2}{6})$	26 $(6^{-}\overset{2}{6}^{-}\overset{2}{6})$	43 $(1\ 3^{\sim}\overset{2}{6}^{\sim}\overset{2}{6})$	44 $(1\ 3^{\sim}\overset{2}{6}^{-}\overset{2}{6})$	46 $(1\ 3^{-}\overset{2}{6}^{-}\overset{2}{6})$	49 $(3^{\sim}3^{-}3^{\sim}\overset{2}{6})$
3. $(8^{\sim}8)_e$	×	×	×	8.882795	×	8.6889662	8.8818546	
6. $(\underline{16\ 8})_e$	10.8607401	×	×	11.8983597	×	×	11.8910222	
21. $(8^{\sim}8^{\sim}8)_e$	×	10.3524003	10.8501906	9.9956146	9.9807992	9.6722502	9.9908882	
23. $(8^{-}8^{\sim}8)_e$	9.0672901	10.5354925	11.0018745	10.2144695	10.1381525	9.8686203	10.2093005	
38. $(16^{-}\overset{2}{16}^{\sim}8)_e$	12.1221909	12.8989864	×	13.0766292	×	12.3956657	13.0685993	
125. $(8^{\sim}8^{\sim}\overset{2}{16}^{\sim}8)_e$	11.28925	×	×	12.318954	×	11.723516	12.311191	
127. $(8^{\sim}8^{-}\overset{2}{16}^{\sim}8)_e$	11.356805	×	×	12.38601	×	11.780471 ϕ (green) = 179.98 deg	12.37820	
129. $(8^{-}8^{\sim}\overset{2}{16}^{\sim}8)_e$	11.547512	×	×	12.558301	×	11.934321	12.55041	
131. $(8^{-}8^{-}\overset{2}{16}^{\sim}8)_e$	11.662907	×	×	12.670114	×	12.029898	12.6621505	
P. 3	50 $(3^{\sim}3^{-}\overset{2}{3}^{-}\overset{2}{6})$	55 $(3^{\sim}3^{\sim}\overset{2}{6}^{\sim}\overset{2}{6})$	56 $(3^{\sim}3^{\sim}\overset{2}{6}^{-}\overset{2}{6})$	58 $(3^{\sim}3^{-}\overset{2}{6}^{-}\overset{2}{6})$	62 $(3^{-}3^{-}\overset{2}{6}^{-}\overset{2}{6})$	71. $(3^{\sim}\overset{2}{6}^{\sim}\overset{2}{6}^{\sim}\overset{2}{6})$	72 $(3^{\sim}\overset{2}{6}^{-}\overset{2}{6}^{-}\overset{2}{6})$	73 $(3^{\sim}\overset{2}{6}^{-}\overset{2}{6}^{\sim}\overset{2}{6})$
3. $(8^{\sim}8)_e$	×	×	8.7038065	×	×	×	×	×
6. $(\underline{16\ 8})_e$	×	×	×	12.0243373	×	×	12.0274778	×
21. $(8^{\sim}8^{\sim}8)_e$	×	10.0158649	9.6907753	10.03811	×	11.163364	10.038722	11.7043891
23. $(8^{-}8^{\sim}8)_e$	×	10.1726622	9.8877511	10.2671123	×	11.312533	10.2694029	11.851446
38. $(16^{-}\overset{2}{16}^{\sim}8)_e$	12.4093546	×	12.4258705	13.2499909	13.5733046	×	13.2512899	×

125. $(8\ {}^{\sim}8\ {}^{\sim}\overset{2}{16}\ {}^{\sim}8)_e$	11.476459	×	11.750681	12.464363	12.696285	×	12.467032	×
127. $(8\ {}^{\sim}8\ {}^{-}\overset{2}{16}\ {}^{\sim}8)_e$	11.554332	×	11.808001	12.53469	12.776685	×	12.537264	×
129. $(8\ {}^{-}8\ {}^{\sim}\overset{2}{16}\ {}^{\sim}8)_e$	11.769107	×	11.962477	12.713264	12.9767	×	12,71555	×
131. $(8\ {}^{-}8\ {}^{-}\overset{2}{16}\ {}^{\sim}8)_e$	11.90086	×	12.058652	12.830237	13.109396	×	12.832335	×
P. 4	74. $(3\ {}^{\sim}\overset{2}{6}\ {}^{-}\overset{2}{6}\ {}^{-}\overset{2}{6})$	76. $(3\ {}^{-}\overset{2}{6}\ {}^{\sim}\overset{2}{6}\ {}^{-}\overset{2}{6})$	86. $(\underline{6\ 3}\ {}^{-}\overset{2}{6}\ {}^{-}\overset{2}{6})$	103. $(6\ {}^{\sim}6\ {}^{\sim}6\ {}^{-}\overset{2}{6})$	113. $(6\ {}^{-}\overset{2}{6}\ {}^{-}\overset{2}{6}\ {}^{\sim}\overset{2}{6})$	114 $(6\ {}^{-}\overset{2}{6}\ {}^{-}\overset{2}{6}\ {}^{-}\overset{2}{6})$		
3. $(8\ {}^{\sim}8)_e$	×	8.8464017	×	9.9250093	×	×		
6. $(\underline{16\ 8})_e$	12.1698766	12.041679	×	×	×	×		
21. $(8\ {}^{\sim}8\ {}^{\sim}8)_e$	×	10.042	×	10.8906999	12.8902956	×		
23. $(8\ {}^{-}8\ {}^{\sim}8)_e$	×	10.2734665	×	11.0779262	13.0319696	×		
38. $(16\ {}^{-}\overset{2}{16}\ {}^{\sim}8)_e$	13.48768	13.2705436	13.6235251	13.5633131	×	13.5773653		
125. $(8\ {}^{\sim}8\ {}^{\sim}\overset{2}{16}\ {}^{\sim}8)_e$	12.64766	12.483422	×	×	×	×		
127. $(8\ {}^{\sim}8\ {}^{-}\overset{2}{16}\ {}^{\sim}8)_e$	12.724102	12.554027	×	×	×	×		
129. $(8\ {}^{-}8\ {}^{\sim}\overset{2}{16}\ {}^{\sim}8)_e$	12.915307	12.73296	×	13.130814	×	×		

End of 48E’ Group (10): 6(8M)e + 8(6A) [x = y = E8M, z = E6A]

48E' Group (11): 6(8M)e + 12(4A) [x = y = E8M, z = E4A]	2	5	8	13	14	18	19	20	21	26	28	29	36	46	68	69	96	97	114	116
3	•																			
6																				
21		•	•		•	•	•				•			•	•		•			
23		•	•		•	•	•				•			•	•		•			
38					•		•		•	•		•		•					•	
125			•						•											
127			•				•		•											
129							•		•					•						
131							•		•	•				•						
170																				

48E' Group (11): (8M)e +(4A) [x = y = E8M, z = E4A]		2. (1 4)	5. (2 4)	8. (4 4)	14. (1 4 4)	18. (2 4 4)	19. (2 4 4)	21. (2 4 4)
3. (8 8)e	R =	6.74884	×	×	×	×	×	×
21. (8 8 8)e	R =	×	8.5517746	9.7487035	10.0247063	10.9681134	10.0431075	×
23. (8 8 8)e	R =	×	8.7570291	9.9410873	10.2495208	11.1506887	10.2788797	×
38. (16 16 8)e	R =	×	×	×	13.1034652	×	13.3197892	13.6056685
125. (8 8 16 8)e	R =	×	×	11.575311	×	×	×	12.703986
127. (8 8 16 8)e	R =	×	×	×	×	×	12.604305	12.78661
129. (8 8 16 8)e	R =	×	×	×	×	×	12.7834	12.992275
131. (8 8 16 8)e	R =	×	×	×	×	×	12.900605	13.128866
		26. (4 4 4)	28 (4 4 4)	29. (4 4 4)	46 (1 4 4 4)	68 (2 4 4 4)	96 (4 4 4 4)	114 (4 8 4 4)
21. (8 8 8)e	R =	×	12.202095	×	11.2783151	13.445903	14.696447	×
23. (8 8 8)e	R =	×	12.3772	×	11.5042204	13.615207	14.861243	×
38. (16 16 8)e	R =	14.3618788	×	13.6298966	14.4978548	×	×	16.048478
129. (8 8 16 8)e	R =	×	×	×	13.995861	×	×	×
131. (8 8 16 8)e	R =	×	×	×	14.1139	×	×	×

End of 48E' Group (11): 6(8M)e + 12(4A) [x = y = E8M, z = E6A]

48E' Family Group (12): 6(8M)e,1 + 8(6A) [x = E8M, y = 1, z = E6A]

P. 1-1	1	2	5	7	8	9	10	11	12	13	14
1	•	•	•	•	•	•	•	•		•	•
2	•	•	•	•	•	•	•	•		•	•
5			•		•		•		•	•	•
7	•	•	•	•	•	•	•		•	•	•
8		•	•		•		•		•	•	•
9		•	•		•		•		•	•	•
10			•		•		•		•	•	•
11		•	•		•		•		•	•	•
12			•		•		•		•	•	•
13					•		•	•	•		•
14					•		•		•		•
15			•		•		•		•	•	•
16			•		•		•		•		•
17			•		•		•		•	•	•
20									•		•
27									•		•
28									•		
41		•	•		•		•	•	•	•	•
42			•		•		•		•	•	•

P. 1-2	15	16	17	18	23	25	26	43	44	45	46
1		•	•			•		•	•		
2			•		•	•		•			
5	•	•		•	•	•	•	•	•	•	•
7		•	•		•	•	•	•	•		•
8		•			•	•	•	•	•	•	•
9		•			•	•	•	•	•		•
10	•	•		•	•	•	•	•	•	•	•
11		•		•	•	•	•	•	•	•	•
12	•	•		•	•	•	•	•	•	•	•
13		•		•	•	•	•	•	•	•	•
14		•		•	•		•		•		•
15	•	•		•	•		•	•	•	•	•
16		•		•	•		•		•		•
17	•	•		•	•		•	•	•		•
20				•	•		•		•		•
27				•	•		•		•		•
28				•			•		•		
41		•		•	•	•	•	•	•		•
42	•	•		•	•	•	•	•	•	•	•

P. 1-3	49	50	55	56	57	58	61	62	71	72	73
1	•		•	•					•		•
2	•		•						•		•
5			•	•	•				•	•	•
7			•	•					•		•
8			•	•	•	•			•	•	•
9			•	•	•	•			•	•	•
10			•	•	•	•	•		•	•	•
11			•	•	•	•			•	•	•
12			•	•	•	•	•		•	•	•
13			•	•		•			•	•	•
14		•		•		•		•		•	
15		•	•	•	•	•		•		•	
16		•		•		•		•		•	
17		•		•		•		•		•	
20		•		•		•				•	
27		•		•		•		•		•	
28		•		•		•		•		•	
41			•	•		•	•		•	•	•
42			•	•	•	•	•	•	•	•	•

P. 1-4	74	75	76	81	82	85	86	103	113	114	
1		•		•				•	•		
2		•						•	•		
5	•	•	•			•		•	•		
7		•						•	•		
8		•	•					•	•		
9		•	•					•	•		
10	•	•	•			•		•	•		
11		•	•			•		•	•		
12		•	•			•		•	•		
13		•	•					•	•		
14	•		•				•	•			
15	•		•					•			
16	•		•				•	•			
17	•		•					•			
20	•		•				•	•		•	
27	•		•				•	•			
28	•		•		•		•			•	
41		•	•					•	•		
42	•	•	•			•		•	•		

End of P. 1 of 3

P. 2-1	1	2	5	7	8	9	10	11	12	13	14	15	16	17	18	23	25	26	43	44	45	46
45							•		•		•		•		•	•		•		•		•
46							•		•		•				•	•		•		•		•
47							•		•		•		•		•	•		•		•		•
53							•		•		•		•		•	•		•		•		•
54									•		•				•	•		•		•		•
55									•		•				•	•		•		•		•
60									•						•			•		•		
65							•		•		•		•		•	•		•		•		•
66									•		•				•	•		•		•		•
67									•		•				•			•		•		•
69							•		•		•		•		•	•		•		•		•
70									•		•				•			•		•		•
71																						
74									•		•				•			•		•		•
78									•		•				•					•		
81									•													

P. 2-2	49	50	55	56	57	58	61	62	71	72	73	74	75	76	81	82	85	86	103	113	114	
45		•		•		•		•		•		•		•				•	•			
46		•		•		•		•		•		•		•				•	•			
47		•		•		•		•		•		•		•				•	•			
53		•		•		•		•		•		•		•				•	•			
54		•		•		•		•		•		•		•				•	•		•	
55		•		•		•		•		•		•		•				•	•			
60		•		•		•		•		•		•		•				•				
65		•		•		•		•		•		•		•				•	•		•	
66		•		•		•		•		•		•		•				•	•		•	
67		•		•		•		•		•		•		•				•	•			
69		•		•		•		•		•		•		•				•	•			
70		•		•		•		•		•		•		•				•	•		•	
71																						
74		•		•		•		•		•		•		•				•	•			
78				•																		
81																						

End of P. 2 of 3

P. 3	1	2	5	7	8	9	10	11	12	13	14	15	16	17	18	23	25	26	43	44	45	46
82															•			•				
83																						
84																		•				
85																						
86															•			•				
92																						
113															•			•				
115															•			•				
122																		•				
124																						
	49	50	55	56	57	58	61	62	71	72	73	74	75	76	81	82	85	86	103	113	114	
82		•						•		•		•		•		•		•			•	
83																						
84		•						•		•								•				
85																						
86		•						•		•								•				
92		•								•								•				
113		•						•		•								•				
115		•						•		•		•		•		•		•			•	
122		•						•		•								•				
124								•														

End of P. 3 of 3

48E' Family Group (12): 6(8M)e,1 + 8(6A) [x = E8M, y = 1, z = E6A]											P. 1
P. 1-1	1 $(3\ {}^{\sim2}6)$	2 $(3\ {}^{-2}6)$	5 $(6\ {}^{-2}6)$	7 $(1\ 3\ {}^{\sim2}6)$	8 $(1\ 3\ {}^{-2}6)$	9 $(3\ {}^{\sim}3\ {}^{\sim2}6)$	10 $(3\ {}^{\sim}3\ {}^{-2}6)$	11 $(3\ {}^{-}3\ {}^{\sim2}6)$	12 $(3\ {}^{-}3\ {}^{-2}6)$	13 $(3\ {}^{\sim2}6\ {}^{\sim2}6)$	14 $(3\ {}^{\sim2}6\ {}^{-2}6)$
1 $(4\ {}^{\sim2}8)$ R =	5.872808	5.8109818	6.990429	6.707363	6.2697425	7.0280245	6.3806693	7.8730405	×	7.838871	7.472152
2 $(4\ {}^{-2}8)$	5.89716	5.891069	7.0723893	6.7263992	6.379906	7.0451838	6.502949	7.886067	×	7.9217722	7.5849674
5 $(8\ {}^{-2}8)$	×	×	7.7088906	×	7.4099104	×	7.640725	×	8.0900522	8.517512	8.5535423
7 $(1\ 4\ {}^{\sim2}8)$	5.9284118	6.072365	7.2326352	6.737023	6.615818	7.0497044	6.760825	×	6.955104	8.073203	7.800715
8 $(1\ 4\ {}^{-2}8)$	×	6.250388	7.40646	×	6.8963965	×	7.0423595	×	7.3244365	8.244719	8.0542695
9 $(2\ 4\ {}^{\sim2}8)$	×	6.2635014	7.4021756	×	6.8728396	×	7.0423185	×	7.3170608	8.233244	8.0388808
10 $(2\ 4\ {}^{-2}8)$	×	×	7.6566374	×	7.296306	×	7.5161474	×	7.9352476	8.4756294	8.455629
11 $(4\ {}^{\sim}4\ {}^{\sim2}8)$	×	6.305624	7.437656	×	6.92952	×	7.1041415	×	7.395242	8.2658047	8.089684
12 $(4\ {}^{\sim}4\ {}^{-2}8)$	×	×	7.7075266	×	7.4037535	×	7.6364844	×	8.0904965	8.5188213	8.554044
13 $(4\ {}^{-}4\ {}^{\sim2}8)$	×	×	×	×	7.0827549	×	7.27054	7.383311	7.6026557	×	8.2223725
14 $(4\ {}^{-}4\ {}^{-2}8)$	×	×	×	×	×	×	8.0784296	×	8.678985	×	8.9003396
15 $(4\ {}^{\sim2}8\ {}^{\sim2}8)$	×	×	7.7846859	×	7.6390198	×	7.9019305	×	8.4326559	8.5714573	8.760572
16 $(4\ {}^{\sim2}8\ {}^{-2}8)$	×	×	7.7924469	×	7.2795449	×	8.0094486	×	8.5815424	×	8.8469557
17 $(4\ {}^{-2}8\ {}^{\sim2}8)$	×	×	7.7908155	×	7.6833965	×	7.9534815	×	8.5015216	8.5712648	8.8006119
20 $(\underline{8\ 4}\ {}^{-2}8)$	×	×	×	×	×	×	×	×	9.2437109	×	9.1237835
27 $(8\ {}^{-2}8\ {}^{\sim2}8)$	×	×	×	×	×	×	×	×	8.9454361	×	9.0268512
28 $(8\ {}^{-2}8\ {}^{-2}8)$	×	×	×	×	×	×	×	×	9.5723605	×	×
41 $(1\ 4\ {}^{\sim}4\ {}^{\sim2}8)$	×	6.333455	7.4598395	×	6.966144	×	7.14381	7.6784757	7.444708	8.2855697	8.12115
42 $(1\ 4\ {}^{\sim}4\ {}^{-2}8)$	×	×	7.7374224	×	7.4768	×	7.718269	×	8.1952834		8.6186661

P. 1-2	15 $(3\ {}^{-2}6\ {}^{\sim2}6)$	16 $(3\ {}^{-2}6\ {}^{-2}6)$	17 $(\underline{6\ 3}\ {}^{\sim2}6)$	18 $(\underline{6\ 3}\ {}^{-2}6)$	23 $(6\ {}^{\sim}6\ {}^{-2}6)$	25 $(6\ {}^{-2}6\ {}^{\sim2}6)$	26 $(6\ {}^{-2}6\ {}^{-2}6)$	43. $(1\ 3\ {}^{\sim2}6\ {}^{\sim2}6)$	44. $(1\ 3\ {}^{\sim2}6\ {}^{-2}6)$	45. $(1\ 3\ {}^{-2}6\ {}^{\sim2}6)$	46. $(1\ 3\ {}^{-2}6\ {}^{-2}6)$
1 $(4\ {}^{\sim2}8)$ R =	×	7.586655	8.2138029	×	×	9.0334906	×	8.153697	9.5932363	×	×
2 $(4\ {}^{-2}8)$	×	×	8.225321	×	8.45569	9.1165442	×	8.236236	×	×	×
5 $(8\ {}^{-2}8)$	8.8083129	8.780331	×	8.2029843	9.3959265	9.667355	9.285926	8.815904	8.7860703	9.643957	9.274591
7 $(1\ 4\ {}^{\sim2}8)$	×	7.946327	8.2132668	×	8.6636634	9.2586228	8.198545	8.384059	7.9532068	×	8.1947732
8 $(1\ 4\ {}^{-2}8)$	×	8.2268204	×	×	8.9184528	9.4261535	8.5666786	8.5536552	8.233601	9.400957	8.559641
9 $(2\ 4\ {}^{\sim2}8)$	×	8.2075438	×	×	8.894316	9.408463	8.5363736	8.5402732	8.214264	×	8.529664
10 $(2\ 4\ {}^{-2}8)$	8.7692856	8.674427	×	8.0337587	9.309178	9.635497	9.1505266	8.7772402	8.6804528	9.6117825	9.1398
11 $(4\ {}^{\sim}4\ {}^{\sim2}8)$	×	8.2632517	×	7.4309977	8.9426208	9.4380536	8.6076531	8.78154354	8.2698989	9.4130188	8.6004262
12 $(4\ {}^{\sim}4\ {}^{-2}8)$	9.8102079	8.7859406	×	8.2049479	9.4035087	9.6702825	9.2975326	8.817801	8.7917	9.6468816	9.2860541
13 $(4\ {}^{-}4\ {}^{\sim2}8)$	×	8.4084535	×	7.661652	9.066029	9.5097527	8.7914243	8.6500288	8.4148468	9.4853272	9.7830349
14 $(4\ {}^{-}4\ {}^{-2}8)$	×	9.1948352	×	8.8532605	9.73190834	×	9.8609309	×	9.1994215	×	9.8467899
15 $(4\ {}^{\sim2}8\ {}^{\sim2}8)$	8.8555723	9.0236221	×	8.579859	9.5965657	×	9.6138731	8.8626476	9.0287014	9.674316	9.6009803

16 (4 ~$\overset{2}{8}$ -$\overset{2}{8}$)	×	9.1283402	×	8.7449499	9.6806902	×	9.3633538	×	9.1331219	×	9.7497049
17 (4 -$\overset{2}{8}$ ~$\overset{2}{8}$)	×	9.0712342	×	8.655602	9.6344552	×	9.6797624	8.8604246	9.0761704	×	9.6665673
20 (8 4 -$\overset{2}{8}$)	×	×	×	9.4791656	9.91571	×	10.386815	×	9.5239768	×	10.37005
27 (8 -$\overset{2}{8}$ ~$\overset{2}{8}$)	×	×	×	9.1391206	9.8341441	×	10.075755	×	9.3565421	×	10.061014
28 (8 -$\overset{2}{8}$ -$\overset{2}{8}$)	×	×	×	9.8545671	×	×	10.6860948	×	9.6192221	×	×
41 (1 4 ~4 ~$\overset{2}{8}$)	×	8.2976194	×	7.4862444	9.9718662	9.4554706	8.651176	8.5907397	8.3042141	×	8.64366
42 (1 4 ~4 -$\overset{2}{8}$)	8.8316483	8.8595013	×	8.31993	9.4642405	9.6867548	9.3944421	8.8390327	8.8650593	9.6635499	9.3825164

48E' Family Group (12): 6(8M)e,1 + 8(6A)
[x = E8M, y = 1, z = E6A] P. 1-1,2

P. 1-3	49 $(3^{\sim}3^{-}\overset{2}{3}{}^{\sim}\overset{2}{6})$	50 $(3^{\sim}3^{-}\overset{2}{3}{}^{-}\overset{2}{6})$	55 $(3^{\sim}3^{\sim}\overset{2}{6}{}^{\sim}\overset{2}{6})$	56 $(3^{\sim}3^{\sim}\overset{2}{6}{}^{-}\overset{2}{6})$	57 $(3^{\sim}3^{-}\overset{2}{6}{}^{\sim}\overset{2}{6})$	58 $(3^{\sim}3^{-}\overset{2}{6}{}^{-}\overset{2}{6})$	61 $(3^{-}3^{-}\overset{2}{6}{}^{\sim}\overset{2}{6})$	62 $(3^{-}3^{-}\overset{2}{6}{}^{-}\overset{2}{6})$	71 $(3^{\sim}\overset{2}{6}{}^{\sim}\overset{2}{6}{}^{\sim}\overset{2}{6})$	72 $(3^{\sim}\overset{2}{6}{}^{-}\overset{2}{6}{}^{-}\overset{2}{6})$
1 $(4^{\sim}\overset{2}{8})$	9.0582462	×	8.194413	7.6070625	×	×	×	×	9.362115	×
2 $(4^{-}\overset{2}{8})$	9.0667548	×	8.2766634	×	×	×	×	×	9.444583	×
5 $(8^{-}\overset{2}{8})$	×	×	8.8529497	8.8131004	9.958197	×	×	×	9.9822278	9.409485
7 $(1\,4^{\sim}\overset{2}{8})$	×	×	8.4236512	7.9709738	×	×	×	×	9.5837436	×
8 $(1\,4^{-}\overset{2}{8})$	×	×	8.592565	8.254527	9.7222496	8.613987	×	×	9.7492471	8.6175857
9 $(2\,4^{\sim}\overset{2}{8})$	×	×	8.579002	8.2347074	9.703372	8.5799547	×	×	9.730473	8.5830503
10 $(2\,4^{-}\overset{2}{8})$	×	×	8.8147483	8.706386	9.9281163	9.252653	10.778283	×	9.952592	9.262095
11 $(4^{\sim}4^{\sim}\overset{2}{8})$	×	×	8.6103101	8.2908921	9.7325253	8.6580291	×	×	9.7592617	8.6621185
12 $(4^{\sim}4^{-}\overset{2}{8})$	×	×	8.8549084	8.8189884	9.9613061	9.4129033	10.8072018	×	9.9853215	9.4232069
13 $(4^{-}4^{\sim}\overset{2}{8})$	×	×	8.6879669	8.4371868	×	8.858396	×	×	9.8284745	8.864594
14 $(4^{-}4^{-}\overset{2}{8})$	×	9.1194205	×	9.2330194	×	10.0293851	×	10.35527	×	10.041716
15 $(4^{\sim}\overset{2}{8}{}^{\sim}\overset{2}{8})$	×	8.7619691	8.8987488	9.059114	9.9830718	9.757197	×	10.0024447	×	9.7687498
16 $(4^{\sim}\overset{2}{8}{}^{-}\overset{2}{8})$	×	8.9780832	×	9.1654	×	9.9216638	×	10.2156777	×	9.9337211
17 $(4^{-}\overset{2}{8}{}^{\sim}\overset{2}{8})$	×	8.860331	×	9.1073422	×	9.8292546	×	10.095136	×	9.841014
20 $(\underline{8\,4}^{-}\overset{2}{8})$	×	9.919672	×	9.567226	×	10.61346	×	×	×	10.6265053
27 $(8^{-}\overset{2}{8}{}^{\sim}\overset{2}{8})$	×	9.4688383	×	9.3909018	×	10.260887	×	10.639072	×	10.263426
28 $(8^{-}\overset{2}{8}{}^{-}\overset{2}{8})$	×	10.410235	×	9.6750331	×	10.959884	×	11.5713926	×	10.973262
41 $(1\,4^{\sim}4^{\sim}\overset{2}{8})$	×	×	8.629128	8.3255219	×	8.7055745	10.60545	×	9.776095	8.710214
42 $(1\,4^{\sim}4^{-}\overset{2}{8})$	×	×	8.8758436	8.8932557	9.9764667	9.5183491	10.8189464	9.6943346	10.0002227	9.5291048

48E' Family Group (12): 6(8M)e,1 + 8(6A)
[x = E8M, y = 1, z = E6A] P. 1-3

P. 1-4	73 $(3\,\tilde{6}^{2}\,\bar{6}^{2}\,\tilde{6}^{2})$	74 $(3\,\tilde{6}^{2}\,\bar{6}^{2}\,\bar{6}^{2})$	75 $(3\,\bar{6}^{2}\,\tilde{6}^{2}\,\tilde{6}^{2})$	76 $(3\,\bar{6}^{2}\,\tilde{6}^{2}\,\bar{6}^{2})$	81 $(\underline{6\,3}\,\tilde{3}\,\tilde{6}^{2})$	82 $(\underline{6\,3}\,\bar{3}\,\bar{6}^{2})$	85 $(\underline{6\,3}\,\bar{6}^{2}\,\tilde{6}^{2})$	86 $(\underline{6\,3}\,\bar{6}^{2}\,\bar{6}^{2})$	103 $(6\,\tilde{6}\,\tilde{6}\,\bar{6}^{2})$	113 $(6\,\bar{6}^{2}\,\bar{6}^{2}\,\tilde{6}^{2})$	114 $(6\,\bar{6}^{2}\,\bar{6}^{2}\,\bar{6}^{2})$
1 $(4\,\tilde{8}^{2})$	9.8975444	×	9.410915	×	10.257852	×	×	×	8.84018	11.09851	×
2 $(4\,\bar{8}^{2})$	9.9808106	×	9.483023	×	×	×	×	×	8.9677472	11.181666	×
5 $(8\,\bar{8}^{2})$	10.508943	9.527417	10.027199	9.4248137	×	×	11.1307397	×	10.008875	11.684415	×
7 $(1\,4\,\tilde{8}^{2})$	10.1184945	×	9.631327	×	×	×	×	×	9.19358	11.3144565	×
8 $(1\,4\,\bar{8}^{2})$	10.284044	×	9.7960247	8.620869	×	×	×	×	9.4773898	11.477259	×
9 $(2\,4\,\tilde{8}^{2})$	10.2634244	×	9.777163	8.5856271	×	×	×	×	9.445655	11.453918	×
10 $(2\,4\,\bar{8}^{2})$	10.482045	9.3542056	9.9979405	9.2576	×	×	11.106706	×	9.9137149	11.662535	×
11 $(\tilde{4}\,4\,\tilde{8}^{2})$	10.29157	×	9.8057542	8.6661496	×	×	10.9207262	×	9.4981942	11.480521	×
12 $(\tilde{4}\,4\,\bar{8}^{2})$	10.51225	9.5446848	10.0303044	9.4388298	×	×	11.1341243	×	10.0208391	11.687746	×
13 $(\bar{4}\,4\,\tilde{8}^{2})$	10.358417	×	9.874421	8.8719717	×	×	×	×	9.631208	11.5422874	×
14 $(\bar{4}\,4\,\bar{8}^{2})$	×	10.270828	×	10.06492	×	×	×	10.406916	10.4135	×	×
15 $(4\,\tilde{8}^{2}\,\tilde{8}^{2})$	×	9.9490848	×	9.7885136	×	×	×	×	10.243095	×	×
16 $(4\,\tilde{8}^{2}\,\bar{8}^{2})$	×	10.1434074	×	9.9555545	×	×	×	10.25121	10.3477	×	×
17 $(4\,\bar{8}^{2}\,\tilde{8}^{2})$	×	10.0336595	×	9.8616338	×	×	×	×	10.288919	×	×
20 $(\underline{8\,4}\,\bar{8}^{2})$	×	10.960511	×	10.656915	×	×	×	11.235004	10.707835	×	11.360765
27 $(8\,\bar{8}^{2}\,\tilde{8}^{2})$	×	10.534188	×	10.298487	×	×	×	10.718167	10.540395	×	×
28 $(8\,\bar{8}^{2}\,\bar{8}^{2})$	×	11.383727	×	11.009107	×	10.791557	×	11.744598	×	×	11.95971
41 $(1\,\tilde{4}\,4\,\tilde{8}^{2})$	10.307861	×	9.8224558	8.715085	×	×	×	×	9.529673	11.4956365	×
42 $(1\,\tilde{4}\,4\,\bar{8}^{2})$	10.525064	9.669042	10.044955	9.5460448	×	×	11.144715	×	10.089929	11.696471	×

48E' Family Group (12): 6(8M)e,1 + 8(6A)
[x = E8M, y = 1, z = E6A] P. 1-4

P. 2-1	10 $(3\ {}^{\sim}3\ {}^{-}\overset{2}{6})$	12 $(3\ {}^{-}3\ {}^{-}\overset{2}{6})$	14 $(3\ {}^{\sim}\overset{2}{6}\ {}^{-}\overset{2}{6})$	16 $(3\ {}^{-}\overset{2}{6}\ {}^{-}\overset{2}{6})$	18 $(\underline{6\ 3}\ {}^{-}\overset{2}{6})$	23 $(6\ {}^{\sim}6\ {}^{-}\overset{2}{6})$	26 $(6\ {}^{-}\overset{2}{6}\ {}^{-}\overset{2}{6})$
45 $(1\ 4\ {}^{\sim}\overset{2}{8}\ {}^{\sim}\overset{2}{8})$	8.044331	8.6200395	8.8663183	9.1496406	8.7848944	9.6939836	9.78715
46 $(1\ 4\ {}^{\sim}\overset{2}{8}\ {}^{-}\overset{2}{8})$	8.228481	8.884214	9.001093	×	9.078566	9.820786	10.049287
47 $(1\ 4\ {}^{-}\overset{2}{8}\ {}^{\sim}\overset{2}{8})$	8.142092	8.7525715	8.9363617	9.2364145	8.9303422	9.7580351	9.9109578
53 $(2\ 4\ {}^{\sim}\overset{2}{8}\ {}^{\sim}\overset{2}{8})$	8.1361185	8.7413774	8.9302592	9.227954	8.9172348	9.7511496	9.89671
54 $(2\ 4\ {}^{\sim}\overset{2}{8}\ {}^{-}\overset{2}{8})$	×	9.1006933	9.0868097	×	9.3172178	9.8910897	10.250238
55 $(2\ 4\ {}^{-}\overset{2}{8}\ {}^{\sim}\overset{2}{8})$	×	8.9181075	9.0150524	×	9.11025	9.825314	10.0554996
60 $(4\ {}^{\sim}4\ {}^{-}4\ {}^{-}\overset{2}{8})$	×	9.574545	×	×	9.858804	×	10.69248
65 $(4\ {}^{\sim}4\ {}^{\sim}\overset{2}{8}\ {}^{\sim}\overset{2}{8})$		8.7652583	8.9422842	9.242925	8.943173	9.761641	9.9177099
66 $(4\ {}^{\sim}4\ {}^{\sim}\overset{2}{8}\ {}^{-}\overset{2}{8})$	×	9.145925	9.1007019	×	9.3683744	9.9016728	10.29137
67 $(4\ {}^{\sim}4\ {}^{-}\overset{2}{8}\ {}^{\sim}\overset{2}{8})$	×	8.946971	9.0275607	×	9.14104	×	10.077935
69 $(4\ {}^{-}4\ {}^{\sim}\overset{2}{8}\ {}^{\sim}\overset{2}{8})$	8.1993299	8.6236232	8.9708685	9.282557	9.006248	9.7858815	9.9674849
70 $(4\ {}^{-}4\ {}^{\sim}\overset{2}{8}\ {}^{-}\overset{2}{8})$	×	9.2609849	9.1269498	×	9.4958654	×	10.393303
74 $(4\ {}^{\sim}\overset{2}{8}\ {}^{\sim}4\ {}^{-}\overset{2}{8})$	×	9.3259959	9.1339555	×	9.5709596	×	10.4595027
78 $(4\ {}^{-}\overset{2}{8}\ {}^{\sim}4\ {}^{-}\overset{2}{8})$ [Rc3< 9.805]	×	9.3306555	9.1342234	×	9.5757431	×	×
81 $(4\ {}^{\sim}\overset{2}{8}\ {}^{\sim}\overset{2}{8}\ {}^{\sim}\overset{2}{8})$	×	8.9774088	×	×	×	×	×

P. 2-2	44. $(1\ 3\ {}^{\sim}\overset{2}{6}\ {}^{-}\overset{2}{6})$	46. $(1\ 3\ {}^{-}\overset{2}{6}\ {}^{-}\overset{2}{6})$	50 $(3\ {}^{\sim}3\ {}^{-}\overset{2}{3}\ {}^{-}\overset{2}{6})$	56 $(3\ {}^{\sim}3\ {}^{\sim}\overset{2}{6}\ {}^{-}\overset{2}{6})$	58 $(3\ {}^{\sim}3\ {}^{-}\overset{2}{6}\ {}^{-}\overset{2}{6})$	62 $(3\ {}^{-}3\ {}^{-}\overset{2}{6}\ {}^{-}\overset{2}{6})$	72 $(3\ {}^{\sim}\overset{2}{6}\ {}^{\sim}\overset{2}{6}\ {}^{-}\overset{2}{6})$
45 $(1\ 4\ {}^{\sim}\overset{2}{8}\ {}^{\sim}\overset{2}{8})$	9.1543232	9.7735181	9.0253015	9.1867185	9.9462075	10.2436852	9.9582067
46 $(1\ 4\ {}^{\sim}\overset{2}{8}\ {}^{-}\overset{2}{8})$	9.327651	10.03433	9.4059205	9.363936	10.236241	10.61872	10.248875
47 $(1\ 4\ {}^{-}\overset{2}{8}\ {}^{\sim}\overset{2}{8})$	9.2408156	9.8967919	9.210813	9.274838	10.0818029	10.416385	10.0940546
53 $(2\ 4\ {}^{\sim}\overset{2}{8}\ {}^{\sim}\overset{2}{8})$	9.2323737	9.8826395	9.1926503	9.2661472	10.0657076	10.3948082	10.0778965
54 $(2\ 4\ {}^{\sim}\overset{2}{8}\ {}^{-}\overset{2}{8})$	9.4519286	10.234299	9.7110877	9.4918971	10.4592624	10.9059493	10.4721294
55 $(2\ 4\ {}^{-}\overset{2}{8}\ {}^{\sim}\overset{2}{8})$	9.341433	10.040788	9.434771	9.3775135	10.2393	10.613378	10.2516837
60 $(4\ {}^{\sim}4\ {}^{-}4\ {}^{-}\overset{2}{8})$	9.619281	×	10.419785	9.675265	10.968591	11.586401	10.982034
65 $(4\ {}^{\sim}4\ {}^{\sim}\overset{2}{8}\ {}^{\sim}\overset{2}{8})$	9.2472901	9.903566	9.2250982	9.2813429	10.088567	10.4235148	10.1007829
66 $(4\ {}^{\sim}4\ {}^{\sim}\overset{2}{8}\ {}^{-}\overset{2}{8})$	9.4755321	10.2752147	9.7746882	9.5163996	10.5051427	10.9639153	10.51804
67 $(4\ {}^{\sim}4\ {}^{-}\overset{2}{8}\ {}^{\sim}\overset{2}{8})$	9.357703	10.063177	9.471644	9.394111	10.26336	10.64232	10.275723
69 $(4\ {}^{-}4\ {}^{\sim}\overset{2}{8}\ {}^{\sim}\overset{2}{8})$	9.2828907	9.9531646	9.3031526	9.3175979	10.1425852	10.4907632	10.154844
70 $(4\ {}^{-}4\ {}^{\sim}\overset{2}{8}\ {}^{-}\overset{2}{8})$	9.5305419	10.3766188	9.9352593	9.573961	10.6190505	11.1098838	10.6320033
74 $(4\ {}^{\sim}\overset{2}{8}\ {}^{\sim}4\ {}^{-}\overset{2}{8})$	9.5588394	10.4423222	10.03604	9.6042447	10.6954236	11.2131696	10.708507
78 $(4\ {}^{-}\overset{2}{8}\ {}^{\sim}4\ {}^{-}\overset{2}{8})$ [Rc3< 9.805]	9.5605197	×	×	9.6060157	×	×	×

P. 2-3	74 (3 ~$\overset{2}{6}$ -$\overset{2}{6}$ -$\overset{2}{6}$)	76 (3 -$\overset{2}{6}$ ~$\overset{2}{6}$ -$\overset{2}{6}$)	86 (<u>6 3</u> -$\overset{2}{6}$ -$\overset{2}{6}$)	103 (6 ~6 ~6 -$\overset{2}{6}$)	86 (<u>6 3</u> -$\overset{2}{6}$ -$\overset{2}{6}$)	103 (6 ~6 ~6 -$\overset{2}{6}$)	
45 (1 4 ~$\overset{2}{8}$ ~$\overset{2}{8}$)	10.169789	9.980123	10.2819868	10.3604236	10.2819868	10.3604236	
46 (1 4 ~$\overset{2}{8}$ -$\overset{2}{8}$)	10.51275	10.274441	10.698166	10.530615	10.698166	10.530615	
47 (1 4 -$\overset{2}{8}$ ~$\overset{2}{8}$)	10.3282227	10.117518	10.473664	10.4413056	10.473664	10.4413056	
53 (2 4 ~$\overset{2}{8}$ ~$\overset{2}{8}$)	10.3086647	10.101111	10.4496	10.431333	10.4496	10.431333	
54 (2 4 ~$\overset{2}{8}$ -$\overset{2}{8}$)	10.7756267	10.500414	×	11.012837	×	11.012837	
55 (2 4 -$\overset{2}{8}$ ~$\overset{2}{8}$)	10.51012	10.276772	×	10.6902957	×	10.6902957	
60 (4 ~4 -4 -$\overset{2}{8}$)	11.396545	11.01818	×	11.761949	×	11.761949	
65 (4 ~4 ~$\overset{2}{8}$ ~$\overset{2}{8}$)	10.335114	10.1242395	×	10.481293	×	10.481293	
66 (4 ~4 ~$\overset{2}{8}$ -$\overset{2}{8}$)	10.829824	10.5468955	×	11.077425	×	11.077425	
67 (4 ~4 -$\overset{2}{8}$ ~$\overset{2}{8}$)	10.53716	10.301	×	10.721751	×	10.721751	
69 (4 -4 ~$\overset{2}{8}$ ~$\overset{2}{8}$)	12.3972506	10.178838	×	10.555313	×	10.555313	
70 (4 -4 ~$\overset{2}{8}$ -$\overset{2}{8}$)	10.964007	10.662256	×	11.2368655	×	11.2368655	
74 (4 ~$\overset{2}{8}$ ~4 -$\overset{2}{8}$)	11.0579934	10.739985	×	11.350857	×	11.350857	

48E' Family Group (12): 6(8M)e,1 + 8(6A)
[x = E8M, y = 1, z = E6A] P. 2-1, 2, 3

P. 3	18 $(\underline{6\ 3}\ \bar{6}^2)$	26 $(6\ \bar{6}^2\ \bar{6}^2)$	50 $(3\ \tilde{3}\ \bar{3}^2\ \bar{6}^2)$	62 $(3\ \bar{3}\ \bar{6}^2\ \bar{6}^2)$	72 $(3\ \tilde{6}^2\ \tilde{6}^2\ \bar{6}^2)$	74 $(3\ \tilde{6}^2\ \bar{6}^2\ \bar{6}^2)$	76 $(3\ \bar{6}^2\ \tilde{6}^2\ \bar{6}^2)$	82 $(\underline{6\ 3}\ \bar{3}\ \bar{6}^2)$	86. $(\underline{6\ 3}\ \bar{6}^2\ \bar{6}^2)$	114 $(6\ \bar{6}^2\ \bar{6}^2\ \bar{6}^2)$
82 $(4\ \tilde{8}^2\ \tilde{8}^2\ \bar{8}^2)$	10.077635	10.847568	10.714705	11.8583159	11.1738935	11.6418576	11.2140565	11.221961	12.0614677	12.3308234
84 $(4\ \tilde{8}^2\ \bar{8}^2\ \bar{8}^2)$	×	10.9118734	10.857873	11.9957605	11.26212	×	×	×	12.215074	×
86 $(4\ \bar{8}^2\ \tilde{8}^2\ \bar{8}^2)$	10.1208882	10.8761839	10.77586	11.9150947	11.211916	×	×	×	12.125215	×
92 $(\underline{8\ 4}\ \bar{4}\ \bar{8}^2)$			11.42437	×	11.5005888	×	×	×	12.8108726	×
113 $(8\ \tilde{8}\ \tilde{8}\ \bar{8}^2)$	9.953155	10.7508978	10.53071	11.6671377	11.048283	×	×	×	11.8474117	×
115 $(8\ \bar{8}\ \tilde{8}\ \bar{8}^2)$	10.042916	10.813807	10.65021	11.775561	11.125868	11.571013	11.16432	11.116051	11.96632	12.214221
122 $(8\ \bar{8}^2\ \tilde{8}^2\ \bar{8}^2)$	×	10.996141	11.125777	12.2295404	11.4029698	×	×	×	12.475391	×
124 $(8\ \bar{8}^2\ \bar{8}^2\ \bar{8}^2)$	×	×	×	12.7389598	×	×	×	×	×	×

End of 48E' Family Group (12):
6(8M)e,1 + 8(6A) [x = E8M, y = 1, z = E6A]

48E' Group (13): 6(8M)e,1 + 12(4A) [x = 1, y = E8M, z = E4A] ---- Outline

P.1	2	4	5	8	13	14	18	19	20	21	26	28	29	36	46	68	69	96	97	114	116
1	•	•	•	•			•									•		•			
2	•	•	•	•			•									•		•			
5			•	•	•	•	•	•	•			•			•	•		•			
7	•		•	•		•	•									•		•			
8			•	•		•	•		•							•		•			
9			•	•		•	•									•		•			
10			•	•	•	•	•	•	•			•			•	•		•			
11			•	•		•	•									•		•			
12			•	•	•	•	•	•	•			•			•	•		•			
13			•	•	•	•	•	•										•			
14			•	•		•		•		•	•									•	
15			•	•	•	•		•			•										
16			•			•		•	•	•	•	•									
17			•	•		•		•		•	•										
20						•		•		•	•		•							•	
27						•		•		•	•		•							•	
28								•		•			•	•	•					•	
41			•	•	•	•	•								•	•		•			
42			•	•	•	•	•	•	•		•	•			•	•		•			
45			•			•		•		•	•				•						
P.2	2	4	5	8	13	14	18	19	20	21	26	28	29	36	46	68	69	96	97	114	116
46						•		•		•	•		•		•					•	•
47						•		•		•	•			•	•						
53						•		•		•	•			•	•						
54						•		•		•	•		•		•					•	•
55						•		•		•	•		•		•					•	•
60								•		•			•	•						•	•
65						•		•		•	•				•					•	
66						•		•		•	•		•		•					•	•
67						•		•		•	•		•		•					•	•
69						•		•		•	•		•		•					•	•
70						•		•		•	•		•		•					•	•
71																					
74						•		•		•	•		•		•					•	•
78						•															
P.3	2	4	5	8	13	14	18	19	20	21	26	28	29	36	46	68	69	96	97	114	116
81																					
82										•			•	•	•		•			•	•
83																					
84										•			•	•	•		•			•	•
85																					
86										•			•	•			•			•	•
92													•								
113										•			•	•						•	•
115										•			•	•							
122										•			•	•			•			•	•
124																	•				

48E' Group (13): 6(8M)e,1 + 12(4A) [x = 1, y = E8M, z = E4A]									
P. 1-1	$2\,(1\,\overset{2}{4})$	$4\,(2\,{}^{\sim}\overset{2}{4})$	$5\,(2\,{}^{-}\overset{2}{4})$	$8\,(4\,{}^{-}\overset{2}{4})$	$13\,(1\,\overset{2}{4}\,{}^{\sim}\overset{2}{4})$	$14\,(1\,\overset{2}{4}\,{}^{-}\overset{2}{4})$	$18\,(2\,{}^{\sim}\overset{2}{4}\,{}^{\sim}\overset{2}{4})$	$19\,(2\,{}^{\sim}\overset{2}{4}\,{}^{-}\overset{2}{4})$	$20\,(2\,{}^{-}\overset{2}{4}\,{}^{\sim}\overset{2}{4})$
$1\,(4\,{}^{\sim}\overset{2}{8})$	5.681422	6.8708164	6.2876503	7.5178305	×	×	8.7621933	×	×
$2\,(4\,{}^{-}\overset{2}{8})$	5.7121728	6.8926103	6.3860572	7.61816	×	×	8.8631616	×	×
$5\,(8\,{}^{-}\overset{2}{8})$	×	×	7.2010305	8.3501763	9.2086148	8.8754363	9.5379873	8.9729447	10.411678
$7\,(1\,4\,{}^{\sim}\overset{2}{8})$	5.7563049	×	6.5927685	7.8050387	×	7.966387	9.0376688	×	×
$8\,(1\,4\,{}^{-}\overset{2}{8})$	×	×	6.8056595	8.0135379	×	8.277392	9.2417407	×	10.1362
$9\,(2\,4\,{}^{\sim}\overset{2}{8})$	×	×	6.809588	8.0017133	×	8.255466	9.2212253	×	×
$10\,(2\,4\,{}^{-}\overset{2}{8})$	×	×	7.1275327	8.2975106	9.1658797	8.7638468	9.49836	8.8448511	10.37845
$11\,(4\,{}^{\sim}4\,{}^{\sim}\overset{2}{8})$	×	×	6.8558643	8.0418888	×	8.316626	9.2545498	×	×
$12\,(4\,{}^{\sim}4\,{}^{-}\overset{2}{8})$	×	×	7.1959432	8.351253	9.2116131	8.882011	9.5413843	8.9813457	10.4156575
$13\,(4\,{}^{-}4\,{}^{\sim}\overset{2}{8})$	×	×	6.977465	8.143057	9.012214	8.4748435	9.3458	8.5146385	×
$14\,(4\,{}^{-}4\,{}^{-}\overset{2}{8})$	×	×	×	8.3619593	×	9.294763	×	9.480369	×
$15\,(4\,{}^{\sim}\overset{2}{8}\,{}^{\sim}\overset{2}{8})$	×	×	7.310695	8.419602	9.2549568	9.126612	×	9.269857	×
$16\,(4\,{}^{\sim}\overset{2}{8}\,{}^{-}\overset{2}{8})$	×	×	7.331241	×	×	9.2307648	×	9.39863	10.3878
$17\,(4\,{}^{-}\overset{2}{8}\,{}^{\sim}\overset{2}{8})$	×	×	7.3232067	8.4201233	×	9.174208	×	9.3278307	×
$20\,(\underline{8\,4}\,{}^{-}\overset{2}{8})$	×	×	×	×	×	9.545689	×	9.8721815	×
$27\,(8\,{}^{-}\overset{2}{8}\,{}^{\sim}\overset{2}{8})$	×	×	×	×	×	9.4338317	×	9.6635263	×
$28\,(8\,{}^{-}\overset{2}{8}\,{}^{-}\overset{2}{8})$	×	×	×	×	×	×	×	9.9852214	×
$41\,(1\,4\,{}^{\sim}4\,{}^{\sim}\overset{2}{8})$	×	×	6.8854917	8.066426	8.9429755	8.3542056	9.278999	×	×
$42\,(1\,4\,{}^{\sim}4\,{}^{-}\overset{2}{8})$	×	×	7.2379525	8.380728	9.2344385	8.9588457	9.5620494	9.0708576	10.4316172
$45\,(1\,4\,{}^{\sim}\overset{2}{8}\,{}^{\sim}\overset{2}{8})$	×	×	7.33396	×	×	9.2506375	×	9.422049	×
P. 1-2	$21\,(2\,{}^{-}\overset{2}{4}\,{}^{-}\overset{2}{4})$	$26\,(4\,{}^{\sim}4\,{}^{-}\overset{2}{4})$	$28\,(4\,{}^{-}\overset{2}{4}\,{}^{\sim}\overset{2}{4})$	$29\,(4\,{}^{-}\overset{2}{4}\,{}^{-}\overset{2}{4})$	$36\,(\underline{8\,4}\,{}^{-}\overset{2}{4})$	$46\,(1\,4\,{}^{\sim}4\,{}^{-}\overset{2}{4})$	$68\,(2\,{}^{\sim}\overset{2}{4}\,{}^{-}\overset{2}{4}\,{}^{\sim}\overset{2}{4})$	$96\,(2\,{}^{\sim}\overset{2}{4}\,{}^{-}\overset{2}{4}\,{}^{-}\overset{2}{4})$	$114\,(4\,{}^{\sim}\overset{2}{8}\,{}^{\sim}4\,{}^{-}\overset{2}{4})$
$1\,(4\,{}^{\sim}\overset{2}{8})$	×	×	×	×	×	×	11.27335	12.535209	×
$2\,(4\,{}^{-}\overset{2}{8})$	×	×	×	×	×	×	11.37449	12.63629	×
$5\,(8\,{}^{-}\overset{2}{8})$	×	×	×	×	×	10.19729	11.979679	13.219722	×
$7\,(1\,4\,{}^{\sim}\overset{2}{8})$	×	×	×	×	×	×	11.53513	12.792845	×
$8\,(1\,4\,{}^{-}\overset{2}{8})$	×	×	×	×	×	×	11.732445	12.987727	×
$9\,(2\,4\,{}^{\sim}\overset{2}{8})$	×	×	×	×	×	×	11.70352	12.956659	×
$10\,(2\,4\,{}^{-}\overset{2}{8})$	×	×	×	×	×	10.082026	11.954379	13.19852	×
$11\,(4\,{}^{\sim}4\,{}^{\sim}\overset{2}{8})$	×	×	×	×	×	×	11.735515	12.987374	×
$12\,(4\,{}^{\sim}4\,{}^{-}\overset{2}{8})$	×	×	10.4150556	×	×	10.212881	11.983686	13.223403	×
$13\,(4\,{}^{-}4\,{}^{\sim}\overset{2}{8})$	×	×	×	×	×	×	11.809407	13.057075	×
$14\,(4\,{}^{-}4\,{}^{-}\overset{2}{8})$	9.722	10.553039	×	×	×	×	×	×	12.22285
$15\,(4\,{}^{\sim}\overset{2}{8}\,{}^{\sim}\overset{2}{8})$	×	10.3566174	×	×	×	×	×	×	×

16 (4 ~$\overset{2}{8}$ -$\overset{2}{8}$)	9.6022448	10.4775748	10.3877997	×	×	×	×	×	×
17 (4 -$\overset{2}{8}$ ~$\overset{2}{8}$)	9.5002245	10.4094444	×	×	×	×	×	×	×
20 ($\underline{8\ 4}$ -$\overset{2}{8}$)	10.3690687	10.868836	×	10.461868	×	×	×	×	12.905555
27 (8 -$\overset{2}{8}$ ~$\overset{2}{8}$)	9.9897171	10.6928706	×	10.0296835	×	×	×	×	12.441933
28 (8 -$\overset{2}{8}$ -$\overset{2}{8}$)	10.735343	×	×	10.893248	13.16214	11.0988	×	×	13.301751
41 (1 4 ~4 ~$\overset{2}{8}$)	×	×	×	×	×	9.60935	11.75364	13.00454	×
42 (1 4 ~4 -$\overset{2}{8}$)	×	10.177901	10.4316172	×	×	10.296879	11.99348	13.22977	×
45 (1 4 ~$\overset{2}{8}$ ~$\overset{2}{8}$)	9.63338	10.491404	×	×	×	10.623841	×	×	×
P. 2	14 (1 $\overset{2}{4}$ -$\overset{2}{4}$)	19 (2 ~$\overset{2}{4}$ -$\overset{2}{4}$)	21 (2 -$\overset{2}{4}$ -$\overset{2}{4}$)	26 (4 ~4 -$\overset{2}{4}$)	29 (4 -$\overset{2}{4}$ -$\overset{2}{4}$)	36 ($\underline{8\ 4}$ -$\overset{2}{4}$)	46 (1 4 ~4 -$\overset{2}{4}$)	114 (4 ~$\overset{2}{8}$ ~4 -$\overset{2}{4}$)	116 (4 ~$\overset{2}{8}$ ~4 -$\overset{2}{4}$)
46 (1 4 ~$\overset{2}{8}$ -$\overset{2}{8}$)	9.410659	9.635007	9.954664	10.684	9.9916	×	10.830033	12.46082	12.461741
47 (1 4 -$\overset{2}{8}$ ~$\overset{2}{8}$)	9.332461	9.526813	9.785955	10.583317	×	12.154713	10.721042	×	×
53 (2 4 ~$\overset{2}{8}$ ~$\overset{2}{8}$)	9.32445	9.51584	9.7688385	10.57183	×	12.131761	10.7086265	×	×
54 (2 4 ~$\overset{2}{8}$ -$\overset{2}{8}$)	9.5069404	9.7848238	10.201693	10.80502	10.2703547	×	10.963463	12.720182	12.720819
55 (2 4 -$\overset{2}{8}$ ~$\overset{2}{8}$)	9.421369	9.646377	9.9644856	10.6808525	10.001625	×	10.8248515	12.4241023	12.42505
60 (4 ~4 -4 -$\overset{2}{8}$)	×	9.985306	10.743355	×	10.9038	13.178014	×	13.31885	13.31873
65 (4 ~4 ~$\overset{2}{8}$ ~$\overset{2}{8}$)	9.33817	9.533628	9.79482	10.586656	×	×	10.7243057	12.2571931	×
66 (4 ~4 ~$\overset{2}{8}$ -$\overset{2}{8}$)	9.5217882	9.8130546	10.252183	10.825972	10.327701	×	10.98747	12.7722172	12.772815
67 (4 ~4 -$\overset{2}{8}$ ~$\overset{2}{8}$)	9.434823	9.665085	9.992349	10.694403	10.032656	×	10.839217	12.44486	12.445791
69 (4 -4 ~$\overset{2}{8}$ ~$\overset{2}{8}$)	9.37029	9.57571	9.856463	10.6207405	9.8808417	×	10.760355	12.3137908	12.314844
70 (4 -4 ~$\overset{2}{8}$ -$\overset{2}{8}$)	9.5479093	9.878304	10.3771605	10.870094	10.4702582	×	11.0401597	12.896957	12.8974
74 (4 ~$\overset{2}{8}$ ~4 -$\overset{2}{8}$)	9.553207	9.9133678	10.4581323	10.892301	10.5643878	×	11.0699725	12.995025	12.99535
P. 3	21 (2 -$\overset{2}{4}$ -$\overset{2}{4}$)	29 (4 -$\overset{2}{4}$ -$\overset{2}{4}$)	36 ($\underline{8\ 4}$ -$\overset{2}{4}$)	46 (1 4 ~4 -$\overset{2}{4}$)	69 (2 ~$\overset{2}{4}$ -$\overset{2}{4}$ -$\overset{2}{4}$)	114 (4 ~$\overset{2}{8}$ ~4 -$\overset{2}{4}$)	116 (4 -$\overset{2}{8}$ ~4 -$\overset{2}{4}$)		
82 (4 ~$\overset{2}{8}$ ~$\overset{2}{8}$ -$\overset{2}{8}$)	10.932722	11.1432032	13.0368909	10.9305443	11.462583	13.5255568	13.522128		
84 (4 ~$\overset{2}{8}$ -$\overset{2}{8}$ -$\overset{2}{8}$)	11.0116848	11.253554	13.4565354	10.6208244	11.6550297	13.6191668	13.618556		
86 (4 -$\overset{2}{8}$ ~$\overset{2}{8}$ -$\overset{2}{8}$)	10.9677482	11.1907172	13.4060597	×	11.543309	13.5631003	13.562591		
92 ($\underline{8\ 4}$ -4 -$\overset{2}{8}$)	×	11.5647893	×	×	12.410379	×	×		
113 (8 ~8 ~8 -$\overset{2}{8}$)	10.8137794	10.987477	13.22131	×	×	13.3633304	13.36316		
115 (8 -8 ~8 -$\overset{2}{8}$)	10.89035	11.084	13.297296	×	×	13.44407	13.443767		
122 (8 -$\overset{2}{8}$ ~$\overset{2}{8}$ -$\overset{2}{8}$)	11.115891	11.4315105	13.562515	×	11.99278	13.742904	13.742026		
124 (8 -$\overset{2}{8}$ -$\overset{2}{8}$ -$\overset{2}{8}$)	×	×	×	×	12.778135	×	×		

End of [DATA 2B – 7] 48E' Group (13): 6(8M)e,1 + 12(4A

48E' Group (14): 8(6M)e + 12(4A) [x = y = E6M, z = E4A]																					
	2	4	5	8	13	14	18	19	20	21	26	28	29	36	46	68	69	96	97	114	116
3						•															
6										•											
19			•	•	•	•	•	•	•			•			•	•		•			
115										•			•							•	
117								•		•			•		•					•	
119								•		•			•		•					•	
121						•		•		•	•		•		•						
152																					

48E' Group (14): 8(6M)e + 12(4A) [x = y = E6M, z = E4A]								
19 $(6\overset{\sim}{6}\overset{\sim}{6})$e	5 $(2\overset{-2}{4})$	8 $(4\overset{-2}{4})$	13 $(1\overset{2}{4}\overset{\sim 2}{4})$	14 $(1\overset{2}{4}\overset{-2}{4})$	18 $(2\overset{\sim 2}{4}\overset{\sim 2}{4})$	19 $(2\overset{\sim 2}{4}\overset{-2}{4})$	20 $(2\overset{-2}{4}\overset{\sim 2}{4})$	28 $(4\overset{-2}{4}\overset{\sim 2}{4})$
	8.7632279	10.3062845	11.44787	10.726404	11.8781456	10.7697358	13.029737	13.46675
	46 $(1\,4\overset{\sim}{4}\overset{-2}{4})$	68 $(2\overset{\sim 2}{4}\overset{-2}{4}\overset{\sim 2}{4})$	96 $(4\overset{-2}{4}\overset{-2}{4}\overset{\sim 2}{4})$					
	12.3526664	15.06568	16.67128					
	14 $(1\overset{2}{4}\overset{-2}{4})$	19 $(2\overset{\sim 2}{4}\overset{-2}{4})$	21 $(2\overset{-2}{4}\overset{-2}{4})$	26 $(4\overset{\sim}{4}\overset{-2}{4})$	29 $(4\overset{-2}{4}\overset{-2}{4})$	46 $(1\,4\overset{\sim}{4}\overset{-2}{4})$	96 $(4\overset{-2}{4}\overset{-2}{4}\overset{\sim 2}{4})$	114 $(4\overset{\sim 2}{8}\overset{\sim}{4}\overset{-2}{4})$
3 $(6\overset{\sim}{6})$e	9.9156652	×	×	×	×	×	×	×
6 $(4\overset{\sim}{4})$e	×	×	12.894073	×	×	×	×	×
115 $(6\overset{\sim}{6}\overset{\sim 2}{12}\overset{-}{6})$e	×	×	×	×	13.667171	×	×	16.828797
117 $(6\overset{\sim}{6}\overset{-2}{12}\overset{\sim}{6})$e	×	13.328425	13.63418	×	13.779191	14.87927	×	16.926061
119 $(6\overset{-}{6}\overset{\sim 2}{12}\overset{\sim}{6})$e	×	13.335554	13.738618	×	13.788041	14.883931	×	16.93008
121 $(6\overset{-}{6}\overset{-2}{12}\overset{\sim}{6})$e	12.9987826	13.2191934	13.746937	14.6160349	13.692997	14.7891144	×	×

48E' Group (15): 8(6M)e + 6(8A) [x = z = E6M, y = E8A]	2	5	8	10	12	14	15	16	17	18	20	25	27	28	35	42	45	46	47	48	53	54	55	56	60	65
3				•	•	•					•			•		•		•		•		•		•	•	
6											•														•	
19		•		•	•	•	•	•	•		•	•	•	•		•	•	•	•	•	•	•	•	•	•	•
115											•			•	•					•				•	•	
117											•			•	•									•	•	
119											•			•	•									•	•	
121																										
152																										
	66	67	68	69	70	71	72	73	74	78	81	82	83	84	85	86	92	95	96	113	115	121	122	123	124	173
3	•		•		•		•	?	•	•		•		•		•			•	•	•		•			•
6							•	?	•	•		•		•		•			•				•		•	•
19	•	•	•	•	•	•	•	?	•		•	•	•	•	•	•			•	•	•	•	•	•		•
115							•	?	•			•		•		•	•		•				•		•	•
117					•		•	?	•			•		•		•	•		•				•		•	•
119					•		•	?	•			•		•		•	•		•		•		•		•	•
121								?																		
152								?									•								•	

48E' Group (15): 8(6M)e + 6(8A) [x = z = E6M, y = E8A]									
3 (6 $\tilde{6}$)e	10 (2 4 $\bar{8}^{2}$)	12 (4 $\tilde{4}$ $\bar{8}^{2}$)	14 (4 $\bar{4}$ $\bar{8}^{2}$)	20 ($\underline{8\ 4}$ $\bar{8}^{2}$)	28 (8 $\bar{8}^{2}$ $\bar{8}^{2}$)	42 (1 4 $\tilde{4}$ $\bar{8}^{2}$)	46 (1 4 $\tilde{8}^{2}$ $\bar{8}^{2}$)	48 (1 4 $\bar{8}^{2}$ $\bar{8}^{2}$)	54 (2 4 $\tilde{8}^{2}$ $\bar{8}^{2}$)
	7.3566397	7.4729202	7.8899525	8.227243	9.3956266	7.551856	8.663664	8.9184528	8.8943161
	56 (2 4 $\bar{8}^{2}$ $\bar{8}^{2}$)	60 (4 $\tilde{4}$ $\bar{4}$ $\bar{8}^{2}$)	66 (4 $\tilde{4}$ $\tilde{8}^{2}$ $\bar{8}^{2}$)	68 (4 $\tilde{4}$ $\bar{8}^{2}$ $\bar{8}^{2}$)	70 (4 $\bar{4}$ $\tilde{8}^{2}$ $\bar{8}^{2}$)	72 (4 $\bar{4}$ $\bar{8}^{2}$ $\bar{8}^{2}$)	74 (4 $\tilde{8}^{2}$ $\tilde{4}$ $\bar{8}^{2}$)	78 (4 $\bar{8}^{2}$ $\tilde{4}$ $\bar{8}^{2}$)	82 (4 $\tilde{8}^{2}$ $\tilde{8}^{2}$ $\bar{8}^{2}$)
	9.3091778	8.3226324	8.9426207	9.4035086	9.0660253	9.7319083	8.2648845	8.26761	9.596561
	84 (4 $\tilde{8}^{2}$ $\bar{8}^{2}$ $\bar{8}^{2}$)	86 (4 $\bar{8}^{2}$ $\tilde{8}^{2}$ $\bar{8}^{2}$)	96 ($\underline{8\ 4}$ $\bar{8}^{2}$ $\bar{8}^{2}$)	113 (8 $\tilde{8}$ $\tilde{8}$ $\bar{8}^{2}$)	115 (8 $\bar{8}$ $\tilde{8}$ $\bar{8}^{2}$)	122 (8 $\bar{8}^{2}$ $\tilde{8}^{2}$ $\bar{8}^{2}$)	173 (16 $\bar{16}^{2}$ $\tilde{8}$ $\bar{8}^{2}$)		
	9.6806875	9.6344533	9.91571	10.3524003	10.535495	9.8341441	12.8989864		
6 ($\underline{12\ 6}$)e	20 ($\underline{8\ 4}$ $\bar{8}^{2}$)	60 (4 $\tilde{4}$ $\bar{4}$ $\bar{8}^{2}$)	72 (4 $\bar{4}$ $\bar{8}^{2}$ $\bar{8}^{2}$)	74 (4 $\tilde{8}^{2}$ $\tilde{4}$ $\bar{8}^{2}$)	78 (4 $\bar{8}^{2}$ $\tilde{4}$ $\bar{8}^{2}$)	82 (4 $\tilde{8}^{2}$ $\tilde{8}^{2}$ $\bar{8}^{2}$)	84 (4 $\bar{8}^{2}$ $\bar{8}^{2}$ $\bar{8}^{2}$)	86 (4 $\bar{8}^{2}$ $\bar{8}^{2}$ $\bar{8}^{2}$)	96 ($\underline{8\ 4}$ $\bar{8}^{2}$ $\bar{8}^{2}$)
	10.5062863	10.854535	11.9272704	10.587501	10.59028	11.6785045	11.82933	11.7416177	12.4140134
	122 (8 $\bar{8}^{2}$ $\tilde{8}^{2}$ $\bar{8}^{2}$)	124 (8 $\bar{8}^{2}$ $\bar{8}^{2}$ $\bar{8}^{2}$)	173 (16 $\bar{16}^{2}$ $\tilde{8}$ $\bar{8}^{2}$)						
	12.0899987	12.6808877	15.076496						
19 (6 $\tilde{6}$ $\tilde{6}$)e	5 (8 $\bar{8}^{2}$)	10 (2 4 $\bar{8}^{2}$)	12 (4 $\tilde{4}$ $\bar{8}^{2}$)	14 (4 $\bar{4}$ $\bar{8}^{2}$)	15 (4 $\tilde{8}^{2}$ $\tilde{8}^{2}$)	16 (4 $\tilde{8}^{2}$ $\bar{8}^{2}$)	17 (4 $\bar{8}^{2}$ $\tilde{8}^{2}$)	20 ($\underline{8\ 4}$ $\bar{8}^{2}$)	25 (8 $\tilde{8}$ $\bar{8}^{2}$)
	8.2575832	7.9406626	8.068317	8.5407339	8.811682	8.8401818	8.9444746	8.9566122	9.9250093
	27 (8 $\bar{8}^{2}$ $\tilde{8}^{2}$)	28 (8 $\bar{8}^{2}$ $\bar{8}^{2}$)	42 (1 4 $\tilde{4}$ $\bar{8}^{2}$)	45 (1 4 $\tilde{8}^{2}$ $\tilde{8}^{2}$)	46 (1 4 $\tilde{8}^{2}$ $\bar{8}^{2}$)	47 (1 4 $\bar{8}^{2}$ $\tilde{8}^{2}$)	48 (1 4 $\bar{8}^{2}$ $\bar{8}^{2}$)	53 (2 4 $\tilde{8}^{2}$ $\tilde{8}^{2}$)	54 (2 4 $\tilde{8}^{2}$ $\bar{8}^{2}$)
	10.258478	10.0088734	8.1540989	9.1928	9.19358	9.5298209	9.477388	9.49015	9.4456544
	55 (2 4 $\bar{8}^{2}$ $\tilde{8}^{2}$)	56 (2 4 $\bar{8}^{2}$ $\bar{8}^{2}$)	60 (4 $\tilde{4}$ $\bar{4}$ $\bar{8}^{2}$)	65 (4 $\tilde{4}$ $\tilde{8}^{2}$ $\tilde{8}^{2}$)	66 (4 $\tilde{4}$ $\tilde{8}^{2}$ $\bar{8}^{2}$)	67 (4 $\tilde{4}$ $\bar{8}^{2}$ $\tilde{8}^{2}$)	68 (4 $\tilde{4}$ $\bar{8}^{2}$ $\bar{8}^{2}$)	69 (4 $\bar{4}$ $\tilde{8}^{2}$ $\tilde{8}^{2}$)	70 (4 $\bar{4}$ $\tilde{8}^{2}$ $\bar{8}^{2}$)
	10.117349	9.9137149	9.1477847	9.5550695	9.4981943	10.278524	10.0208335	9.7241253	9.631208
	71 (4 $\bar{4}$ $\bar{8}^{2}$ $\tilde{8}^{2}$)	72 (4 $\bar{4}$ $\bar{8}^{2}$ $\bar{8}^{2}$)	74 (4 $\tilde{8}^{2}$ $\tilde{4}$ $\bar{8}^{2}$)	78 (4 $\bar{8}^{2}$ $\tilde{4}$ $\bar{8}^{2}$)	81 (4 $\tilde{8}^{2}$ $\tilde{8}^{2}$ $\tilde{8}^{2}$)	82 (4 $\tilde{8}^{2}$ $\tilde{8}^{2}$ $\bar{8}^{2}$)	83 (4 $\tilde{8}^{2}$ $\bar{8}^{2}$ $\tilde{8}^{2}$)	84 (4 $\tilde{8}^{2}$ $\bar{8}^{2}$ $\bar{8}^{2}$)	85 (4 $\bar{8}^{2}$ $\tilde{8}^{2}$ $\tilde{8}^{2}$)
	10.9678808	10.4135	9.011004	9.0143456	11.6422058	10.243297	10.835943	10.3477005	10.723825
	86 (4 $\bar{8}^{2}$ $\tilde{8}^{2}$ $\bar{8}^{2}$)	96 ($\underline{8\ 4}$ $\bar{8}^{2}$ $\bar{8}^{2}$)	113 (8 $\tilde{8}$ $\tilde{8}$ $\bar{8}^{2}$)	115 (8 $\bar{8}$ $\tilde{8}$ $\bar{8}^{2}$)	121 (8 $\bar{8}^{2}$ $\tilde{8}^{2}$ $\tilde{8}^{2}$)	122 (8 $\bar{8}^{2}$ $\tilde{8}^{2}$ $\bar{8}^{2}$)	123 (8 $\bar{8}^{2}$ $\bar{8}^{2}$ $\tilde{8}^{2}$)	173 (16 $\bar{16}^{2}$ $\tilde{8}$ $\bar{8}^{2}$)	
	10.288917	10.7078375	10.8906942	11.0779262	11.22654	10.540398	12.299484	13.563313	
	20 ($\underline{8\ 4}$ $\bar{8}^{2}$)	28 (8 $\bar{8}^{2}$ $\bar{8}^{2}$)	35 ($\underline{16\ 8}$ $\bar{8}^{2}$)	48 (1 4 $\bar{8}^{2}$ $\bar{8}^{2}$)	56 (2 4 $\bar{8}^{2}$ $\bar{8}^{2}$)	60 (4 $\tilde{4}$ $\bar{4}$ $\bar{8}^{2}$)	70 (4 $\bar{4}$ $\tilde{8}^{2}$ $\bar{8}^{2}$)	72 (4 $\bar{4}$ $\bar{8}^{2}$ $\bar{8}^{2}$)	74 (4 $\tilde{8}^{2}$ $\tilde{4}$ $\bar{8}^{2}$)

115 (6 ~6 ~$\overset{2}{12}$ -6)e	11.0314594	11.8111958	14.385945	11.0636315	11.6842742	11.4647623	×	12.4305205	11.12013
117 (6 ~6 -$\overset{2}{12}$ ~6)e	11.115437	11.879742	14.44718	×	11.75171	11.514265	113.14669	12.505965	11.204946
119 (6 -6 ~$\overset{2}{12}$ ~6)e	11.128165	11.88826	14.45137	×	11.76037	11.52646	11.323456	12.514008	11.217496
	78 (4 -$\overset{2}{8}$ ~4 -$\overset{2}{8}$)	82 (4 ~$\overset{2}{8}$ ~$\overset{2}{8}$ -$\overset{2}{8}$)	84 (4 ~$\overset{2}{8}$ -$\overset{2}{8}$ -$\overset{2}{8}$)	86 (4 -$\overset{2}{8}$ ~$\overset{2}{8}$ -$\overset{2}{8}$)	92 ($\underline{8\ 4}$ -4 -$\overset{2}{8}$)	96 ($\underline{8\ 4}$ -$\overset{2}{8}$ -$\overset{2}{8}$)	115 (8 -8 ~8 -$\overset{2}{8}$)	122 (8 -$\overset{2}{8}$ ~$\overset{2}{8}$ -$\overset{2}{8}$)	124 (8 -$\overset{2}{8}$ -$\overset{2}{8}$ -$\overset{2}{8}$)
115 (6 ~6 ~$\overset{2}{12}$ -6)e	11.1224437	12.1553255	12.321961	12.2243677	12.0677143	12.9729657	×	12.6021008	13.2874637
117 (6 ~6 -$\overset{2}{12}$ ~6)e	11.207175	12.22753	12.39612	12.297286	12.168223	13.054865	×	12.678315	13.374865
119 (6 -6 ~$\overset{2}{12}$ ~6)e	11.2197	12.235759	12.404218	12.30546	12.19091	13.06253	12.716888	12.68611	13.38247
	173 (16 -$\overset{2}{16}$ ~8 -$\overset{2}{8}$)								
115 (6 ~6 ~$\overset{2}{12}$ -6)e	15.560618								
117 (6 ~6 -$\overset{2}{12}$ ~6)e	15.624772								
119 (6 -6 ~$\overset{2}{12}$ ~6)e	15.628203								
152 ($\underline{12\ 6}$ ~$\overset{2}{12}$ -6)e	92 ($\underline{8\ 4}$ -4 -$\overset{2}{8}$)	124 (8 -$\overset{2}{8}$ -$\overset{2}{8}$ -$\overset{2}{8}$)							
	13.894297	14.861809							

End of 48E' Group (15): 8(6M)e + 6(8A) [x = z = E6M, y = E8A]

48E' Group (16): 8(6M)e,1 + 12(4A) [z = E6M, y = E4A, x = 1] ---- Outline

	2	4	5	8	13	14	18	19	20	21	26	28	29	36	46	68	69	96	97	114	116
1			•	•	•	•	•		•			•				•		•			
2			•	•		•	•	•	•			•			•	•		•			
5						•		•		•	•				•						
7			•	•	•	•	•	•	•			•			•	•		•			
8						•		•		•	•				•						
9			•	•	•	•	•	•	•			•			•	•		•			
10								•		•	•		•		•					•	•
12										•			•	•						•	•
13								•		•	•		•		•					•	•
14										•			•	•						•	•
15								•		•	•		•		•					•	•
18										•			•	•			•			•	•
26																	•				
43								•		•	•		•		•					•	•
44										•			•	•						•	•
50																	•				
55								•		•	•		•		•					•	•
56										•			•	•						•	•
72																	•				
74																	•		•		
76																	•				
82																			•		
103													•	•			•			•	•
114																			•		

48E' Group (16): 8(6M)e,1 + 12(4A) [z = E6M, y = E4A, x = 1]

	5 (2 $\overset{-1}{4}$)	8 (4 $\overset{-2}{4}$)	13 (1 $\overset{2}{4}$ $\overset{\sim 2}{4}$)	14 (1 $\overset{2}{4}$ $\overset{-2}{4}$)	18 (2 $\overset{\sim 2}{4}$ $\overset{\sim 2}{4}$)	19 (2 $\overset{\sim 2}{4}$ $\overset{-2}{4}$)	20 (2 $\overset{-2}{4}$ $\overset{\sim 2}{4}$)	21 (2 $\overset{-2}{4}$ $\overset{-2}{4}$)	26 (4 $\overset{\sim}{4}$ $\overset{-2}{4}$)
1 (3 $\overset{\sim 2}{6}$)	6.7564009	8.3171954	9.4662695	9.5691991	9.9032225	×	11.0589395	×	
2 (3 $\overset{-2}{6}$)	6.8584707	8.401055	×	8.810679	9.9755445	8.8538089	11.125695	×	
5 (6 $\overset{-2}{6}$)				9.229425		9.4430929		9.6955247	10.8345265
7 (1 3 $\overset{\sim 2}{6}$)	6.8659528	8.4029928	9.541825	8.8163228	9.97559	8.8605494	11.1251846		
8 (1 3 $\overset{-2}{6}$)				9.2230595		9.4322963		9.6738802	10.8282465
9 (3 $\overset{\sim}{3}$ $\overset{\sim 2}{6}$)	6.8763969	8.40908	9.5462746	8.8434816	9.9795795	8.894409	11.1283106		

	28 (4 $\overset{-2}{4}$ $\overset{\sim 2}{4}$)	46 (1 4 $\overset{\sim}{4}$ $\overset{-2}{4}$)	68 (2 $\overset{\sim 2}{4}$ $\overset{-2}{4}$ $\overset{\sim 2}{4}$)	96 (4 $\overset{-2}{4}$ $\overset{-2}{4}$ $\overset{\sim 2}{4}$)					
1 (3 $\overset{\sim 2}{6}$)	11.50261	×	13.109727	14.721637					
2 (3 $\overset{-2}{6}$)	11.567399	10.44701	13.16937	14.77765					
5 (6 $\overset{-2}{6}$)		11.0019012							
7 (1 3 $\overset{\sim 2}{6}$)	11.56677	10.4452738	13.168524	14.77675					
8 (1 3 $\overset{-2}{6}$)		10.9948068							
9 (3 $\overset{\sim}{3}$ $\overset{\sim 2}{6}$)	11.569627	10.475244	13.1706892	14.77847					

	19 (2 $\overset{\sim 2}{4}$ $\overset{-2}{4}$)	21 (2 $\overset{-2}{4}$ $\overset{-2}{4}$)	26 (4 $\overset{\sim}{4}$ $\overset{-2}{4}$)	29 (4 $\overset{-2}{4}$ $\overset{-2}{4}$)	36 ($\underline{8\ 4}$ $\overset{-2}{4}$)	46 (1 4 $\overset{\sim}{4}$ $\overset{-2}{4}$)	69 (2 $\overset{\sim 2}{4}$ $\overset{-2}{4}$ $\overset{-2}{4}$)	114 (4 $\overset{\sim 2}{8}$ $\overset{\sim}{4}$ $\overset{-2}{4}$)	116 (4 $\overset{-2}{8}$ $\overset{\sim}{4}$ $\overset{-2}{4}$)
10 (3 $\overset{\sim}{3}$ $\overset{-2}{6}$)	9.5410704	9.9415803	10.9009986	9.9845623	×	11.086521	×	13.144278	13.144414
12 (3 $\overset{-}{3}$ $\overset{-2}{6}$)	×	10.3892369	×	10.59273	13.5174652	×	×	13.693439	13.69266
13 (3 $\overset{\sim 2}{6}$ $\overset{\sim 2}{6}$)	9.54763	9.960034	10.9026847	10.0069203	×	11.0890517	×	13.14688	13.147012

$14\,(3\,\overset{\sim 2}{6}\,\overset{-2}{6})$	×	10.281115	×	10.4248395	13.3859926	×	×	13.54552	13.545044
$15\,(3\,\overset{-2}{6}\,\overset{\sim 2}{6})$	9.5571237	9.9935871	10.906137	10.0483213	×	11.0954022	×	13.17933	13.179406
$18\,(\underline{6\,3}\,\overset{-2}{6})$	×	10.4392763	×	10.7197346	13.5901304	×	11.147163	13.786549	13.785518
$26\,(6\,\overset{-2}{6}\,\overset{-2}{6})$	×	×	×	×	×	×	11.60099	×	×
$43\,(1\,3\,\overset{\sim 2}{6}\,\overset{\sim 2}{6})$	9.557166	9.9936022	10.9061245	10.0483183	×	11.0953565	×	13.178862	13.17848
$44\,(1\,3\,\overset{\sim 2}{6}\,\overset{-2}{6})$	×	10.391315	×	10.59584	13.515385	×	×	13.690811	13.690029
$50\,(3\,\overset{\sim}{3}\,\overset{-2}{3}\,\overset{-2}{6})$	×	×	×	×	×	×	11.77772	×	×
$55\,(3\,\overset{\sim}{3}\,\overset{\sim 2}{6}\,\overset{\sim 2}{6})$	9.5576768	9.995423	10.96625	10.05055	×	11.095625	×	13.180033	13.180118

	$21\,(2\,\overset{-2}{4}\,\overset{-2}{4})$	$29\,(4\,\overset{-2}{4}\,\overset{-2}{4})$	$36\,(\underline{8\,4}\,\overset{-2}{4})$	69 $(2\,\overset{\sim 2}{4}\,\overset{-2}{4}\,\overset{-2}{4})$	97 $(4\,\overset{-2}{4}\,\overset{-2}{4}\,\overset{-2}{4})$	114 $(4\,\overset{\sim 2}{8}\,\overset{\sim}{4}\,\overset{-2}{4})$	116 $(4\,\overset{-2}{8}\,\overset{\sim}{4}\,\overset{-2}{4})$		
$56\,(3\,\overset{\sim}{3}\,\overset{\sim 2}{6}\,\overset{-2}{6})$	10.400182	10.612073	13.525719	×	×	13.702887	13.180118		
$72\,(3\,\overset{\sim 2}{6}\,\overset{\sim 2}{6}\,\overset{-2}{6})$	×	×	×	11.78675	×	×	×		
$74\,(3\,\overset{\sim 2}{6}\,\overset{-2}{6}\,\overset{-2}{6})$	×	×	×	11.9766174	12.53563	×	×		
$76\,(3\,\overset{-2}{6}\,\overset{\sim 2}{6}\,\overset{-2}{6})$	×	×	×	11.8069987	×	×	×		
$82\,(\underline{6\,3}\,\overset{-}{3}\,\overset{-2}{6})$	×	×	×	×	12.9789	×	×		
$103\,(6\,\overset{\sim}{6}\,\overset{\sim}{6}\,\overset{-2}{6})$	×	10.745726	13.595241	11.236717	×	13.796026	13.794944		
$114\,(6\,\overset{-2}{6}\,\overset{-2}{6}\,\overset{-2}{6})$	×	×	×	×	13.198532	×	×		

End of 48E’ Group (16): 8(6M)e,1 + 12(4A) [z = E6M, y = E4A, x = 1]

48E' Group (17): 8(6M)e,1 + +6(8A) [x = E6M, y = E8A, z = 1]

	P. 1-1										P. 1-2								P. 1-3							
	2	5	8	10	12	14	15	16	17	18	20	27	28	35	42	45	46	47	48	53	54	55	56	60	65	66
1		•	•	•	•	•	•	•		•	•	•	•		•	•	•		•	•	•	•	•	•	•	•
2		•		•	•	•	•	•				•	•		•	•	•	•	•	•	•	•	•	•	•	•
5						•					•		•	•			•				•		•	•		•
7		•		•	•	•	•	•			•	•	•		•	•	•	•	•	•	•	•	•	•	•	•
8						•					•		•	•			•		•		•		•	•		•
9		•		•	•	•		•			•	•	•	•	•	•	•	•	•	•	•	•	•	•	•	•

	P. 1-4										P. 1-5								P. 1-6							
			67	68	69	70	71	72	74	78	81	82	83	84	85	86	92	95	96	113	115	121	122	123	124	173
1				•	•	•		•			•	•	•	•	•	•			•	•	•	•	•	•		•
2				•	•	•		•			•	•	•	•	•	•				•	•	•		•		
5								•	•	•		•	•	•			•		•	•	•		•		•	•
7			•	•	•	•	•	•	•	•	•	•	•	•	•	•			•	•	•	•	•	•	•	•
8						•		•	•	•		•		•					•	•	•		•			•
9			•	•	•	•	•	•	•	•	•	•	•	•	•	•		•	•	•	•	•	•	•	•	•

	P. 2-1																									
	2	5	8	10	12	14	15	16	17	18	20	27	28	35	42	45	46	47	48	53	54	55	56	60	65	66
10											•		•	•							•		•	•		•
12														•												
13											•		•	•							•			•		•
14													•	•										•		
15											•		•	•							•			•		•
18																										
26																										
43											•		•	•							•			•		•
44																										
50																										
55											•		•	•							•			•		•
56																										

	P. 2-2														P. 2-3											
	67	68	69	70	71	72	74	78	81	82	83	84	85	86	92	95	96	113	115	121	122	123	124	173		
10				•		•	•	•		•		•		•			•	•	•		•		•	•		
12						•						•									•		•	•		
13				•		•	•	•		•		•		•	•		•	•	•		•		•	•		
14										•		•		•	•		•				•		•	•		
15				•		•	•	•		•		•		•	•		•	•	•		•		•	•		
18															•						•		•	•		
26															•								•	•		
43				•		•	•	•		•		•		•	•		•	•	•		•		•	•		
44										•		•		•	•						•		•	•		
50																							•			
55				•		•	•	•		•		•		•	•		•	•	•		•		•	•		
56										•		•		•	•						•		•	•		
72																							•			
74																										
76																							•			
82																										
103															•						•		•	•		
114																										

48E' Group (17): 8(6M)e,1 + +6(8A) [x = E6M, y = E8A, z = 1]
P. 1

P. 1-1	5 $(8\,\overset{-2}{8})$	8 $(1\,4\,\overset{-2}{8})$	10 $(2\,4\,\overset{-2}{8})$	12 $(4\,\overset{\sim}{4}\,\overset{-2}{8})$	14 $(4\,\bar{4}\,\overset{-2}{8})$	15 $(4\,\overset{\sim 2}{8}\,\overset{\sim 2}{8})$	16 $(4\,\overset{\sim 2}{8}\,\overset{-2}{8})$	18 $(4\,\overset{-2}{8}\,\overset{-2}{8})$
1 $(3\,\overset{\sim 2}{6})$	6.953623	6.0323393	6.499544	6.6191142	7.043444	7.5024775	7.4721525	7.584969
2 $(3\,\overset{-2}{6})$	7.0178144	×	6.6736196	6.805401	7.2843998	7.5627586	7.586655	×

5 $(6\,\bar{6}^2)$	×	×	×	×	7.8443996	×	×	×
7 $(1\,3\,\tilde{6}^2)$	7.0220745	×	6.6919391	6.823679	7.302448	7.565492	×	×
8 $(1\,3\,\bar{6}^2)$	×	×	×	×	7.8262155	×	×	×
9 $(3\,\tilde{3}\,\tilde{6}^2)$	7.028445	×	6.7177847	6.850919	7.3357094	×	7.6070622	×
P. 1-2	20 $(\underline{8\,4}\,\bar{8}^2)$	27 $(8\,\bar{8}^2\,\tilde{8}^2)$	28 $(8\,\bar{8}^2\,\bar{8}^2)$	35 $(\underline{16\,8}\,\bar{8}^2)$	42 $(1\,4\,\tilde{4}\,\bar{8}^2)$	45 $(1\,4\,\tilde{8}^2\,\tilde{8}^2)$	46 $(1\,4\,\tilde{8}^2\,\bar{8}^2)$	47 $(1\,4\,\bar{8}^2\,\tilde{8}^2)$
1 $(3\,\tilde{6}^2)$	7.404095	8.9737478	8.5505423	×	6.7027455	7.8928611	7.80072	×
2 $(3\,\bar{6}^2)$	×	9.0244154	8.7803302	×	6.8973624	7.9500525	7.946325	8.2800651
5 $(6\,\bar{6}^2)$	8.4594464	×	9.2859254	11.898363	×	×	8.198549	×
7 $(1\,3\,\tilde{6}^2)$	7.7405421	9.025026	8.7860666	×	6.9155219	7.951988	7.953207	8.2815003
8 $(1\,3\,\bar{6}^2)$	8.4380012	×	9.2745891	11.8910202	×	×	8.194774	×
9 $(3\,\tilde{3}\,\tilde{6}^2)$	7.781827	9.0283141	8.813101	11.386583	6.9436677	7.9565585	7.97097	8.2856432
P. 1-3	48 $(1\,4\,\bar{8}^2\,\bar{8}^2)$	53 $(2\,4\,\tilde{8}^2\,\tilde{8}^2)$	54 $(2\,4\,\tilde{8}^2\,\bar{8}^2)$	55 $(2\,4\,\bar{8}^2\,\tilde{8}^2)$	56 $(2\,4\,\bar{8}^2\,\bar{8}^2)$	60 $(4\,\tilde{4}\,\bar{4}\,\bar{8}^2)$	65 $(4\,\tilde{4}\,\tilde{8}^2\,\tilde{8}^2)$	66 $(4\,\tilde{4}\,\tilde{8}^2\,\bar{8}^2)$
1 $(3\,\tilde{6}^2)$	8.0542694	8.2009635	8.038881	8.8197799	8.455629	7.5204857	8.2696453	8.089684
2 $(3\,\bar{6}^2)$	8.22682	8.2557512	8.207544	8.8717875	8.6744227	×	8.3238493	8.2632517
5 $(6\,\bar{6}^2)$	×	×	8.536373	×	9.1505303	8.8178257	×	8.6076531
7 $(1\,3\,\tilde{6}^2)$	8.233601	8.2571931	8.214264	8.8725371	8.6804528	7.941705	8.3251996	8.2698979
8 $(1\,3\,\bar{6}^2)$	8.5596413	×	8.5296645	×	9.1398005	8.7938171	×	8.600427
9 $(3\,\tilde{3}\,\tilde{6}^2)$	8.2545276	8.2613081	8.2347074	8.8760047	8.706386	7.9915164	8.32921	8.2908921
P. 1-4	67 $(4\,\tilde{4}\,\bar{8}^2\,\tilde{8}^2)$	68 $(4\,\tilde{4}\,\bar{8}^2\,\bar{8}^2)$	69 $(4\,\bar{4}\,\tilde{8}^2\,\tilde{8}^2)$	70 $(4\,\bar{4}\,\tilde{8}^2\,\bar{8}^2)$	71 $(4\,\bar{4}\,\bar{8}^2\,\tilde{8}^2)$	72 $(4\,\bar{4}\,\bar{8}^2\,\bar{8}^2)$	74 $(4\,\tilde{8}^2\,\tilde{4}\,\bar{8}^2)$	78 $(4\,\bar{8}^2\,\tilde{4}\,\bar{8}^2)$
1 $(3\,\tilde{6}^2)$	×	8.5540443	8.4537206	8.2223725	×	8.9003397	×	×
2 $(3\,\bar{6}^2)$	×	8.7859406	8.5061831	8.4084535	×	9.194838	×	×
5 $(6\,\bar{6}^2)$	×	×	×	8.4914272	×	9.860931	8.5525765	8.5593752
7 $(1\,3\,\tilde{6}^2)$	9.0358071	8.7917	8.5072884	8.4148468	9.7263542	9.1994243	7.8019603	7.80737
8 $(1\,3\,\bar{6}^2)$	×	×	×	8.783634	×	9.846788	8.5306935	8.5374942
9 $(3\,\tilde{3}\,\tilde{6}^2)$	9.0391155	8.8189893	8.5110364	8.4371877	9.7291179	9.2330194	7.8449203	7.8504305
P. 1-5	81 $(4\,\tilde{8}^2\,\tilde{8}^2\,\tilde{8}^2)$	82 $(4\,\tilde{8}^2\,\tilde{8}^2\,\bar{8}^2)$	83 $(4\,\tilde{8}^2\,\bar{8}^2\,\tilde{8}^2)$	84 $(4\,\tilde{8}^2\,\bar{8}^2\,\bar{8}^2)$	85 $(4\,\bar{8}^2\,\tilde{8}^2\,\tilde{8}^2)$	86 $(4\,\bar{8}^2\,\tilde{8}^2\,\bar{8}^2)$	92 $(\underline{8\,4}\,\bar{4}\,\bar{8}^2)$	95 $(\underline{8\,4}\,\bar{8}^2\,\tilde{8}^2)$
1 $(3\,\tilde{6}^2)$	9.3621107	8.760572	9.5501848	8.8469557	9.444582	8.8006119	×	×

2 $(3\,{}^{-}6^{2})$	9.4109206	9.02362	9.5985075	9.1283365	9.4930215	9.0712342	×	×
5 $(6\,{}^{-}6^{2})$	×	9.6138758	9.4001536	9.7633538	×	×	9.386826	×
7 $(1\;3\,{}^{\sim}6^{2})$	9.4112449	9.0287014	9.5987267	9.1331219	9.4932991	9.0761724	×	×
8 $(1\;3\,{}^{-}6^{2})$	×	9.600985	×	9.7497049	×	×	×	×
9 $(3\,{}^{\sim}3\,{}^{\sim}6^{2})$	9.4142042	9.0591139	9.601567	9.1654	9.4961991	9.1073422	×	10.511405

P. 1-6	96 $(\underline{8\;4}\,{}^{-}8^{2}\,{}^{-}8^{2})$	113 $(8\,{}^{\sim}8\,{}^{\sim}8\,{}^{-}8^{2})$	115 $(8\,{}^{-}8\,{}^{\sim}8\,{}^{-}8^{2})$	121 $(8\,{}^{-}8^{2}\,{}^{\sim}8^{2}\,{}^{\sim}8^{2})$	122 $(8\,{}^{-}8^{2}\,{}^{\sim}8^{2}\,{}^{-}8^{2})$	123 $(8\,{}^{-}8^{2}\,{}^{-}8^{2}\,{}^{\sim}8^{2})$	124 $(8\,{}^{-}8^{2}\,{}^{-}8^{2}\,{}^{-}8^{2})$	173 $(16\,{}^{-}16^{2}\,{}^{\sim}8\,{}^{-}8^{2})$
1 $(3\,{}^{\sim}6^{2})$	9.123783	9.499025	9.68887	9.982225	9.0268512	11.027725	×	12.089725
2 $(3\,{}^{-}6^{2})$	×	9.671191	9.868118	10.0272	×	11.07078	×	×
5 $(6\,{}^{-}6^{2})$	10.38182	9.995615	10.21447	×	10.075756	×	10.6861	13.07663
7 $(1\;3\,{}^{\sim}6^{2})$	9.52398	9.672249	9.8686203	10.027203	9.356543	11.070457	9.619223	12.39567
8 $(1\;3\,{}^{-}6^{2})$	10.370054	9.990893	10.2093	×	10.0610156	×	×	13.0686
9 $(3\,{}^{\sim}3\,{}^{\sim}6^{2})$	9.567228	9.690775	9.887745	10.02967	9.3929	11.0725	9.675033	12.42587

P. 2-1	20 $(\underline{8\;4}\,{}^{-}8^{2})$	28 $(8\,{}^{-}8^{2}\,{}^{-}8^{2})$	35 $(\underline{16\;8}\,{}^{-}8^{2})$	54 $(2\;4\,{}^{\sim}8^{2}\,{}^{-}8^{2})$	56 $(2\;4\,{}^{-}8^{2}\,{}^{-}8^{2})$	60 $(4\,{}^{\sim}4\,{}^{-}4\,{}^{-}8^{2})$	66 $(4\,{}^{\sim}4\,{}^{\sim}8^{2}\,{}^{-}8^{2})$
10 $(3\,{}^{\sim}3\,{}^{-}6^{2})$	8.6438231	9.3993334	12.024339	8.5799547	9.2526535	9.0456995	8.6580291
12 $(3\,{}^{-}3\,{}^{-}6^{2})$	×	×	12.193377	×	×	×	×
13 $(3\,{}^{\sim}6^{2}\,{}^{\sim}6^{2})$	8.667845	9.409486	12.027478	8.58305	×	9.071241	8.662121
14 $(3\,{}^{\sim}6^{2}\,{}^{-}6^{2})$	×	9.527414	12.169875	×	×	9.408105	×
15 $(3\,{}^{-}6^{2}\,{}^{\sim}6^{2})$	8.697615	9.424814	12.04168	8.585624	×	9.106768	8.666148
43 $(1\;3\,{}^{\sim}6^{2}\,{}^{\sim}6^{2})$	8.698126	9.42491	12.041503	8.58566	×	9.10715	8.666216
55 $(3\,{}^{\sim}3\,{}^{\sim}6^{2}\,{}^{\sim}6^{2})$	8.700154	9.42581	12.0421	8.585782	×	9.109402	8.666433

P. 2-2	70 $(4\,{}^{-}4\,{}^{\sim}8^{2}\,{}^{-}8^{2})$	72 $(4\,{}^{-}4\,{}^{-}8^{2}\,{}^{-}8^{2})$	74 $(4\,{}^{\sim}8^{2}\,{}^{\sim}4\,{}^{-}8^{2})$	78 $(4\,{}^{-}8^{2}\,{}^{\sim}4\,{}^{-}8^{2})$	82 $(4\,{}^{\sim}8^{2}\,{}^{\sim}8^{2}\,{}^{-}8^{2})$	84 $(4\,{}^{\sim}8^{2}\,{}^{-}8^{2}\,{}^{-}8^{2})$	86 $(4\,{}^{-}8^{2}\,{}^{\sim}8^{2}\,{}^{-}8^{2})$
10 $(3\,{}^{\sim}3\,{}^{-}6^{2})$	8.858396	10.0293851	8.7457948	8.752824	9.7571969	9.9216638	9.8292528
12 $(3\,{}^{-}3\,{}^{-}6^{2})$	×	10.35527	×	×	10.0024428	10.2156777	×
13 $(3\,{}^{\sim}6^{2}\,{}^{\sim}6^{2})$	8.864594	10.041714	8.770037	8.777021	9.7687498	9.933721	9.8410144
14 $(3\,{}^{\sim}6^{2}\,{}^{-}6^{2})$	×	×	×	×	9.949085	10.1434074	10.03366
15 $(3\,{}^{-}6^{2}\,{}^{\sim}6^{2})$	8.871975	10.06492	8.80097	8.807949	9.788511	9.9555545	9.861634
43 $(1\;3\,{}^{\sim}6^{2}\,{}^{\sim}6^{2})$	8.87207	10.064917	8.80144	8.808438	9.78856	9.9555717	9.861667
44 $(1\;3\,{}^{\sim}6^{2}\,{}^{-}6^{2})$	×	×	×	×	10.003627	12.01731	10.096536
55 $(3\,{}^{\sim}3\,{}^{\sim}6^{2}\,{}^{\sim}6^{2})$	8.872524	10.066167	8.803508	8.810495	9.789664	9.95677	9.861808
56 $(3\,{}^{\sim}3\,{}^{\sim}6^{2}\,{}^{-}6^{2})$	×	×	×	×	10.00727	10.223405	10.101258

P. 2-3	92	96	113	115	122	124	173

	$(\underline{8\,4}\,{}^{-}4\,{}^{-}\overset{2}{8})$	$(\underline{8\,4}\,{}^{-}\overset{2}{8}\,{}^{-}\overset{2}{8})$	$(8\,{}^{\sim}8\,{}^{\sim}8\,{}^{-}\overset{2}{8})$	$(8\,{}^{-}8\,{}^{\sim}8\,{}^{-}\overset{2}{8})$	$(8\,{}^{-}\overset{2}{8}\,{}^{\sim}\overset{2}{8}\,{}^{-}\overset{2}{8})$	$(8\,{}^{-}\overset{2}{8}\,{}^{-}\overset{2}{8}\,{}^{-}\overset{2}{8})$	$(16\,{}^{-}\overset{2}{16}\,{}^{\sim}8\,{}^{-}\overset{2}{8})$
$10\,(3\,{}^{\sim}3\,{}^{-}\overset{2}{6})$	×	10.613459	10.038116	10.267114	10.260887	10.959885	13.25
$12\,(3\,{}^{-}3\,{}^{-}\overset{2}{6})$	×	×	×	×	10.639072	11.571353	13.573305
$13\,(3\,{}^{\sim}\overset{2}{6}\,{}^{\sim}\overset{2}{6})$	9.8086686	10.626508	10.039722	10.2694	10.27325	10.97326	13.25129
$14\,(3\,{}^{\sim}\overset{2}{6}\,{}^{-}\overset{2}{6})$	10.372719	10.960512	×	×	10.534186	11.383733	13.48768
$15\,(3\,{}^{-}\overset{2}{6}\,{}^{\sim}\overset{2}{6})$	9.863428	10.656915	10.042	10.27346	10.298485	11.00911	13.270545
$18\,(\underline{6\,3}\,{}^{-}\overset{2}{6})$	10.89063	×	×	×	10.718176	11.7446	13.623523
$26\,(6\,{}^{-}\overset{2}{6}\,{}^{-}\overset{2}{6})$	11.24501	×	×	×	×	11.9597	×
$43\,(1\,3\,{}^{\sim}\overset{2}{6}\,{}^{\sim}\overset{2}{6})$	9.863516	10.65677	10.042	10.273456	10.298438	11.008866	13.270265
$44\,(1\,3\,{}^{\sim}\overset{2}{6}\,{}^{-}\overset{2}{6})$	10.640343	×	×	×	10.640931	11.572068	13.57191
$50\,(3\,{}^{\sim}3\,{}^{-}\overset{2}{3}\,{}^{-}\overset{2}{6})$	×	×	×	×	×	12.028836	×
$55\,(3\,{}^{\sim}3\,{}^{\sim}\overset{2}{6}\,{}^{\sim}\overset{2}{6})$	9.866567	10.658271	10.04209	10.273645	10.299746	11.01056	13.270955
$56\,(3\,{}^{\sim}3\,{}^{\sim}\overset{2}{6}\,{}^{-}\overset{2}{6})$	10.668631	×	×	×	10.651056	11.591096	13.578752
$72\,(3\,{}^{\sim}\overset{2}{6}\,{}^{\sim}\overset{2}{6}\,{}^{-}\overset{2}{6})$	×	×	×	×	×	12.031484	×
$76\,(3\,{}^{-}\overset{2}{6}\,{}^{\sim}\overset{2}{6}\,{}^{-}\overset{2}{6})$	×	×	×	×	×	12.037765	×
$103\,(6\,{}^{\sim}6\,{}^{\sim}6\,{}^{-}\overset{2}{6})$	10.96205	×	×	×	10.734194	11.785093	13.62772

48E' Group (18): 12(4M)e + 6(8A) [y = z = E4M, x = E8A]

	5	8	10	12	14	16	18	20	28	35	42	46	48	54	56	60
6					•			•								•
9																
10								•	•		•				•	•
30								•								•
103																
153																

	66	68	70	72	74	78	82	84	86	92	96	113	115	122	124	173
6				•	•	•	•	•	•		•			•		•
9										•	•				•	
10		•	•	•	•	•	•	•	•		•		•	•		•
30				•	•			•		•	•			•	•	•
103																
153																

48E' Group (18): 12(4M)e + 6(8A) [y = z = E4M, x = E8A]

	$14\,(4\,{}^{-}4\,{}^{-}\overset{2}{8})$	$20\,(\underline{8\,4}\,{}^{-}\overset{2}{8})$	$28\,(8\,{}^{-}\overset{2}{8}\,{}^{-}\overset{2}{8})$	$42\,(1\,4\,{}^{\sim}4\,{}^{-}\overset{2}{8})$	$56\,(2\,4\,{}^{-}\overset{2}{8}\,{}^{-}\overset{2}{8})$	$60\,(4\,{}^{\sim}4\,{}^{-}4\,{}^{-}\overset{2}{8})$	$68\,(4\,{}^{\sim}4\,{}^{-}\overset{2}{8}\,{}^{-}\overset{2}{8})$
$6\,(4\,{}^{\sim}4)e$	8.269111	8.759184	×	×	×	8.967428	×
$10\,(1\,4\,{}^{\sim}4)e$	×	8.393035	10.19729	7.913838	10.083031	9.143239	10.212882
$30\,(4\,{}^{\sim}\overset{2}{8}\,{}^{\sim}4)e$	×	10.50931	×	×	×	11.007753	×
	$70\,(4\,{}^{-}4\,{}^{\sim}\overset{2}{8}\,{}^{-}\overset{2}{8})$	$72\,(4\,{}^{-}4\,{}^{-}\overset{2}{8}\,{}^{-}\overset{2}{8})$	$74\,(4\,{}^{\sim}\overset{2}{8}\,{}^{\sim}4\,{}^{-}\overset{2}{8})$	$78\,(4\,{}^{-}\overset{2}{8}\,{}^{\sim}4\,{}^{-}\overset{2}{8})$	$82\,(4\,{}^{\sim}\overset{2}{8}\,{}^{\sim}\overset{2}{8}\,{}^{-}\overset{2}{8})$	$84\,(4\,{}^{\sim}\overset{2}{8}\,{}^{-}\overset{2}{8}\,{}^{-}\overset{2}{8})$	$86\,(4\,{}^{-}\overset{2}{8}\,{}^{\sim}\overset{2}{8}\,{}^{-}\overset{2}{8})$
$6\,(4\,{}^{\sim}4)e$	×	10.553034	8.8214536	8.82547	10.356616	10.477575	10.40944
$10\,(1\,4\,{}^{\sim}4)e$	9.733433	10.690112	8.9750046	8.9791703	10.482645	10.610041	10.537977

30 $(4 \sim \overset{2}{8} \sim 4)$e	×	12.22285	10.623843	×	×	12.08661	×
	92 $(\underline{8\ 4} - 4 - \overset{2}{8})$	96 $(\underline{8\ 4} - \overset{2}{8} - \overset{2}{8})$	115 $(8 - 8 \sim 8 - \overset{2}{8})$	122 $(8 - \overset{2}{8} \sim \overset{2}{8} - \overset{2}{8})$	124 $(8 - \overset{2}{8} - \overset{2}{8} - \overset{2}{8})$	173 $(16 - \overset{2}{16} \sim 8 - \overset{2}{8})$	
6 $(4 \sim 4)$e	×	10.868836	×	10.692871	×	14.361875	
9 $(\underline{8\ 4})$e	11.661952	12.782976	×	×	13.16214	×	
10 $(1\ 4 \sim 4)$e	×	11.039071	11.504225	10.837547	×	14.49785	
30 $(4 \sim \overset{2}{8} \sim 4)$e	11.84825	12.905554	×	12.441933	13.301753	16.048469	

End of 48e' Group (18): 12(4M)e + 6(8A) [y = z = E4M, x = E8A]

48e' Group (19): 12(4M)e + 8(6A) [y = z = E4M, x = E6A]	5	8	10	12	14	16	18	23	26	44	46	50	56	58	62	72	74	76	82	86	103	114
6				•			•		•		•			•		•		•				
9												•			•		•			•		
10			•	•		•	•		•	•	•		•			•		•			•	
30				•			•					•		•	•	•	•	•		•		
103																			•			•
153																						

48e' Group (19): 12(4M)e + 8(6A) [y = z = E4M, x = E6A]

	10 (3 3 6)	12 (3 3 6)	16 (3 6 6)	18 (6 3 6)	26 (6 6 6)	44 (1 3 6 6)	46 (1 3 6 6)	50 (3 3 3 6)	56 (3 3 6 6)
6 (4 4)e	×	8.951982	×	9.0083556	10.834526	×	10.8282465	×	×
9 (8 4)e	×	×	×	×	×	×	×	11.6481796	×
10 (1 4 4)e	8.615705	9.128776	10.447013	9.207023	11.001901	10.445274	10.994808	×	10.475244
30 (4 8 4)e	×	11.02444	×	11.327	×	×	×	11.8361	×

	58 (3 3 6 6)	62 (3 3 6 6)	72 (3 6 6 6)	74 (3 6 6 6)	76 (3 6 6 6)	82 (6 3 3 6)	86 (6 3 6 6)	103 (6 6 6 6)	114 (6 6 6 6)
6 (4 4)e	10.900993	×	10.902685	×	10.906138	×	×	×	×
9 (8 4)e	×	13.517465	×	13.386	×	×	13.590236	×	×
10 (1 4 4)e	×	×	11.089051	×	11.095401	×	×	12.352661	×
30 (4 8 4)e	13.144278	13.69344	13.1468796	13.545537	13.1793274	×	13.786549	×	×
103 (4 4 8 4)e	×	×	×	×	×	14.353736	×	×	16.086357

End of 48E' Group (19): 12(4M)e + 8(6A) [y = z = E4M, x = E6A]

48E' Group (20): 12(4M)e,1 + 6(8A) [y = E4M, x = E8A, z = 1]																
	5	8	10	12	14	16	18	20	28	35	42	46	48	54	56	60
2			•	•	•			•	•	•	•	•	•	•	•	•
4			•	•	•			•	•	•	•				•	•
5								•								•
8								•								•
19																
	66	68	70	72	74	78	82	84	86	92	96	113	115	122	124	173
2	•	•	•	•	•	•	•	•	•		•	•	•	•		•
4		•	•	•	•	•	•	•	•		•		•	•	•	•
5				•	•	•		•	•	•	•			•	•	•
8				•	•	•				•	•			•	•	•
19										•						

48E' Group (20): 12(4M)e,1 + 6(8A) [y = E4M, x = E8A, z = 1]							
	10 (2 4 $\overset{-2}{8}$)	12 (4 $\overset{\sim}{4}$ $\overset{-2}{8}$)	14 (4 $\overset{-}{4}$ $\overset{-2}{8}$)	20 ($\underline{8\ 4}$ $\overset{-2}{8}$)	28 (8 $\overset{-2}{8}$ $\overset{-2}{8}$)	35 ($\underline{16\ 8}$ $\overset{-2}{8}$)	42 (1 4 $\overset{\sim}{4}$ $\overset{-2}{8}$)
2 (1 $\overset{1}{4}$)	6.3862044	6.5338199	7.0550936	7.4964095	8.8754364	11.9613643	6.63711
4 (2 $\overset{\sim 1}{4}$)	6.4092196	6.5748665	7.1717995	7.7194083	8.9729447	12.085651	6.69104
5 (2 $\overset{-1}{4}$)	×	×	×	8.0329985	×	×	×
8 (4 $\overset{-2}{4}$)	×	×	×	8.0723004	×	×	×
	46 (1 4 $\overset{\sim 2}{8}$ $\overset{-2}{8}$)	48 (1 4 $\overset{-2}{8}$ $\overset{-2}{8}$)	54 (2 4 $\overset{\sim 2}{8}$ $\overset{-2}{8}$)	56 (2 4 $\overset{-2}{8}$ $\overset{-2}{8}$)	60 (4 $\overset{\sim}{4}$ $\overset{-}{4}$ $\overset{-2}{8}$)	66 (4 $\overset{\sim}{4}$ $\overset{\sim 2}{8}$ $\overset{-2}{8}$)	68 (4 $\overset{\sim}{4}$ $\overset{-2}{8}$ $\overset{-2}{8}$)
2 (1 $\overset{2}{4}$)	7.966387	8.2773957	8.255465	8.7638468	7.6416838	8.3166261	8.882011
4 (2 $\overset{\sim 1}{4}$)	×	×	×	8.84485	7.9744307	×	8.891346
5 (2 $\overset{-1}{4}$)	×	×	×	×	8.4759098	×	×
8 (4 $\overset{-2}{4}$)	×	×	×	×	8.5750652	×	×
	70 (4 $\overset{-}{4}$ $\overset{\sim 2}{8}$ $\overset{-2}{8}$)	72 (4 $\overset{-}{4}$ $\overset{-2}{8}$ $\overset{-2}{8}$)	74 (4 $\overset{\sim 2}{8}$ $\overset{\sim}{4}$ $\overset{-2}{8}$)	78 (4 $\overset{-2}{8}$ $\overset{\sim}{4}$ $\overset{-2}{8}$)	82 (4 $\overset{\sim 2}{8}$ $\overset{\sim 2}{8}$ $\overset{-2}{8}$)	84 (4 $\overset{\sim 2}{8}$ $\overset{-2}{8}$ $\overset{-2}{8}$)	86 (4 $\overset{-2}{8}$ $\overset{\sim 2}{8}$ $\overset{-2}{8}$)
2 (1 $\overset{2}{4}$)	8.4748445	9.294763	7.5503724	7.5552108	9.126612	9.2307648	9.1742081
4 (2 $\overset{\sim 1}{4}$)	8.5146394	9.4803672	7.7965355	7.8033473	9.269857	9.3986354	9.3278307
5 (2 $\overset{-1}{4}$)	×	9.7219996	8.1515663	8.1612839	×	9.60224	9.5002227
8 (4 $\overset{-2}{4}$)	×	9.731154	8.2050404	8.2157636	×	×	×
	92 ($\underline{8\ 4}$ $\overset{-}{4}$ $\overset{-2}{8}$)	96 ($\underline{8\ 4}$ $\overset{-2}{8}$ $\overset{-2}{8}$)	113 (8 $\overset{\sim}{8}$ $\overset{\sim}{8}$ $\overset{-2}{8}$)	115 (8 $\overset{-}{8}$ $\overset{\sim}{8}$ $\overset{-2}{8}$)	122 (8 $\overset{-2}{8}$ $\overset{\sim 2}{8}$ $\overset{-2}{8}$)	124 (8 $\overset{-2}{8}$ $\overset{-2}{8}$ $\overset{-2}{8}$)	173 (16 $\overset{-2}{16}$ $\overset{\sim}{8}$ $\overset{-2}{8}$)
2 (1 $\overset{2}{4}$)	×	9.545692	10.0247063	10.2495227	9.433829	×	13.1024652
4 (2 $\overset{\sim 1}{4}$)	×	9.8721814	×	10.2788797	9.663527	9.9852232	13.314776
5 (2 $\overset{-1}{4}$)	9.2071186	10.3690687	×	×	9.9897208	10.7353453	13.6056543
8 (4 $\overset{-2}{4}$)	9.5215592	10.4618717	×	×	10.0296854	10.89325	13.6298938
19 (2 $\overset{\sim 2}{4}$ $\overset{-2}{4}$)	10.08264	×	×	×	×	×	×

End of 48E' Group (20): 12(4M)e,1 + 6(8A)
[y = E4M, x = E8A, z = 1]

48E' Group (21): 12(4M)e,1 + 8(6A) [y = E4M, x = E6A, z = 1]

	5	8	10	12	14	16	18	23	26	44	46	50	56	58	62	72	74	76	82	86	103	114
2		•	•	•	•	•		•	•	•	•		•								•	
4			•	•		•	•		•	•	•		•	•		•		•			•	
5							•		•		•	•		•	•	•	•	•		•		
8							•					•		•	•	•	•	•		•		
19																			•	•		•
29																						
69																						
97																						

48E' Group (21): 12(4M)e,1 + 8(6A) [y = E4M, x = E6A, z = 1]

	$8\,(1\ 3\ \overset{-2}{6})$	$10\,(3\ \overset{\sim}{3}\ \overset{-2}{6})$	$12\,(3\ \overset{-}{3}\ \overset{-2}{6})$	$14\,(3\ \overset{\sim 2}{6}\ \overset{-2}{6})$	$16\,(3\ \overset{-2}{6}\ \overset{-2}{6})$	$18\,(\underline{6\ 3}\ \overset{-2}{6})$	$23\,(6\ \overset{\sim}{6}\ \overset{-2}{6})$
$2\,(1\ \overset{1}{4})$	6.7274177	6.976249	7.344163	8.5691982	8.810679	×	9.91567
$4\,(2\ \overset{\sim 1}{4})$	×	7.0418794	7.5788547	×	8.853809	7.6748497	×
$5\,(2\ \overset{-1}{4})$	×	×	7.8749535	×	×	8.1420694	×
$8\,(4\ \overset{-2}{4})$	×	×	×	×	×	8.2127004	×

	$26\,(6\ \overset{-2}{6}\ \overset{-2}{6})$	$44\,(1\ 3\ \overset{\sim 2}{6}\ \overset{-2}{6})$	$46\,(1\ 3\ \overset{-2}{6}\ \overset{-2}{6})$	$50\,(3\ \overset{\sim}{3}\ \overset{-2}{3}\ \overset{-2}{6})$	$56\,(3\ \overset{\sim}{3}\ \overset{\sim 2}{6}\ \overset{-2}{6})$	$58\,(3\ \overset{\sim}{3}\ \overset{-2}{6}\ \overset{-2}{6})$	$62\,(3\ \overset{-}{3}\ \overset{-2}{6}\ \overset{-2}{6})$
$2\,(1\ \overset{1}{4})$	9.229423	8.8163229	9.2239595	×	8.8434816	×	×
$4\,(2\ \overset{\sim 1}{4})$	9.4430937	8.8605495	9.4322963	×	8.8944107	9.5410705	×
$5\,(2\ \overset{-1}{4})$	9.6955247	×	9.6738802	8.5297037	×	9.9415822	10.3892322
$8\,(4\ \overset{-2}{4})$	×	×	×	8.7636104	×	9.9845605	10.59273

	$72\,(3\ \overset{\sim 2}{6}\ \overset{\sim 2}{6}\ \overset{-2}{6})$	$74\,(3\ \overset{\sim 2}{6}\ \overset{-2}{6}\ \overset{-2}{6})$	$76\,(3\ \overset{-2}{6}\ \overset{\sim 2}{6}\ \overset{-2}{6})$	$82\,(\underline{6\ 3}\ \overset{-}{3}\ \overset{-2}{6})$	$86\,(\underline{6\ 3}\ \overset{-2}{6}\ \overset{-2}{6})$	$103\,(6\ \overset{\sim}{6}\ \overset{\sim}{6}\ \overset{-2}{6})$	$114\,(6\ \overset{-2}{6}\ \overset{-2}{6}\ \overset{-2}{6})$
$2\,(1\ \overset{1}{4})$	×	×	×	×	×	10.7264095	×
$4\,(2\ \overset{\sim 1}{4})$	9.54793	×	9.5571237	×	×	10.7697358	×
$5\,(2\ \overset{-1}{4})$	9.9600338	10.281115	9.993587	×	10.4392763	×	×
$8\,(4\ \overset{-2}{4})$	10.00692	10.4248395	10.0483213	×	10.7197337	×	×
$19\,(2\ \overset{\sim 2}{4}\ \overset{-2}{4})$	×	×	×	9.972042	11.147163	×	11.60099

End of 48E' Group (21): 12(4M)e,1 + 8(6A)
[y = E4M, x = E6A, z = 1]

48E' Group (22): 6(8S)e,1 + 8(6S)e,1 [x = 1, y = E8M, z = E6M] ---- Outline

P. 1	1	2	5	7	8	9	10	12	13	14	15	18	26	43	44	50	55	56	72	74	76	82	103	114
1	•		•	•		•			•	•	•		•	•	•		•	•	•	•	•		•	•
2						•								•			•						•	
5	•		•	•		•			•	•	•		•	•	•		•	•	•	•	•		•	•
7	•		•	•		•			•	•	•		•	•	•		•	•	•	•	•		•	•
8						•								•			•						•	
9	•		•	•		•			•	•	•		•	•	•		•	•	•	•	•		•	•
10						•								•			•						•	
11	•		•	•		•			•	•	•		•	•	•		•	•	•	•	•		•	•
12						•								•	•		•						•	
13	•	•	•	•	•	•	•	•	•	•	•	•	•	•	•	•	•	•	•	•	•	•	•	•
14						•					•			•			•						•	
15	•		•	•		•			•	•	•		•	•	•		•	•	•	•	•		•	•
16	•		•	•		•			•	•	•		•	•	•		•	•	•	•	•		•	•
17	•		•	•		•			•	•	•		•	•	•		•	•	•	•	•		•	•
20						•											•						•	
27	•	•	•	•	•	•	•	•	•	•	•	•	•	•	•	•	•	•	•	•	•	•	•	•
28	•		•	•		•			•	•	•		•	•	•		•	•	•	•	•		•	•
41	•	•	•	•	•	•	•	•	•	•	•	•	•	•	•	•	•	•	•	•	•	•	•	•
42						•			•					•			•						•	
45	•		•	•		•			•	•	•		•	•	•		•	•	•	•	•		•	•
46	•		•	•		•			•	•	•		•	•	•		•	•	•	•	•		•	•
47	•		•	•		•			•	•	•		•	•	•		•	•	•	•	•		•	•
53	•		•	•		•			•	•	•		•	•	•		•	•	•	•	•		•	•
54	•		•	•		•			•	•	•		•	•	•		•	•	•	•	•		•	•
55	•	•	•	•	•	•	•	•	•	•	•	•	•	•	•	•	•	•	•	•	•	•	•	•
60						•											•						•	
65	•		•	•		•			•	•	•		•	•	•		•	•	•	•	•		•	•
66	•		•	•		•			•	•	•		•	•	•		•	•	•	•	•		•	•
67	•	•	•	•	•	•	•	•	•	•	•	•	•	•	•	•	•	•	•	•	•	•	•	•
69	•	•	•	•	•	•	•	•	•	•	•	•	•	•	•	•	•	•	•	•	•	•	•	•
70	•		•	•		•			•	•	•		•	•	•		•	•	•	•	•		•	•
P.4	1	2	5	7	8	9	10	12	13	14	15	18	26	43	44	50	55	56	72	74	76	82	103	114
71																								
74						•											•						•	
78						•																		
81		•				•																		
82	•		•	•		•			•	•	•		•	•	•		•	•	•	•	•		•	•
83																								
84	•		•	•		•			•	•	•		•	•	•		•	•	•	•	•		•	•
85																								
86	•		•	•		•			•	•	•		•	•	•		•	•	•	•	•		•	•
92						•					•						•						•	
113	•		•	•		•			•	•	•		•	•	•		•	•	•	•	•		•	•
115	•		•	•		•			•	•	•		•	•	•		•	•	•	•	•		•	•
122	•		•	•		•			•	•	•		•	•	•		•	•	•	•	•		•	•
124	•		•	•		•			•	•	•		•	•	•		•	•	•	•	•		•	•

48E' Group (22): 6(8S)e,1 + 8(6S)e,1 [x = 1, y = E8M, z = E6M] **P. 1**

P. 1-1	1 (3 $\tilde{6}^2$)	5 (6 $\bar{6}^2$)	7 (1 3 $\tilde{6}^2$)	9 (3 $\tilde{3}$ $\tilde{6}^2$)	13 (3 $\tilde{6}^2$ $\tilde{6}^2$)	14 (3 $\tilde{6}^2$ $\bar{6}^2$)	15 (3 $\bar{6}^2$ $\tilde{6}^2$)	26 (6 $\bar{6}^2$ $\bar{6}^2$)	43 (1 3 $\bar{6}^2$ $\tilde{6}^2$)
1 (4 $\tilde{8}^2$)	5.9495441	7.1532938	6.282395	6.333762	7.5024779	8.0143885	7.5627605	9.220445	7.565492
2 (4 $\bar{8}^2$)	×	×	×	6.4555052	×	×	×	×	7.6966
5 (8 $\bar{8}^2$)	7.4469692	8.64197	7.640822	7.806323	9.9737496	9.5053448	9.0244154	10.707073	9.0250267
7 (1 4 $\tilde{8}^2$)	6.3512433	7.549733	6.6780397	6.7262256	7.8928611	8.450458	7.9500525	9.6140409	2.9519876
8 (1 4 $\bar{8}^2$)	×	×	×	7.0424073	×	×	×	×	8.281497

9 (2 4 $\overset{\sim 2}{8}$)	6.6764032	7.8636224	6.9966405	7.0423567	8.2009644	8.7203374	8.25575	9.9185004	8.257194
10 (2 4 $\overset{-2}{8}$)	×	×	×	7.636896	×	×	×	×	8.8725353
11 (4 $\tilde{4}$ $\overset{\sim 2}{8}$)	6.7517805	7.9340441	7.0698163	7.1149113	8.2696453	8.7886682	8.3238501	9.9844775	8.3251987
12 (4 $\tilde{4}$ $\overset{-2}{8}$)	×	×	×	7.8028577	×	×	×	×	9.0358062
P. 1-2	44 (1 3 $\overset{\sim 2}{6}$ $\overset{-2}{6}$)	55 (3 $\tilde{3}$ $\overset{\sim 2}{6}$ $\overset{\sim 2}{6}$)	56 (3 $\tilde{3}$ $\overset{\sim 2}{6}$ $\overset{-2}{6}$)	72 (3 $\overset{\sim 2}{6}$ $\overset{\sim 2}{6}$ $\overset{-2}{6}$)	74 (3 $\overset{\sim 2}{6}$ $\overset{-2}{6}$ $\overset{-2}{6}$)	76 (3 $\overset{-2}{6}$ $\overset{\sim 2}{6}$ $\overset{-2}{6}$)	103 (6 $\tilde{6}$ $\tilde{6}$ $\overset{-2}{6}$)	114 (6 $\overset{-2}{6}$ $\overset{-2}{6}$ $\overset{-2}{6}$)	
1 (4 $\overset{\sim 2}{8}$)	8.3317504	7.5707372	8.3722587	9.550185	10.0909647	9.5985093	8.8111682	11.2980475	
2 (4 $\overset{-2}{8}$)	×	7.6996921	×	×	×	×	8.9444747	×	
5 (8 $\overset{-2}{8}$)	9.8155208	9.0283113	9.8519493	11.0277248	11.5785375	11.0707755	10.2584758	12.783173	
7 (1 4 $\overset{\sim 2}{8}$)	9.7252106	7.9565585	8.764381	9.9407505	10.4842792	9.987432	9.1928	11.6899992	
8 (1 4 $\overset{-2}{8}$)	×	8.2856431	×	×	×	×	9.5298209	×	
9 (2 4 $\overset{\sim 2}{8}$)	9.0321869	8.2613081	9.070313	10.2423091	10.7865758	10.2877403	9.49015	11.9892993	
10 (2 4 $\overset{-2}{8}$)	×	8.8760047	×	×	×	×	10.117349	×	
11 (4 $\tilde{4}$ $\overset{\sim 2}{8}$)	9.0995507	8.3292107	9.1734211	10.3074327	10.8514561	10.35257	9.555067	12.0529142	
12 (4 $\tilde{4}$ $\overset{-2}{8}$)	×	9.0391192	×	×	×	×	10.2785239	×	

(continue) **48E' Group (22): 6(8S)e,1 + 8(6S)e,1 [x = 1, y = E8M, z = E6M]** **P. 2**

P. 2-1	1 (3 $\overset{\sim 2}{6}$)	2 (3 $\overset{-2}{6}$)	5 (6 $\overset{-2}{6}$)	7 (1 3 $\overset{\sim 2}{6}$)	8 (1 3 $\overset{-2}{6}$)	9 (3 $\tilde{3}$ $\overset{\sim 2}{6}$)	10 (3 $\tilde{3}$ $\overset{-2}{6}$)	12 (3 $\bar{3}$ $\overset{-2}{6}$)
13 (4 $\bar{4}$ $\overset{\sim 2}{8}$)	6.9624474	7.2490256	8.1243187	7.2725366	8.0906616	7.3157384	8.4095744	9.2599568
14 (4 $\bar{4}$ $\overset{-2}{8}$)	×	×	×	×	×	8.493169	×	×
15 (4 $\overset{\sim 2}{8}$ $\overset{\sim 2}{8}$)	7.8388308	×	9.03349	8.1536966	×	8.194412	×	×
16 (4 $\overset{\sim 2}{8}$ $\overset{-2}{8}$)	8.014388	×	9.2204433	8.3317504	×	8.3722587	×	×
17 (4 $\overset{-2}{8}$ $\overset{\sim 2}{8}$)	7.921772	×	9.1165471	8.2362361	×	8.276665	×	×
20 (<u>8 4</u> $\overset{-2}{8}$)	×	×	×	×	×	9.28725	×	×
27 (8 $\overset{\sim 2}{8}$ $\overset{\sim 2}{8}$)	8.5175119	8.8083128	9.66736	8.8159023	9.6439567	8.8529497	9.9581914	10.8038325
28 (8 $\overset{\sim 2}{8}$ $\overset{-2}{8}$)	9.5053373	×	10.7070756	9.8155208	×	9.85195	×	×
41 (1 4 $\tilde{4}$ $\overset{\sim 2}{8}$)	6.801481	7.0913399	7.978454	7.1174791	7.943157	7.1620347	8.2655321	9.1228195
42 (1 4 $\tilde{4}$ $\overset{-2}{8}$)	×	×	×	×	×	7.919282	×	×
45 (1 4 $\overset{\sim 2}{8}$ $\overset{\sim 2}{8}$)	8.073203	×	9.2586226	8.3840585	×	8.4236512	×	×
P. 2-2	13 (3 $\overset{\sim 2}{6}$ $\overset{\sim 2}{6}$)	14 (3 $\overset{\sim 2}{6}$ $\overset{-2}{6}$)	15 (3 $\overset{-2}{6}$ $\overset{\sim 2}{6}$)	18 (<u>6 3</u> $\overset{-2}{6}$)	26 (6 $\overset{-2}{6}$ $\overset{-2}{6}$)	43 (1 3 $\overset{\sim 2}{6}$ $\overset{\sim 2}{6}$)	44 (1 3 $\overset{\sim 2}{6}$ $\overset{-2}{6}$)	50 (3 $\tilde{3}$ $\overset{-2}{3}$ $\overset{-2}{6}$)
13 (4 $\bar{4}$ $\overset{\sim 2}{8}$)	8.4537206	8.96936	8.5061832	9.594235	10.1546474	8.5072884	9.2764698	10.4450717
14 (4 $\bar{4}$ $\overset{-2}{8}$)	×	×	9.7262101 $\hat{\phi}$ (red)=179.9997	×	×	9.726355	×	×
15 (4 $\overset{\sim 2}{8}$ $\overset{\sim 2}{8}$)	9.3621107	9.8975425	9.41091775	×	11.0985065	9.411245	10.6083642	×
16 (4 $\overset{\sim 2}{8}$ $\overset{-2}{8}$)	9.5501866	10.0909628	9.5985093	×	11.2980475	9.5987258	10.4010578	×
17 (4 $\overset{-2}{8}$ $\overset{\sim 2}{8}$)	9.4445835	9.9808107	9.4930215	×	11.1816659	9.4933	10.2894062	×
27 (8 $\overset{\sim 2}{8}$ $\overset{\sim 2}{8}$)	9.9822259	10.508943	10.0271992	11.13074	11.6844105	10.027204	10.8089805	11.9794994

28 $(8\ {}^{-2}8\ {}^{-2}8)$	11.0277244	11.5785393	11.0717755	×	12.783173	11.070457	11.8847719	×
41 $(1\ 4\ {}^{\sim}4\ {}^{\sim 2}8)$	8.3125463	8.83077	8.3663345	9.4600062	10.02416	8.3676209	9.1407638	10.3154502
42 $(1\ 4\ {}^{\sim}4\ {}^{-2}8)$	×	×	×	×	×	9.147429	×	×
45 $(1\ 4\ {}^{\sim 2}8\ {}^{\sim 2}8)$	9.583744	10.1184907	9.631327	×	11.3144565	9.63152	10.425311	×
P. 2-3	55 $(3\ {}^{\sim}3\ {}^{\sim 2}6\ {}^{\sim 2}6)$	56 $(3\ {}^{\sim}3\ {}^{\sim 2}6\ {}^{-2}6)$	72 $(3\ {}^{\sim 2}6\ {}^{\sim 2}6\ {}^{-2}6)$	74 $(3\ {}^{\sim 2}6\ {}^{-2}6\ {}^{-2}6)$	76 $(3\ {}^{-2}6\ {}^{\sim 2}6\ {}^{-2}6)$	82 $(\underline{6\ 3}\ {}^{-}3\ {}^{-2}6)$	103 $(6\ {}^{\sim}6\ {}^{\sim}6\ {}^{-2}6)$	114 $(6\ {}^{-2}6\ {}^{-2}6\ {}^{-2}6)$
13 $(4\ {}^{-}4\ {}^{\sim 2}8)$	8.5110363	9.3135504	10.47446	11.0163121	10.518744	11.64243	9.7241245	12.2117676
14 $(4\ {}^{-}4\ {}^{-2}8)$	9.7291179	×	×	×	×	×	10.9678845	×
15 $(4\ {}^{\sim 2}8\ {}^{\sim 2}8)$	9.4142042	10.2420363	11.4173131	11.9700904	11.4592739	×	10.6422058	13.1740965
16 $(4\ {}^{\sim 2}8\ {}^{-2}8)$	9.6015658	10.43656	11.6173807	12.1733183	11.659	×	10.8359436	13.3809295
17 $(4\ {}^{-2}8\ {}^{\sim 2}8)$	9.4961991	10.3249058	11.5001053	12.0532823	11.5418562	×	10.723825	13.2571693
20 $(\underline{8\ 4}\ {}^{-2}8)$	10.5114055	×	×	×	×	×	11.7414994	×
27 $(8\ {}^{-2}8\ {}^{\sim 2}8)$	10.029673	10.8428444	11.9954992	12.5425096	12.0353985	13.167756	11.2265439	13.7308697
28 $(8\ {}^{-2}8\ {}^{-2}8)$	11.0724902	11.9181161	13.0972123	13.6589198	13.1357035	×	12.2994904	14.865006
41 $(1\ 4\ {}^{\sim}4\ {}^{\sim 2}8)$	8.371571	9.1784413	10.3463941	10.8899618	10.3913369	11.518401	9.5944362	12.0901094
42 $(1\ 4\ {}^{\sim}4\ {}^{-2}8)$	9.1506258	×	×	×	×	×	10.3867097	×
45 $(1\ 4\ {}^{\sim 2}8\ {}^{\sim 2}8)$	9.6543027	10.4604191	11.631349	12.1835943	11.6726422	×	10.8558164	13.3846268

(continue) **48E' Group (22): 6(8S)e,1 + 8(6S)e,1 [x = 1, y = E8M, z = E6M]**								**P. 3**
P. 3-1	1 $(3\ {}^{\sim 2}6)$	2 $(3\ {}^{-2}6)$	5 $(6\ {}^{-2}6)$	7 $(1\ 3\ {}^{\sim 2}6)$	8 $(1\ 3\ {}^{-2}6)$	9 $(3\ {}^{\sim}3\ {}^{\sim 2}6)$	10 $(3\ {}^{\sim}3\ {}^{-2}6)$	12 $(3\ {}^{-}3\ {}^{-2}6)$
46 $(1\ 4\ {}^{\sim 2}8\ {}^{-2}8)$	8.41045	×	9.614044	8.72521	×	8.764385	×	×
47 $(1\ 4\ {}^{-2}8\ {}^{\sim 2}8)$	8.2447191	×	9.3261536	8.553656	×	8.592565	×	×
53 $(2\ 4\ {}^{\sim 2}8\ {}^{\sim 2}8)$	8.2332443	×	9.4084658	8.5402732	×	8.579	×	×
54 $(2\ 4\ {}^{\sim 2}8\ {}^{-2}8)$	8.7203374	×	9.918502	9.0321851	×	9.070313	×	×
55 $(2\ 4\ {}^{-2}8\ {}^{\sim 2}8)$	8.4756294	8.769285	9.6354985	8.7772402	9.611779	8.814749	9.9281126	10.778286
60 $(4\ {}^{\sim}4\ {}^{-}4\ {}^{-2}8)$	×	×	×	×	×	9.85889	×	×
65 $(4\ {}^{\sim}4\ {}^{\sim 2}8\ {}^{\sim 2}8)$	8.2658047	×	9.4380537	8.5717835	×	8.610301	×	×
66 $(4\ {}^{\sim}4\ {}^{\sim 2}8\ {}^{-2}8)$	8.7886682	×	9.9844775	9.0995507	×	9.1374211	×	×
67 $(4\ {}^{\sim}4\ {}^{-2}8\ {}^{\sim 2}8)$	8.51882	8.81021	9.670282	8.817802	9.64688	8.854908	9.961309	10.80721
69 $(4\ {}^{-}4\ {}^{\sim 2}8\ {}^{\sim 2}8)$	8.34743	8.641051	9.50975	8.6500288	9.485326	8.6879687	9.802635	10.65507
70 $(4\ {}^{-}4\ {}^{\sim 2}8\ {}^{-2}8)$	8.9693617	×	10.1546474	9.2764706	×	9.3135504	×	×
P. 3-2	13 $(3\ {}^{\sim 2}6\ {}^{\sim 2}6)$	14 $(3\ {}^{\sim 2}6\ {}^{-2}6)$	15 $(3\ {}^{-2}6\ {}^{\sim 2}6)$	18 $(\underline{6\ 3}\ {}^{-2}6)$	26 $(6\ {}^{-2}6\ {}^{-2}6)$	43 $(1\ 3\ {}^{\sim 2}6\ {}^{\sim 2}6)$	44 $(1\ 3\ {}^{\sim 2}6\ {}^{-2}6)$	50 $(3\ {}^{\sim}3\ {}^{-2}3\ {}^{-2}6)$
46 $(1\ 4\ {}^{\sim 2}8\ {}^{-2}8)$	9.94075	10.48428	9.98743	×	11.69	9.98745	10.79302	×
47 $(1\ 4\ {}^{-2}8\ {}^{\sim 2}8)$	9.7492471	10.2840447	9.7960239	×	11.4772744	9.7961325	10.5896822	×

53 (2 4 8 8)	9.7304721	10.2634243	9.7771611	×	11.453917	9.7772822	10.5682998	×
54 (2 4 8 8)	10.24231	10.7865757	10.2877402	×	11.9892993	10.287644	11.0937952	×
55 (2 4 8 8)	9.9525942	10.4820409	9.99679415	11.1067062	11.662537	9.9979605	10.783605	11.958569
65 (4 4 8 8)	9.7592616	10.291572	9.805755	×	11.480521	9.8058591	10.5959334	×
66 (4 4 8 8)	10.3074327	10.8514553	10.3525692	×	12.0529142	10.3524557	11.1581654	×
67 (4 4 8 8)	9.98532	10.512251	10.03031	11.13412	11.687749	10.03031	10.81233	11.98282
69 (4 4 8 8)	9.8284698	10.358417	9.874422	10.984737	11.5422873	9.874494	11.6611437	11.8386741
70 (4 4 8 8)	10.4744602	11.0163115	10.5187468	×	12.2117676	10.5185751	11.3209873	×

P. 3-3	55 (3 3 6 6)	56 (3 3 6 6)	72 (3 6 6 6)	74 (3 6 6 6)	76 (3 6 6 6)	82 (6 3 3 6)	103 (6 6 6 6)	114 (6 6 6 6)
46 (1 4 8 8)	9.990031	10.82784	12.00764	12.56511	12.04831	×	11.221569	13.771896
47 (1 4 8 8)	9.7987962	10.6244379	11.792933	12.3450293	11.833756	×	11.0167194	13.5441843
53 (2 4 8 8)	9.779945	10.6029964	11.7691149	12.3202528	11.809936	×	10.9946704	13.7580154
54 (2 4 8 8)	10.2900376	11.1280752	12.3053003	12.8631138	12.345244	×	11.5171904	14.0680957
55 (2 4 8 8)	10.0004685	10.8176598	11.9747577	12.5232997	12.014835	13.15014	11.203315	13.7145037
60 (4 4 4 8)	11.0889	×	×	×	×	×	12.32259	×
65 (4 4 8 8)	9.8085008	10.6305432	11.7953047	12.3460602	11.83602	×	11.0213457	13.5429253
66 (4 4 8 8)	10.354813	11.192326	12.3684077	12.9260498	12.4081846	×	11.5802798	14.1302496
67 (4 4 8 8)	10.032777	10.846185	11.99878	12.5457	12.038675	13.17082	11.229864	13.733755
69 (4 4 8 8)	9.8770757	10.6954856	11.8557725	12.4051548	11.8962197	13.0333	11.0837826	13.599114
70 (4 4 8 8)	10.5208335	11.3547397	12.5254836	13.0817255	12.5647874	×	11.7389502	14.2822195

(continue) **48E' Group (22): 6(8S)e,1 + 8(6S)e,1 [x = 1, y = E8M, z = E6M]** **P. 4**

P. 4-1	1 (3 6)	2 (3 6)	5 (6 6)	7 (1 3 6)	9 (3 3 6)	13 (3 6 6)	14 (3 6 6)	15 (3 6 6)	26 (6 6 6)
74 (4 8 4 8)	×	×	×	×	9.4182686	×	×	×	×
78 (4 8 4 8) Rc3< 9.805	×	×	×	×	9.4255872	×	×	×	×
81 (4 8 8 8) [Rc<8.91739]	8.57145	8.8555722	×	8.8626476	8.8987488	×	×	×	×
82 (4 8 8 8)	9.8975425	×	11.0985094	10.2063042	10.242035	11.417313	11.9700904	11.4592786	×
84 (4 8 8 8)	10.0909628	×	11.2980466	10.4010597	10.4365573	11.6173806	12.17332	11.659	13.3809247
86 (4 8 8 8)	9.9808106	×	11.181666	10.2894062	10.324905	11.50011	12.053281	11.5418497	13.2571693
92 (8 4 4 8)	×	×	×	×	11.3458294	×	×	12.567037	×
113 (8 8 8 8)	9.6804175	×	10.850185	9.9807992	10.015865	11.1633648	11.7043806	11.205162	12.8902955
115 (8 8 8 8)	9.840649	×	11.001878	10.138154	10.17266	11.31253	11.85144	11.3537	13.03197
122 (8 8 8 8)	10.508945	×	11.684405	10.8089814	10.8428444	11.9955105	12.5425	12.035397	13.730863
124 (8 8 8 8)	11.5785393	×	12.783173	11.8847794	11.9181161	13.097217	13.6588892	13.1356987	14.864989

P. 4-2	$43\,(1\ 3\ \overset{\sim 2}{6}\ \overset{\sim 2}{6})$	$44\,(1\ 3\ \overset{\sim 2}{6}\ \overset{-2}{6})$	$55\,(3\ \overset{\sim}{3}\ \overset{\sim 2}{6}\ \overset{\sim 2}{6})$	$56\,(3\ \overset{\sim}{3}\ \overset{\sim 2}{6}\ \overset{-2}{6})$	$72\,(3\ \overset{\sim 2}{6}\ \overset{\sim 2}{6}\ \overset{-2}{6})$	$74\,(3\ \overset{\sim 2}{6}\ \overset{-2}{6}\ \overset{-2}{6})$	$76\,(3\ \overset{-2}{6}\ \overset{\sim 2}{6}\ \overset{-2}{6})$	$103\,(6\ \overset{\sim}{6}\ \overset{\sim}{6}\ \overset{-2}{6})$	$114\,(6\ \overset{-2}{6}\ \overset{-2}{6}\ \overset{-2}{6})$
$74\,(4\ \overset{\sim 2}{8}\ \overset{\sim}{4}\ \overset{-2}{8})$	×	×	10.6356168	×	×	×	×	11.8606706	×
$82\,(4\ \overset{\sim 2}{8}\ \overset{\sim 2}{8}\ \overset{-2}{8})$	11.4588905	12.2755246	11.460771	12.3084225	13.4869934	14.0498043	13.5247942	12.6860308	15.255442
$84\,(4\ \overset{\sim 2}{8}\ \overset{-2}{8}\ \overset{-2}{8})$	11.6585794	12.4794393	11.6604075	12.512237	13.694111	14.258921	13.7316766	12.8892437	15.466844
$86\,(4\ \overset{-2}{8}\ \overset{\sim 2}{8}\ \overset{-2}{8})$	11.5414577	12.3585563	11.5433077	12.3913627	13.5698257	14.1328855	13.6974968	12.7682166	15.33839
$92\,(\underline{8\ 4}\ \bar{4}\ \overset{-2}{8})$	×	×	12.5680603	×	×	×	×	13.7950645	×
$113\,(8\ \overset{\sim}{8}\ \overset{\sim}{8}\ \overset{-2}{8})$	11.204826	12.0051303	11.2067656	12.0377803	13.1999918	13.7554192	13.2378173	12.41211	14.9497075
$115\,(8\ \bar{8}\ \overset{\sim}{8}\ \overset{-2}{8})$	11.35334	12.15046	11.355209	12.182799	13.34103	13.894121	13.37759	12.554117	15.084878
$122\,(8\ \overset{-2}{8}\ \overset{\sim 2}{8}\ \overset{-2}{8})$	12.0349393	12.84275	12.0366054	12.874655	14.0390583	14.597978	14.075645	13.2427736	15.7934089
$124\,(8\ \overset{-2}{8}\ \overset{-2}{8}\ \overset{-2}{8})$	13.13514	13.962734	13.1365666	13.9942555	15.1748394	15.4732522	15.2104086	14.363005	16.9501677

End of 48E' Group (22): 6(8S)e,1 + 8(6S)e,1
[x = 1, y = E8M, z = E6M] P. 4

48E' Group (23): 12(4S)e,1 + 8(6S)e,1 [x = E6M, y = E4M, z = 1]

	1	5	7	9	13	14	15	26	43	44	55	56	72	74	76	103	114
2				•					•		•					•	
4	•	•	•	•	•	•	•	•	•	•	•	•	•	•	•	•	•
5				•							•					•	
8	•	•	•	•	•	•	•	•	•	•	•	•	•	•	•	•	•
19	•	•	•	•	•	•	•	•	•	•	•	•	•	•	•	•	•
29	•	•	•	•	•	•	•	•	•	•	•	•	•	•	•	•	•
69	•	•	•	•	•	•	•	•	•	•	•	•	•	•	•	•	•
97	•	•	•	•	•	•	•	•	•	•	•	•	•	•	•	•	•

48E' Group (23): 12(4S)e,1 + 8(6S)e,1 [x = E6M, y = E4M, z = 1]

	$1(3\,\overset{\sim 2}{6})$	$5(6\,\overset{-2}{6})$	$7(1\,3\,\overset{\sim 2}{6})$	$9(3\,\overset{\sim}{3}\,\overset{\sim 2}{6})$	$13(3\,\overset{\sim 2}{6}\,\overset{\sim 2}{6})$	$14(3\,\overset{\sim 2}{6}\,\overset{-2}{6})$	$15(3\,\overset{-2}{6}\,\overset{\sim 2}{6})$	$26(6\,\overset{-2}{6}\,\overset{-2}{6})$	$43(1\,3\,\overset{\sim 2}{6}\,\overset{\sim 2}{6})$
$2(1\,\overset{1}{4})$	×	×	×	76.44414	×	×	×	×	9.5418249
$4(2\,\overset{\sim 1}{4})$	7.524367	9.3953672	8.019438	8.0836998	9.9032206	10.7478632	9.9755482	12.6215702	9.9755912
$5(2\,\overset{-1}{4})$	×	×	×	9.2274537	×	×	×	×	×
$8(4\,\overset{-2}{4})$	9.1338505	11.0065742	9.6185631	9.6758275	11.5026135	12.3646493	11.5673995	14.238995	11.56677
$19(2\,\overset{\sim 2}{4}\,\overset{-2}{4})$	10.7478604	12.6215768	11.226315	11.2793536	13.1097265	13.9832807	13.1693758	15.8580184	13.1685234
$29(4\,\overset{-2}{4}\,\overset{-2}{4})$	12.3646474	14.238995	12.8390402	12.88934	14.7216467	15.6031546	14.7776495	17.4781485	14.77675
$69(2\,\overset{\sim 2}{4}\,\overset{-2}{4}\,\overset{-2}{4})$	13.983277	15.8580193	14.45488	14.503305	16.3366862	17.2239904	16.3900195	19.0991488	16.3891173
$97(4\,\overset{-2}{4}\,\overset{-2}{4}\,\overset{-2}{4})$	15.6031413	17.4781485	16.072753	16.1198282	17.9538394	18.8453894	18.005237	20.7208147	18.0044088

	$44(1\,3\,\overset{\sim 2}{6}\,\overset{-2}{6})$	$55(3\,\overset{\sim}{3}\,\overset{\sim 2}{6}\,\overset{\sim 2}{6})$	$56(3\,\overset{\sim}{3}\,\overset{\sim 2}{6}\,\overset{-2}{6})$	$72(3\,\overset{\sim 2}{6}\,\overset{\sim 2}{6}\,\overset{-2}{6})$	$74(3\,\overset{\sim 2}{6}\,\overset{-2}{6}\,\overset{-2}{6})$	$76(3\,\overset{-2}{6}\,\overset{\sim 2}{6}\,\overset{-2}{6})$	$103(6\,\overset{\sim}{6}\,\overset{\sim}{6}\,\overset{-2}{6})$	$114(6\,\overset{-2}{6}\,\overset{-2}{6}\,\overset{-2}{6})$	
$2(1\,\overset{1}{4})$	×	9.5462738	×	×	×	×	11.4479766	×	
$4(2\,\overset{\sim 1}{4})$	11.226315	9.9795795	11.2793536	13.1097265	13.9832844	13.169372	11.8781438	15.8580184	
$5(2\,\overset{-1}{4})$	×	11.1283069	×	×	×	×	13.0297398	×	
$8(4\,\overset{-2}{4})$	12.8390402	11.5696235	12.8893352	14.7216362	15.6031404	14.7776446	13.4667469	17.4781485	
$19(2\,\overset{\sim 2}{4}\,\overset{-2}{4})$	14.45488	13.1706892	14.5033098	16.3366442	17.2239904	16.3900195	15.065679	19.0991488	
$29(4\,\overset{-2}{4}\,\overset{-2}{4})$	16.072753	14.7784734	16.1198282	17.9538394	18.8454313	18.0052367	16.671296	20.7208013	
$69(2\,\overset{\sim 2}{4}\,\overset{-2}{4}\,\overset{-2}{4})$	17.6921014	16.3905669	17.7382117	19.5725489	20.4673643	19.6223784	18.281394	22.342902	
$97(4\,\overset{-2}{4}\,\overset{-2}{4}\,\overset{-2}{4})$	19.3124608	18.0056028	19.3578138	21.1923934	22.0897684	21.2410517	19.89473	23.9653654	

End of 48E' Group (23): 12(4S)e,1 + 8(6S)e,1 [x = E6M, y = E4M, z = 1]

48E' Group (24): 12(4S)e,1 + 6(8S)e,1 [x = E8M, y = 1, z = E4M]

	1	5	7	9	11	13	15	16	17	27	28	41	45	46	47	53
2						•				•		•				
4	•	•	•	•	•	•	•	•	•	•	•	•	•	•	•	•
5						•				•		•				
8	•	•	•	•	•	•	•	•	•	•	•	•	•	•	•	•
9	•	•	•	•	•	•	•	•	•	•	•	•	•	•	•	•
19	•	•	•	•	•	•	•	•	•	•	•	•	•	•	•	•
29	•	•	•	•	•	•	•	•	•	•	•	•	•	•	•	•
69	•	•	•	•	•	•	•	•	•	•	•	•	•	•	•	•
97	•	•	•	•	•	•	•	•	•	•	•	•	•	•	•	•

	54	55	65	66	67	69	70	82	84	86	113	115	122	124		
2		•			•	•										
4	•	•	•	•	•	•	•	•	•	•	•	•	•	•		
5					•	•										
8	•	•	•	•	•	•	•	•	•	•	•	•	•	•		
9	•	•	•	•	•	•	•	•	•	•	•	•	•	•		
19	•	•	•	•	•	•	•	•	•	•	•	•	•	•		
29	•	•	•	•	•	•	•	•	•	•	•	•	•	•		
69	•	•	•	•	•	•	•	•	•	•	•	•	•	•		
97	•	•	•	•	•	•	•	•	•	•	•	•	•	•		

48E' Group (24): 12(4S)e,1 + 6(8S)e,1 [x = E8M, y = 1, z = E4M]

	1 (4 $\overset{\sim 2}{8}$)	5 (8 $\overset{-2}{8}$)	7 (1 4 $\overset{\sim 2}{8}$)	9 (2 4 $\overset{\sim 2}{8}$)	11 (4 $\overset{\sim}{4}$ $\overset{\sim 2}{8}$)	13 (4 $\overset{-}{4}$ $\overset{\sim 2}{8}$)	15 (4 $\overset{\sim 2}{8}$ $\overset{\sim 2}{8}$)	16 (4 $\overset{\sim 2}{8}$ $\overset{-2}{8}$)
2 (1 $\overset{2}{4}$)	×	×	×	×	×	7.3345887	×	×
4 (2 $\overset{\sim 2}{4}$)	6.470197	8.2859644	6.955871	7.3430702	7.4307616	7.6700665	8.7621933	8.9872098
5 (2 $\overset{-2}{4}$)	×	×	×	×	×	8.5422835	×	×
8 (4 $\overset{-2}{4}$)	7.7262843	9.538672	8.2083173	8.5857254	8.669005	8.889229	10.015079	10.2511906
19 (2 $\overset{\sim 2}{4}$ $\overset{-2}{4}$)	8.98721	10.797212	9.4668803	9.8378758	9.918253	10.125417	11.2733493	11.5172095
29 (4 $\overset{-2}{4}$ $\overset{\sim 2}{4}$)	10.251187	12.059448	10.72924	11.095815	11.174209	11.372131	12.535209	12.784663
69 (2 $\overset{\sim 2}{4}$ $\overset{-2}{4}$ $\overset{-2}{4}$)	11.5172095	13.3241763	11.99412	12.3575234	12.4344925	12.625687	13.799526	14.053166
97 (4 $\overset{-2}{4}$ $\overset{-2}{4}$ $\overset{-2}{4}$)	12.7846647	14.590635	13.2607055	13.621795	13.69772	13.8838553	15.065638	15.322459

	17 (4 $\overset{-2}{8}$ $\overset{\sim 2}{8}$)	27 (8 $\overset{-2}{8}$ $\overset{\sim 2}{8}$)	28 (8 $\overset{-2}{8}$ $\overset{-2}{8}$)	41 (1 4 $\overset{\sim}{4}$ $\overset{\sim 2}{8}$)	45 (1 4 $\overset{\sim 2}{8}$ $\overset{\sim 2}{8}$)	46 (1 4 $\overset{\sim 2}{8}$ $\overset{-2}{8}$)	47 (1 4 $\overset{-2}{8}$ $\overset{\sim 2}{8}$)	53 (2 4 $\overset{\sim 2}{8}$ $\overset{\sim 2}{8}$)
2 (1 $\overset{2}{4}$)	×	9.208615	×	7.145704	×	×	×	×
4 (2 $\overset{\sim 2}{4}$)	8.8631645	9.537989	10.797212	7.48693	9.0376696	9.466889	9.2417407	9.221226
5 (2 $\overset{-2}{4}$)	×	10.411676	×	8.3691507	×	×	×	×
8 (4 $\overset{-2}{4}$)	10.116223	10.750666	12.059448	8.720407	10.28254	10.72924	10.48286	10.457197
19 (2 $\overset{\sim 2}{4}$ $\overset{-2}{4}$)	11.3744894	11.9796796	13.3241764	9.9666104	11.535129	11.9941	11.7324404	11.703521
29 (4 $\overset{-2}{4}$ $\overset{\sim 2}{4}$)	12.6362873	13.619724	14.590635	11.220509	12.792845	13.260724	12.9877267	12.956655
69 (2 $\overset{\sim 2}{4}$ $\overset{-2}{4}$ $\overset{-2}{4}$)	13.90052	14.467426	15.858355	12.4793505	14.0540556	14.528579	14.247029	14.2144455
97 (4 $\overset{-2}{4}$ $\overset{-2}{4}$ $\overset{-2}{4}$)	15.1665397	15.7206234	17.126998	13.7414984	15.3177294	15.796396	15.50917	15.475511

	54 (2 4 $\overset{\sim 2}{8}$ $\overset{-2}{8}$)	55 (2 4 $\overset{-2}{8}$ $\overset{\sim 2}{8}$)	65 (4 $\overset{\sim}{4}$ $\overset{\sim 2}{8}$ $\overset{\sim 2}{8}$)	66 (4 $\overset{\sim}{4}$ $\overset{\sim 2}{8}$ $\overset{-2}{8}$)	67 (4 $\overset{\sim}{4}$ $\overset{-2}{8}$ $\overset{\sim 2}{8}$)	69 (4 $\overset{-}{4}$ $\overset{\sim 2}{8}$ $\overset{\sim 2}{8}$)	70 (4 $\overset{-}{4}$ $\overset{\sim 2}{8}$ $\overset{-2}{8}$)	82 (4 $\overset{\sim 2}{8}$ $\overset{-2}{8}$ $\overset{-2}{8}$)
2 (1 $\overset{2}{4}$)	×	9.165876	×	×	9.211615	9.012214	×	×
4 (2 $\overset{\sim 2}{4}$)	9.8378758	9.4983573	9.257547	9.9182525	9.541393	9.3458056	10.1254116	11.2733493

$5(2\,\overset{-2}{4})$	×	10.37845	×	×	10.415653	10.228137	×	×
$8(4\,\overset{-2}{4})$	11.095815	10.7195564	10.4909654	11.1742053	10.754712	11.8094077	11.37213	12.535209
$19(2\,\overset{\sim 2}{4}\,\overset{-2}{4})$	12.3575234	11.9543767	11.735514	12.4344906	11.983691	11.8094077	12.625685	13.7995266
$29(4\,\overset{-2}{4}\,\overset{-2}{4})$	13.621795	13.1985193	12.9873747	13.69772	13.2234	13.057074	13.883864	15.065639
$69(2\,\overset{\sim 2}{4}\,\overset{-2}{4}\,\overset{-2}{4})$	14.887878	14.449224	14.2441993	14.963003	14.470679	14.310825	15.145272	16.3329974
$97(4\,\overset{-2}{4}\,\overset{-2}{4}\,\overset{-2}{4})$	16.15532	15.7046677	15.5045657	16.229806	15.723433	15.568829	16.4090395	17.601453
	$84(4\,\overset{\sim 2}{8}\,\overset{-2}{8}\,\overset{-2}{8})$	$86(4\,\overset{-2}{8}\,\overset{\sim 2}{8}\,\overset{-2}{8})$	$113(8\,\tilde{8}\,\tilde{8}\,\overset{-2}{8})$	$115(8\,\bar{8}\,\tilde{8}\,\overset{-2}{8})$	$122(8\,\overset{-2}{8}\,\overset{\sim 2}{8}\,\overset{-2}{8})$	$124(8\,\overset{-2}{8}\,\overset{-2}{8}\,\overset{-2}{8})$		
$4(2\,\overset{\sim 2}{4})$	11.5172095	11.3744894	10.968107	11.150683	11.9796777	13.3241763		
$8(4\,\overset{-2}{4})$	12.784661	12.63629	12.2021022	12.377201	13.219724	14.5906348		
$19(2\,\overset{\sim 2}{4}\,\overset{-2}{4})$	14.053159	13.90052	13.4459	13.61521	14.4674344	15.858355		
$29(4\,\overset{-2}{4}\,\overset{-2}{4})$	15.3224397	15.1665397	14.696447	14.861244	15.7206054	17.126998		
$69(2\,\overset{\sim 2}{4}\,\overset{-2}{4}\,\overset{-2}{4})$	16.592347	16.433875	15.951838	16.113047	16.97782	18.3964414		
$97(4\,\overset{-2}{4}\,\overset{-2}{4}\,\overset{-2}{4})$	17.8627405	17.70218	17.2107554	17.369079	18.2379466	19.666375		

End of 48E' Group (24): 12(4S)e,1 + 6(8S)e,1
[x = E8M, y = 1, z = E4M]

------------- END OF [DATA 2B-7]48E' and FAMILY -----------

[DATA 2B – 8] The 60E'1 and Family

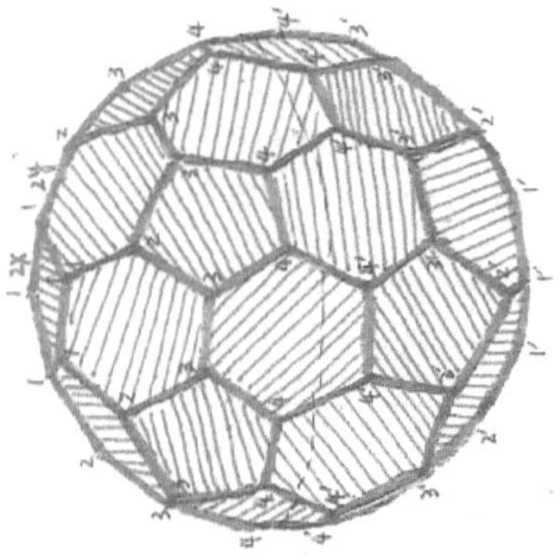

$$\mathbf{Ro} = \sqrt{\frac{1}{2}(29+9\sqrt{5})} \doteq 4.9560373$$

Family groups:

(1) 12 (5 S)e (2) 20 (6 S)e (3) 20 (6 S)5-e (4) 20 (6 S)6-e

(5) 12 (5 M)e + 20 (6 A)

60E'1 Group (1): 12 (5 S)e

2 $(5 \sim 5)$	4 $(\underline{10\ 5})$	5 $(1\ 5 \sim 5)$	11 $(5 \sim \overset{2}{10} \sim 5)$	13 $(5 - \overset{2}{10} \sim 5)$	25 $(1\ 5 \sim \overset{2}{10} \sim 5)$	27 $(1\ 5 - \overset{2}{10} \sim 5)$	42 $(5 - 5 \sim \overset{2}{10} - 5)$	86 $(10 - \overset{2}{10} \sim 10 - 5)$
R=7.483074	10.61393	7.719369	10.83272	10.869322	11.00067	11.06748	13.3784	16.5016

60E'1 Group (2): 20 (6 S)e

3 $(6 \sim 6)$e	6 $(\underline{12\ 6})$e	19 $(6 \sim 6 \sim 6)$e	115 $(6 \sim 6 \sim \overset{2}{12} - 6)$e	117 $(6 \sim 6 - \overset{2}{12} \sim 6)$e	119 $(6 - 6 \sim \overset{2}{12} \sim 6)$e	121 $(6 - 6 - \overset{2}{12} \sim 6)$e	152 $(\underline{12\ 6} \sim \overset{2}{12} - 6)$e
R=8.41954	13.3106	× d < 1	× d < 1	× d < 1	× d < 1	× d < 1	18.210712

60E'1-Group (3): 20(6 S)5-e

1 $(3 \sim \overset{2}{6})$	5 $(6 - \overset{2}{6})$	7 $(1\ 3 \sim \overset{2}{6})$	9 $(3 \sim 3 \sim \overset{2}{6})$	13 $(3 \sim \overset{2}{6} \sim \overset{2}{6})$	14 $(3 \sim \overset{2}{6} - \overset{2}{6})$	15 $(3 - \overset{2}{6} \sim \overset{2}{6})$	26 $(6 - \overset{2}{6} - \overset{2}{6})$	43 $(1\ 3 \sim \overset{2}{6} \sim \overset{2}{6})$
R=7.154226	10.308786	7.989644	8.097217	11.14812	12.5973	11.25883	15.758918	11.257924
44 $(1\ 3 \sim \overset{2}{6} - \overset{2}{6})$	55 $(3 \sim 3 \sim \overset{2}{6} \sim \overset{2}{6})$	56 $(3 \sim 3 \sim \overset{2}{6} - \overset{2}{6})$	72 $(3 \sim \overset{2}{6} \sim \overset{2}{6} - \overset{2}{6})$	74 $(3 \sim \overset{2}{6} - \overset{2}{6} - \overset{2}{6})$	76 $(3 - \overset{2}{6} \sim \overset{2}{6} - \overset{2}{6})$	103 $(6 \sim 6 \sim 6 - \overset{2}{6})$	114 $(6 - \overset{2}{6} - \overset{2}{6} - \overset{2}{6})$	
13.395614	11.26298	13.47914	16.5653	18.063285	16.65447	14.434686	21.226563	

60E'1 Group (4): 20(6 S)6-e

1. $(3 \sim \overset{2}{6})$	2. $(3 - \overset{2}{6})$	5. $(6 - \overset{2}{6})$	7. $(1\ 3 \sim \overset{2}{6})$	8. $(1\ 3 - \overset{2}{6})$	9. $(3 \sim 3 \sim \overset{2}{6})$	10. $(3 \sim 3 - \overset{2}{6})$	12. $(3 - 3 - \overset{2}{6})$
6.175793	6.605467	7.935095	d < 1	7.88175	d < 1	8.362795	9.638955
13. $(3 \sim \overset{2}{6} \sim \overset{2}{6})$	14. $(3 \sim \overset{2}{6} - \overset{2}{6})$	15. $(3 - \overset{2}{6} \sim \overset{2}{6})$	18. $(\underline{6\ 3} - \overset{2}{6})$	26. $(6 - \overset{2}{6} - \overset{2}{6})$	43. $(1\ 3 \sim \overset{2}{6} \sim \overset{2}{6})$	44. $(1\ 3 \sim \overset{2}{6} - \overset{2}{6})$	50. $(3 \sim 3 - \overset{2}{3} - \overset{2}{6})$
d < 1	9.201887	d < 1 ×	10.135685	10.967685	d < 1	d < 1 ×	11.4055925
55. $(3 \sim 3 \sim \overset{2}{6} \sim \overset{2}{6})$	56. $(3 \sim 3 \sim \overset{2}{6} - \overset{2}{6})$	72. $(3 \sim \overset{2}{6} - \overset{2}{6} - \overset{2}{6})$	74. $(3 \sim \overset{2}{6} - \overset{2}{6} - \overset{2}{6})$	76. $(3 - \overset{2}{6} \sim \overset{2}{6} - \overset{2}{6})$	82. $(\underline{6\ 3} - 3 - \overset{2}{6})$	103. $(6 \sim 6 \sim 6 - \overset{2}{6})$	114. $(6 - \overset{2}{6} - \overset{2}{6} - \overset{2}{6})$
d < 1	9.7125497	11.435891	12.248316	11.49726	13.179444	d < 1	14.016173

60E'1 Group (5): 12 (5M)e + 20 (6A)									
P. 1-1	1 $(3\,{}^{\sim 2}6)$	2 $(3\,{}^{-2}6)$	5 $(6\,{}^{-2}6)$	7 $(1\ 3\,{}^{\sim 2}6)$	8 $(1\ 3\,{}^{-2}6)$	9 $(3\,{}^{\sim}3\,{}^{\sim 2}6)$	10 $(3\,{}^{\sim}3\,{}^{-2}6)$	11 $(3\,{}^{-}3\,{}^{\sim 2}6)$	12 $(3\,{}^{-}3\,{}^{-2}6)$
1 (1 5)	6.807106	6.52513	8.28701	8.082265	×	8.56237	×	9.838121	×
2 $(5\,{}^{\sim}5)$	×	7.72157	9.4521	×	8.636065	×	×	×	×
3 $(5\,{}^{-}5)$	×	×	9.913559	×	9.317582	×	9.62487	×	×
4 $(\underline{10\ 5})$	×	×	×	×	×	×	10.86521	×	11.7676
5 $(1\ 5\,{}^{\sim}5)$	×	7.83925	9.565145	×	8.79593	×	×	×	×
8 $(5\,{}^{\sim}5\,{}^{-}5)$	×	×	×	×	×	×	11.0898	×	12.083823
11 $(5\,{}^{\sim 2}10\,{}^{\sim}5)$	×	×	×	×	×	×	10.97613	×	11.91935
13 $(5\,{}^{-2}10\,{}^{\sim}5)$	×	×	×	×	×	×	10.99422	×	11.9444
P. 1-2	13 $(3\,{}^{\sim 2}6\,{}^{\sim 2}6)$	14 $(3\,{}^{\sim 2}6\,{}^{-2}6)$	15 $(3\,{}^{-2}6\,{}^{\sim 2}6)$	16 $(3\,{}^{-2}6\,{}^{-2}6)$	17 $(3\,{}^{-2}6\,{}^{-2}6)$	18 $(\underline{6\ 3}\,{}^{-2}6)$	23 $(6\,{}^{\sim}6\,{}^{-2}6)$	25 $(6\,{}^{-2}6\,{}^{\sim 2}6)$	26 $(6\,{}^{-2}6\,{}^{-2}6)$
1 (1 5)	9.55665	×	10.01187	×	10.333611	×	×	11.323115	×
2 $(5\,{}^{\sim}5)$	10.712805	10.38262	11.1675645	10.62244	×	×	11.659791	12.46643	×
3 $(5\,{}^{-}5)$	11.167995	11.055325	11.621561	11.35932	×	×	12.3304	12.91395	×
4 $(\underline{10\ 5})$	×	12.10835	×	12.55705	×	×	13.304509	×	13.50521
5 $(1\ 5\,{}^{\sim}5)$	10.824034	10.538863	×	10.7925	×	×	11.8148493	12.575547	×
8 $(5\,{}^{\sim}5\,{}^{-}5)$	11.24599	12.28793	×	×	×	12.35791	13.543772	12.87741	13.27212
11 $(5\,{}^{\sim 2}10\,{}^{\sim}5)$	×	12.19674	×	12.6658	×	×	13.45157	×	13.65624
13 $(5\,{}^{-2}10\,{}^{\sim}5)$	×	12.2109	×	12.6835	×	×	13.46547	×	13.68115
16 $(\underline{10\ 5}\,{}^{-}5)$	×	×	×	×	×	14.23131	×	×	15.4257
P. 1-3	43 $(1\ 3\,{}^{\sim 2}6\,{}^{\sim 2}6)$	44 $(1\ 3\,{}^{\sim 2}6\,{}^{-2}6)$	45 $(1\ 3\,{}^{-2}6\,{}^{\sim 2}6)$	46 $(1\ 3\,{}^{-2}6\,{}^{-2}6)$	49 $(3\,{}^{\sim}3\,{}^{-2}3\,{}^{\sim 2}6)$	50 $(3\,{}^{\sim}3\,{}^{-2}3\,{}^{-2}6)$	55 $(3\,{}^{\sim}3\,{}^{\sim 2}6\,{}^{\sim 2}6)$	56 $(3\,{}^{\sim}3\,{}^{\sim 2}6\,{}^{-2}6)$	57 $(3\,{}^{\sim}3\,{}^{-2}6\,{}^{\sim 2}6)$
1 (1 5)	10.012032	×	×	×	11.60375	×	10.064897	×	11.774396
2 $(5\,{}^{\sim}5)$	11.160133	10.620465	×	×	×	×	11.209886	×	12.9189945
3 $(5\,{}^{-}5)$	11.611789	11.354105	12.89423	×	×	×	11.66045	11.38426	13.36624
4 $(\underline{10\ 5})$	×	12.54438	×	13.494826	×	×	×	12.581185	×
5 $(1\ 5\,{}^{\sim}5)$	11.27041	10.789843	12.554709	×	×	×	11.31988	10.815269	13.028034
8 $(5\,{}^{\sim}5\,{}^{-}5)$	11.657662	12.768256	12.858724	13.816161	×	×	×	12.815	13.3043
11 $(5\,{}^{\sim 2}10\,{}^{\sim}5)$	×	12.65216	×	13.645545	×	×	×	12.69668	×

13 $(5\,{}^{-}\overset{2}{10}\,{}^{\sim}5)$	×	12.66969	×	13.670404	×	×	×	12.71453	×
16 $(\underline{10\ 5}\,{}^{-}5)$	×	×	×	×	×	15.20866	×	×	×

P.1-4	58 $(3\,{}^{\sim}3\,{}^{-}\overset{2}{6}\,{}^{-}\overset{2}{6})$	61 $(3\,{}^{-}3\,{}^{-}\overset{2}{6}\,{}^{\sim}\overset{2}{6})$	62 $(3\,{}^{-}3\,{}^{-}\overset{2}{6}\,{}^{-}\overset{2}{6})$	71 $(3\,{}^{\sim}\overset{2}{6}\,{}^{\sim}\overset{2}{6}\,{}^{\sim}\overset{2}{6})$	72 $(3\,{}^{\sim}\overset{2}{6}\,{}^{\sim}\overset{2}{6}\,{}^{-}\overset{2}{6})$	73 $(3\,{}^{\sim}\overset{2}{6}\,{}^{-}\overset{2}{6}\,{}^{\sim}\overset{2}{6})$	74 $(3\,{}^{\sim}\overset{2}{6}\,{}^{-}\overset{2}{6}\,{}^{-}\overset{2}{6})$	75 $(3\,{}^{-}\overset{2}{6}\,{}^{\sim}\overset{2}{6}\,{}^{\sim}\overset{2}{6})$	76 $(3\,{}^{-}\overset{2}{6}\,{}^{\sim}\overset{2}{6}\,{}^{-}\overset{2}{6})$
1 (1 5)	×	13.05891	×	11.78947	×	12.604815	×	11.849601	×
2 $(5\,{}^{\sim}5)$	×	14.198965	×	12.924951	×	13.743736	×	12.9814	×
3 $(5\,{}^{-}5)$	×	14.64291	×	13.36921	×	14.187909	×	×	×
4 $(\underline{10\ 5})$	×	×	×	×	×	×	×	×	×
5 $(1\ 5\,{}^{\sim}5)$	×	14.307119	×	13.03322	×	13.851972	×	13.08935	×
8 $(5\,{}^{\sim}5\,{}^{-}5)$	14.094735	×	×	×	14.0941	×	×	×	14.122605
11 $(5\,{}^{\sim}\overset{2}{10}\,{}^{\sim}5)$	13.907095	×	×	×	13.90701	×	×	×	×
13 $(5\,{}^{-}\overset{2}{10}\,{}^{\sim}5)$	13.93423	×	×	×	13.934095	×	×	×	13.961064
16 $(\underline{10\ 5}\,{}^{-}5)$	×	×	16.93398	×	15.904	×	16.616901	×	15.9533

P.1-5	81. $(\underline{6\ 3}\,{}^{-}3\,{}^{\sim}\overset{2}{6})$	82. $(\underline{6\ 3}\,{}^{-}3\,{}^{-}\overset{2}{6})$	85. $(\underline{6\ 3}\,{}^{-}\overset{2}{6}\,{}^{\sim}\overset{2}{6})$	86. $(\underline{6\ 3}\,{}^{-}\overset{2}{6}\,{}^{-}\overset{2}{6})$	103. $(6\,{}^{\sim}6\,{}^{\sim}6\,{}^{-}\overset{2}{6})$
1 (1 5)	13.376756	×	13.5453	×	×
2 $(5\,{}^{\sim}5)$	×	×	1.68155	×	12.423003
3 $(5\,{}^{-}5)$	×	×	15.12369	×	13.15768
4 $(\underline{10\ 5})$	×	×	×	×	14.33466
5 $(1\ 5\,{}^{\sim}5)$	×	×	14.789233	×	12.59153
8 $(5\,{}^{\sim}5\,{}^{-}5)$	×	×	×	×	14.563656
11 $(5\,{}^{\sim}\overset{2}{10}\,{}^{\sim}5)$	×	×	×	×	14.442767
13 $(5\,{}^{-}\overset{2}{10}\,{}^{\sim}5)$	×	×	×	1.446033	×
16 $(\underline{10\ 5}\,{}^{-}5)$	×	×	×	17.2118	14.9323

End of 60E'1-Group (5): 12 (5 M) + 20 (6 A) P. 1-1,2,3,4, 5

(continue) 60E'1-Group (5): 12(5 M) + 20(6 A) **P. 2**

P.2-1	10 (3 ~3 -$\overset{2}{6}$)	12 (3 -3 -$\overset{2}{6}$)	14 (3 ~$\overset{2}{6}$ -$\overset{2}{6}$)	16 (3 -$\overset{2}{6}$ -$\overset{2}{6}$)	18 ($\underline{6\ 3}$ -$\overset{2}{6}$)	23 (6 ~6 -$\overset{2}{6}$)	26 (6 -$\overset{2}{6}$ -$\overset{2}{6}$)
22 (1 5 ~5 -5)	×	12.23496	12.365119	×	12.523225	13.61785	13.97495
25 (1 5 ~$\overset{2}{10}$ ~5)	11.05785	12.029497	12.25772	12.74174	12.29605	13.50972	13.76145
27 (1 5 -$\overset{2}{10}$ ~5)	11.08971	12.073294	12.28133	×	12.343725	13.53239	13.80375
42 (5 -5 ~$\overset{2}{10}$ -5)	×	13.39575	×	×	13.820159	×	15.0966
46 (5 ~$\overset{2}{10}$ ~5 -5)	×	×	×	×	×	×	15.50323
50 (5 -$\overset{2}{10}$ ~5 -5)	×	×	×	×	×	×	15.51545

P. 2-2	44 (1 3 ~$\overset{2}{6}$ -$\overset{2}{6}$)	46 (1 3 -$\overset{2}{6}$ -$\overset{2}{6}$)	50 (3 ~3 -$\overset{2}{3}$ -$\overset{2}{6}$)	56 (3 ~3 ~$\overset{2}{6}$ -$\overset{2}{6}$)	58 (3 ~3 -$\overset{2}{6}$ -$\overset{2}{6}$)	62 (3 -3 -$\overset{2}{6}$ -$\overset{2}{6}$)	72 (3 ~$\overset{2}{6}$ ~$\overset{2}{6}$ -$\overset{2}{6}$)
22 (1 5 ~5 -5)	11.86873	13.96359	×	12.911745	14.25628	×	14.255105
25 (1 5 ~$\overset{2}{10}$ ~5)	12.72742	13.75057	×	12.773166	14.021335	×	14.02091
27 (1 5 -$\overset{2}{10}$ ~5)	12.75712	13.79279	×	12.803405	14.06735	×	14.066809
42 (5 -5 ~$\overset{2}{10}$ -5)	×	×	14.65333	13.47788	15.5131	16.384979	15.50657
46 (5 ~$\overset{2}{10}$ ~5 -5)	15.50323	×	15.35659	×	×	17.079237	16.00392
50 (5 -$\overset{2}{10}$ ~5 -5)	×	×	15.380927	×	×	17.103134	16.02008
86 (10 -$\overset{2}{10}$ ~10 5)	×	×	16.54085	×	×	×	16.5648

P. 2-3	74 (3 ~$\overset{2}{6}$ -$\overset{2}{6}$ -$\overset{2}{6}$)	76 (3 -$\overset{2}{6}$ ~$\overset{2}{6}$ -$\overset{2}{6}$)	82 ($\underline{6\ 3}$ -3 -$\overset{2}{6}$)	86 ($\underline{6\ 3}$ -$\overset{2}{6}$ -$\overset{2}{6}$)	103 (6 ~6 ~6 -$\overset{2}{6}$)	114 (6 -$\overset{2}{6}$ -$\overset{2}{6}$ -$\overset{2}{6}$)	
22 (1 5 ~5 -5)	×	14.28503	×	×	14.66281	×	
25 (1 5 ~$\overset{2}{10}$ ~5)	×	14.04862	×	×	14.51562	×	
27 (1 5 -$\overset{2}{10}$ ~5)	×	14.094896	×	×	14.54465	×	
42 (5 -5 ~$\overset{2}{10}$ -5)	16.117512	15.549444	×	×	×	×	
46 (5 ~$\overset{2}{10}$ ~5 -5)	16.7478	16.05527	×	17.371005	×	×	
50 (5 -$\overset{2}{10}$ ~5 -5)	16.769293	16.07175	×	17.397199	×	×	
86 (10 -$\overset{2}{10}$ ~10 -5)	17.731795	16.65053	17.78645	18.650845	×	19.247556	

--------------- End of [DATA 2B -- 8] 60E'1 Family -------------

[DATA 2B – 9] The 60E'2 and Family

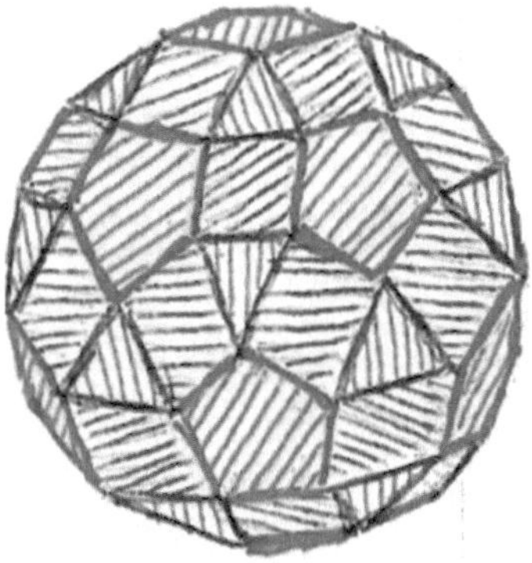

$$\mathbf{Ro} = \sqrt{11+4\sqrt{5}} \doteq 4.465901$$

Family groups:

(1) 12 (5 S)e
(2) 20 (3 S)e
(3) 30 (4 S)c
(4) 30 (4 S)5-e
(5) 30 (4 S)3-e
(6) 12 (5 M)e + 30 (4 A)
(7) 20 (3 M)e + 30 (4 A)
(8) 12 (5 S)e1 +20 (3 S)e2

Group (1): 12(5S)e	1 (1 5)	2 (5~5)	3 (5‾5)	4 (10 5)	5 (1 5~5)	8 (5~5‾5)	11 (5~$\overset{2}{10}$~5)	13 (5‾$\overset{2}{10}$~5)
	R=4.8747465	6.605464	7.4830723	9.2782707	6.8071013	9.6389504		
	16 (10 5‾5)	22 (1 5~5‾5)	25 (1 5‾$\overset{1}{10}$~5)	27 (1 5‾$\overset{2}{10}$~5)	37 (5~5~$\overset{2}{10}$~5)	42 (5‾5~$\overset{2}{10}$‾5)	46 (5~$\overset{2}{10}$~5‾5)	50 (5‾$\overset{2}{10}$~5‾5)
		9.5344992	9.607784	9.665487	9.871979	11.62158	12.47713	12.50885
	86 (10‾$\overset{2}{10}$~10‾5)							
	14.2865							
Group (2): 20(3S)e	1 (1 3)	2 (3~3)	3 (3‾3)	4 (6 3)	8 (3~3‾3)	16 (6 3‾3)		
	R=5.7294405	6.2180293	7.483074	7.9949026	9.250972	11.028006		
Group (3): 30(4S)e	6 (4~4)e	9 (8 4)e	10 (1 4~4)e	30 (4~$\overset{2}{8}$~4)e	32 (4‾$\overset{2}{8}$~4)e [Rc<9.805]	54 (1 4~$\overset{2}{8}$~4)e [Rc<6.191]	103 (4‾4~$\overset{2}{8}$‾4)e	153 (8‾$\overset{2}{8}$~8‾$\overset{2}{4}$)
	8.5326381	12.9456802	8.8719449	13.2913995	×	×	17.36493	21.7873564
Group (4): 30(4S)5-e	2 (1 $\overset{1}{4}$)	4 (2~$\overset{1}{4}$)	5 (2‾$\overset{1}{4}$)	8 (4‾$\overset{2}{4}$)	19 (2~$\overset{2}{4}$‾$\overset{2}{4}$)	29 (4‾$\overset{2}{4}$‾$\overset{2}{4}$)	69 (2~$\overset{2}{4}$‾$\overset{2}{4}$‾$\overset{2}{4}$)	97 (4‾$\overset{2}{4}$‾$\overset{2}{4}$‾$\overset{2}{4}$)
	6.390918	7.1542282	9.10574245	9.8705854	12.5973	15.328921	18.0632886	20.7993516
Group (5): 30(4S)3-e	2 (1 $\overset{1}{4}$)	4 (2~$\overset{1}{4}$)	5 (2‾$\overset{1}{4}$)	8 (4‾$\overset{2}{4}$)	19 (2~$\overset{2}{4}$‾$\overset{2}{4}$)	29 (4‾$\overset{2}{4}$‾$\overset{2}{4}$)	69 (2~$\overset{2}{4}$‾$\overset{2}{4}$‾$\overset{2}{4}$)	97 (4‾$\overset{2}{4}$‾$\overset{2}{4}$‾$\overset{2}{4}$)
	5.7127975	6.2176962	7.4830737	7.9955654	9.784494	11.57911	13.3770443	15.1770692

60E'2 Group (6): 12(5M)e + 30(4A) [x = E5M, y = E4A]

	2	4	5	8	13	14	18	19	20	21	26	28	29	36	46	68	69	96	97	114	116
1	•	•	•	•			•					•				•		•			
2			•	•		•	•					•				•		•			
3			•	•	•	•	•	•				•			•	•		•			
4						•		•		•	•				•						
5						•	•														
8						•		•		•	•				•						
11						•		•		•	•				•						
13						•		•		•	•				•						
16										•			•	•						•	•
22								•		•	•		•		•					•	•
25										•					•						
27										•			•							•	
37																					
42													•	•						•	•
46										•			•	•			•			•	•
50										•			•	•			•			•	•

60E'2 Group (6): 12(5M)e + 30(4A) [x = E5M, y = E4A]

P. 1	2 (1 $\overset{1}{4}$)	4 (2~$\overset{1}{4}$)	5 (2‾$\overset{1}{4}$)	8 (4‾$\overset{2}{4}$)	13 (1 $\overset{2}{4}$~$\overset{2}{4}$)	14 (1 $\overset{2}{4}$‾$\overset{2}{4}$)	18 (2~$\overset{2}{4}$~$\overset{2}{4}$)
1 (1 5)	5.9508088	7.7193683	6.6291868	8.4118486	×	×	12.203014
2 (5~5)	×	×	8.0301424	9.7837653	×	10.1256465	11.557757
(5‾5)	×	×	8.5851721	10.3242897	11.6064391	10.9206404	12.0884642
4 (10 5)	×	×	×	×	×	12.1636948	×

5 (1 5 ~5)	×	×	×	9.916183	×	10.310255	11.6873479
8 (5 ~5 -5)	×	×	×	×	×	12.376059	×
11 (5 ~2 10 ~5)	×	×	×	×	×	12.2682043	×
13 (5 -2 10 ~5)	×	×	×	×	×	12.2849155	×
P. 2	19 (2 ~2 4 -2 4)	21 (2 -2 4 -2 4)	26 (4 ~ 4 -2 4)	28 (4 -2 4 ~2 4)	46 (1 4 ~ 4 -2 4)	68 (2 ~2 4 -2 4 ~2 4)	96 (4 -2 4 -2 4 ~2 4)
1 (1 5)	×	×	×	11.9988504	×	13.797545	15.598054
2 (5 ~5)	×	×	×	13.342668	×	15.133991	16.92935
3 (5 -5)	11.0067439	×	×	13.866593	12.7873148	15.653089	17.444884
4 (10 5)	12.4300695	12.762735	13.992565	×	14.1839748	×	×
8 (5 ~5 -5)	12.6979366	13.140167	14.2522502	×	14.4544587	×	×
11 (5 ~2 10 ~5)	12.5590376	12.9411788	14.1175045	×	14.3118262	×	×
13 (5 -2 10 ~5)	12.580048	12.9706073	14.137279	×	14.3326116	×	×

60E'2 Group (6): 12(5M)e + 30(4A) [x = E5M, y = E4A]
P. 3

P. 3	19 (2 ~2 4 -2 4)	21 (2 -2 4 -2 4)	26 (4 ~ 4 -2 4)	29 (4 -2 4 -2 4)	36 (8 4 -2 4)	46 (1 4 ~ 4 -2 4)	69 (2 ~2 4 -2 4 -2 4)	114 (4 ~2~ 8 4 -2 4)	116 (4 -2~ 8 4 -2 4)
16 (10 5 -5)	×	15.043344	×	15.354868 4	18.648831 4	×	×	18.855791	18.847955
22 (1 5 ~5 -5)	12.818782 4	13.315858 4	14.362709 5	13.372644	×	14.5718016	×	16.9304342	16.926326
25 (1 5 ~2 10 ~5)	×	13.117793	×	×	×	14.3981138	×	×	×
27 (1 5 -2 10 ~5)	×	×	×	13.154531	×	×	×	16.693102	×
42 (5 -5 ~2 10 -5)	×	×	×	14.877228	18.203199	×	×	18.38859	18.381895
46 (5 ~2 10 ~5 -5)	×	15.136359	×	15.475838	18.757894	×	15.99909	18.971753	18.963506
50 (5 -2 10 ~5 -5)	×	15.151067 2	×	15.495442	18.775382	×	16.0304614	18.99039	18.982136

60E'2 Group (7): 20 (3M)e + 30 (4A)

	2	4	5	8	13	14	18	19	20	21	26	28	29	36	46	68	69	96	97	114	116
1			•	•		•	•					•				•		•			
2			•	•			•	•							•						
3						•		•		•	•				•						
4													•							•	
8										•			•	•						•	•
16																	•				

60E'₂ Group (7): 20(3M)e + 30(4A)								
	5. (2 $\overset{-2}{4}$)	8. (4 $\overset{-2}{4}$)	14 (1 $\overset{2}{4}$ $\overset{-2}{4}$)	18. (2 $\overset{\sim 2}{4}$ $\overset{\sim 2}{4}$)	19. (2 $\overset{\sim 2}{4}$ $\overset{-2}{4}$)	21 (2 $\overset{-2}{4}$ $\overset{-2}{4}$)	26 (4 $\overset{\sim}{4}$ $\overset{-2}{4}$)	28 (4 $\overset{-2}{4}$ $\overset{\sim 2}{4}$)
1. (1 3)	7.9896442	10.6806845	11.1017	13.3956933	×	×	×	16.120985
2. (3 $^{\sim}$3)	8.097217	10.771818	×	13.479154	11.56399	×	×	×
3. (3 $^{-}$3)	×	×	12.020377	×	12.38986	12.815858	14.8061824	×
4. ($\underline{6\ 3}$)	×	×	×	×	×	×	×	×
8. (3 $^{\sim}$3 $^{-}$3)	×	×	×	×	×	13.76027	×	×
	29 (4 $\overset{-2}{4}$ $\overset{-2}{4}$)	36 ($\underline{8\ 4}$ $\overset{-2}{4}$)	46. (1 4 $\overset{\sim}{4}$ $\overset{-2}{4}$)	68. (2 $\overset{\sim 2}{4}$ $\overset{-2}{4}$ $\overset{\sim 2}{4}$)	69. (2 $\overset{\sim 2}{4}$ $\overset{-2}{4}$ $\overset{-2}{4}$)	96. (4 $\overset{-2}{4}$ $\overset{-2}{4}$ $\overset{\sim 2}{4}$)	114. (4 $\overset{\sim 2}{8}$ $\overset{\sim}{4}$ $\overset{-2}{4}$)	116. (4 $\overset{-2}{8}$ $\overset{\sim}{4}$ $\overset{-2}{4}$)
1. (1 3)	×	×	×	18.851539	×	21.585123	×	×
2. (3 $^{\sim}$3)	×	×	14.2727265	×	×	×	×	×
3. (3 $^{-}$3)	×	×	15.0828872	×	×	×	×	×
4. ($\underline{6\ 3}$)	13.293084	×	×	×	×	×	18.686795	×
8. (3 $^{\sim}$3 $^{-}$3)	14.1295763	19.2109252	×	×	×	×	19.495056	19.48351
16. ($\underline{6\ 3}$ $^{-}$3)	×	×	×	×	15.887529	×	×	×

60E'₂ Group (8) 12 (5S)e1 + 20 (3S)e2								
P. 1	1. (1 5)	2. (5 $^{\sim}$5)	3. (5 $^{-}$5)	4. ($\underline{10\ 5}$)	5. (1 5 $^{\sim}$5)	8. (5 $^{\sim}$5 $^{-}$5)	11. (5 $^{\sim}$ $\overset{2}{10}$ $^{\sim}$5)	13. (5 $^{-}$ $\overset{2}{10}$ $^{\sim}$5)
1. (1 3)	× Cri-b	7.8817556	× Cri-b	10.5433278	8.0822521	× Cri-b	10.7292352	10.7603364
2. (3 $^{\sim}$3)	6.6492517	8.3627925	9.2509713	11.015695	8.562368	11.405599	11.202642	11.2339253
3. (3 $^{-}$3)	× Cri-b	9.6389508	× Cri-b	12.287581	9.838121	× Cri-b	12.4771428	12.5088505
4. ($\underline{6\ 3}$)	8.434029	10.1356828	11.028006	12.772185	10.33361	13.1794572	12.9619545	12.9937
8. (3 $^{\sim}$3 $^{-}$3)	× Cri-b	11.4055925	× Cri-b	14.0437446	11.6037298	× Cri-b	14.2352137	14.2672228
16. ($\underline{6\ 3}$ $^{-}$3)	× Cri-b	13.1794572	× Cri-b	15.8085943	13.376756	× Cri-b	16.0010996	16.0332613
P. 2	16. ($\underline{10\ 5}$ $^{-}$5)	22. (1 5 $^{\sim}$5 $^{-}$5)	25. (1 5 $^{\sim}$ $\overset{2}{10}$ $^{\sim}$5)	27. (1 5 $^{-}$ $\overset{2}{10}$ $^{\sim}$5)	42. (5 $^{-}$5 $^{\sim}$ $\overset{2}{10}$ $^{-}$5)	46. (5 $^{\sim}$ $\overset{2}{10}$ $^{\sim}$5 $^{-}$5)	50. (5 $^{-}$ $\overset{2}{10}$ $^{\sim}$5 $^{-}$5)	86. (10 $^{-}$ $\overset{2}{10}$ $^{\sim}$10 $^{-}$5)
1. (1 3)	× Cri-b	× Cri-b	10.872578	10.929526	12.894247	× Cri-b	× Cri-b	15.554033
2. (3 $^{\sim}$3)	14.043741	11.603743	11.3442565	11.400851	13.366231	14.235206	14.2672228	16.022742
3. (3 $^{-}$3)	× Cri-b	× Cri-b	12.6154285	12.671325	14.642904	× Cri-b	× Cri-b	17.295702
4. ($\underline{6\ 3}$)	15.808594	13.3767553	13.0988346	13.15435	15.123671	16.001098	16.0332613	17.771669
8. (3 $^{\sim}$3 $^{-}$3)	× Cri-b	× Cri-b	14.3699747	14.424935	16.4	× Cri-b	× Cri-b	19.046246
16. ($\underline{6\ 3}$ $^{-}$3)	× Cri-b	× Cri-b	16.1333284	16.187541	18.1639315	× Cri-b	× Cri-b	20.80392

---------------- End of 60E'2 Family ----------------

[DATA 2B – 10] 60E'3 and Family

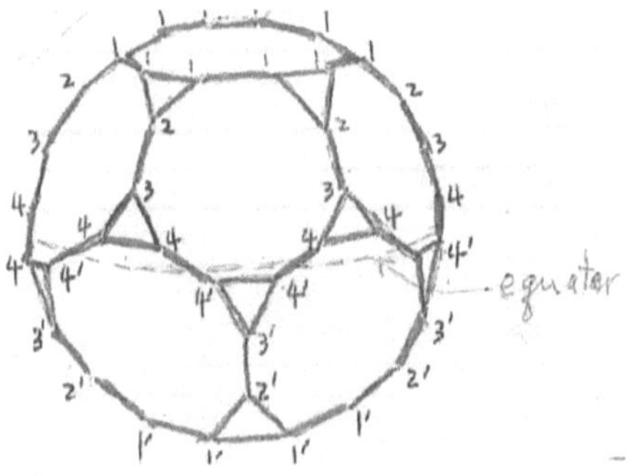

$$Ro = \sqrt{\frac{1}{2}(37+15\sqrt{5}} \doteq 5.938898$$

Family groups:

(1) 20 (3S)e (2) 12 (10S)e (3) 12 (10S)10-e

(4) 12 (10S)3-e (5) 20 (3M)e +12 (10A)

60E'3 Family (1): 20 (3S)	2. (3 3) R=9.1057429	4. (6 3) 12.4804255

60E'3 Family (2) 12 (10S)e						
3. (10 10)e R=8.5236525	6. (20 10)e 14.3829302	19. (10 10 10)e 10.4920278	21. (10 10 10) d<1 ×	36 (20 20 10)e d<1 ×	96. (10 10 10 10)e d<1 ×	100. (10 10 10 10)e d<1 ×

60E'3 Family Group (3): 12 (10S)10-e						
1. (5 10) R=6.175792	2. (5 10) 6.2180293	5. (10 10) 8.835113	7. (1 5 10) 6.5251315	8. (1 5 10) 6.6492531	9. (5 5 10) 7.7215716	10. (5 5 10) 8.3628029
11. (5 5 10) d < 1 ×	12. (5 5 10) R=9.250975	13. (5 10 10) 9.1352696	14. (5 10 10) 9.2018867	15. (5 10 10) 9.1884	18. (10 5 10) 11.015695	23. (10 10 10) [R<13.64…] R=11.037327
25. (10 10 10) d < 1 ×	26. (10 10 10) 11.839649	39. (1 5 5 10) R=7.839245	40 . (1 5 5 10) 8.562368	43. (1 5 10 10) 9.412697	44. (1 5 10 10) 9.5566506	45. (1 5 10 10) 9.530657
50. (5 5 5 10) 11.405591	55. (5 5 10 10) 10.22953	56. (5 5 10 10) 10.712807	57. (5 5 10 10) R=10.646745	59. (5 5 10 10) 10.512126	60. (5 5 10 10) 11.167995	61. (5 5 10 10) d < 1 ×
64. (5 10 5 10) 11.202645	68. (5 10 5 10) 11.23393	71. (5 10 10 10) d < 1 ×	72. (5 10 10 10) 12.16809	73. (5 10 10 10) d < 1 ×	74. (5 10 10 10) R=12.248308	75. (5 10 10 10) d < 1 ×
76. (5 10 10 10) 12.227	82. (10 5 5 10) 14.043745	103. (10 10 10 10) 12.54268	105. (10 10 10 10) d < 1 ×	108. (10 10 10 10) [R<13.64] R=14.04… ×	112. (10 10 10 10) R=13.83114	114. (10 10 10 10) R=14.875524

60E'3 Family (4) 12 (10S)3-e							
1. (5 10) R=6.2176965	5. (10 10) 9.3404003	7. (1 5 10) 6.6291855	9. (5 5 10) 8.0301424	11. (5 5 10) 8.58517	13. (5 10 10) 9.701965	14. (5 10 10) 9.784493	15. (5 10 10) 9.766336
23 (10 10 10) [R<13 64] R=11.931261	25. (10 10 10) 11.715738	26. (10 10 10) 12.886197	39. (1 5 5 10) 8.166985	43. (1 5 10 10) 10.02684	44. (1 5 10 10) 10.203013	45. (1 5 10 10) 10.167855	55. (5 5 10 10) 10.97328
56. (5 5 10 10) 11.557757	57. (5 5 10 10) 11.462173	59. (5 5 10 10) 11.297109	60. (5 5 10 10) 12.0884604	61. (5 5 10 10) 11.957417	71. (5 10 10 10) 11.888871	72. (5 10 10 10) 13.279034	73. (5 10 10 10) 11.92729
74 (5 10 10 10) R=13.377044	75 (5 10 10 10) 11.917388	76 (5 10 10 10) 13.3497879	103 (10 10 10 10) 13.687669	105 (10 10 10 10) 14.905234	108 (10 10 10 10) [R<13.64] R=15.48… ×	112. (10 10 10 10) 15.2124	114 (10 10 10 10) 16.4684076

---------- End of [DATA 2B—10] 60E'3 and Family ----------

[DATA 2B – 11] The 120E' and Family

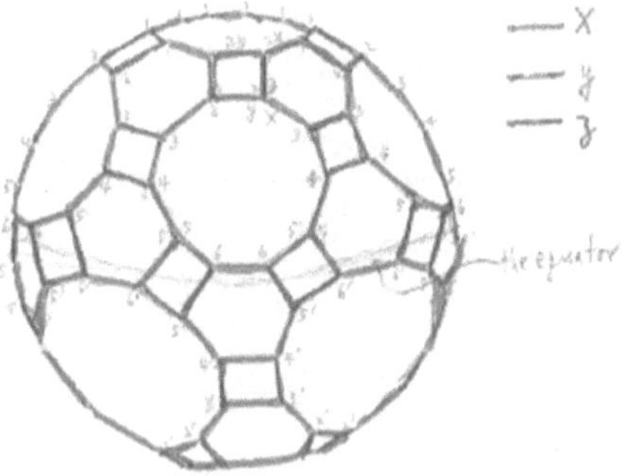

$$Ro = \sqrt{31+12\sqrt{5}} \doteq 7.604789$$

Family groups:

(1) 12(10S)e
(2) 12(10S)6-e
(3) 12(10S)4-e

(4) 20(6S)e
(5) 20(6S)4-e
(6) 20(6S)10-e

(7) 30(4S)e
(8) 30(4S)10-e
(9) 30(4S)6-e

(10) 12(10S1)e1 + 20(6S2)e2
(11) 20(6S1)e1 + 30(4S2)e2
(12) 30(4S1)e1 = 12(10S2)e2

(13) 12(10M)e +20(6A)
(14) 12(10M)e,1 +20(6A)
(15) 12(10M)e + 30(4A)
(16) 12(10M)e,1 +30(4A)

(17)20(6M)e +30(4A)
(18)20(6M)e,1 +30(4A)
(19)20(6M)e +12(10A)
(20) 20(6M)e,1 + 12(10A)

(21) 30(4M)e + 12(10A)
(22) 30(4M)e,1 +12(10A)
(23) 30(4M)e +20(6A)
(24) 30(4M)e,1 +20(6A)

120E' Group (1): 12(10S)e						
R	3. (10 10)e 10.267778	6 (20 10)e 16.091412	19. (10 10 10)e 12.230452	21. (10 10 10) e 12.67489	36. (20 20 10)e 18.489265	115. (10 10 20 10)e 16.175606[R>13.645]
R	117. (10 10 20 10)e 16.1863165[R>13.645]	119. (10 10 20 10)e 16.6682654	121. (10 10 20 10)e 16.747627[R>8.9983]	147. (20 10 20 10)e 19.151196	148. (20 10 20 10)e 20.71465	149. (20 10 20 10)e [Rc<16.141308] ×

120E' Group (2): 12(10S)6-e						
1 (5 10) R=7.995566	5 (10 10) 11.10587	7. (1 5 10) 8.4118485	9. (5 5 10) 9.78377	11. (5 5 10) 10.3242898	13. (5 10 10) 11.4871898	14 (5 10 10) 11.57913
15 (5 10 10) R=11.555541	23 (10 10 10) [R<13.64] 13.704045 ×	25 (10 10 10) 13.453574	26 (10 10 10) 14.674802	39 (1 5 5 10) 9.9161849	43 (1 5 10 10) 11.80872	44 (1 5 10 10) 11.9988523
45 (1 5 10 10) R=11.95409	55 (5 5 10 10) 12.7307973	56 (5 5 10 10) 13.342665	57 (5 5 10 10) 13.218193	59 (5 5 10 10) 13.042374	60 (5 5 10 10) 13.866953	61 (5 5 10 10) 13.6902871
71 (5 10 10 10) R=13.621947	72. (5 10 10 10) 15.074928	73. (5 10 10 10) 13.6599832	74. (5 10 10 10) 15.1770692	75 (5 10 10 10) 13.64948	76 (5 10 10 10) 15.147229	103 (10 10 10 10) 15.44706
105 (10 10 10 10) R=15.822474	108 (10 10 10 10) [R<13 64] ×	112 (10 10 10 10) 16.984551	114 (10 10 10 10) 18.2652178			

120E' Group (3): 12(10S)4-e						
1. (5 10) R=7.935095	2. (5 10) 7.994901	5. (10 10) 1.057276	7. (1 5 10) 8.28701	8. (1 5 10) 8.434029	9. (5 5 10) 9.452105	10. (5 5 10) 10.1356807
11. (5 5 10) 9.91356	12. (5 5 10) 11.028	13. (5 10 10) 10.892039	14. (5 10 10) 10.967685	15. (5 10 10) 10.949059	18. (10 5 10) 12.772184	23. (10 10 10) [R<13.64] 12.778005
25. (10 10 10) R=12.578657	26. (10 10 10) 13.595599	39. (1 5 5 10) 9.5651355	40. (1 5 5 10) 10.333612	43. (1 5 10 10) 11.165628	44. (1 5 10 10) 12.323113	45. (1 5 10 10) 11.28782
50. (5 5 5 10) 13.17945	55. (5 5 10 10) 11.956132	56. (5 5 10 10) 12.43346	57. (5 5 10 10) 14.3714242	59. (5 5 10 10) 12.225515	60. (5 5 10 10) 12.913954	61. (5 5 10 10) 12.782635
64. (5 10 5 10) 13.171203	68. (5 10 5 10) 12.9936967	71. (5 10 10 10) 12.724173	72. (5 10 10 10) 13.931616	73. (5 10 10 10) 12.756802	74. (5 10 10 10) 14.0161738	75. (5 10 10 10) 12.748074
76. (5 10 10 10) 13.99217	82. (10 5 5 10) 15.808595	103. (10 10 10 10) 14.266831	105. (10 10 10 10) 14.58911	108. (10 10 10 10) [R<13 64] ×	112. (10 10 10 10) 15.567496	114. (10 10 10 10) 16.638156

120E' Group (4) 20(6S)e						
3. (6 6)e 11.112296	6. (12 6)e 15.98626	19. (6 6 6)e 12.490737	115. (6 6 12 6)e 17.2256727	117. (6 6 12 6)e 17.385608	119. (6 6 12 6)e 17.391682	152. (12 6 12 6)e 20.8762

120E' Group (5) 20(6S)4-e							
1. (3 6) 8.835113	2. (3 6) 9.278272	5. (6 6) 10.57276	7. (1 3 6) 9.28483	8. (1 3 6) 10.543323	9. (3 3 6) 9.338979	10. (3 3 6) 11.015695	12. (3 3 6) 12.28759
13. (3 6 6)	14. (3 6 6)	15. (3 6 6)	18. (6 3 6)	26. (6 6 6)	43. (1 3 6 6)	44. (1 3 6 6)	50. (3 3 3 6)

11.03818	11.839648	11.100465	12.772184	13.595596	11.100002	12.284725	14.043734
55. $(3\,\tilde{3}\,\overset{\sim2}{6}\,\overset{\sim2}{6})$ 11.102924	56. $(3\,\tilde{3}\,\overset{\sim2}{6}\,\overset{-2}{6})$ 12.3326	72. $(3\,\overset{\sim2}{6}\,\overset{\sim2}{6}\,\overset{-2}{6})$ 14.050516	74. $(3\,\overset{\sim2}{6}\,\overset{-2}{6}\,\overset{-2}{6})$ 14.87553	76. $(3\,\overset{-2}{6}\,\overset{\sim2}{6}\,\overset{-2}{6})$ 14.10439	82. $(\underline{6\,3}\,\bar{3}\,\overset{-2}{6})$ 15.808595	103. $(6\,\tilde{6}\,\tilde{6}\,\overset{-2}{6})$ 12.88021	114. $(6\,\overset{-2}{6}\,\overset{-2}{6}\,\overset{-2}{6})$ 16.63817

120E' Group (6) 20(6S)10-e

1. $(3\,\overset{\sim2}{6})$ R=9.870597	5. $(6\,\overset{-2}{6})$ 13.030105	7. $(1\,3\,\overset{\sim2}{6})$ 10.680679	9. $(3\,\tilde{3}\,\overset{\sim2}{6})$ 10.771819	13. $(3\,\overset{\sim2}{6}\,\overset{\sim2}{6})$ 13.848369	14. $(3\,\overset{\sim2}{6}\,\overset{-2}{6})$ 15.32893	15. $(3\,\overset{-2}{6}\,\overset{\sim2}{6})$ 13.9454646	26. $(6\,\overset{-2}{6}\,\overset{-2}{6})$ 18.491625	43. $(1\,3\,\overset{\sim2}{6}\,\overset{\sim2}{6})$ 13.943966
44. $(1\,3\,\overset{\sim2}{6}\,\overset{-2}{6})$ 16.120985	55. $(3\,\tilde{3}\,\overset{\sim2}{6}\,\overset{\sim2}{6})$ 13.9472336	56. $(3\,\tilde{3}\,\overset{\sim2}{6}\,\overset{-2}{6})$ 16.2002897	72. $(3\,\overset{\sim2}{6}\,\overset{\sim2}{6}\,\overset{-2}{6})$ 10.2906409	74. $(3\,\overset{\sim2}{6}\,\overset{-2}{6}\,\overset{-2}{6})$ 20.79935	76. $(3\,\overset{-2}{6}\,\overset{\sim2}{6}\,\overset{-2}{6})$ 19.374934	103. $(6\,\tilde{6}\,\tilde{6}\,\overset{-2}{6})$ 17.12498	114. $(6\,\overset{-2}{6}\,\overset{-2}{6}\,\overset{-2}{6})$ 23.962995	

120E' Group (7) 30(4S)e

6. $(4\,\tilde{4})$e 11.6918207	9. $(\underline{8\,4})$e 16.1004286	10. $(1\,4\,\tilde{4})$e 12.0132498	30. $(4\,\overset{\sim2\sim}{8}\,4)$e 16.4308404	103. $(4\,\bar{4}\,\overset{\sim2-}{8}\,4)$e 20.515921	153. $(8\,\overset{-2\sim}{8}\,8\,\overset{-2}{4})$ 24.9351243		

120E' Group (8) 30(4S)10-e

2. $(1\,\overset{2}{4})$ 9.564609	4. $(2\,\overset{\sim2}{4})$ 10.30878	5. $(2\,\overset{-1}{4})$ 12.2804094	8. $(4\,\overset{-2}{4})$ 13.030105	19. $(2\,\overset{\sim2}{4}\,\overset{-2}{4})$ 15.758917	29. $(4\,\overset{-2}{4}\,\overset{-2}{4})$ 18.491609	69. $(2\,\overset{\sim2}{4}\,\overset{-2}{4}\,\overset{-2}{4})$ 21.226563	97. $(4\,\overset{-2}{4}\,\overset{-2}{4}\,\overset{-2}{4})$ 23.963

120E' Group (9) 30(4S)6-e

2. $(1\,\overset{2}{4})$ 8.85845	4. $(2\,\overset{\sim2}{4})$ 9.340402	5. $(2\,\overset{-1}{4})$ 10.613928	8. $(4\,\overset{-2}{4})$ 11.10587	19. $(2\,\overset{\sim2}{4}\,\overset{-2}{4})$ 12.88619	29. $(4\,\overset{-2}{4}\,\overset{-2}{4})$ 14.674804	69. $(2\,\overset{\sim2}{4}\,\overset{-2}{4}\,\overset{-2}{4})$ 16.468407	97. $(4\,\overset{-2}{4}\,\overset{-2}{4}\,\overset{-2}{4})$ 18.265235

120E' Group (10): 12(10 S_1) + 20(6 S_2) [x = 1, y = E10M, z = E6M]

	P. 1								P. 2								P. 3							
	1	2	5	7	8	9	10	12	13	14	15	18	26	43	44	50	55	56	72	74	76	82	103	114
1	•		•	•		•			•	•	•		•	•	•		•	•	•	•	•		•	•
2			•			•											•						•	
5	•		•	•		•			•	•	•		•	•	•		•	•	•	•	•		•	•
7	•		•	•		•			•	•	•		•	•	•		•	•	•	•	•		•	•
8						•											•						•	
9	•		•	•		•			•	•	•		•	•	•		•	•	•	•	•		•	•
10			•			•											•						•	
11	•	•	•	•	•	•	•	•	•	•	•	•	•	•	•	•	•	•	•	•	•	•	•	•
12						•											•						•	
13	•		•	•		•			•	•	•		•	•	•			•	•	•	•		•	•
14	•		•	•		•			•	•	•		•	•	•			•	•	•	•		•	•
15	•		•	•		•			•	•	•		•	•	•			•	•	•	•		•	•
18						•																	•	
23																								
25	•	•	•	•	•	•	•	•	•	•	•	•	•	•	•	•	•	•	•	•	•	•	•	•
26	•		•	•		•			•	•	•		•	•	•			•	•	•	•		•	•
39	•		•	•		•			•	•	•		•	•	•			•	•	•	•		•	•
40						•					•												•	
43	•	•	•	•		•			•	•	•		•	•	•		•	•	•	•	•		•	•
44	•		•	•		•			•	•	•		•	•	•		•	•	•	•	•		•	•
45	•		•	•		•			•	•	•		•	•	•		•	•	•	•	•		•	•
50						•					•						•							
55	•		•	•		•			•	•	•		•	•	•		•	•	•	•	•		•	•
56	•		•	•		•			•	•	•		•	•	•		•	•	•	•	•		•	•
57	•		•	•	•	•	•	•	•	•	•	•	•	•	•	•	•	•	•	•	•	•	•	•
59	•		•	•		•			•	•	•		•	•	•		•	•	•	•	•		•	•
60	•		•	•		•			•	•	•		•	•	•		•	•	•	•	•		•	•
61	•	•	•	•	•	•	•	•	•	•	•	•	•	•	•	•	•	•	•	•	•	•	•	•
64						•					•						•						•	

68						•											•						•	
71	•	•	•	•	•	•	•	•	•	•	•	•	•	•	•	•	•	•	•	•	•	•	•	•
72	•		•	•		•			•	•	•		•	•	•		•	•	•	•	•		•	•
73	•	•	•	•	•	•	•	•	•	•	•	•	•	•	•	•	•	•	•	•	•	•	•	•
74			•	•		•			•	•	•		•	•	•		•	•	•	•	•		•	•
75	•	•	•	•	•	•	•	•	•	•	•	•	•	•	•	•	•	•	•	•	•	•	•	•
76	•		•	•		•			•	•	•		•	•	•		•	•	•	•	•		•	•
82						•																	•	
103	•		•	•		•			•	•	•		•	•	•		•	•	•	•	•		•	•
105	•		•	•		•			•	•	•		•	•	•		•	•	•	•	•		•	•
108																								
112	•		•	•		•			•	•	•		•	•	•		•	•	•	•	•		•	•
114	•		•	•		•			•	•	•		•	•	•		•	•	•	•	•		•	•

120E' Group (10): $12(10\ S_1)e_1 + 20(6\ S_2)e_2$								
P. 1	1. $(3 \sim 6^{2})$	2. $(3 - 6^{2})$	5. $(6 - 6^{2})$	7. $(1\ 3 \sim 6^{2})$	8. $(1\ 3 - 6^{2})$	9. $(3 \sim 3 \sim 6^{2})$	10. $(3 \sim 3 - 6^{2})$	12. $(3 - 3 - 6^{2})$
1. $(5 \sim 10^{2})$ R =	9.2018895	Ûa ×	10.967688	9.658657	Ûa ×	9.712548	Ûa ×	Ûa ×
2. $(5 - 10^{2})$	Ûa ×	Ûa ×	Ûa ×	Ûa ×	Ûa ×	9.781091	Ûa ×	Ûa ×
5. $(10 - 10^{2})$	11.839646	Ûa ×	13.595596	12.284727	Ûa ×	12.332605	Ûa ×	Ûa ×
7. $(1\ 5 \sim 10^{2})$	9.5566506	Ûa ×	11.323113	10.012033	Ûa ×	10.064897	Ûa ×	Ûa ×
8. $(1\ 5 - 10^{2})$	Ûa ×	Ûa ×	Ûa ×	Ûa ×	Ûa ×	10.22367	Ûa ×	Ûa ×
9. $(5 \sim 5 \sim 10^{2})$	10.712807	Ûa ×	12.466415	11.160135	Ûa ×	11.209888	Ûa ×	Ûa ×
10. $(5 \sim 5 - 10^{2})$	Ûa ×	Ûa ×	Ûa ×	Ûa ×	Ûa ×	11.916911	Ûa ×	Ûa ×
11. $(5 - 5 \sim 10^{2})$	11.167997	11.621561	12.913955	11.611787	12.864247	11.66045	13.366231	14.64291
12. $(5 - 5 - 10^{2})$	Ûa ×	Ûa ×	Ûa ×	Ûa ×	Ûa ×	12.810789	Ûa ×	Ûa ×
13. $(5 \sim 10^{2} \sim 10^{2})$	12.16809	Ûa ×	13.931617	12.614692	Ûa ×	12.662337	Ûa ×	Ûa ×
14. $(5 \sim 10^{2} - 10^{2})$	12.24831	Ûa ×	14.016173	12.969014	Ûa ×	12.743666	Ûa ×	Ûa ×
15. $(5 - 10^{2} \sim 10^{2})$	12.227	Ûa ×	13.99217	12.673978	Ûa ×	12.7215772	Ûa ×	Ûa ×
18. $(\underline{10\ 5} - 10^{2})$	Ûa ×	Ûa ×	Ûa ×	Ûa ×	Ûa ×	14.540291	Ûa ×	Ûa ×
25. $(10 - 10^{2} \sim 10^{2})$	13.831145	14.286497	15.567495	14.267079	15.554032	14.312067	16.022743	17.295702
26. $(10 - 10^{2} - 10^{2})$	14.875529	Ûa ×	16.638156	15.317594	Ûa ×	15.3624	Ûa ×	Ûa ×
39. $(1\ 5 \sim 5 \sim 10^{2})$	10.824034	Ûa ×	12.575553	11.270412	Ûa ×	11.319875	Ûa ×	Ûa ×
40. $(1\ 5 \sim 5 - 10^{2})$	Ûa ×	Ûa ×	Ûa ×	Ûa ×	Ûa ×	12.113599	Ûa ×	Ûa ×
43. $(1\ 5 \sim 10^{2} \sim 10^{2})$	12.440347	12.604811	14.201953	15.88585	Ûa ×	12.933065	Ûa ×	Ûa ×
44. $(1\ 5 \sim 10^{2} - 10^{2})$	12.604811	Ûa ×	14.37291	13.051883	Ûa ×	13.099083	Ûa ×	Ûa ×
45. $(1\ 5 - 10^{2} \sim 10^{2})$	12.5647	Ûa ×	14.328033	13.01047	Ûa ×	13.057676	Ûa ×	Ûa ×
50. $(5 \sim 5 - 5 - 10^{2})$	Ûa ×	Ûa ×	Ûa ×	Ûa ×	Ûa ×	14.958561	Ûa ×	Ûa ×
55. $(5 \sim 5 \sim 10^{2} \sim 10^{2})$	13.219186	Ûa ×	14.967625	13.65948	Ûa ×	13.70541	Ûa ×	Ûa ×
56. $(5 \sim 5 \sim 10^{2} - 10^{2})$	13.743755	Ûa ×	15.505444	14.187033	Ûa ×	14.2328	Ûa ×	Ûa ×
57. $(5 \sim 5 - 10^{2} \sim 10^{2})$	13.63345	Ûa ×	15.379651	14.072504	15.36577	14.11795	15.836648	17.114403
59. $(5 - 5 \sim 10^{2} \sim 10^{2})$	13.482196	Ûa ×	15.2237575	13.920115	Ûa ×	13.965574	Ûa ×	Ûa ×

60. (5 -5 ~10^{2} -10^{2})	14.187925	Ûa ×	15.945445	14.629405	Ûa ×	14.67465	Ûa ×	Ûa ×
61. (5 -5 -10^{2} ~10^{2})	14.031975	14.486404	15.764077	14.466515	15.750982	14.511196	16.218646	17.489172
64. (5 ~10^{2} ~5 -10^{2})	Ûa ×	Ûa ×	Ûa ×	Ûa ×	Ûa ×	14.731475	Ûa ×	Ûa ×
68. (5 -10^{2} ~5 -10^{2})	Ûa ×	Ûa ×	Ûa ×	Ûa ×	Ûa ×	14.7634505	Ûa ×	Ûa ×
71. (5 ~10^{2} ~10^{2} ~10^{2})	13.973847	14.4283895	15.706725	14.408645	15.693522	14.45341	16.16139	17.432466
72. (5 ~10^{2} ~10^{2} -10^{2})	15.21573	Ûa ×	16.9821175	15.658495	Ûa ×	15.70315	Ûa ×	Ûa ×
73. (5 ~10^{2} -10^{2} ~10^{2})	14.006159	14.460615	15.73855	14.440784	15.725393	14.485501	16.193152	17.463873
74. (5 ~10^{2} -10^{2} -10^{2})	15.30264	Ûa ×	17.071402	15.745967	Ûa ×	15.790628	Ûa ×	Ûa ×
75. (5 -10^{2} ~10^{2} ~10^{2})	13.997155	14.451515	15.729316	14.431729	15.71616	14.476444	16.183843	17.454537
76. (5 -10^{2} ~10^{2} -10^{2})	15.277188	Ûa ×	17.044431	15.720114	Ûa ×	15.764769	Ûa ×	Ûa ×
82. (10 5 -5 -10^{2})	Ûa ×	Ûa ×	Ûa ×	Ûa ×	Ûa ×	17.57986	Ûa ×	Ûa ×
103. (10 ~10 ~10 -10^{2})	15.5304226	Ûa ×	17.275216	15.966777	Ûa ×	16.01065	Ûa ×	Ûa ×
105. (10 -10 ~10 -10^{2})	16.48587	Ûa ×	17.591119	16.923713	Ûa ×	16.945351	Ûa ×	Ûa ×
112. (10 -10^{2} ~10^{2} -10^{2})	16.8382807	Ûa ×	18.5888295	17.275089	Ûa ×	17.31835	Ûa ×	Ûa ×
114. (10 -10^{2} -10^{2} -10^{2})	17.924025	Ûa ×	19.689649	18.364243	Ûa ×	18.40732	Ûa ×	Ûa ×

(continue) 120E' Group (10): 12(10-S)e1 + 20(6-S)e2
P. 2

	13. (3 ~6^{2} ~6^{2})	14. (3 ~6^{2} -6^{2})	15. (3 -6^{2} ~6^{2})	18. (6 3 -6^{2})	26. (6 -6^{2} -6^{2})	43. (1 3 ~6^{2} ~6^{2})	44. (1 3 ~6^{2} -6^{2})	50. (3 ~3 -3^{2} -6^{2})
1. (5 ~10^{2}) R =	11.43589	12.24831	11.497269 5	Ûa ×	14.0161597	11.496689	12.696014	Ûa ×
5. (10 -10^{2})	14.05516	14.875542	14.10439	Ûa ×	16.638156	14.10355	15.317599	Ûa ×
7. (1 5 ~10^{2})	11.789469	12.604815	11.849601	Ûa ×	14.37291	11.848947	13.05188	Ûa ×
8. (1 5 -10^{2})	Ûa ×	Ûa ×	12.01517 Ûa × =180.0003	Ûa ×	Ûa ×	Ûa ×	Ûa ×	Ûa ×
9. (5 ~5 ~10^{2})	17.92495	13.743736	12.9814	Ûa ×	15.505444	12.9806	14.18704	Ûa ×
11. (5 -5 ~10^{2})	13.369213	14.187924	13.424377	15.12369	15.945444	13.423565	14.629406	16.39999
12. (5 -5 -10^{2})	Ûa ×	Ûa ×	Ûa ×	Ûa ×	Ûa ×	Ûa ×	Ûa ×	Ûa ×
13. (5 ~10^{2} ~10^{2})	14.386979	15.215729	14.440354	Ûa ×	16.982115	14.4395	15.65848	Ûa ×
14. (5 ~10^{2} -10^{2})	14.472156	15.302649 5	14.52545	Ûa ×	17.071374	14.5246	15.7459674	Ûa ×
15. (5 -10^{2} ~10^{2})	14.447685	15.277188	14.500975	Ûa ×	17.0444307	14.500125	15.720113	Ûa ×
25. (10 -10^{2} ~10^{2})	16.013122	16.838281	16.063198	17.771669	18.588825	16.06237	17.27509	19.046246
26. (10 -10^{2} -10^{2})	17.0870677	17.92403	17.136354	Ûa ×	19.689666	17.135535	18.364243	Ûa ×
39. (1 5 ~5 ~10^{2})	13.03322	13.851974	13.089349	Ûa ×	15.612556	13.08855	14.294789	Ûa ×
40. (1 5 ~5 -10^{2})	Ûa ×	Ûa ×	13.89836	Ûa ×	Ûa ×	Ûa ×	Ûa ×	Ûa ×
43. (1 5 ~10^{2} ~10^{2})	14.656135	15.4853	14.708949	Ûa ×	17.2505694	14.708099 7	15.927495	Ûa ×

44. $(1\ 5\,{}^{\sim}10^{2}\,{}^{-}10^{2})$	14.827961	15.659914	14.880621	Ûa ×	17.42878	14.879757	16.102917	Ûa ×
45. $(1\ 5\,{}^{-}10^{2}\,{}^{\sim}10^{2})$	14.782208	15.612400 5	14.834839	Ûa ×	17.3785734	14.833994	16.0546848	Ûa ×
50. $(5\,{}^{\sim}5\,{}^{-}5\,{}^{-}10^{2})$	Ûa ×	Ûa ×	16.741291	Ûa ×	Ûa ×	Ûa ×	Ûa ×	Ûa ×
55. $(5\,{}^{\sim}5\,{}^{\sim}10^{2}\,{}^{\sim}10^{2})$	15.417141	16.244664	15.468372	Ûa ×	18.002498	15.467529	16.68399	Ûa ×
56. $(5\,{}^{\sim}5\,{}^{\sim}10^{2}\,{}^{-}10^{2})$	15.9564	16.790000	16.007138	Ûa ×	18.55522	16.006266 6	17.230952	Ûa ×
57. $(5\,{}^{\sim}5\,{}^{-}10^{2}\,{}^{\sim}10^{2})$	15.827687	16.655588	15.878237	17.59205	18.411732	15.877407	17.094106	18.870079
59. $(5\,{}^{-}5\,{}^{\sim}10^{2}\,{}^{\sim}10^{2})$	15.671251	16.497263	15.721955	Ûa ×	18.251141	15.721117	16.935263	Ûa ×
60. $(5\,{}^{-}5\,{}^{\sim}10^{2}\,{}^{-}10^{2})$	16.394644	17.22804	16.444644	Ûa ×	18.990772	1644382	17.667965	Ûa ×
61. $(5\,{}^{-}5\,{}^{-}10^{2}\,{}^{\sim}10^{2})$	16.208353	17.032371	16.25802	17.96409	18.77984	16.257196	17.468195	19.23674
64. $(5\,{}^{\sim}10^{2}\,{}^{\sim}5\,{}^{-}10^{2})$	Ûa ×	Ûa ×	16.507616	Ûa ×	Ûa ×	Ûa ×	Ûa ×	Ûa ×
71. $(5\,{}^{\sim}10^{2}\,{}^{\sim}10^{2}\,{}^{\sim}10^{2})$	16.151287	16.975554 5	16.201098	17.907678	18.72375	16.200277	17.411605	19.1807835
72. $(5\,{}^{\sim}10^{2}\,{}^{\sim}10^{2}\,{}^{-}10^{2})$	17.431206	18.27027	17.48013	Ûa ×	20.037996	17.479345	18.710793	Ûa ×
73. $(5\,{}^{\sim}10^{2}\,{}^{-}10^{2}\,{}^{\sim}10^{2})$	16.182897	17.007033	16.23267	17.938923	18.754826	16.231827	17.442961	19.211804
74. $(5\,{}^{\sim}10^{2}\,{}^{-}10^{2}\,{}^{-}10^{2})$	17.5208	18.360871 4	17.569708	Ûa ×	20.130074	17.568871	18.801757	Ûa ×
75. $(5\,{}^{-}10^{2}\,{}^{\sim}10^{2}\,{}^{\sim}10^{2})$	16.173687	16.997728	16.223420 1	17.929557	18.745399	16.2226	17.433624	19.20239
76. $(5\,{}^{-}10^{2}\,{}^{\sim}10^{2}\,{}^{-}10^{2})$	17.493582	18.333019	17.542473	Ûa ×	20.1013002	17.541663	18.773676	Ûa ×
103. $(10\,{}^{\sim}10\,{}^{\sim}10\,{}^{-}10^{2})$	17.7193403	18.551438	17.767581	Ûa ×	20.306195	17.766695	18.988244	Ûa ×
105. $(10\,{}^{-}10\,{}^{\sim}10\,{}^{-}10^{2})$	18.59719	19.408589 4	18.643106	Ûa ×	21.124701	18.642401	19.834343	Ûa ×
112. $(10\,{}^{-}10^{2}\,{}^{\sim}10^{2}\,{}^{-}10^{2})$	19.03223	19.868794	19.07927	Ûa ×	21.626606	19.078519	20.503735	Ûa ×
114. $(10\,{}^{-}10^{2}\,{}^{-}10^{2}\,{}^{-}10^{2})$	20.13495	20.978704	20.181426	Ûa ×	22.7459917	20.180702	21.417682	Ûa ×

(continue)**120E' Group (10): 12(10-S)e_1 + 20(6-S)e_2**
P. 3

	55. $(3\,{}^{\sim}3\,{}^{\sim}6^{2}\,{}^{\sim}6^{2})$	56. $(3\,{}^{\sim}3\,{}^{\sim}6^{2}\,{}^{-}6^{2})$	72. $(3\,{}^{\sim}6^{2}\,{}^{\sim}6^{2}\,{}^{-}6^{2})$	74. $(3\,{}^{\sim}6^{2}\,{}^{-}6^{2}\,{}^{-}6^{2})$	76. $(3\,{}^{-}6^{2}\,{}^{\sim}6^{2}\,{}^{-}6^{2})$	82. $(\underline{6\ 3}\,{}^{-}3\,{}^{-}6^{2})$	103. $(6\,{}^{\sim}6\,{}^{\sim}6\,{}^{-}6^{2})$	114. $(6\,{}^{-}6^{2}\,{}^{-}6^{2}\,{}^{-}6^{2})$
1. $(5\,{}^{\sim}10^{2})$	11.4994312	12.743666	14.472156	15.302641	14.525445	Ûa ×	13.290284	17.071402
2. $(5\,{}^{-}10^{2})$	11.572862	Ûa ×	Ûa ×	Ûa ×	Ûa ×	Ûa ×	13.366727	Ûa ×
5. $(10\,{}^{-}10^{2})$	14.10534	15.362418	17.087068	17.924025	17.136355	Ûa ×	15.885495	19.689645
7. $(1\ 5\,{}^{\sim}10^{2})$	11.851515	13.099083	14.827952	15.659915	14.880621	Ûa ×	13.642045	17.428786
8. $(1\ 5\,{}^{-}10^{2})$	12.016999	Ûa ×	Ûa ×	Ûa ×	Ûa ×	Ûa ×	13.811711	Ûa ×
9. $(5\,{}^{\sim}5\,{}^{\sim}10^{2})$	12.982707	14.23281	15.9564	16.79	16.007136	Ûa ×	14.763637	18.55522
10. $(5\,{}^{\sim}5\,{}^{-}10^{2})$	13.703637	Ûa ×	Ûa ×	Ûa ×	Ûa ×	Ûa ×	15.49343	Ûa ×
11. $(5\,{}^{-}5\,{}^{\sim}10^{2})$	13.42553	14.674645	16.394646	17.22804	16.444644	18.1639315	15.200859	18.99078
12. $(5\,{}^{-}5\,{}^{-}10^{2})$	Ûa ×	Ûa ×	Ûa ×	Ûa ×	Ûa ×	Ûa ×	16.387467	Ûa ×

13. (5 ~$\overset{2}{10}$ ~$\overset{2}{10}$)	14.4412198	15.703151	17.4312	18.27027	17.48013	Ûa ×	16.225246	20.038
14. (5 ~$\overset{2}{10}$ -$\overset{2}{10}$)	14.526301	15.790628	17.5207968	18.360858	17.569708	Ûa ×	16.312887	20.1300783
15. (5 -$\overset{2}{10}$ ~$\overset{2}{10}$)	14.5018258	15.764769	17.49357	18.333053	17.5424737	Ûa ×	16.286757	20.101266
18. (10 5 -$\overset{2}{10}$)	16.316165	Ûa ×	Ûa ×	Ûa ×	Ûa ×	Ûa ×	18.097226	Ûa ×
25. (10 -$\overset{2}{10}$ ~$\overset{2}{10}$)	16.063755	17.318343	19.03223	19.868798	19.079291	20.80392	17.827353	21.626606
26. (10 -$\overset{2}{10}$ -$\overset{2}{10}$)	17.136786	18.4073195	20.13494	20.978704	20.181426	Ûa ×	18.9163007	22.745903
39. (1 5 ~5 ~$\overset{2}{10}$)	13.090619	14.340431	16.063075	16.8966094	16.113603	Ûa ×	14.870072	18.661206
40. (1 5 ~5 -$\overset{2}{10}$)	13.899349	Ûa ×	Ûa ×	Ûa ×	Ûa ×	Ûa ×	15.68838	Ûa ×
43. (1 5 ~$\overset{2}{10}$ ~$\overset{2}{10}$)	14.709754	15.971908	17.69901	18.53823	12.74764	Ûa ×	16.492103	20.305336
44. (1 5 ~$\overset{2}{10}$ -$\overset{2}{10}$)	14.88139	16.147355	17.877639	18.71857	17.926201	Ûa ×	16.667597	20.487806
45. (1 5 -$\overset{2}{10}$ ~$\overset{2}{10}$)	14.835617	16.099054	17.826968	18.666774	17.8755	Ûa ×	16.618803	20.434374
50. (5 ~5 -5 -$\overset{2}{10}$)	16.741729	Ûa ×	Ûa ×	Ûa ×	Ûa ×	Ûa ×	18.527833	Ûa ×
55. (5 ~5 ~$\overset{2}{10}$ ~$\overset{2}{10}$)	15.469027	16.7277575	18.448233	19.286465	18.4959555	Ûa ×	17.241715	21.048865
56. (5 ~5 ~$\overset{2}{10}$ -$\overset{2}{10}$)	16.007688	17.2745696	19.001761	19.843498	19.049176	Ûa ×	17.788112	21.6105
57. (5 ~5 -$\overset{2}{10}$ ~$\overset{2}{10}$)	15.878836	17.137605	18.856544	19.694774	18.903851	20.631774	17.64909	21.455999
59. (5 -5 ~$\overset{2}{10}$ ~$\overset{2}{10}$)	15.7225604	16.97879	18.695673	19.532886	18.7431	Ûa ×	17.490351	21.292892
60. (5 -5 ~$\overset{2}{10}$ -$\overset{2}{10}$)	16.445158	17.7112895	19.436225	20.277871	19.483217	Ûa ×	18.2222	22.043239
61. (5 -5 -$\overset{2}{10}$ ~$\overset{2}{10}$)	16.258559	17.511263	19.222282	20.057966	19.26911	20.99204	18.018497	21.81357
64. (5 ~$\overset{2}{10}$ ~5 -$\overset{2}{10}$)	16.508098	Ûa ×	Ûa ×	Ûa ×	Ûa ×	Ûa ×	18.289498	Ûa ×
68. (5 -$\overset{2}{10}$ ~5 -$\overset{2}{10}$)	16.54018	Ûa ×	Ûa ×	Ûa ×	Ûa ×	Ûa ×	18.3215975	Ûa ×
71. (5 ~$\overset{2}{10}$ ~$\overset{2}{10}$ ~$\overset{2}{10}$)	16.201627	17.454727	19.166486	20.0023852	19.213366	20.936695	17.962442	21.758603
72. (5 ~$\overset{2}{10}$ ~$\overset{2}{10}$ -$\overset{2}{10}$)	17.48058	18.753823	20.483329	21.3284	20.529636	Ûa ×	19.262094	23.00688
73. (5 ~$\overset{2}{10}$ -$\overset{2}{10}$ ~$\overset{2}{10}$)	16.233177	17.496019	19.197396	20.033165	19.244209	20.967327	17.993475	21.789049
74. (5 ~$\overset{2}{10}$ -$\overset{2}{10}$ -$\overset{2}{10}$)	17.570122	18.844777	20.575576	21.421306	20.621855	Ûa ×	19.3531805	23.190725
75. (5 -$\overset{2}{10}$ ~$\overset{2}{10}$ ~$\overset{2}{10}$)	16.223955	17.476723	19.188011	20.023764	19.234859	20.9579455	17.984188	21.77963
76. (5 -$\overset{2}{10}$ ~$\overset{2}{10}$ -$\overset{2}{10}$)	17.54285	18.816681	20.546646	21.391983	20.592926	Ûa ×	19.32488	23.160776
82. (10 5 -5 -$\overset{2}{10}$)	19.356154	Ûa ×	Ûa ×	Ûa ×	Ûa ×	Ûa ×	21.13616	Ûa ×
103. (10 ~10 ~10 -$\overset{2}{10}$)	17.7678918	19.0307894	20.748489	21.589113	20.794314	Ûa ×	19.53434	23.349182
105. (10 -10 ~10 -$\overset{2}{10}$)	18.643394	19.875584	21.557899	22.385238	21.602447	Ûa ×	20.365167	24.119892
112. (10 -$\overset{2}{10}$ ~$\overset{2}{10}$ -$\overset{2}{10}$)	19.0796	20.347906	22.06833	22.911681	22.113438	Ûa ×	20.84822	24.673545
114. (10 -$\overset{2}{10}$ -$\overset{2}{10}$ -$\overset{2}{10}$)	20.181646	21.459749	23.188699	24.036854	23.23358	Ûa ×	21.95989	25.804904

End of 120E' Group (10): 12(10S1)e1 + 20(6S2)e2 (P. 3)

120E' Group (11): 20(6S_1)e_1 + 30(4S_2)e_2

	2. (1 $\overset{1}{4}$)	4. (2 $\overset{\sim 1}{4}$)	5. (2 $\overset{-1}{4}$)	8. (4 $\overset{-2}{4}$)	19. (2 $\overset{\sim 2}{4}$ $\overset{-2}{4}$)	29. (4 $\overset{-2}{4}$ $\overset{-2}{4}$)	69. (2 $\overset{\sim 2}{4}$ $\overset{-2}{4}$ $\overset{-2}{4}$)	97. (4 $\overset{-2}{4}$ $\overset{-2}{4}$ $\overset{-2}{4}$)
1. (3 $\overset{\sim 2}{6}$) R =	Ûa ×	12.5973	Ûa ×	15.32893	18.063288	20.799035	23.536465	26.274321
5. (6 $\overset{-2}{6}$)	Ûa ×	15.7589	Ûa ×	1.849163	21.226563	23.962976	26.7003431	29.438334
7. (1 3 $\overset{\sim 2}{6}$)	Ûa ×	13.395675	Ûa ×	16.12099	18.85154	21.585123	24.320555	27.057254
9. (3 $\tilde{3}$ $\overset{\sim 2}{6}$)	12.742229	13.47914	15.458004	16.200307	18.92837	21.660299	24.35459	27.130507
13. (3 $\overset{\sim 2}{6}$ $\overset{\sim 2}{6}$)	Ûa ×	16.565274	Ûa ×	19.2906408	22.02076	24.75378	27.488742	30.22485
14. (3 $\overset{\sim 2}{6}$ $\overset{-2}{6}$)	Ûa ×	18.063286	Ûa ×	20.799305	23.536444	26.274326	29.01259	31.751365
15. (3 $\overset{-2}{6}$ $\overset{\sim 2}{6}$)	Ûa ×	16.6545	18.636747*	19.374934	20.101811	24.832556	27.565891	30.300878
26. (6 $\overset{-2}{6}$ $\overset{-2}{6}$)	Ûa ×	21.22659	Ûa ×	23.962995	26.700343	29.438472	32.1766035	34.915906
43. (1 3 $\overset{\sim 2}{6}$ $\overset{\sim 2}{6}$)	Ûa ×	16.653	Ûa ×	19.37358	22.10057	24.8315	27.56478	30.29999
44. (1 3 $\overset{\sim 2}{6}$ $\overset{-2}{6}$)	Ûa ×	18.851539	Ûa ×	21.58517	24.320574	27.057254	29.794783	32.53302
55. (3 $\tilde{3}$ $\overset{\sim 2}{6}$ $\overset{\sim 2}{6}$)	15.92204	16.65535	18.637302	19.37537	22.102037	24.83274	27.566042	30.30088
56. (3 $\tilde{3}$ $\overset{\sim 2}{6}$ $\overset{-2}{6}$)	Ûa ×	18.92837	Ûa ×	21.660299	24.394743	27.13052	29.8676	32.6054058
72. (3 $\overset{\sim 2}{6}$ $\overset{\sim 2}{6}$ $\overset{-2}{6}$)	Ûa ×	22.02076	Ûa ×	24.75378	27.488742	30.224848	32.9621	35.69986
74. (3 $\overset{\sim 2}{6}$ $\overset{-2}{6}$ $\overset{-2}{6}$)	Ûa ×	23.5364946	Ûa ×	26.27432	29.0128	31.75137	34.49045	37.22993
76. (3 $\overset{-2}{6}$ $\overset{\sim 2}{6}$ $\overset{-2}{6}$)	Ûa ×	22.10181	Ûa ×	24.83157	27.565858	30.300878	33.03715	35.77407
103. (6 $\tilde{6}$ $\tilde{6}$ $\overset{-2}{6}$)	19.102366	19.833697	21.817295	22.552735	25.27803	28.00747	30.73945	$\hat{V}_{21}$ ×
114. (6 $\overset{-2}{6}$ $\overset{-2}{6}$ $\overset{-2}{6}$)	Ûa ×	26.700343	Ûa ×	29.4383344	32.1766034	34.91555	37.655	40.39445

* 15. (3 $\overset{-2}{6}$ $\overset{\sim 2}{6}$) + 5. (2 $\overset{-1}{4}$) Ûa = 179.9997 deg.

120E' Group (12): 12(10 S_1) + 30(4 S_2)

	2. (1 $\overset{1}{4}$)	4. (2 $\overset{\sim 1}{4}$)	5. (2 $\overset{-1}{4}$)	8. (4 $\overset{-2}{4}$)	19. (2 $\overset{\sim 2}{4}$ $\overset{-2}{4}$)	29. (4 $\overset{-2}{4}$ $\overset{-2}{4}$)	69. (2 $\overset{\sim 2}{4}$ $\overset{-2}{4}$ $\overset{-2}{4}$)	97. (4 $\overset{-2}{4}$ $\overset{-2}{4}$ $\overset{-2}{4}$)
1. (5 $\overset{\sim 2}{10}$) R =	Ûa ×	9.784493	Ûa ×	11.579116	13.377045	15.17707	16.978505	18.78096
5. (10 $\overset{-2}{10}$)	Ûa ×	12.88619	Ûa ×	14.674802 1	16.46844	18.265244	20.0642	21.86475
7. (1 5 $\overset{\sim 2}{10}$)	Ûa ×	10.20302	Ûa ×	11.998865	13.79755	15.598055	17.399806	19.202467
9. (5 $\tilde{5}$ $\overset{\sim 2}{10}$)	Ûa ×	11.557755	Ûa ×	13.342665	15.13399	16.92935	18.7273	20.52714
11. (5 $\overset{-}{5}$ $\overset{\sim 2}{10}$)	11.60644	12.088464	13.3784	13.86658	15.65309	17.444884	10.24019	21.03824
13. (5 $\overset{\sim 2}{10}$ $\overset{\sim 2}{10}$)	Ûa ×	13.27905	Ûa ×	15.074928	16.873435	18.673694	20.47517	22.277658
14. (5 $\overset{\sim 2}{10}$ $\overset{-2}{10}$)	Ûa ×	13.377044	Ûa ×	15.17707	16.978522	18.780914	20.584039	22.38779
15. (5 $\overset{-2}{10}$ $\overset{\sim 2}{10}$)	Ûa ×	13.349786	Ûa ×	15.147222	16.94682	18.74783	20.549896	22.35282
25. (10 $\overset{-2}{10}$ $\overset{\sim 2}{10}$)	14.734863 0	15.212399	16.50163	16.984551	18.765457	20.552355	22.34353	24.137795

26. $(10^{-}\overset{2}{10}^{-}\overset{2}{10})$	Ûa ×	16.46844	Ûa ×	18.26525	20.064195	21.864757	23.6665	25.46885
39 $(1\ 5^{\sim}5^{\sim}\overset{2}{10})$	Ûa ×	11.68735	Ûa ×	13.47041	15.26042	17.0548	18.85209	20.651408
43. $(1\ 5^{\sim}\overset{2}{10}^{\sim}\overset{2}{10})$	Ûa ×	13.598271	Ûa ×	15.392555	17.1898946	18.989291	20.7901	23.592016
44. $(1\ 5^{\sim}\overset{2}{10}^{-}\overset{2}{10})$	Ûa ×	13.79755	Ûa ×	15.598053	17.3998	19.202498	21.005784	22.8096342
45. $(1\ 5^{-}\overset{2}{10}^{\sim}\overset{2}{10})$	Ûa ×	13.74625	Ûa ×	15.542215	17.3406744	19.1408759	20.94237	22.744707
55. $(5^{\sim}5^{\sim}\overset{2}{10}^{\sim}\overset{2}{10})$	Ûa ×	14.603956	Ûa ×	16.286705	18.075653	19.868757	21.6648	23.462887
56. $(5^{\sim}5^{\sim}\overset{2}{10}^{-}\overset{2}{10})$	Ûa ×	15.133991	Ûa ×	16.92935	18.7273	20.52714	22.32828	24.1303
57. $(5^{\sim}5^{-}\overset{2}{10}^{\sim}\overset{2}{10})$	14.508756	14.989353	16.284996	16.770185	18.557449	20.34917	22.144009	23.94114
59. $(5^{-}5^{\sim}\overset{2}{10}^{\sim}\overset{2}{10})$	Ûa ×	14.807047	Ûa ×	16.58374	18.36819	20.157945	21.95133	23.7474
60. $(5^{-}5^{\sim}\overset{2}{10}^{-}\overset{2}{10})$	Ûa ×	15,652088	Ûa ×	17.4448844	19.24019	21.03804	22.837556	24.63835
61. $(5^{-}5^{-}\overset{2}{10}^{\sim}\overset{2}{10})$	14.9682	15.4443385	16.730699	17.212353	18.989824	20.77393	22.56278	24.355145
71. $(5^{\sim}\overset{2}{10}^{\sim}\overset{2}{10}^{\sim}\overset{2}{10})$	14.900356	15.376765	16.663707	17.14564	18.92394	20.7088	22.498301	24.29113
72. $(5^{\sim}\overset{2}{10}^{\sim}\overset{2}{10}^{-}\overset{2}{10})$	Ûa ×	16.87345	Ûa ×	18.67366	20.47517	22.277658	24.080795	25.884405
73. $(5^{\sim}\overset{2}{10}^{-}\overset{2}{10}^{\sim}\overset{2}{10})$	14.938003	15.4142455	16.70084	17.18259	18.9604	20.74484	22.533995	24.32653
74. $(5^{\sim}\overset{2}{10}^{-}\overset{2}{10}^{-}\overset{2}{10})$	Ûa ×	16.97853	Ûa ×	18.78092	20.58401	22.38771	24.191799	25.996258
75. $(5^{-}\overset{2}{10}^{\sim}\overset{2}{10}^{\sim}\overset{2}{10})$	14.92726	15.40346	16.68991	17.17164	18.94939	20.733798	22.52296	24.315545
76. $(5^{-}\overset{2}{10}^{\sim}\overset{2}{10}^{-}\overset{2}{10})$	Ûa ×	16.9468	Ûa ×	18.74783	20.5499	22.35282	24.1563005	25.96006
103. $(10^{\sim}10^{\sim}10^{-}\overset{2}{10})$	Ûa ×	17.219544	Ûa ×	19.00069	20.78778	22.57903	24.3733	26.170046
105. $(10^{-}10^{\sim}10^{-}\overset{2}{10})$	Ûa ×	18.26564	Ûa ×	19.992434	21.7378	23,496575	25.26496	27.040662
112. $(10^{-}\overset{2}{10}^{\sim}\overset{2}{10}^{-}\overset{2}{10})$	Ûa ×	18.26545	Ûa ×	20.552384	22,34353	24.137795	25.934355	27.732693
114. $(10^{-}\overset{2}{10}^{-}\overset{2}{10}^{-}\overset{2}{10})$	Ûa ×	20.0642	Ûa ×	21.864757	23.66646	25.46885	27.272	29.075753

End of 120E' Group (12) 30(4S1)e1 + 12(10S2)e2

120E' Group (13): 12(10M) + 20(6A) [x = E10M, y = x, z = E6A]	2	5	8	10	12	13	14	15	16		18	23	25	26	33	43	44	45	46
3	•	•	•	•			•		•								•		
6									•		•			•			•		•
19	•	•	•	•	•	•	•		•			•	•	•		•	•		•
21	•	•	•	•	•	•	•		•			•	•	•		•	•		•
36				•	•		•		•		•	•		•	•		•		•
115				•	•				•		•			•			•		•
117				•	•				•		•			•			•		•
119				•	•				•		•			•			•		•
121				•	•				•		•			•			•		•
147				•	•		•		•		•	•		•	•		•		•
148					•						•				•				
149																			

	50	55	56	57	58	61	62	71	72	73	74	75	76	82	85	86	103	113	114
3			•														•		
6			•		•				•		•						•		
19		•	•					•		•		•					•	•	
21		•	•					•		•		•					•	•	
36	•		•		•		•		•		•						•		
115			•		•		•		•		•						•		
117			•		•				•		•						•		
119			•		•				•		•						•		
121			•		•				•		•						•		
147	•		•		•		•		•		•					•	•		
148	•				•		•		•		•					•			
149																			

120E' Group (13): 12(10M)e + 20(6A)

P. 1	2. (3 ¯$\overset{2}{6}$)	5. (6 ¯$\overset{2}{6}$)	8. (1 3 ¯$\overset{2}{6}$)	10. (3 ~3 ¯$\overset{2}{6}$)	12. (3 ¯3 ¯$\overset{2}{6}$)	13. (3 ~$\overset{2}{6}$ ~$\overset{2}{6}$)	14. (3 ~$\overset{2}{6}$ ¯$\overset{2}{6}$)	16. (3 ¯$\overset{2}{6}$ ¯$\overset{2}{6}$)
3. (10 ~10)$_e$ R =	11.03753	12.77802	11.742179	11.903099	×	×	13.49659	13.660586
6. (20 10)$_e$	×	×	×	16.86463	17.550456	×	×	18.570465
19. (10 ~10 ~10)$_e$	16.542694	14.266834	13.42805	13.654656	13.98332	15.530435	15.167969	15.396523
21. (10 ¯10 ~10)	12.870462	14.58911	13.797804	14.038545	14.411923	15.84985	15.53299	15.77557
36. (20 ¯$\overset{2}{20}$ ~10)$_e$	×	×	×	18.705217	19.61359	×	19.93851	20.395799
115. (10 ~10 ~$\overset{2}{20}$ ~10)$_e$	×	×	×	16.945611	17.6531096	×	×	18.6757917
117. (10 ~10 ¯$\overset{2}{20}$ ~10)$_e$	×	×	×	16.9558992	17.66624	×	×	18.6893014
119. (10 ¯10 ~$\overset{2}{20}$ ~10)$_e$	×	×	×	17.3310462	18.0828	×	×	19.0568227
121. (10 ¯10 ¯$\overset{2}{20}$ ~10)$_e$	×	×	×	17.3944884	18.1558141	×	×	19.236257
147. (20 10 ~$\overset{2}{20}$ ~10)$_e$	×	×	×	19.1871043	20.1497697	×	20.3695956	20.85583
148. (20 10 ~$\overset{2}{20}$ ¯10)$_e$	×	×	×	×	21.4286719	×	×	×
P. 2	18. (6 3 ¯$\overset{2}{6}$)	23. (6 ~6 ¯$\overset{2}{6}$)	25. (6 ¯$\overset{2}{6}$ ~$\overset{2}{6}$)	26. (6 ¯$\overset{2}{6}$ ¯$\overset{2}{6}$)	33. (12 6 ¯$\overset{2}{6}$)	43. (1 3 ~$\overset{2}{6}$ ~$\overset{2}{6}$)	44. (1 3 ~$\overset{2}{6}$ ¯$\overset{2}{6}$)	46. (1 3 ¯$\overset{2}{6}$ ¯$\overset{2}{6}$)
3. (10 ~10)$_e$	×	×	×	×	×	×	13.655206	×
6. (20 10)$_e$	17.79618	×	×	19.28122	×	×	18.551	19.27766
19. (10 ~10 ~10)$_e$	×	16.443923	17.21521	15.75151	×	15.966777	15.386644	15.750345

21. (10¯10˜10)	×	16.80652	17.59111	16.174609	×	16.284937	15.76465	16.172904
36. (20¯20˜10)e	19.849845	21.184901	×	21.33134	24.333775	×	20.369115	21.32725
115. (10˜10¯20˜10)e	17.81565	×	×	19.4080439	×	×	18.6556787	19.4044333
117. (10˜10¯20˜10)e	17.82969	×	×	19.4243689	×	×	18.669107	19.4207354
119. (10¯10˜20˜10)e	18.262359	×	×	19.833549	×	×	19.035327	1.9829823
121. (10¯10˜20˜10)e	18.3387671	×	×	19.909934	×	×	19.1018457	19.9061232
147. (20 10˜20˜10)e	20.404102	21.60255	×	21.846548	24.82209	×	20.8271665	21.842406
148. (20 10˜20¯10)e	×	×	×	×	26.1527262	×	×	×

P. 3	50. (3˜3¯3¯6)	55. (3˜3˜6˜6)	56. (3˜3˜6¯6)	58. (3˜3¯6¯6)	62. (3¯3¯6¯6)	71. (3˜6˜6˜6)	72. (3˜6¯6¯6)	73. (3˜6¯6˜6)
3. (10˜10)e	×	×	13.668934	×	×	×	×	×
6. (20 10)e	×	×	18.581263	19.4394	×	×	19.43394	×
19. (10˜10˜10)e	×	16.010647	15.406271	×	×	17.7193041	×	18.551439
21. (10¯10˜10)	×	16.328483	15.785501	×	×	18.034002	×	18.865496
36. (20¯20˜10)e	20.207645	×	20.407236	21.568435	21.94946	×	21.557996	×
115. (10˜10¯20˜10)e	×	×	18.636736	19.5732412	×	×	19.5673895	×
117. (10˜10¯20˜10)e	×	×	18.7002534	19.5905	×	×	19.5846414	×
119. (10¯10˜20˜10)e	×	×	19.067863	20.0152387	×	×	20.00841	×
121. (10¯10˜20˜10)e	×	×	19.134656	20.094896	×	×	20.0879145	×
147. (20 10˜20˜10)e	20.820101	×	20.867392	22.1015215	22.5387926	×	22.08983	×
148. (20 10˜20¯10)e	22.3168745	×	×	×	24.0590487	×	23.4462042	×

P. 4	74. (3˜6¯6¯6)	75. (3¯6˜6˜6)	76. (3¯6˜6¯6)	86. (6 3¯6¯6)	103. (6˜6˜6¯6)	113. (6¯6¯6˜6)		
3. (10˜10)e	×	×	×	×	15.444588	×		
6. (20 10)e	19.57671	×	19.447182	×	20.316744	×		
19. (10˜10˜10)e	×	17.767521	×	×	17.17034	20.30619		
21. (10¯10˜10)	×	18.081768	×	×	17.545785	20.617906		
36. (20¯20˜10)e	21.861356	×	21.578697	×	22.12908	×		
115. (10˜10¯20˜10)e	19.7230215	×	19.5813036	×	20.4400359	×		
117. (10˜10¯20˜10)e	19.7419157	×	19.5986147	×	20.4559489	×		
119. (10¯10˜20˜10)e	2.01991454	×	20.0239249	×	20.8175555	×		
121. (10¯10˜20˜10)e	20.285266	×	20.1037855	×	20.8868342	×		
147. (20 10˜20˜10)e	22.427928	×	22.1120978	22.586998	22.5715704	×		
148. (20 10˜20¯10)e	23.8878222	×	23.4736413	24.1764956	×	×		

End of 120E' Group (13) 12(10M)e + 20(6A)

120E' Group (14): 12(10M)e,1 + 20(6A) [x = E10M, y = 1, z =E6A]

P. 1	1	2	5	7	8	9	10	11	12	13	14	15	16	17	18	23	25	26	43	44	45	46	49	50
1	•	•	•	•	•	•	•	•		•	•		•	•			•		•	•			•	
2	•	•	•	•		•		•		•				•			•		•				•	
5			•		•		•		•	•	•	•	•		•	•	•	•	•	•	•	•		
7	•	•	•	•	•	•	•	•		•	•		•	•			•		•	•			•	
8	•	•	•	•		•		•		•				•			•		•				•	
9	•	•	•		•		•		•	•	•		•				•	•	•	•				
10		•	•		•		•			•	•	•	•			•	•		•	•	•			
11		•	•		•		•		•	•	•		•			•	•	•	•	•		•		
12			•		•		•			•	•	•	•			•	•		•	•	•			
13			•		•		•		•	•	•	•	•		•	•	•	•	•	•	•	•		
P. 2	1	2	5	7	8	9	10	11	12	13	14	15	16	17	18	23	25	26	43	44	45	46	49	50
14			•		•		•		•	•	•	•	•		•	•	•	•	•	•	•	•		
15			•		•		•		•	•	•	•	•		•	•	•	•	•	•	•	•		
18							•		•		•		•		•	•		•		•		•		
25							•		•		•		•		•	•		•		•		•		•
26									•		•				•	•		•		•		•		•
39							•		•	•	•		•			•	•	•	•	•		•		
40							•			•	•	•	•			•	•		•	•	•			
43							•		•	•	•	•	•		•	•	•	•	•	•	•	•		
44							•		•	•	•	•	•		•	•	•	•	•	•	•	•		
45							•		•	•	•	•	•		•	•	•	•	•	•	•	•		
P. 3	1	2	5	7	8	9	10	11	12	13	14	15	16	17	18	23	25	26	43	44	45	46	49	50
50									•		•				•	•		•		•		•		
55							•		•	•	•	•	•		•	•	•	•	•	•	•	•		
56							•		•		•		•		•	•		•		•		•		•
57							•		•		•		•		•	•		•		•		•		•
59							•		•		•		•		•	•		•		•		•		•
60									•		•		•		•	•		•		•		•		•
61							•		•		•		•		•	•		•		•		•		•
64							•		•		•		•		•	•		•		•		•		•
68									•		•				•	•		•		•		•		•
71							•		•		•		•		•	•		•		•		•		•
72									•		•				•	•		•		•		•		•
P. 4	1	2	5	7	8	9	10	11	12	13	14	15	16	17	18	23	25	26	43	44	45	46	49	50
73							•		•		•		•		•	•		•		•		•		•
74									•		•				•	•		•		•		•		•
75							•		•		•		•		•	•		•		•		•		•
76									•		•				•	•		•		•		•		•
82																		•						•
103															•			•		•		•		•
105															•			•						•
112																		•						•
114																		•						•

(continue) **120E' Group (14) 12(10M)e,1 + 20(6A)**

	55	56	57	58	61	62	71	72	73	74	75	76	81	82	85	86	103	113	114
1	•	•					•		•		•		•				•	•	
2	•						•		•		•		•				•	•	
5	•	•	•	•	•		•	•	•		•	•			•		•	•	
7	•	•					•		•		•		•				•	•	
8	•						•		•		•		•				•	•	
9	•	•		•			•		•		•						•	•	
10	•	•	•		•		•		•		•				•		•	•	
11	•	•					•		•	•	•				•		•	•	
12	•	•	•		•		•		•		•				•		•	•	
13	•	•	•	•	•		•	•	•		•	•			•		•	•	
14	•	•	•	•	•		•	•	•		•	•			•		•	•	

15	•	•	•	•	•		•	•	•		•	•			•		•	•	
18		•		•				•				•					•		
25		•		•		•		•		•		•					•	•	
26		•		•		•		•		•		•				•	•		
39	•	•					•		•		•				•		•	•	
40	•	•	•		•		•		•		•				•		•	•	
43	•	•	•	•	•		•	•	•	•	•	•			•		•	•	
44	•	•	•	•	•		•	•	•	•	•	•			•		•	•	
45	•	•	•	•	•		•	•	•	•	•	•			•		•	•	
50		•		•				•				•					•		
55	•	•	•	•	•	•	•	•	•	•	•	•			•		•	•	
56		•		•		•		•		•		•					•		
57		•		•		•		•		•		•					•		
59		•		•		•		•		•		•					•		
60		•		•		•		•		•		•				•	•		
61		•		•		•		•		•		•				•	•		
64																			
68		•		•		•		•		•		•				•	•		
71		•		•		•		•		•		•				•	•		
72		•		•		•		•		•		•				•	•		•
73		•		•		•		•		•		•				•	•		
74		•		•		•		•		•		•				•	•		•
75		•		•		•		•		•		•				•	•		
76		•		•		•		•		•		•				•	•		•
82						•		•		•		•				•			
103		•		•		•		•		•		•				•	•		•
105				•		•		•		•		•				•			•
112						•		•		•		•		•		•			•
114						•		•		•		•		•		•			•

120E' Group (14) 12(10M)e,1 + 20(6A)

P. 1-1	1. $(3\,{}^{\sim 2}6)$	2. $(3\,{}^{-2}6)$	5. $(6\,{}^{-2}6)$	7. $(1\ 3\,{}^{\sim 2}6)$	8. $(1\ 3\,{}^{-2}6)$	9. $(3\,{}^{\sim}3\,{}^{\sim 2}6)$	10. $(3\,{}^{\sim}3\,{}^{-2}6)$	11. $(3\,{}^{-}3\,{}^{\sim 2}6)$
1. $(5\,{}^{\sim 2}10)$	9.45925	9.13527	10.89104	10.72924	9.6715313	11.20265	9.7632518	12.477145
2. $(5\,{}^{-2}10)$	9.48949	9.188401	10.94906	10.76035	×	11.23393	×	12.508855
5. $(10\,{}^{-2}10)$	×	×	12.578657	×	11.95714	×	12.250656	×
7. $(1\ 5\,{}^{\sim 2}10)$	9.60778	9.412987	11.165627	10.872587	10.0215698	11.344256	10.144749	12.615425
8. $(1\ 5\,{}^{-2}10)$	9.66548	9.530658	11.28782	10.929526	×	11.400846	×	12.6713204
9. $(5\,{}^{\sim}5\,{}^{\sim 2}10)$	9.871774	10.229535	11.956137	×	11.065247	×	11.27702	×
10. $(5\,{}^{\sim}5\,{}^{-2}10)$	×	10.646745	12.371423	×	11.6379	×	11.9050479	×
11. $(5\,{}^{-}5\,{}^{\sim 2}10)$	×	10.512126	12.225517	×	11.43955	×	11.68316	×
12. $(5\,{}^{-}5\,{}^{-2}10)$	×	×	12.782626	×	12.293688	×	12.6251	×
13. $(5\,{}^{\sim 2}10\,{}^{\sim 2}10)$	×	×	12.724174	×	12.19287	×	12.511799	×
P. 1-2	12. $(3\,{}^{-}3\,{}^{-2}6)$	13. $(3\,{}^{\sim 2}6\,{}^{\sim 2}6)$	14. $(3\,{}^{\sim 2}6\,{}^{-2}6)$	15. $(3\,{}^{-2}6\,{}^{\sim 2}6)$	16. $(3\,{}^{-2}6\,{}^{-2}6)$	17. $(\underline{6\ 3}\,{}^{\sim 2}6)$	18. $(\underline{6\ 3}\,{}^{-2}6)$	23. $(6\,{}^{\sim}6\,{}^{-2}6)$
1. $(5\,{}^{\sim 2}10)$	×	12.16808	11.4453884	×	11.544537	12.96195	×	×
2. $(5\,{}^{-2}10)$	×	12.226997	×	×	×	12.993695	×	×
5. $(10\,{}^{-2}10)$	12.7780022	13.83114	13.68908	14.2865	13.98405	×	12.8678774	14.961467

7. $(1\ 5\ \tilde{10}^{2})$	×	12.44035	11.791467	×	11.920257	13.098837	×	×
8. $(1\ 5\ \bar{10}^{2})$	×	12.564688	×	×	×	13.15435	×	×
9. $(5\ \tilde{5}\ \tilde{10}^{2})$	11.5571122	13.219184	12.8106808	×	13.0245224	×	×	×
10. $(5\ \tilde{5}\ \bar{10}^{2})$	×	13.633453	13.388308	14.0915207	13.655394	×	×	14.668669
11. $(5\ \bar{5}\ \tilde{10}^{2})$	12.0655798	13.4821963	13.17356	×	13.418607	×	×	14.4462
12. $(5\ \bar{5}\ \bar{10}^{2})$	×	14.03198	14.029425	14.486409	14.362824	×	×	15.304051
13. $(5\ \tilde{10}^{2}\ \tilde{10}^{2})$	13.1080812	13.97384	13.924855	14.428387	14.245463	×	13.2282997	15.197459

P. 1-3	25. $(6\ \bar{6}^{2}\ \tilde{6}^{2})$	26. $(6\ \bar{6}^{2}\ \bar{6}^{2})$	43. $(1\ 3\ \tilde{6}^{2}\ \tilde{6}^{2})$	44. $(1\ 3\ \tilde{6}^{2}\ \bar{6}^{2})$	45. $(1\ 3\ \bar{6}^{2}\ \tilde{6}^{2})$	46. $(1\ 3\ \bar{6}^{2}\ \bar{6}^{2})$	49. $(3\ \tilde{3}\ \bar{3}^{2}\ \tilde{6}^{2})$	55. $(3\ \tilde{3}\ \tilde{6}^{2}\ \tilde{6}^{2})$
1. $(5\ \tilde{10}^{2})$	13.93162	×	12.6146979	11.543095	×	×	14.235209	12.662336
2. $(5\ \bar{10}^{2})$	13.99217	×	12.67398	×	×	×	14.267235	12.72158
5. $(10\ \bar{10}^{2})$	15.567486	14.5452953	14.267079	13.97299	15.554033	14.54098	×	14.312065
7. $(1\ 5\ \tilde{10}^{2})$	14.201938	×	12.885845	11.9177534	×	×	14.26997	12.93307
8. $(1\ 5\ \bar{10}^{2})$	14.325033	×	13.01047	×	×	×	14.424966	13.057574
9. $(5\ \tilde{5}\ \tilde{10}^{2})$	14.967631	13.339313	13.65948	13.018005	×	13.338313	×	13.705412
10. $(5\ \tilde{5}\ \bar{10}^{2})$	15.379652	×	12.074504	13.645964	15.36578	×	×	14.117944
11. $(5\ \bar{5}\ \tilde{10}^{2})$	15.22375	13.833837	13.9201	13.410374	×	13.8312736	×	13.965574
12. $(5\ \bar{5}\ \bar{10}^{2})$	15.764092	×	14.466515	14.349546	15.750982	×	×	14.511204
13. $(5\ \tilde{10}^{2}\ \tilde{10}^{2})$	15.706727	14.876617	14.408645	14.232924	15.693531	14.871629	×	14.45341

P. 1-4	56 $(3\ \tilde{3}\ \tilde{6}^{2}\ \bar{6}^{2})$	57 $(3\ \tilde{3}\ \bar{6}^{2}\ \tilde{6}^{2})$	58 $(3\ \tilde{3}\ \bar{6}^{2}\ \bar{6}^{2})$	61 $(3\ \bar{3}\ \bar{6}^{2}\ \tilde{6}^{2})$	71. $(3\ \tilde{6}^{2}\ \tilde{6}^{2}\ \tilde{6}^{2})$	72 $(3\ \tilde{6}^{2}\ \tilde{6}^{2}\ \bar{6}^{2})$	73 $(3\ \tilde{6}^{2}\ \bar{6}^{2}\ \tilde{6}^{2})$	74 $(3\ \tilde{6}^{2}\ \bar{6}^{2}\ \bar{6}^{2})$
1. $(5\ \tilde{10}^{2})$	11.5503134	×	×	×	14.386975	×	15.215729	×
2. $(5\ \bar{10}^{2})$	×	×	×	×	14.44768	×	15.277189	×
5. $(10\ \bar{10}^{2})$	13.999794	16.022744	14.642351	17.295702	16.013127	14.641765	16.838281	×
7. $(1\ 5\ \tilde{10}^{2})$	11.9282803	×	×	×	14.656133	×	15.485299	×
8. $(1\ 5\ \bar{10}^{2})$	×	×	×	×	14.782219	×	15.612401	×
9. $(5\ \tilde{5}\ \tilde{10}^{2})$	13.0372376	×	13.522596	×	15.417141	×	16.244664	×
10. $(5\ \tilde{5}\ \bar{10}^{2})$	13.6702332	15.836671	×	17.114403	15.827687	×	16.6556	×
11. $(5\ \bar{5}\ \tilde{10}^{2})$	13.432586	×	×	×	15.67125	×	16.497276	×
12. $(5\ \bar{5}\ \bar{10}^{2})$	14.379795	16.218646	×	17.490172	16.208353	×	17.032382	×
13.	14.26203	16.1613	15.0020776	17.432466	16.151287	15.00088	16.975555	×

(5 ~$\overset{2}{10}$ ~$\overset{2}{10}$)								
P. 1-5	75 (3 -$\overset{2}{6}$ ~$\overset{2}{6}$ ~$\overset{2}{6}$)	76. (3 -$\overset{2}{6}$ ~$\overset{2}{6}$ -$\overset{2}{6}$)	81. (6 3 -3 ~$\overset{2}{6}$)	85 (6 3 -$\overset{2}{6}$ ~$\overset{2}{6}$)	103 (6 ~6 ~6 -$\overset{2}{6}$)	113 (6 -$\overset{2}{6}$ -$\overset{2}{6}$ ~$\overset{2}{6}$)		
1. (5 ~$\overset{2}{10}$)	14.440354	×	16.0011	×	13.341005	16.982117		
2. (5 -$\overset{2}{10}$)	14.520975	×	16.033262	×	13.4241834	17.04443		
5. (10 -$\overset{2}{10}$)	16.0632	14.6491342	×	17.771678	15.763393	18.588819		
7. (1 5 ~$\overset{2}{10}$)	14.7089505	×	16.133328	×	13.7165207	17.250579		
8. (1 5 -$\overset{2}{10}$)	14.834838	×	16.187541	×	13.8918375	17.378573		
9. (5 ~5 ~$\overset{2}{10}$)	15.468365	×	×	×	14.80946	18.002475		
10. (5 ~5 -$\overset{2}{10}$)	15.878236	×	×	17.59205	15.444915	1.841177		
11. (5 -5 ~$\overset{2}{10}$)	15.721955	×	×	×	15.197095	18.2511415		
12. (5 -5 -$\overset{2}{10}$)	16.25802	×	×	17.96409	16.146419	18.779804		
13. (5 ~$\overset{2}{10}$ ~$\overset{2}{10}$)	16.201104	15.011679	×	17.90769	16.025853	18.723749		

(continue) **120E' Group (14) 12(10M)e,1 + 20(6A)** **P. 2**

P. 2-1	5. (6 -$\overset{2}{6}$)	8. (1 3 -$\overset{2}{6}$)	10. (3 ~3 -$\overset{2}{6}$)	12. (3 -3 -$\overset{2}{6}$)	13. (3 ~$\overset{2}{6}$ ~$\overset{2}{6}$)	14. (3 ~$\overset{2}{6}$ -$\overset{2}{6}$)	15. (3 -$\overset{2}{6}$ ~$\overset{2}{6}$)	16. (3 -$\overset{2}{6}$ -$\overset{2}{6}$)	18. (6 3 -$\overset{2}{6}$)
14. (5 ~$\overset{2}{10}$ -$\overset{2}{10}$)	12.7568	12.24861	12.57426	13.1881947	14.00616	1.398214	14.460611	14.309605	13.3158268
15. (5 -$\overset{2}{10}$ ~$\overset{2}{10}$)	12.748068	12.233881	12.557477	13.1661344	13.997152	13.965995	14.45151	14.291339	13.29162
18. (10 5 -$\overset{2}{10}$)	×	×	13.780339	14.757049	×	14.988566	×	15.478094	15.02353
25. (10 -$\overset{2}{10}$ ~$\overset{2}{10}$)	×	×	13.654951	14.572245	×	14.882677	×	15.34225	14.81646
26. (10 -$\overset{2}{10}$ -$\overset{2}{10}$)	×	×	×	15.3668122	×	15.263456	×	×	15.6910815
39. (1 5 ~5 ~$\overset{2}{10}$)	×	×	11.378078	12.684053	×	12.900945	×	13.1224198	×
40. (1 5 ~5 -$\overset{2}{10}$)	×	×	12.070476	×	×	13.5361343	×	13.8173184	×
43 (1 5 ~$\overset{2}{10}$ ~$\overset{2}{10}$)	×	×	12.7184244	13.3678737	×	14.105436	×	14.4464508	13.5107923
44 (1 5 ~$\overset{2}{10}$ -$\overset{2}{10}$)	×	×	12.840551	13.52507	×	14.2150444	×	14.5701957	13.6821494
45 (1 5 -$\overset{2}{10}$ ~$\overset{2}{10}$)	×	×	12.8106666	13.485336	×	14.1867557	×	14.53781	13.6385789
P. 2-2	23. (6 ~6 -$\overset{2}{6}$)	25 (6 -$\overset{2}{6}$ ~$\overset{2}{6}$)	26. (6 -$\overset{2}{6}$ -$\overset{2}{6}$)	43 (1 3 ~$\overset{2}{6}$ ~$\overset{2}{6}$)	44. (1 3 ~$\overset{2}{6}$ -$\overset{2}{6}$)	45. (1 3 -$\overset{2}{6}$ ~$\overset{2}{6}$)	46. (1 3 -$\overset{2}{6}$ -$\overset{2}{6}$)	50. (1 3 -$\overset{2}{6}$ -$\overset{2}{6}$)	55. (3 ~3 ~$\overset{2}{6}$ ~$\overset{2}{6}$)
14. (5 ~$\overset{2}{10}$ -$\overset{2}{10}$)	15.255609	15.738523	14.9589267	14.440784	14.29667	15.725393	14.9537682	×	14.485489
15. (5 -$\overset{2}{10}$ ~$\overset{2}{10}$)	15.23869	15.729317	14.935082	14.431729	14.27853	15.716161	14.9299875	×	14.476443
18. (10 5 -$\overset{2}{10}$)	16.240512	×	16.498947	×	15.456104	×	16.4913795	×	×
25. (10 -$\overset{2}{10}$ ~$\overset{2}{10}$)	16.130786	×	16.3	×	15.32184	×	16.292082	15.1913972	×
26. (10 -$\overset{2}{10}$ -$\overset{2}{10}$)	16.4977923	×	17.0954428	×	16.8416562	×	17.0868998	16.2911405	×
39. (1 5 ~5 ~$\overset{2}{10}$)	14.1774258	×	13.462501	×	13.1154886	×	13.4610668	×	×

40. (1 5~5-10^2)	14.8149763	×	×	×	13.8070665	×	×	×	×
43 (1 5~10^2~10^2)	15.3752312	×	15.1311327	×	14.4327263	×	15.12569	×	×
44 (1 5~10^2-10^2)	15.485604	×	15.290915	×	14.5556559	×	15.2851805	×	×
45 (1 5-10^2~10^2)	15.4560723	×	15.2481514	×	14.5235037	×	15.242509	×	×
P. 2-3	56. (3~3~6^2-6^2)	57. (3~3-6^2~6^2)	58. (3~3-6^2-6^2)	61. (3-3-6^2~6^2)	62. (3-3-6^2-6^2)	71. (3~6^2~6^2~6^2)	72. (3~6^2~6^2-6^2)	73. (3~6^2-6^2~6^2)	74. (3~6^2-6^2-6^2)
14. (5~10^2-10^2)	14.326388	16.193152	15.0914884	17.463873	×	16.182897	15.090113	17.007033	×
15. (5-10^2~10^2)	14.30805	16.183843	15.0655218	17.45452	×	16.173689	15.064189	16.99773	×
18. (10 5-10^2)	15.499494	×	16.762099	×	×	×	16.755725	×	×
25. (10-10^2~10^2)	15.36271	×	16.542048	×	16.9507646	×	16.53658	×	16.852289
26. (10-10^2-10^2)	15.89404	×	17.4177962	×	18.052224	×	17.4086732	×	17.8679686
39. (1 5~5~10^2)	13.135467	×	×	×	×	×	×	×	×
40 (1 5~5-10^2)	13.8326364	×	×	×	×	×	×	×	×
43. (1 5~10^2~10^2)	14.4636254	×	15.2776895	×	×	×	15.275908	×	15.3926683
44. (1 5~10^2-10^2)	14.5878062	×	15.4510827	×	×	×	15.4488892	×	15.5952697
45. (1 5-10^2~10^2)	14.55528	×	15.404491	×	×	×	15.402407	×	15.5407176
P. 2-4	75. (3-6^2~6^2~6^2)	76. (3-6^2~6^2-6^2)	85. (6 3-6^2~6^2)	86. (6 3-6^2-6^2)	103. (6~6~6-6^2)	113. (6-6^2-6^2~6^2)	114. (6-6^2-6^2-6^2)		
14. (5~10^2-10^2)	16.23266	15.10174	17.93893	×	16.091429	18.754787	×		
15. (5-10^2~10^2)	16.223414	15.07558	17.929557	×	16.072012	18.745405	×		
18. (10 5-10^2)	×	16.781299	×	×	17.239304	×	×		
25. (10-10^2~10^2)	×	16.5601	×	×	17.095138	×	16.8056784		
26. (10-10^2-10^2)	×	17.4399442	×	18.1873216	17.6175795	×	×		
39. (1 5~5~10^2)	×	×	×	×	14.9057	×	×		
40. (1 5~5-10^2)	×	×	×	×	15.60512	×	×		
43. (1 5~10^2~10^2)	×	15.289141	×	×	16.2236815	×	×		
44. (1 5~10^2-10^2)	×	15.4636732	×	×	16.3490572	×	×		
45. (1 5-10^2~10^2)	×	15.416755	×	×	16.314763	×	×		

(continue) **120E' Group (14) 12(10M)e,1 + 20(6A)**

P. 3

P. 3-1	10. (3~3-6^2)	12. (3-3-6^2)	13. (3~6^2~6^2)	14. (3~6^2-6^2)	15. (3-6^2~6^2)	16. (3-6^2-6^2)	18. (6 3-6^2)	23. (6~6-6^2)
50. (5~5-5-10^2)	×	15.0763803	×	15.1489568	×	×	15.37219	16.3944557
55. (5~5~10^2~10^2)	13.2679142	14.0654683	14.24009	14.571385	14.68641	14.9741788	14.2670246	15.8298468
56. (5~5~10^2-10^2)	13.6008655	14.5112062	×	14.8491483	×	15.3048082	14.7530695	16.105277

57. $(5^{\sim}5^{-}\overset{2}{10}^{\sim}\overset{2}{10})$	13.5342432	14.4147602	×	14.7900587	×	15.2313513	14.6462826	16.0432462
59. $(5^{-}5^{\sim}\overset{2}{10}^{\sim}\overset{2}{10})$	14.4393793	14.28681	×	14.7109504	×	15.136559	14.5066657	15.9645954
60. $(5^{-}5^{\sim}\overset{2}{10}^{-}\overset{2}{10})$	×	14.8609089	×	15.0434318	×	15.5479818	15.13411	16.291768
61. $(5^{-}5^{-}\overset{2}{10}^{\sim}\overset{2}{10})$	13.7724987	14.7309652	×	14.9724774	×	15.4532685	14.9888845	16.21685
68. $(5^{-}\overset{2}{10}^{\sim}5^{-}\overset{2}{10})$	×	14.9294657	×	15.0786357	×	×	15.0100692	16.3270007
71. $(5^{\sim}\overset{2}{10}^{\sim}\overset{2}{10}^{\sim}\overset{2}{10})$	13.739039	14.6852212	×	14.947042	×	15.4214434	14.9391179	16.1924123
72. $(5^{\sim}\overset{2}{10}^{\sim}\overset{2}{10}^{-}\overset{2}{10})$	×	15.5985647	×	15.319609	×	×	15.9493108	16.5425196
P. 3-2	25. $(6^{-}\overset{2}{6}^{\sim}\overset{2}{6})$	26. $(6^{-}\overset{2}{6}^{-}\overset{2}{6})$	43. $(1\ 3^{\sim}\overset{2}{6}^{\sim}\overset{2}{6})$	44. $(1\ 3^{\sim}\overset{2}{6}^{-}\overset{2}{6})$	45. $(1\ 3^{-}\overset{2}{6}^{\sim}\overset{2}{6})$	46. $(1\ 3^{-}\overset{2}{6}^{-}\overset{2}{6})$	49. $(3^{\sim}3^{-}\overset{2}{3}^{\sim}\overset{2}{6})$	50. $(3^{\sim}3^{-}\overset{2}{3}^{-}\overset{2}{6})$
50. $(5^{\sim}5^{-}5^{-}\overset{2}{10})$	×	16.8173589	×	15.6681028	×	16.8092738	×	×
55. $(5^{\sim}5^{\sim}\overset{2}{10}^{\sim}\overset{2}{10})$	15.94412	15.8087944	14.666389	14.9568925	15.931485	15.8023037	×	×
56. $(5^{\sim}5^{\sim}\overset{2}{10}^{-}\overset{2}{10})$	×	16.2565375	×	15.28461	×	16.2493581	×	15.117652
57. $(5^{\sim}5^{-}\overset{2}{10}^{\sim}\overset{2}{10})$	×	16.1510362	×	15.211922	×	16.1440019	×	14.980575
59. $(5^{-}5^{\sim}\overset{2}{10}^{\sim}\overset{2}{10})$	×	16.0219316	×	15.1180025	×	16.0151244	×	14.804505
60. $(5^{-}5^{\sim}\overset{2}{10}^{-}\overset{2}{10})$	×	16.5970773	×	15.5252022	×	16.5894	×	15.593042
61. $(5^{-}5^{-}\overset{2}{10}^{\sim}\overset{2}{10})$	×	16.4523668	×	15.4317413	×	16.444953	×	15.4055
68. $(5^{-}\overset{2}{10}^{\sim}5^{-}\overset{2}{10})$	×	16.6688128	×	15.572003	×	16.6609869	×	15.6897085
71. $(5^{\sim}\overset{2}{10}^{\sim}\overset{2}{10}^{\sim}\overset{2}{10})$	×	16.4079923	×	15.4002395	×	16.4006339	×	15.3436504
72. $(5^{\sim}\overset{2}{10}^{\sim}\overset{2}{10}^{-}\overset{2}{10})$	×	17.3035044	×	15.9641796	×	17.3145266	×	16.617743
P. 3-3	55. $(3^{\sim}3^{\sim}\overset{2}{6}^{\sim}\overset{2}{6})$	56. $(3^{\sim}3^{\sim}\overset{2}{6}^{-}\overset{2}{6})$	57. $(3^{\sim}3^{-}\overset{2}{6}^{\sim}\overset{2}{6})$	58. $(3^{\sim}3^{-}\overset{2}{6}^{-}\overset{2}{6})$	61. $(3^{-}3^{-}\overset{2}{6}^{\sim}\overset{2}{6})$	62. $(3^{-}3^{-}\overset{2}{6}^{-}\overset{2}{6})$	71. $(3^{\sim}\overset{2}{6}^{\sim}\overset{2}{6}^{\sim}\overset{2}{6})$	72. $(3^{\sim}\overset{2}{6}^{\sim}\overset{2}{6}^{-}\overset{2}{6})$
50. $(5^{\sim}5^{-}5^{-}\overset{2}{10})$	×	15.715858	16.392941	17.1116763	17.649219	×	16.3822	17.1039075
55. $(5^{\sim}5^{\sim}\overset{2}{10}^{\sim}\overset{2}{10})$	14.710137	14.9930825	×	16.0106096	×	16.2863284	×	16.0068674
56. $(5^{\sim}5^{\sim}\overset{2}{10}^{-}\overset{2}{10})$	×	15.3251472	×	16.4977217	×	16.8996859	×	16.4923583
57. $(5^{\sim}5^{-}\overset{2}{10}^{\sim}\overset{2}{10})$	×	15.2513233	×	16.3816748	×	16.7518797	×	16.3767252

59. $(5\,{}^{-}5\,{}^{\sim}\overset{2}{10}\,{}^{\sim}\overset{2}{10})$	×	1.51561017	×	16.2411719	×	16.5752973	×	16.2367297
60. $(5\,{}^{-}5\,{}^{\sim}\overset{2}{10}\,{}^{-}\overset{2}{10})$	×	15.5697583	×	16.8687792	×	17.362687	×	16.8620367
61. $(5\,{}^{-}5\,{}^{-}\overset{2}{10}\,{}^{\sim}\overset{2}{10})$	×	15.474335	×	16.7083692	×	17.1574428	×	16.7022982
68. $(5\,{}^{-}\overset{2}{10}\,{}^{\sim}5\,{}^{-}\overset{2}{10})$	×	15.6175625	×	16.9479476	×	17.4629984	×	16.9408541
71. $(5\,{}^{\sim}\overset{2}{10}\,{}^{\sim}\overset{2}{10}\,{}^{\sim}\overset{2}{10})$	×	15.4423127	×	16.660098	×	17.0974302	×	16.654193
72. $(5\,{}^{\sim}\overset{2}{10}\,{}^{\sim}\overset{2}{10}\,{}^{-}\overset{2}{10})$	×	16.02155	×	17.6724082	×	18.3755942	×	17.6620111

P. 3-4	73 $(3\,{}^{\sim}\overset{2}{6}\,{}^{-}\overset{2}{6}\,{}^{\sim}\overset{2}{6})$	74. $(3\,{}^{\sim}\overset{2}{6}\,{}^{-}\overset{2}{6}\,{}^{-}\overset{2}{6})$	75. $(3\,{}^{-}\overset{2}{6}\,{}^{\sim}\overset{2}{6}\,{}^{\sim}\overset{2}{6})$	76. $(3\,{}^{-}\overset{2}{6}\,{}^{\sim}\overset{2}{6}\,{}^{-}\overset{2}{6})$	85 $(\underline{6\,3}\,{}^{-}\overset{2}{6}\,{}^{\sim}\overset{2}{6})$	86. $(\underline{6\,3}\,{}^{-}\overset{2}{6}\,{}^{-}\overset{2}{6})$	103. $(6\,{}^{\sim}6\,{}^{\sim}6\,{}^{-}\overset{2}{6})$	113. $(6\,{}^{-}\overset{2}{6}\,{}^{-}\overset{2}{6}\,{}^{\sim}\overset{2}{6})$
50. $(5\,{}^{\sim}5\,{}^{-}5\,{}^{-}\overset{2}{10})$	×	×	×	17.1325136	×	×	17.4511557	×
55. $(5\,{}^{\sim}5\,{}^{\sim}\overset{2}{10}\,{}^{\sim}\overset{2}{10})$	17.197052	16.2417155	16.431059	16.0261697	18.119457	×	16.73846	18.928
56. $(5\,{}^{\sim}5\,{}^{\sim}\overset{2}{10}\,{}^{-}\overset{2}{10})$	×	16.8043375	×	16.5157462	×	×	17.069207	×
57. $(5\,{}^{\sim}5\,{}^{-}\overset{2}{10}\,{}^{\sim}\overset{2}{10})$	×	16.6688815	×	16.3990298	×	×	16.9903632	×
59. $(5\,{}^{-}5\,{}^{\sim}\overset{2}{10}\,{}^{\sim}\overset{2}{10})$	×	16.5066939	×	16.25787	×	×	16.8952913	×
60. $(5\,{}^{-}5\,{}^{\sim}\overset{2}{10}\,{}^{-}\overset{2}{10})$	×	17.23137	×	16.8884383	×	17.4361983	17.3051385	×
61. $(5\,{}^{-}5\,{}^{-}\overset{2}{10}\,{}^{\sim}\overset{2}{10})$		17.043086	×	16.7271529	×	17.2109595	17.2021736	×
68. $(5\,{}^{-}\overset{2}{10}\,{}^{\sim}5\,{}^{-}\overset{2}{10})$	×	17.3236647	×	16.9680167	×	17.54609	17.3539362	×
71. $(5\,{}^{\sim}\overset{2}{10}\,{}^{\sim}\overset{2}{10}\,{}^{\sim}\overset{2}{10})$	×	16.9876604	×	16.6786623	×	17.1452494	17.1712695	×
72. $(5\,{}^{\sim}\overset{2}{10}\,{}^{\sim}\overset{2}{10}\,{}^{-}\overset{2}{10})$	×	18.1662569	×	17.6957522	×	18.5385733	17.7372007	×

End of P. 3----1, 2, 3, 4

(continue) **120E' Group (14) 12(10M)e,1 + 20(6A)**
P. 4

P. 4	10. $(3 \sim 3 - 6^{2})$	12. $(3 - 3 - 6^{2})$	14. $(3 \sim 6^{2} - 6^{2})$	16. $(3 - 6^{2} - 6^{2})$	18. $(\underline{6\ 3} - 6^{2})$	23. $(6 \sim 6 - 6^{2})$	26 $(6 - 6^{2} - 6^{2})$	44. $(1\ 3 \sim 6^{2} - 6^{2})$
73. $(5 \sim 10^{2} - 10^{2} \sim 10^{2})$	13.7577	14.71065	14.9612317	15.4391474	14.9667774	16.2060214	16.4326354	15.4177708
74. $(5 \sim 10^{2} - 10^{2} - 10^{2})$	×	15.6558567	15.32624	×	16.01379	16.5452966	17.3805533	15.992204
75. $(5 - 10^{2} \sim 10^{2} \sim 10^{2})$	13.7525582	14.7034393	14.957192	15.4340606	14.958898	16.2020692	16.4254236	15.4127326
76. $(5 - 10^{2} \sim 10^{2} - 10^{2})$	×	15.9390585	15.3246694	×	15,9947057	16.5449323	17.3633375	15.9833788
82. $(\underline{10\ 5} - 5 - 10^{2})$	×	×	×	×	×	×	18.287898	×
103. $(10 \sim 10 \sim 10 - 10^{2})$	×	15.7983404	×	×	16.1701346	×	17.5051547	16.0494986
105. $(10 - 10 \sim 10 - 10^{2})$	×	×	×	×	16.7892466	×	17.999	×
112. $(10 - 10^{2} \sim 10^{2} - 10^{2})$	×	×	×	×	×	×	18.1851588	×
114. $(10 - 10^{2} - 10^{2} - 10^{2})$	×	×	×	×	×	×	18.4835673	×

End of 120E' Group (14) 12(10M)e,1 + 20(6A)

120E' Group (15): 12(10M)e +30(4A)

	2	4	5	8	13	14	18	19	20	21	26	28	29	36	46	68	69	96	97	114	116
3			•																		
6								•							•						
19			•	•		•	•		•							•		•			
21			•	•		•	•		•			•				•		•			
36						•				•	•				•						
115								•							•						
117								•							•						
119								•							•						•
121								•							•						
147								•		•	•				•					•	
148								•					•							•	•
149																					

120E' Group (15): 12(10M)e +30(4A)

P. 1	5. (2 4)	8. (4 4)	14. (1 4 4)	18. (2 4 4)	19 (2 4 4)	20 (2 4 4)	21 (2 4 4)	26. (4 4 4)
3. (10 10)e R =	11.93127	×	×	×	×	×	×	×
6. (20 10)e	×	×	×	×	19.47578	×	×	×
19. (10 10 10)e	13.68767	15.44705	15.75861	17.219545	×	18.517168	×	×
21. (10 10 10)	14.067836	15.82247	16.18683	17.591199	×	18.886803	×	×
36. (20 20 10)e	×	×	21.335384	×	21.610844	×	21.960333	23.174285
115. (10 10 20 10)e	×	×	×	×	19.6115962	×	×	×
117. (10 10 20 10)e	×	×	×	×	19.629161	×	×	×
119. (10 10 20 10)e	×	×	×	×	20.0579253	×	×	×
121. (10 10 20 10)e	×	×	×	×	20.1383714	×	×	21.7493982
147. (20 10 20 10)e	×	×	21.82274	×	22.1352415	×	22.5573434	23.6742945
148. (20 10 20 10)e	×	×	×	×	23.4395856	×	×	×

P. 2	28 (4 4 4)	29 (4 4 4)	46 (1 4 4 4)	68 (2 4 4 4)	96 (4 4 4 4)	114 (4 8 4 4)	11 (4 8 4 4)	
6. (20 10)e	×	×	21.234937	×	×	×	×	
19. (10 10 10)e	19.000684	×	×	20.787738	22.579009	×	×	
21. (10 10 10)	19.369355	×	×	21.154077	22.9433852	×	×	
36. (20 20 10)e	×	×	23.359047	×	×	×	×	
115 (10 10 20 10)e	×	×	21.389151	×	×	×	×	
117. (10 10 20 10)e	×	×	21.4091578	×	×	×	×	
119. (10 10 20 10)e	×	×	21.8323069	×	×	×	23.7136698	
121. (10 10 20 10)e	×	×	21.9152575	×	×	×	×	
147. (20 10 20 10)e	×	×	23.8665456	×	×	26.085462	×	
148. (20 10 20 10)e	×	24.1956644	×	×	×	27.7342177	27.7251997	

End of 120E' Group (15): 12(10M)e +30(4A)

120E' Group (16) 12(10M)e,1 + 30(4A)

P. 1	2	4	5	8	13	14	18	19	20	21	26	28	29	36	46	68	69	96	97	114	116
1	•	•	•	•			•					•				•		•			
2	•	•	•	•			•					•				•		•			
5			•	•	•	•	•	•	•			•			•	•		•			
7	•	•	•	•			•					•				•		•			
8	•	•	•	•			•					•				•		•			
9	•	•	•	•		•	•					•				•		•			
10			•	•	•	•	•	•	•			•			•	•		•			
P. 2	**2**	**4**	**5**	**8**	**13**	**14**	**18**	**19**	**20**	**21**	**26**	**28**	**29**	**36**	**46**	**68**	**69**	**96**	**97**	**114**	**116**
11		•	•	•		•	•					•				•		•			
12			•	•	•	•	•	•	•			•			•	•		•			
13			•	•	•	•	•	•	•			•			•	•		•			
14			•	•	•	•	•	•	•			•			•	•		•			
15			•	•	•	•	•	•	•			•			•	•		•			
18					•	•		•		•	•		•		•					•	•
23																					
P. 3	**2**	**4**	**5**	**8**	**13**	**14**	**18**	**19**	**20**	**21**	**26**	**28**	**29**	**36**	**46**	**68**	**69**	**96**	**97**	**114**	**116**
25						•		•		•	•				•						
26						•		•		•	•		•	•	•					•	•
39			•	•		•	•		•			•				•		•			
40			•	•	•	•	•	•	•		•	•			•	•		•			
43			•	•	•	•	•	•	•		•	•			•	•		•			
44			•	•	•	•	•	•	•		•	•			•	•		•			
45			•	•	•	•	•	•	•		•	•			•	•		•			
P. 4	**2**	**4**	**5**	**8**	**13**	**14**	**18**	**19**	**20**	**21**	**26**	**28**	**29**	**36**	**46**	**68**	**69**	**96**	**97**	**114**	**116**
50						•		•		•	•		•		•					•	•
55			•	•	•	•	•	•	•		•	•									
56					•	•		•		•	•				•						
57						•		•		•	•				•						
59						•		•		•	•				•						
60						•		•		•	•		•	•	•					•	•
61						•		•		•	•			•	•						
64																					
P. 5	**2**	**4**	**5**	**8**	**13**	**14**	**18**	**19**	**20**	**21**	**26**	**28**	**29**	**36**	**46**	**68**	**69**	**96**	**97**	**114**	**116**
68						•		•		•	•		•	•	•					•	•
71						•		•		•	•				•						
72						•		•		•	•		•	•	•					•	•
73						•		•		•	•				•					•	•
74						•		•		•	•		•	•	•					•	•
75						•		•		•	•				•						
76						•		•		•	•		•	•	•					•	•
P. 6	**2**	**4**	**5**	**8**	**13**	**14**	**18**	**19**	**20**	**21**	**26**	**28**	**29**	**36**	**46**	**68**	**69**	**96**	**97**	**114**	**116**
82										•			•	•			•			•	•
103								•		•	•		•	•	•					•	•
105										•			•	•						•	•
112										•			•	•			•			•	•
114										•			•	•			•			•	•

120E' Group (16) 12(10M)e,1 +30(4A) P. 1

	2. $(1\ \overset{1}{4})$	4. $(\tilde{2}\ \overset{1}{4})$	5. $(\bar{2}\ \overset{1}{4})$	8. $(\bar{4}\ \overset{2}{4})$	13. $(1\ \overset{2}{4}\ \tilde{\overset{2}{4}})$	14. $(1\ \overset{2}{4}\ \bar{\overset{2}{4}})$	18. $(\tilde{2}\ \overset{2}{4}\ \tilde{\overset{2}{4}})$	19. $(\tilde{2}\ \overset{2}{4}\ \bar{\overset{2}{4}})$
1. $(\tilde{5}\ \overset{2}{10})$	9.0681128	10.8327041	9.701957	11.48719	×	×	13.279047	×
2. $(\bar{5}\ \overset{2}{10})$	9.103167	10.869324	9.766337	11.555544	×	×	13.349789	×
5. $(\bar{10}\ \overset{2}{10})$	×	×	11.71575	13.453567	14.734863	14.0187182	15.2123905	14.093038
7. $(1\ \tilde{5}\ \overset{2}{10})$	9.244633	11.00068	10.026842	11.808723	×	×	13.598277	×

8. (1 5 -$\overset{2}{10}$)	9.312488	11.06747	10.167854	11.95409	×	×	13.746257	×
9. (5 ~5 ~$\overset{2}{10}$)	9.566376	11.26687	10.973283	12.730798	×	12.98425	14.503957	×
10. (5 ~5 -$\overset{2}{10}$)	×	×	11.462175	13.218185	14.50876	13.666493	14.989353	13.7048087
	20. (2 -$\overset{2}{4}$ ~$\overset{2}{4}$)	28. (4 -$\overset{2}{4}$ ~$\overset{2}{4}$)	46. (1 4 ~4 -$\overset{2}{4}$)	68 (2 ~$\overset{2}{4}$ -$\overset{2}{4}$ ~$\overset{2}{4}$)	96 (4 -$\overset{2}{4}$ -$\overset{2}{4}$ ~$\overset{2}{4}$)			
1. (5 ~$\overset{2}{10}$)	×	15.0749287	×	16.8734197	18.673665			
2. (5 -$\overset{2}{10}$)	×	15.1472296	×	16.946788	18.7478152			
5. (10 -$\overset{2}{10}$)	16.5016	16.9845362	15.871687	18.765456	20.5523843			
7. (1 5 ~$\overset{2}{10}$)	×	15.3925566	×	17.1898946	18.9892759			
8. (1 5 -$\overset{2}{10}$)	×	15.5422073	×	17.3406744	19.1408759			
9. (5 ~5 ~$\overset{2}{10}$)	×	16.286704	×	18.075659	19.8687754			
10. (5 ~5 -$\overset{2}{10}$)	16.285	16.7701846	15.495501	18.55749	20.3491545			

120E' Group (16) 12(10M)e,1 +30(4A) P. 2

	5. (2 -$\overset{2}{4}$)	8. (4 -$\overset{2}{4}$)	13. (1 $\overset{2}{4}$ ~$\overset{2}{4}$)	14. (1 $\overset{2}{4}$ -$\overset{2}{4}$)	18. (2 ~$\overset{2}{4}$ ~$\overset{2}{4}$)	19. (2 ~$\overset{2}{4}$ -$\overset{2}{4}$)	20 (2 -$\overset{2}{4}$ ~$\overset{2}{4}$)	21. (2 -$\overset{2}{4}$ -$\overset{2}{4}$)
11. (5 -5 ~$\overset{2}{10}$)	11.29711	13.042374	×	13.411134	14.807053	×	×	×
12. (5 -5 -$\overset{2}{10}$)	11.957415	13.690286	14.968201	14.420779	15.444339	14.54145	16.730699	×
13. (5 ~$\overset{2}{10}$ ~$\overset{2}{10}$)	11.88888	13.621956	14.9003568	14.296729	15.37676	14.402239	16.663729	×
14. (5 ~$\overset{2}{10}$ -$\overset{2}{10}$)	11.927292	13.659984	14.938002	14.364604	15.414246	14.478271	16.700834	×
15. (5 -$\overset{2}{10}$ ~$\overset{2}{10}$)	11.916385	13.64948	14.9272608	14.345256	15.403454	14.456526	16.689904	×
18. (10 5 -$\overset{2}{10}$)	×	×	14.8596464	15.540155	×	15.854688	×	16.283815
	26 (4 ~4 -$\overset{2}{4}$)	28. (4 -$\overset{2}{4}$ ~$\overset{2}{4}$)	29. (4 -$\overset{2}{4}$ -$\overset{2}{4}$)	46. (1 4 ~4 -$\overset{2}{4}$)	68 (2 ~$\overset{2}{4}$ -$\overset{2}{4}$ ~$\overset{2}{4}$)	96 (4 -$\overset{2}{4}$ -$\overset{2}{4}$ ~$\overset{2}{4}$)	114 (4 ~$\overset{2}{8}$ ~4 -$\overset{2}{4}$)	116. (4 -$\overset{2}{8}$ ~4 -$\overset{2}{4}$)
11. (5 -5 ~$\overset{2}{10}$)	×	16.5837336	×	×	18.368195	20.1579447	×	×
12. (5 -5 -$\overset{2}{10}$)	×	17.21235	×	16.324736	18.989825	20.7739249	×	×
13. (5 ~$\overset{2}{10}$ ~$\overset{2}{10}$)	×	17.1456366	×	16.181924	18.92397	20.7088041	×	×
14. (5 ~$\overset{2}{10}$ -$\overset{2}{10}$)	×	17.1826143	×	16.259553	18.960399	20.7448359	×	×
15. (5 -$\overset{2}{10}$ ~$\overset{2}{10}$)	×	17.1716338	×	16.236482	18.9493895	20.7337962	×	×
18. (10 5 -$\overset{2}{10}$)	17.412695	×	16.3168879	17.610789	×	×	19.8836242	19.8779

120E' Group (16) 12(10M)e,1 +30(4A) P. 3

	5. (2 -$\overset{2}{4}$)	8. (4 -$\overset{2}{4}$)	13. (1 $\overset{2}{4}$ ~$\overset{2}{4}$)	14. (1 $\overset{2}{4}$ -$\overset{2}{4}$)	18. (2 ~$\overset{2}{4}$ ~$\overset{2}{4}$)	19. (2 ~$\overset{2}{4}$ -$\overset{2}{4}$)	20 (2 -$\overset{2}{4}$ ~$\overset{2}{4}$)	21. (2 -$\overset{2}{4}$ -$\overset{2}{4}$)
25. (10 -$\overset{2}{10}$ ~$\overset{2}{10}$)	×	×	×	15.413256	×	15.691536	×	16.05034
26. (10 -$\overset{2}{10}$ -$\overset{2}{10}$)	×	×	×	15.854936	×	16.3134985	×	16.9888067
39. (1 5 ~5 ~$\overset{2}{10}$)	11.05506	12.809484	×	13.0905	14.5805354	×	15.876095	×
40. (1 5 ~5 -$\overset{2}{10}$)	11.584523	13.33569	14.6237235	13.840459	15.103396	13.8964239	16.397229	×
43. (1 5 ~$\overset{2}{10}$ ~$\overset{2}{10}$)	12.01493	13.738359	15.0115075	14.50846	15.486159	13.6394005	16.76929	×

44. $(1\ 5\ {}^{\sim 2}10\ {}^{-2}10)$	12.080978	13.800931	15.0718321	14.637955	15.545667	14.785936	16.82695	×
45. $(1\ 5\ {}^{-2}10\ {}^{\sim 2}10)$	12.066006	13.785649	15.0565877	14.60409	15.530487	14.747346	16.8119	×
	26. $(4\ {}^{\sim}4\ {}^{-2}4)$	28. $(4\ {}^{-2}4\ {}^{\sim 2}4)$	29. $(4\ {}^{-2}4\ {}^{-2}4)$	46. $(1\ 4\ {}^{\sim}4\ {}^{-2}4)$	68. $(2\ {}^{\sim 2}4\ {}^{-2}4\ {}^{\sim 2}4)$	96. $(4\ {}^{-2}4\ {}^{-2}4\ {}^{\sim 2}4)$	114. $(4\ {}^{\sim 2}8\ {}^{\sim}4\ {}^{-2}4)$	116. $(4\ {}^{-2}8\ {}^{\sim}4\ {}^{-2}4)$
25. $(10\ {}^{-2}10\ {}^{\sim 2}10)$	17.250985	×	×	17.439468	×	×	×	×
26. $(10\ {}^{-2}10\ {}^{-2}10)$	17.82213	×	17.095902	18.05273	×	×	20.6467351	20.6398344
39. $(1\ 5\ {}^{\sim}5\ {}^{\sim 2}10)$	×	16.3617659	×	×	18.1496	19.9419202	×	×
40. $(1\ 5\ {}^{\sim}5\ {}^{-2}10)$	15.548696	16.8816976	×	15.684954	18.667035	20.4572496	×	×
43. $(1\ 5\ {}^{\sim 2}10\ {}^{\sim 2}10)$	16.261021	17.2498465	×	16.415368	19.02425	20.8060636	×	×
44. $(1\ 5\ {}^{\sim 2}10\ {}^{-2}10)$	16.404878	17.3068437	×	16.5636	19.079149	2.085933	×	×
45. $(1\ 5\ {}^{-2}10\ {}^{\sim 2}10)$	16.36553	17.2919015	×	16.522898	19.06447	20.8448878	×	×

120E' Group (16) 12(10M)e,1 +30(4A) **P. 4**

	5. $(2\ {}^{-2}4)$	8. $(4\ {}^{-2}4)$	13. $(1\ {}^{2}4\ {}^{\sim 2}4)$	14. $(1\ {}^{2}4\ {}^{-2}4)$	18. $(2\ {}^{\sim 2}4\ {}^{\sim 2}4)$	19. $(2\ {}^{\sim 2}4\ {}^{-2}4)$	20 $(2\ {}^{-2}4\ {}^{\sim 2}4)$	21. $(2\ {}^{-2}4\ {}^{-2}4)$
50. $(5\ {}^{\sim}5\ {}^{-}5\ {}^{-2}10)$	×	×	×	15.726116	×	16.1078	×	16.6597167
55. $(5\ {}^{\sim}5\ {}^{\sim 2}10\ {}^{\sim 2}10)$	12.2576175	13.941405	15.1929202	15.05214	15.660641	15.26033	16.927987	×
56. $(5\ {}^{\sim}5\ {}^{\sim 2}10\ {}^{-2}10)$	×	×	15.1119895	15.378086	×	15.650813	×	15.997274
57. $(5\ {}^{\sim}5\ {}^{-2}10\ {}^{\sim 2}10)$	×	×	×	15.30695	×	15.56266	×	15.87415
59. $(5\ {}^{-}5\ {}^{\sim 2}10\ {}^{\sim 2}10)$	×	×	×	15.21403	×	15.450695	×	15.7221207
60. $(5\ {}^{-}5\ {}^{\sim 2}10\ {}^{-2}10)$	×	×	×	1.880886	×	15.936089	×	16.400291
61. $(5\ {}^{-}5\ {}^{-2}10\ {}^{\sim 2}10)$	×	×	×	15.5169444	×	15.821545	×	16.231525
	26. $(4\ {}^{\sim}4\ {}^{-2}4)$	28. $(4\ {}^{-2}4\ {}^{\sim 2}4)$	29. $(4\ {}^{-2}4\ {}^{-2}4)$	36. $(\underline{8\ 4}\ {}^{-2}4)$	46. $(1\ 4\ {}^{\sim}4\ {}^{-2}4)$	69. $(2\ {}^{\sim 2}4\ {}^{-2}4\ {}^{-2}4)$	114. $(4\ {}^{\sim 2}8\ {}^{\sim}4\ {}^{-2}4)$	116. $(4\ {}^{-2}8\ {}^{\sim}4\ {}^{-2}4)$
50. $(5\ {}^{\sim}5\ {}^{-}5\ {}^{-2}10)$	17.645565	×	16.731053	×	17.859501	×	20.299205	20.2929048
55. $(5\ {}^{\sim}5\ {}^{\sim 2}10\ {}^{\sim 2}10)$	16.848777	17.4033231	×	×	17.021176	×	×	×
56. $(5\ {}^{\sim}5\ {}^{\sim 2}10\ {}^{-2}10)$	17.2227926	×	×	×	17.4110536	×	×	×
57. $(5\ {}^{\sim}5\ {}^{-2}10\ {}^{\sim 2}10)$	17.1336546	×	×	×	17.317195	×	×	×
59. $(5\ {}^{-}5\ {}^{\sim 2}10\ {}^{\sim 2}10)$	17.0262594	×	×	×	17.2051343	×	×	×
60. $(5\ {}^{-}5\ {}^{\sim 2}10\ {}^{-2}10)$	17.4853354	×	16.4447127	19.865276	17.6874781	×	20.0027132	19.9968386
61. $(5\ {}^{-}5\ {}^{-2}10\ {}^{\sim 2}10)$	17.3701925	×	×	19.658225	17.56453	×	×	×

120E' Group (16) 12(10M)e,1 +30(4A) **P. 5**

	5. $(2\ {}^{-2}4)$	8. $(4\ {}^{-2}4)$	13. $(1\ {}^{2}4\ {}^{\sim 2}4)$	14. $(1\ {}^{2}4\ {}^{-2}4)$	18. $(2\ {}^{\sim 2}4\ {}^{\sim 2}4)$	19. $(2\ {}^{\sim 2}4\ {}^{-2}4)$	20. $(2\ {}^{-2}4\ {}^{\sim 2}4)$	21. $(2\ {}^{-2}4\ {}^{-2}4)$
68. $(5\ {}^{-2}10\ {}^{\sim}5\ {}^{-2}10)$	×	×	×	15.644382	×	16.99259	×	16.149456
71. $(5\ {}^{\sim 2}10\ {}^{\sim 2}10\ {}^{\sim 2}10)$	×	×	×	15.487508	×	15.78456	×	16.179047
72. $(5\ {}^{\sim 2}10\ {}^{\sim 2}10\ {}^{-2}10)$	×	×	×	15.915775	×	16.460657	×	17.2579642
73. $(5\ {}^{\sim 2}10\ {}^{-2}10\ {}^{\sim 2}10)$	×	×	×	15.50394	×	15.805287	×	16.208162
74. $(5\ {}^{\sim 2}10\ {}^{-2}10\ {}^{-2}10)$	×	×	×	15.92189	×	16.493412	×	17.3253011

75. $(5\,\bar{10}^2\,\tilde{10}^2\,\tilde{10}^2)$	×	×	×	15.499224	×	15.799293	×	16.199672
76. $(5\,\bar{10}^2\,\tilde{10}^2\,\bar{10}^2)$	×	×	×	15.92-546	×	16.483855	×	17.3049861

	26. $(4\,\tilde{4}\,\bar{4}^2)$	28. $(4\,\bar{4}^2\,\tilde{4}^2)$	29. $(4\,\bar{4}^2\,\bar{4}^2)$	36. $(\underline{8\ 4}\,\bar{4}^2)$	46. $(1\,4\,\tilde{4}\,\bar{4}^2)$	69. $(2\,\tilde{4}^2\,\bar{4}^2\,\bar{4}^2)$	114. $(4\,\tilde{8}^2\,\tilde{4}\,\bar{4}^2)$	116. $(4\,\bar{8}^2\,\tilde{4}\,\bar{4}^2)$
68. $(5\,\bar{10}^2\,\tilde{5}\,\bar{10}^2)$	17.5393153	×	16.537631	19.961577	17.745219	×	20.1015616	20.01955153
71. $(5\,\tilde{10}^2\,\tilde{10}^2\,\tilde{10}^2)$	17.33591	×	×	×	17.52852	×	×	×
72. $(5\,\tilde{10}^2\,\tilde{10}^2\,\bar{10}^2)$	17.94203	×	17.3977232	20.78321	19.181442	×	20.9451589	20.6376956
73. $(5\,\tilde{10}^2\,\bar{10}^2\,\tilde{10}^2)$	17.354974	×	×	×	17.548507	×	19.7652541	19.7597819
74. $(5\,\bar{10}^2\,\bar{10}^2\,\bar{10}^2)$	17.9667635	×	17.4638474	20.8574458	18.22194	×	21.0217047	21.0140963
75. $(5\,\bar{10}^2\,\tilde{10}^2\,\tilde{10}^2)$	17.3493	×	×	×	17.54257	×	×	×
76. $(5\,\bar{10}^2\,\tilde{10}^2\,\bar{10}^2)$	17.9595479	×	17.45082	20.8342313	18.21289	×	20.9977332	20.9901743

(continue) **120E' Group (16) 12(10M)e,1 +30(4A)** **P. 6**

	5. $(2\,\bar{4}^2)$	8. $(4\,\bar{4}^2)$	13. $(1\,4^2\,\tilde{4}^2)$	14. $(1\,4^2\,\bar{4}^2)$	18. $(2\,\tilde{4}^2\,\tilde{4}^2)$	19. $(2\,\tilde{4}^2\,\bar{4}^2)$	20. $(2\,\bar{4}^2\,\tilde{4}^2)$	21. $(2\,\bar{4}^2\,\bar{4}^2)$
82. $(\underline{10\ 5}\,\bar{5}\,\bar{10}^2)$	×	×	×	×	×	×	×	18.3955564
103 $(10\,\tilde{10}\,\tilde{10}\,\bar{10}^2)$	×	×	×	×	×	16.5620261	×	17.4726382
105 $(10\,\bar{10}\,\tilde{10}\,\bar{10}^2)$	×	×	×	×	×	×	×	18.0535497
112 $(10\,\bar{10}^2\,\tilde{10}^2\,\bar{10}^2)$	×	×	×	×	×	×	×	18.2739974
114. $(10\,\bar{10}^2\,\bar{10}^2\,\bar{10}^2)$	×	×	×	×	×	×	×	18.6274672

	26. $(4\,\tilde{4}\,\bar{4}^2)$	28. $(4\,\bar{4}^2\,\tilde{4}^2)$	29. $(4\,\bar{4}^2\,\bar{4}^2)$	36. $(\underline{8\ 4}\,\bar{4}^2)$	46. $(1\,4\,\tilde{4}\,\bar{4}^2)$	69 $(2\,\tilde{4}^2\,\bar{4}^2\,\bar{4}^2)$	114 $(4\,\tilde{8}^2\,\tilde{4}\,\bar{4}^2)$	116. $(4\,\bar{8}^2\,\tilde{4}\,\bar{4}^2)$
82. $(\underline{10\ 5}\,\bar{5}\,\bar{10}^2)$	×	×	18.76163	22.0384474	×	19.3492555	22.2564364	22.24551
103 $(10\,\tilde{10}\,\tilde{10}\,\bar{10}^2)$	18.0093832	×	17.639672	2.0991884	18.27772	×	21.1595735	21.1517096
105 $(10\,\bar{10}\,\tilde{10}\,\bar{10}^2)$	×	×	18.3141	21.571198	×	×	21.7585288	21.7494574
112 $(10\,\bar{10}^2\,\tilde{10}^2\,\bar{10}^2)$	×	×	18.59683	21.882224	×	19.0796346	22.0886335	22.078433
114. $(10\,\bar{10}^2\,\bar{10}^2\,\bar{10}^2)$	×	×	19.16818	22.362326	×	20.1	22.6218748	22.6087389

End of 120E' group (16) 12(10M)e,1 +30(4A)

120E' Group (17): 20(6M)e +30(4A)

	2	5	8	13	14	18	19	20	21	26	28	29	36	46	68	69	96	97	114	116
3					•															
6									•		•									
19		•	•	•	•	•	•	•						•	•		•			
115									•											
117							•		•			•		•					•	
119							•		•			•		•					•	
152																				

120E' Group (17): 20(6M)e +30(4A)

P. 1	5. $(2\,{}^{-}\overset{2}{4})$	8. $(4\,{}^{-}\overset{2}{4})$	13. $(1\,\overset{2}{4}\,{}^{\sim}\overset{2}{4})$	14. $(1\,\overset{2}{4}\,{}^{-}\overset{2}{4})$	18. $(2\,{}^{\sim}\overset{2}{4}\,{}^{\sim}\overset{2}{4})$	19. $(2\,{}^{\sim}\overset{2}{4}\,{}^{-}\overset{2}{4})$	20. $(2\,{}^{-}\overset{2}{4}\,{}^{\sim}\overset{2}{4})$
3. $(6\,{}^{\sim}6)e$ R =	×	×	×	16.545017	×	×	×
19. $(6\,{}^{\sim}6\,{}^{\sim}6)e$	14.434686	17.124994	19.102351	17.8194457	19.822696	17.8775492	21.817277
117. $(6\,{}^{\sim}6\,{}^{-}\overset{2}{12}\,{}^{\sim}6)e$	×	×	×	×	×	22.002052	×
119. $(6\,{}^{-}6\,{}^{\sim}\overset{2}{12}\,{}^{\sim}6)e$	×	×	×	×	×	22.003718	×
P. 2	21. $(2\,{}^{-}\overset{2}{4}\,{}^{-}\overset{2}{4})$	28. $(4\,{}^{-}\overset{2}{4}\,{}^{\sim}\overset{2}{4})$	29. $(4\,{}^{-}\overset{2}{4}\,{}^{-}\overset{2}{4})$	46. $(1\ 4\,{}^{\sim}4\,{}^{-}\overset{2}{4})$	68. $(2\,{}^{\sim}\overset{2}{4}\,{}^{-}\overset{2}{4}\,{}^{\sim}\overset{2}{4})$	96. $(4\,{}^{-}\overset{2}{4}\,{}^{-}\overset{2}{4}\,{}^{\sim}\overset{2}{4})$	114. $(4\,{}^{\sim}\overset{2}{8}\,{}^{\sim}4\,{}^{-}\overset{2}{4})$
6. $(\underline{12\ 6})e$	21.391568	×	×	×	×	×	×
19. $(6\,{}^{\sim}6\,{}^{\sim}6)e$	×	22.552735	×	20.592495	25.278039	28.007452	×
115 $(6\,{}^{\sim}6\,{}^{\sim}\overset{2}{12}\,{}^{-}6)e$	22.5629244	×	×	×	×	×	×
117. $(6\,{}^{\sim}6\,{}^{-}\overset{2}{12}\,{}^{\sim}6)e$	22.692003	×	22.749116	24.695932	×	×	28.16369
119. $(6\,{}^{-}6\,{}^{\sim}\overset{2}{12}\,{}^{\sim}6)e$	22.694024	×	22.7512545	24.696934	×	×	28.164666

End of 120E' Group (17): 20(6M)e +30(4A)

120E' Group (18): 20(6M)e,1 +30(4A)																					
	2	4	5	8	13	14	18	19	20	21	26	28	29	36	46	68	69	96	97	114	116
1			•	•		•	•	•				•				•		•			
2			•	•		•	•	•				•			•	•		•			
5						•		•			•				•						
7			•	•		•	•	•				•			•	•		•			
8						•		•		•	•				•						
9			•	•	•	•	•	•	•			•			•	•		•			
10								•					•		•					•	•
12										•			•	•						•	•
13								•		•	•		•		•						
14										•			•	•						•	•
15								•		•	•		•		•					•	•
18													•				•			•	•
26																	•				
43								•		•	•		•		•					•	•
44										•			•	•						•	•
50																	•				
55										•			•		•					•	•
56										•			•	•						•	•
72																	•				
74																	•		•		
76																	•				
82																			•		
103										•			•	•			•			•	•
114																					

120E' Group (18): 20(6M)e,1 +30(4A)
P. 1

	5. $(2\ \overset{-1}{4})$	8. $(4\ \overset{-2}{4})$	13. $(1\ \overset{2}{4}\ \overset{\sim 2}{4})$	14. $(1\ \overset{2}{4}\ \overset{-2}{4})$	18. $(2\ \overset{\sim 2}{4}\ \overset{\sim 2}{4})$	19. $(2\ \overset{\sim 2}{4}\ \overset{-2}{4})$	20. $(2\ \overset{-2}{4}\ \overset{\sim 2}{4})$		
1. $(3\ \overset{\sim 2}{6})$	11.148124	13.848368	×	14.245211	16.56527	×	×		
2. $(3\ \overset{-2}{6})$	11.258832	13.945465	×	14.632450	16.654479	14.690193	×		
5. $(6\ \overset{-2}{6})$	×	×	×	15.190725	×	15.547126	×		
7. $(1\ 3\ \overset{\sim 2}{6})$	11.25792	13.943977	×	14.617184	16.6530	14.670515	×		
8. $(1\ 3\ \overset{-2}{6})$	×	×	×	15.189465	×	15.542187	×		
9. $(3\ \overset{\sim}{3}\ \overset{\sim 2}{6})$	11.262989	13.947234	15.922039	14.650545	16.655332	14.713246	18.637307		

120E' Group (18): 20(6M)e,1 +30(4A)
P. 2

1. $(3\ \overset{\sim 2}{6})$	21. $(2\ \overset{-2}{4}\ \overset{-2}{4})$	26. $(4\ \overset{\sim}{4}\ \overset{-2}{4})$	28. $(4\ \overset{-2}{4}\ \overset{\sim 2}{4})$	46. $(1\ 4\ \overset{\sim}{4}\ \overset{-2}{4})$	68. $(2\ \overset{\sim 2}{4}\ \overset{-2}{4}\ \overset{\sim 2}{4})$	96. $(4\ \overset{-2}{4}\ \overset{-2}{4}\ \overset{\sim 2}{4})$			
2. $(3\ \overset{-2}{6})$	×	×	19.290641	×	22.020768	24.753785			
5. $(6\ \overset{-2}{6})$	×	×	19.374934	17.410676	22.101811	24.832574			
7. $(1\ 3\ \overset{\sim 2}{6})$	×	17.974899	×	18.245809	×	×			
8. $(1\ 3\ \overset{-2}{6})$	×	×	19.373555	17.386355	22.100572	24.831386			
9. $(3\ \overset{\sim}{3}\ \overset{\sim 2}{6})$	15.93247	17.972099	×	18.242425	×	×			

120E' Group (18): 20(6M)e,1 +30(4A)
P. 3

	19. $(2\ \tilde{4}^{2}\ \bar{4}^{2})$	21. $(2\ \bar{4}^{2}\ \bar{4}^{2})$	26. $(4\ \tilde{4}\ \bar{4}^{2})$	29. $(4\ \bar{4}^{2}\ \bar{4}^{2})$	36. $(\underline{8\ 4}\ \bar{4}^{2})$	46. $(1\ 4\ \tilde{4}\ \bar{4}^{2})$	69. $(2\ \tilde{4}^{2}\ \bar{4}^{2}\ \bar{4}^{2})$	114. $(4\ \tilde{8}^{2}\ \tilde{4}\ \bar{4}^{2})$	116. $(4\ \bar{8}^{2}\ \tilde{4}\ \bar{4}^{2})$
10. $(3\ \tilde{3}\ \bar{6}^{2})$	15.654746	×	×	16.393485	×	18.343576	×	21.815808	21.805955
12. $(3\ \bar{3}\ \bar{6}^{2})$	×	16.91727	×	17.266606	22.372466	×	×	22.648938	22.634847
13. $(3\ \tilde{6}^{2}\ \tilde{6}^{2})$	15.653896	15.327556	18.039029	16.382911	×	18.342223	×	×	×
14. $(3\ \tilde{6}^{2}\ \bar{6}^{2})$	×	16.802258	×	17.037267	22.186751	×	×	22.434264	22.421788
15. $(3\ \bar{6}^{2}\ \tilde{6}^{2})$	15.658536	16.364944	18.03840	16.430537	×	18.345576	×	21.836290	21.82640
18. $(\underline{6\ 3}\ \bar{6}^{2})$	×	×	×	17.391156	×	×	18.119256	22.750884	22.734876
26. $(6\ \bar{6}^{2}\ \bar{6}^{2})$	×	×	×	×	×	×	18.773006	×	×
43. $(1\ 3\ \tilde{6}^{2}\ \tilde{6}^{2})$	15.658467	16.364299	18.038446	16.429776	×	18.345536	×	21.835649	21.825774
44. $(1\ 3\ \tilde{6}^{2}\ \bar{6}^{2})$	×	16.914804	×	17.258506	22.364432	×	×	22.638963	22.624953
50. $(3\ \tilde{3}\ \bar{3}^{2}\ \bar{6}^{2})$	×	×	×	×	×	×	19.00946	×	×

120E' Group (18): 20(6M)e,1 +30(4A)
P. 4

	21. $(2\ \bar{4}^{2}\ \bar{4}^{2})$	29. $(4\ \bar{4}^{2}\ \bar{4}^{2})$	36. $(\underline{8\ 4}\ \bar{4}^{2})$	46. $(1\ 4\ \tilde{4}\ \bar{4}^{2})$	69. $(2\ \tilde{4}^{2}\ \bar{4}^{2}\ \bar{4}^{2})$	97. $(4\ \bar{4}^{2}\ \bar{4}^{2}\ \bar{4}^{2})$	114. $(4\ \tilde{8}^{2}\ \tilde{4}\ \bar{4}^{2})$	116. $(4\ \bar{8}^{2}\ \tilde{4}\ \bar{4}^{2})$	
55. $(3\ \tilde{3}\ \tilde{6}^{2}\ \tilde{6}^{2})$	16.365306	16.431027	×	18.345601	×	×	21.836452	21.826581	
56. $(3\ \tilde{3}\ \tilde{6}^{2}\ \bar{6}^{2})$	16.91995	17.275334	22.37621	×	×	×	22.653506	22.639337	
72. $(3\ \tilde{6}^{2}\ \tilde{6}^{2}\ \bar{6}^{2})$	×	×	×	×	19.003758	×	×	×	
74. $(3\ \tilde{6}^{2}\ \bar{6}^{2}\ \bar{6}^{2})$	×	×	×	×	19.157756	20.16952	×	×	
76. $(3\ \bar{6}^{2}\ \tilde{6}^{2}\ \bar{6}^{2})$	×	×	×	×	19.022754	×	×	×	
82. $(\underline{6\ 3}\ \bar{3}\ \bar{6}^{2})$	×	×	×	×	×	20.765047	×	×	
103 $(6\ \tilde{6}\ \tilde{6}\ \bar{6}^{2})$	21.882154	17.397435	22.438382	×	18.170838	×	22.752986	22.736951	
114 $(6\ \bar{6}^{2}\ \bar{6}^{2}\ \bar{6}^{2})$	×	×	×	×	×	×	×	×	

End of 120E' Group (18): 20(6M)e,1 + 30(4A)

120E' Group (19) 20(6M)e +12(10A)							
	3	6	19	115	117	119	152
2							
5			•				
8							
10	•						
12	•		•				
14			•				
16							
18	•	•	•	•			
23			•				
26	•		•		•	•	
33			•	•	•	•	
40	•		•				
44			•				
46							
50	•	•	•		•	•	
56	•		•				

120E' Group (19) 20(6M)e +12(10A)							
	3	6	19	115	117	119	152
58	•		•				
60	•		•				
62	•		•				
64	•	•	•	•	•	•	
68	•	•	•	•	•	•	
72	•		•	•	•	•	
74	•		•	•	•	•	
76	•		•	•	•	•	
82		•			•	•	•
86	•	•	•	•	•	•	
103	•		•	•			
105	•		•	•			
112	•	•	•	•	•	•	
114	•	•	•	•	•	•	
161	•	•	•	•	•	•	

120E' Group (19) 20(6M)e +12(10A)							
	3 (6 ~ 6)e	6 ($\underline{12\ 6}$)e	19 (6 ~ 6 ~ 6)e	115 (6 ~ 6 ~ $\overset{2}{12}$ - 6)e	117 (6 ~ 6 - $\overset{2}{12}$ ~ 6)e	119 (6 - 6 ~ $\overset{2}{12}$ ~ 6)e	152 ($\underline{12\ 6}$ ~ $\overset{2}{12}$ - 6)e
5 (10 - $\overset{2}{10}$) R =	×	×	12.880212	×	×	×	×
10 (5 ~ 5 - $\overset{2}{10}$)	11.659791	×	×	×	×	×	×
12 (5 - 5 - $\overset{2}{10}$)	12.3304	×	13.1576648	×	×	×	×
14 (5 ~ $\overset{2}{10}$ - $\overset{2}{10}$)	×	×	13.3410029	×	×	×	×
18 ($\underline{10\ 5}$ - $\overset{2}{10}$)	13.364509	16.5391054	14.3346552	17.2287684	×	×	×
26 (10 - $\overset{2}{10}$ - $\overset{2}{10}$)	14.961467	×	15.4445873	×	18.222992	18.225099	×
33 ($\underline{20\ 10}$ - $\overset{2}{10}$)	×	×	20.3167448	22.8994341	22.964916	22.965931	×
40 (1 5 ~ 5 - $\overset{2}{10}$)	11.814842	×	12.5915485	×	×	×	×
44 (1 5 ~ $\overset{2}{10}$ - $\overset{2}{10}$)	×	×	13.7165142	×	×	×	×
50 (5 ~ 5 - 5 - $\overset{2}{10}$)	13.543773	16.8677585	14.5636564	×	17.676509	17.679702	×
56 (5 ~ 5 ~ $\overset{2}{10}$ - $\overset{2}{10}$)	14.08846	×	14.8094545	×	×	×	×
58 (5 ~ 5 - $\overset{2}{10}$ - $\overset{2}{10}$)	14.668669	×	15.444915	×	×	×	×
60 (5 - 5 ~ $\overset{2}{10}$ - $\overset{2}{10}$)	14.4462	×	15.1970945	×	×	×	×
62 (5 - 5 - $\overset{2}{10}$ - $\overset{2}{10}$)	15.304061	×	16.1464186	×	×	×	×
64 (5 ~ $\overset{2}{10}$ ~ 5 - $\overset{2}{10}$)	13.451575	16.68955	14.442768	17.392056	17.479429	17.482714	×
68 (5 - $\overset{2}{10}$ ~ 5 - $\overset{2}{10}$)	13.46547	16.7143316	14.4603296	17.4189368	17.506583	17.509769	×
72 (5 ~ $\overset{2}{10}$ ~ $\overset{2}{10}$ - $\overset{2}{10}$)	15.197458	×	16.025857	17.5182828	18.587398	18.589497	×
74 (5 ~ $\overset{2}{10}$ - $\overset{2}{10}$ - $\overset{2}{10}$)	15.255608	×	16.0914335	18.6105413	18.680442	18.682541	×
76 (5 - $\overset{2}{10}$ ~ $\overset{2}{10}$ - $\overset{2}{10}$)	15.238699	×	16.0720149	18.5821982	18.6518715	18.653971	×
82 ($\underline{10\ 5}$ - 5 - $\overset{2}{10}$)	×	18.4109735	×	19.3415022	19.45055	19.45344	21.750985

86 $(\underline{10\;5}{}^{-}{\overset{2}{10}}{}^{-}\overset{2}{10})$	16.240512	19.534936	17.2392664	20.2437391	20.326002	20.327855	×
103 $(10^{\sim}10^{\sim}10^{-}\overset{2}{10})$	16.443957	×	17.170331	×	×	×	×
105 $(10^{-}10^{\sim}10^{-}\overset{2}{10})$	16.806553	×	17.545784	×	×	×	×
112 $(10^{-}{\overset{2}{10}}{}^{\sim}{\overset{2}{10}}{}^{-}\overset{2}{10})$	16.130786	19.318325	17.09516	20.0048	20.084784	20.086661	×
114 $(10^{-}{\overset{2}{10}}{}^{-}{\overset{2}{10}}{}^{-}\overset{2}{10})$	16.497787	20.117153	17.61757	20.881394	20.969016	20.970756	×
161 $(20^{-}{\overset{2}{20}}{}^{\sim}10^{-}\overset{2}{10})$	21.184901	24.333736	22.1290807	25.00333	25.074805	25.075554	×

End of 120E' Group (19): 20(6M)e +12(10A)

120E' Group (20) 20(6M)e,1 +12(10A) Outline

P. 1-1,2,3	5	10	12	13	14	15	18	23	25	26	33	40	43	44	45	50	55	56	57	58	59	60	61
1	•	•	•	•	•		•	•	•	•		•	•	•		•	•	•		•	•	•	
2	•		•	•	•		•	•	•	•	•	•	•	•		•	•	•		•	•	•	•
5							•			•	•					•		•				•	
7	•		•	•	•		•	•	•	•	•	•	•	•		•	•	•		•	•	•	
8							•			•	•					•		•				•	•
9	•		•	•	•	•	•	•	•	•	•	•	•	•	•	•	•	•	•	•	•	•	•
P. 1-4,5,6	62	64	68	71	72	73	74		75	76	82	85	86	103	105		108	111	112	113	114	161	
1	•	•	•	•	•	•	•		•	•			•	•	•		•	•	•	•	•	•	
2	•	•	•	•	•	•	•		•	•			•	•	•		•	•	•	•	•	•	
5		•	•		•		•			•	•		•	•	•		•		•		•	•	
7	•	•	•	•	•	•	•		•	•			•	•	•		•	•	•	•	•	•	
8		•	•		•		•			•	•		•	•	•		•		•		•	•	
9	•	•	•	•	•	•	•		•	•		•	•	•	•		•	•	•	•	•	•	
P. 2-1	5	10	12	13	14	15	18	23	25	26	33	40	43	44	45	50	55	56	57	58	59	60	61
10										•	•					•						•	
13										•	•					•						•	
14											•												
15										•	•					•							
43										•	•					•							
55										•	•					•							
P. 2-1	62	64	68	71	72	73	74	75	P.2-2	76	82	85	86	103	105	108	111	112	113	114	161		
10		•	•		•		•			•			•			•		•		•	•		
13	•	•	•		•		•			•	•		•	•	•	•		•		•	•		
14											•					•		•		•	•		
15	•		•		•		•			•	•		•	•	•	•		•		•	•		
43	•		•		•		•			•	•		•	•	•	•		•		•	•		
55			•		•		•			•	•		•	•	•	•		•		•	•		
P. 3	61	62	64	68	71	72	73	74	75	76	82	85	86	103	105	108	111	112	113	114	161		
12											•					•		•		•	•		
18											•					•				•			
44											•					•				•	•		
56											•					•		•		•	•		
103											•					•					•		

120E' Group (20) 20(6M)e,1 + 12(10A)

P. 1-1	5 (10 $^{-}10^{2}$)	10 (5 $^{\sim}5$ $^{-}10^{2}$)	12 (5 $^{-}5$ $^{-}10^{2}$)	13 (5 $^{\sim}10^{2}$ $^{\sim}10^{2}$)	14 (5 $^{\sim}10^{2}$ $^{-}10^{2}$)	15 (5 $^{-}10^{2}$ $^{\sim}10^{2}$)	18 (<u>10 5</u> $^{-}10^{2}$)	23 (10 $^{\sim}10$ $^{-}10^{2}$)
1 (3 $^{\sim}6^{2}$)	11.0381837	10.3826232	11.0553202	11.4358935	11.445393	×	12.108357	13.49659
2 (3 $^{-}6^{2}$)	11.100467	×	11.35932	11.49726	11.544542	×	12.55705	13.660585
5 (6 $^{-}6^{2}$)	×	×	×	×	×	×	13.505222	×
7 (1 3 $^{\sim}6^{2}$)	11.100005	×	11.354109	11.49669	11.543095	×	12.544375	13.655206
8 (1 3 $^{-}6^{2}$)	×	×	×	×	×	×	13.494826	×

9 $(3 \sim 3 \sim 6^{2})$	11.1029245	×	11.384265	11.4994305	11.550312	11.572862	12.58719	13.66894
P. 1-2	25 $(8 \sim 8 - 8^{2})$	26 $(10 - 10^{2} - 10^{2})$	33 $(\underline{20\ 10} - 10^{2})$	40 $(1\ 5 \sim 5 - 10^{2})$	43 $(1\ 5 \sim 10^{2} \sim 10^{2})$	44 $(1\ 5 \sim 10^{2} - 10^{2})$	45 $(1\ 5 - 10^{2} \sim 10^{2})$	50 $(5 \sim 5 - 5 - 10^{2})$
1 $(3 \sim 6^{2})$	14.050516	13.689084	×	10.538863	11.78947	11.791462	×	12.28792
2 $(3 - 6^{2})$	14.104394	13.984047	18.570511	10.792486	11.849601	11.920257	×	12.783055
5 $(6 - 6^{2})$	×	14.54529	19.28122	×	×	×	×	12.827212
7 $(1\ 3 \sim 6^{2})$	14.10355	13.972984	18.550986	10.739841	11.84895	11.917757	×	12.76826
8 $(1\ 3 - 6^{2})$	×	14.5409865	19.277656	×	×	×	×	13.816165
9 $(3 \sim 3 \sim 6^{2})$	14.105342	13.999787	18.58124	10.815267	11.851506	11.9283	12.0169979	12.814995

(continue)120E' Group (20) 20(6M)e,1 + 12(10A)							
P. 1-3	55 $(5 \sim 5 \sim 10^{2} \sim 10^{2})$	56 $(5 \sim 5 \sim 10^{2} - 10^{2})$	57 $(5 \sim 5 - 10^{2} \sim 10^{2})$	58 $(5 \sim 5 - 10^{2} - 10^{2})$	59 $(5 - 5 \sim 10^{2} \sim 10^{2})$	60 $(5 - 5 \sim 10^{2} - 10^{2})$	61 $(5 - 5 - 10^{2} \sim 10^{2})$
1 $(3 \sim 6^{2})$	12.974947	12.81066	×	13.388308	13.369113	13.173556	×
2 $(3 - 6^{2})$	12.9814	13.024524	×	13.655394	13.424367	13.41861	14.597005
5 $(6 - 6^{2})$	×	13.339313	×	×	×	13.833836	×
7 $(1\ 3 \sim 6^{2})$	12.9806	13.018008	×	13.645954	13.423565	13.410372	×
8 $(1\ 3 - 6^{2})$	×	13.338305	×	×	×	13.831271	14.2156044
9 $(3 \sim 3 \sim 6^{2})$	12.982707	13.037237	13.7036367	13.670233	13.425528	13.432575	14.097845
P. 1-4	62 $(5 - 5 - 10^{2} - 10^{2})$	64 $(5 \sim 10^{2} \sim 5 - 10^{2})$	68 $(5 - 10^{2} \sim 5 - 10^{2})$	71 $(5 \sim 10^{2} \sim 10^{2} \sim 10^{2})$	72 $(5 \sim 10^{2} \sim 10^{2} - 10^{2})$	73 $(5 \sim 10^{2} - 10^{2} \sim 10^{2})$	74 $(5 \sim 10^{2} - 10^{2} - 10^{2})$
1 $(3 \sim 6^{2})$	14.02941	12.196731	12.210905	14.386975	13.924855	14.472156	13.98215
2 $(3 - 6^{2})$	14.362816	12.6658	12.683499	14.440354	14.245463	14.52545	14.309605
5 $(6 - 6^{2})$	×	13.656238	13.681141	×	14.876617	×	14.958927
7 $(1\ 3 \sim 6^{2})$	14.349555	12.652138	12.669695	14.4395042	14.232935	14.5246	14.296674
8 $(1\ 3 - 6^{2})$	×	13.645549	13.670401	×	14.8716406	×	14.953767
9 $(3 \sim 3 \sim 6^{2})$	14.379793	12.69668	12.714525	14.4412155	14.262033	14.5263015	14.3264
P. 1-5	75 $(5 - 10^{2} \sim 10^{2} \sim 10^{2})$	76 $(5 - 10^{2} \sim 10^{2} - 10^{2})$	82 $(\underline{10\ 5} - 5 - 10^{2})$	85 $(\underline{10\ 5} - 10^{2} \sim 10^{2})$	86 $(\underline{10\ 5} - 10^{2} - 10^{2})$	10 $(10 \sim 10 \sim 10 - 10^{2})$	105 $(10 - 10 \sim 10 - 10^{2})$
1 $(3 \sim 6^{2})$	14.447685	13.965995	×	×	14.98856	15.167983	15.53299
2 $(3 - 6^{2})$	14.500976	14.291323	×	×	15.478094	15.396522	15.775572
5 $(6 - 6^{2})$	×	14.935086	15.425713	×	16.498947	15.751571	16.1746
7 $(1\ 3 \sim 6^{2})$	14.500126	14.278527	×	×	15.456102	15.386644	15.764654
8 $(1\ 3 - 6^{2})$	×	14.929981	15.409703	×	16.491365	15.750345	16.172905
9 $(3 \sim 3 \sim 6^{2})$	14.50183	14.208061	×	16.316165	15.499492	15.406271	15.785486

P. 1-6	108 $(10\ \tilde{10}\ \bar{10}^2\ \bar{10}^2)$	111 $(10\ \bar{10}^2\ \tilde{10}^2\ \tilde{10}^2)$	112 $(10\ \bar{10}^2\ \tilde{10}^2\ \bar{10}^2)$	113 $(10\ \bar{10}^2\ \bar{10}^2\ \tilde{10}^2)$	114 $(10\ \bar{10}^2\ \bar{10}^2\ \bar{10}^2)$	161 $(20\ \bar{20}^2\ \tilde{10}\ \bar{10}^2)$	
1 $(3\ \tilde{6}^2)$	14.985533	16.01312	14.882676	17.087068	15.263457	19.938544	
2 $(3\ \bar{6}^2)$	15.473078	16.063197	15.342252	17.136355	15.86934	20.395795	
5 $(6\ \bar{6}^2)$	16.488843	×	16.299257	×	17.0954465	21.331303	
7 $(1\ 3\ \tilde{6}^2)$	15.451191	16.062371	15.321838	17.135537	15.841657	20.3691	
8 $(1\ 3\ \bar{6}^2)$	16.481303	×	16.292082	×	17.086901	21.32724	
9 $(3\ \tilde{3}\ \tilde{6}^2)$	15.494397	16.063757	15.36271	17.136786	15.89404	20.407235	

End of P. 1-----1, 2, 3, 4, 5, 6

(continue) **120E' Group (20): 20(6M)e,1 + 12(10A)** **P. 2**

P. 2-1	26 (10 10 10)	33 (20 10 10)	50 (5 5 5 10)	60 (5 5 10 10)	6 (5 5 10 10)	64 (5 10 5 10)	68 (5 10 5 10)
10 (3 3 6)	14.642345	19.439396	14.09475	13.866695	×	13.907092	13.934226
13 (3 6 6)	14.641765	19.43395	14.09409	13.866696	15.16474	13.907004	13.934065
14 (3 6 6)	×	19.576725	×	×	×	×	×
15 (3 6 6)	14.649138	19.447184	14.122609	×	15.17706	×	13.961061
43 (1 3 6 6)	14.649035	19.44698	14.122166	×	15.17685	×	13.96066
5 (3 3 6 6)	14.649245	19.44725	14.123094	×	×	×	13.961547

P. 2-2	72 (5 10 10 10)	74 (5 10 10 10)	76 (5 10 10 10)	82 (10 5 5 10)	86 (10 5 10 10)	103 (10 10 10 10)	105 (10 10 10 10)
10 (3 3 6)	15.00205	15.09148	15.06553	×	16.762099	×	×
13 (3 6 6)	15.00088	15.09012	15.06419	15.904014	16.755725	15.755797	16.2
14 (3 6 6)	×	×	×	16.616853	×	×	×
15 (3 6 6)	15.011679	15.10174	15.07556	15.953308	16.781299	15.751189	16.198224
43 (1 3 6 6)	15.011516	15.101556	15.07539	15.952495	16.780888	15.75128	16.198298
55 (3 3 6 6)	15.011833	15.101908	15.07573	15.9538555	16.781534	15.751125	16.198221

P. 2-3	108 (10 10 10 10)	112 (10 10 10 10)	114 (10 10 10 10)	161 (20 20 10 10)
10 (3 3 6)	16.750295	16.542047	17.417799	21.568407
13 (3 6 6)	16.744004	16.536575	17.408673	21.557955
14 (3 6 6)	17.094431	15.8523053	17.868	21.861356
15 (3 6 6)	16.769396	16.560099	17.39944	21.578687
43 (1 3 6 6)	16.768986	16.559707	17.439426	21.5784
55 (3 3 6 6)	16.769655	16.560321	17.44019	21.578744

End of P. 2-1, 2, 3

(continue) **120E' Group (20) 20(6M)e,1 + 12(10A)**					**P. 3**
	$82\ (\underline{10\ 5}\ \bar{5}\ \bar{10}^{2})$	$108\ (10\ \tilde{10}\ \bar{10}^{2}\ \bar{10}^{2})$	$112\ (10\ \bar{10}^{2}\ \tilde{10}^{2}\ \bar{10}^{2})$	$114\ (10\ \bar{10}^{2}\ \bar{10}^{2}\ \bar{10}^{2})$	$161\ (20\ \bar{20}^{2}\ \tilde{10}\ \bar{10}^{2})$
$12\ (3\ \bar{3}\ \bar{6}^{2})$	16.93397	17.214165	16.950765	18.052202	21.9494687
$18\ (\underline{6\ 3}\ \bar{6}^{2})$	17.21176	17.273682	×	18.187355	×
$44\ (1\ 3\ \tilde{6}^{2}\ \bar{6}^{2})$	16.921254	17.210035	×	18.044508	21.945988
$56\ (3\ \tilde{3}\ \tilde{6}^{2}\ \bar{6}^{2})$	16.953855	16.769655	16.560321	17.44019	21.578744
$103\ (6\ \tilde{6}\ \tilde{6}\ \bar{6}^{2})$	17.242834	17.275777	×	×	21.966723

End of 120E' Group (20): 20(6M)e,1 + 12(10A)

120E' Group (21) 30(4M)e + 12(10A) Outline

	2	5	8	10	12	14	16	18	23	26	33	40	44	46	50	56	58
6								•							•		
9																	
10					•			•		•	•				•		•
30															•		
103																	
153																	

	60	62	64	68	72	74	76	82	86	103	105	108	111	112	114	161	
6			•	•					•			•		•	•	•	
9								•									
10		•	•	•	•	•	•		•			•		•	•	•	
30								•	•			•					
103																	
153																	

120E' Group (21): 30(4M)e + 12(10A)

P. 1	$12\ (5\,^{-}5\,^{-}\overset{2}{10})$	$18\ (\underline{10\ 5}\,^{-}\overset{2}{10})$	$26\ (10\,^{-}\overset{2}{10}\,^{-}\overset{2}{10})$	$33\ (\underline{20\ 10}\,^{-}\overset{2}{10})$	$50\ (5\,^{\sim}5\,^{-}5\,^{-}\overset{2}{10})$	$58\ (5\,^{\sim}5\,^{-}\overset{2}{10}\,^{-}\overset{2}{10})$
6. $(4\,^{\sim}4)$e	×	13.995562	×	×	14.25225	×
10. $(1\ 4\,^{\sim}4)$e	12.7873253	14.183971	15.8716875	21.2349368	14.45445	15.495502
30. $(4\,^{\sim}\overset{2}{8}\,^{\sim}4)$e	×	×	×	×	16.740943	×
P. 2	62 $(5\,^{-}5\,^{-}\overset{2}{10}\,^{-}\overset{2}{10})$	64 $(5\,^{\sim}\overset{2}{10}\,^{\sim}5\,^{-}\overset{2}{10})$	$68\ (5\,^{-}\overset{2}{10}\,^{\sim}5\,^{-}\overset{2}{10})$	$72\ (5\,^{\sim}\overset{2}{10}\,^{\sim}\overset{2}{10}\,^{-}\overset{2}{10})$	$74\ (5\,^{\sim}\overset{2}{10}\,^{-}\overset{2}{10}\,^{-}\overset{2}{10})$	$76\ (5\,^{-}\overset{2}{10}\,^{\sim}\overset{2}{10}\,^{-}\overset{2}{10})$
6. $(4\,^{\sim}4)$e	×	14.117511	14.137279	×	×	×
10. $(1\ 4\,^{\sim}4)$e	16.3247366	14.31181	14.3326039	16.1816239	16.259553	16.2364817
P. 3	$82\ (\underline{10\ 5}\,^{-}5\,^{-}\overset{2}{10})$	86 $(\underline{10\ 5}\,^{-}\overset{2}{10}\,^{-}\overset{2}{10})$	108 $(10\,^{\sim}10\,^{-}\overset{2}{10}\,^{-}\overset{2}{10})$	112 $(10\,^{-}\overset{2}{10}\,^{\sim}\overset{2}{10}\,^{-}\overset{2}{10})$	114 $(10\,^{-}\overset{2}{10}\,^{-}\overset{2}{10}\,^{-}\overset{2}{10})$	161 $(20\,^{-}\overset{2}{20}\,^{\sim}10\,^{-}\overset{2}{10})$
6. $(4\,^{\sim}4)$e	×	17.4126958	17.403259	17.2509814	17.8221426	23.1742983
9. $(\underline{8\ 4})$e	18.6488276	×	×	×	×	×
10. $(1\ 4\,^{\sim}4)$e	×	17.61081	17.60047	17.4394788	18.0527068	23.359046
30. $(4\,^{\sim}\overset{2}{8}\,^{\sim}4)$e	18.8557787	19.883625	19.861958	×	×	×

End of 120E' Group (21): 30(4M)e + 12(10A)

120E' Group (22) 30(4M)e,1 + 12(10A)

	2	5	8	10	12	14	16	18	23	26	33	40	44	46	50	56
2				•	•			•		•		•			•	•
4					•			•		•	•				•	
5								•							•	
8															•	
19																
29																
69																
97																

	58	60	62	64	68	72	74	76	82	86	103	105	108	112	114	161
2	•	•	•	•	•	•	•	•		•	•	•	•	•	•	•
4	•		•	•	•	•	•	•		•			•	•	•	•
5				•	•				•	•			•	•	•	•
8									•	•					•	
19																
29																
69																
97																

120E' Group (22): 30(4M)e,1 + 12(10A)

	$10\ (5\ \tilde{5}\ \bar{\overset{2}{10}})$	$12\ (5\ \bar{5}\ \bar{\overset{2}{10}})$	$18\ (\underline{10\ 5}\ \bar{\overset{2}{10}})$	$26\ (10\ \bar{\overset{2}{10}}\ \bar{\overset{2}{10}})$	$33\ (\underline{20\ 10}\ \bar{\overset{2}{10}})$	$40\ (1\ 5\ \tilde{5}\ \bar{\overset{2}{10}})$	$50\ (5\ \tilde{5}\ \tilde{5}\ \bar{\overset{2}{10}})$	$56\ (5\ \tilde{5}\ \tilde{\overset{2}{10}}\ \bar{\overset{2}{10}})$
$2.\ (1\ \overset{2}{4})$	10.1250503	10.9206367	12.1639648	14.0187221	×	10.3102565	12.3760771	12.9842467
$4.\ (2\ \tilde{\overset{2}{4}})$	×	11.0067440	12.4300657	14.0930390	19.4757738	×	12.6979366	×
$5.\ (2\ \bar{\overset{2}{4}})$	×	×	12.7627349	×	×	×	13.1401697	×
$8.\ (4\ \bar{\overset{2}{4}})$	×	×	×	×	×	×	13.1794525	×
	$58\ (5\ \tilde{5}\ \bar{\overset{2}{10}}\ \bar{\overset{2}{10}})$	$60\ (5\ \bar{5}\ \tilde{\overset{2}{10}}\ \bar{\overset{2}{10}})$	$62\ (5\ \bar{5}\ \bar{\overset{2}{10}}\ \bar{\overset{2}{10}})$	$64\ (5\ \tilde{\overset{2}{10}}\ \tilde{5}\ \bar{\overset{2}{10}})$	$68\ (5\ \bar{\overset{2}{10}}\ \tilde{5}\ \bar{\overset{2}{10}})$	$72\ (5\ \tilde{\overset{2}{10}}\ \tilde{\overset{2}{10}}\ \bar{\overset{2}{10}})$	$74\ (5\ \tilde{\overset{2}{10}}\ \bar{\overset{2}{10}}\ \bar{\overset{2}{10}})$	$76\ (5\ \bar{\overset{2}{10}}\ \tilde{\overset{2}{10}}\ \bar{\overset{2}{10}})$
$2.\ (1\ \overset{2}{4})$	13.6664901	13.4111332	14.4207778	12.2861994	12.2849155	14.2967295	14.3646006	14.3465260
$4.\ (2\ \tilde{\overset{2}{4}})$	13.7048068	×	14.5414509	12.5590330	12.5800453	14.4022402	14.4782663	14.4565262
$5.\ (2\ \bar{\overset{2}{4}})$	×	×	×	12.9411769	12.9706101	×	×	×
$8.\ (4\ \bar{\overset{2}{4}})$	×	×	×	×	×	×	×	×
	82 $(\underline{10\ 5}\ \bar{5}\ \bar{\overset{2}{10}})$	86 $(\underline{10\ 5}\ \bar{\overset{2}{10}}\ \bar{\overset{2}{10}})$	103 $(10\ \tilde{10}\ \tilde{10}\ \bar{\overset{2}{10}})$	105 $(10\ \bar{10}\ \tilde{10}\ \bar{\overset{2}{10}})$	108 $(10\ \tilde{10}\ \bar{\overset{2}{10}}\ \bar{\overset{2}{10}})$	112 $(10\ \bar{\overset{2}{10}}\ \tilde{\overset{2}{10}}\ \bar{\overset{2}{10}})$	114 $(10\ \bar{\overset{2}{10}}\ \bar{\overset{2}{10}}\ \bar{\overset{2}{10}})$	161 $(20\ \bar{\overset{2}{20}}\ \tilde{10}\ \bar{\overset{2}{10}})$
$2.\ (1\ \overset{2}{4})$	×	15.5401283	15.7586226	16.1868638	15.535446	15.4132542	15.8549361	21.3353815
$4.\ (2\ \tilde{\overset{2}{4}})$	×	15.8546870	×	×	15.847828	15.6919445	16.3134985	21.6108198
$5.\ (2\ \bar{\overset{2}{4}})$	15.0433440	16.2838145	×	×	16.272570	16.0503301	16.9888067	21.9603338
$8.\ (4\ \bar{\overset{2}{4}})$	15.3548684	16.3168879	×	×	×	×	17.0959024	×

End of 120E' Group (22): 30(4M)e,1 + 12(10A)

120E’ Group (23) 30(4M)e + 20(6A) Outline																								
	2	5	8	10	12	14	16	18	23	26	33	44	46	50	56	58	62	72	74	76	82	86	103	114
6					•					•			•					•						
9														•			•		•					
10				•	•		•	•		•		•	•		•	•		•		•			•	
30								•						•		•	•		•	•		•		
103																					•			•
153																								

120E’ Group (23): 30(4M)e + 20(6A)									
	$10\,(3\,\tilde{3}\,\overset{-2}{6})$	$12\,(3\,\bar{3}\,\overset{-2}{6})$	$16\,(3\,\overset{-2}{6}\,\overset{-2}{6})$	$18\,(\underline{6\,3}\,\overset{-2}{6})$	$26\,(6\,\overset{-2}{6}\,\overset{-2}{6})$	$44\,(1\,3\,\overset{\sim 2}{6}\,\overset{-2}{6})$	$46\,(1\,3\,\overset{-2}{6}\,\overset{-2}{6})$	$50\,(3\,\tilde{3}\,\overset{-2}{3}\,\overset{-2}{6})$	$56\,(3\,\tilde{3}\,\overset{-2}{6}\,\overset{-2}{6})$
$6\,(4\,\tilde{4})e$	×	14.806201	×	×	17.9748984	×	17.9720908	×	×
$9\,(\underline{8\,4})e$	×	×	×	×	×	×	×	19.2109308	×
$10\,(1\,4\,\tilde{4})e$	14.27273	15.0828881	17.4106761	15.1746793	18.24584	17.3863554	18.2424249	×	17.4269037
$30\,(4\,\overset{\sim 2\sim}{8}\,4)e$	×	×	×	18.6867934	×	×	×	19.4950552	×

	$58\,(3\,\tilde{3}\,\overset{-2}{6}\,\overset{-2}{6})$	$62\,(3\,\bar{3}\,\overset{-2}{6}\,\overset{-2}{6})$	$72\,(3\,\overset{\sim 2}{6}\,\overset{\sim 2}{6}\,\overset{-2}{6})$	$74\,(3\,\overset{\sim 2}{6}\,\overset{-2}{6}\,\overset{-2}{6})$	$76\,(3\,\overset{-2}{6}\,\overset{\sim 2}{6}\,\overset{-2}{6})$	$82\,(\underline{6\,3}\,\bar{3}\,\overset{-2}{6})$	$86\,(\underline{6\,3}\,\overset{-2}{6}\,\overset{-2}{6})$	$103\,(6\,\tilde{6}\,\tilde{6}\,\overset{-2}{6})$	$114\,(6\,\overset{-2}{6}\,\overset{-2}{6}\,\overset{-2}{6})$
$6\,(4\,\tilde{4})e$	×	×	18.0390368	×	×	×	×	×	×
$9\,(\underline{8\,4})e$	×	22.3724508	×	22.1867189	×	×	×	×	×
$10\,(1\,4\,\tilde{4})e$	18.3435584	×	18.3422479	×	18.3455687	×	×	20.5924254	×
$30\,(4\,\overset{\sim 2\sim}{8}\,4)e$	21.8158182	22.6489382	×	22.4342737	21.8362855	×	22.7509241	×	×
$103\,(4\,\bar{4}\,\overset{\sim 2-}{8}\,4)e$	×	×	×	×	×	23.622594	×	×	26.594256

End of 120E’ Group (23): 30(4M)e + 20(6A)

120E' Group (24): 30(4M)e,1 + 20(6A)	2	5	8	10	12	14	16	18	23	26	33	44	46	50	56	58	62	72	74	76	82	86	103	114
2			•	•	•	•	•		•	•		•	•		•								•	
4				•	•		•	•		•		•	•		•	•		•		•			•	
5					•			•		•	•		•	•		•	•	•	•	•				
8								•						•		•	•	•	•	•		•		
19																					•	•		•
29																								
69																								
97																								

120E' Group (24): 30(4M)e,1 + 20(6A)	8. (1 3 6)	10. (3 3 6)	12. (3 3 6)	14. (3 6 6)	16. (3 6 6)	18. (6 3 6)	23. (6 6 6)	26. (6 6 6)
2. (1 4)	11.1016984	11.4924902	12.0203919	14.2452112	14.6324497	×	16.5450163	15.1907249
4. (2 4)	×	11.5639892	12.3898492	×	14.6901927	12.4923348	×	15.5471254
5. (2 4)	×	×	12.8158546	×	×	13.2178894	×	15.9447866
8. (4 4)	×	×	×	×	×	13.2930836	×	×
	33. (12 6 6)	44. (1 3 6 6)	46. (1 3 6 6)	50. (3 3 3 6)	56. (3 3 6 6)	58. (3 3 6 6)	62. (3 3 6 6)	72. (3 6 6 6)
2. (1 4)	×	14.6171841	15.1894708	×	14.6505428	×	×	×
4. (2 4)	×	14.6704994	15.5421872	×	14.7132450	15.6547456	×	15.6538911
5. (2 4)	21.3915681	×	15.9324708	13.7602658	×	16.3357134	16.9172698	16.3275556
8. (4 4)	×	×	×	14.1295762	×	16.3934851	17.2666063	16.3829107
	74. (3 6 6 6)	76. (3 6 6 6)	82. (6 3 3 6)	86. (6 3 6 6)	103. (6 6 6 6)	114. (6 6 6 6)		
2. (1 4)	×	×	×	×	17.8194457	×		
4. (2 4)	×	15.6585355	×	×	17.8775492	×		
5. (2 4)	16.8022584	16.3649435	×	×	×	×		
8. (4 4)	17.0372495	16.4305582	×	17.391166	×	×		
19 (2 4 4)	×	×	15.887529	18.119257	×	18.773029		

End of [DATA 2B--11] 120E' Group (24): 30(4M)e,1 + 20(6A)

******* End of [DATA 2B] All Eleven E's and Families *******

[DATA 3] The TEs and Families

Every TE family has four groups:

(I) P-S

(II) G-S

(III) T-S and (IV) (G-S1) + (T-S2)

(V) (P-S1) + (G-S2)

While z = 1: $\bar{\phi}_1, \bar{\phi}_2, \bar{\phi}_3 \geq 60^{\circ}$

While z > 1: $\bar{\phi}_3 \geq 60^{\circ}$ **and** $\hat{\phi}_{12} (= \hat{\phi}_1 + \hat{\phi}_2) < \pi$.

[The boxed Ds is high Ds while it is > 80 %]

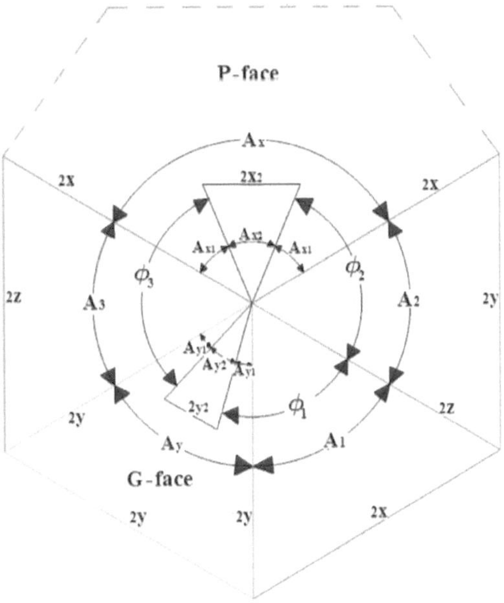

The angles and edges of the general village of a TE

Criteria:

While z = 1: Every $\bar{\phi}_1$, $\bar{\phi}_2$, $\bar{\phi}_3$ $\geq 60^0$

While z > 1: $\bar{\phi}_3 \geq 60^0$ **and** $\hat{\phi}_{12} (= \hat{\phi}_1 + \hat{\phi}_2) < \pi$.

The symbol $\mathbf{z}_{\phi 3}$ means that **z** is lengthened to be $\mathbf{z}_{\phi 3}$ while $\bar{\phi}_3 = 60^0$.

The symbol $\mathbf{z}_{\phi 12}$ means that **z** is lengthened to be $\mathbf{z}_{\phi 12}$ while $\hat{\phi}_{12} = 179.999^0\ 60^0$.

The boxed Ds is while it is > 80 %.

The symbol " 1 🡐 " shows that z = 1; no need to be lengthened.

[DATA 3-1] The T12E and Family

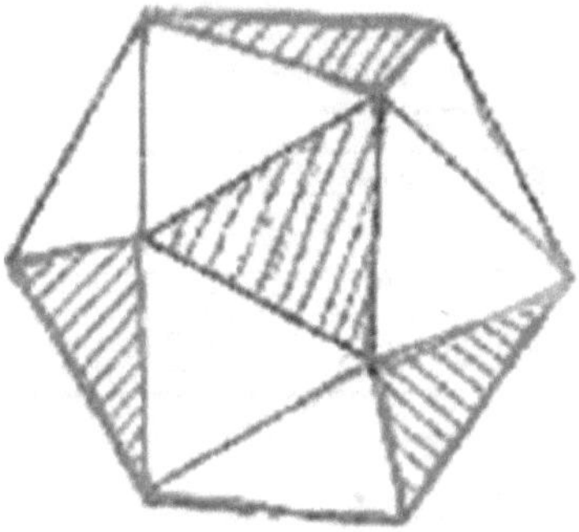

T12E:

$$R = \sqrt{\frac{1}{2}(5+\sqrt{5})} \doteq 1.90211303259.... \quad Ds = \frac{100(12)}{4R^2}\% = 82.91769...\%$$

$$\theta_T = 16.992708^{\circ}$$

Both P-face (red) and G-face (green) of T12E are all 3-gons, so there is no difference between Group (I) and Group (II).

T12E (I): P-S		1. (1 3)	2. $(3^{\sim}3)$	3. $(3^{-}3)$	4. $(\underline{6\ 3})$	8. $(3^{\sim}3^{-}3)$	16. $(\underline{6\ 3}^{-}3)$
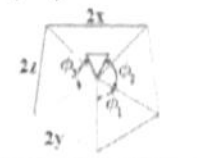	R x y z N, Ds	2.28466 1.557322 1 1 ← 16, 76.633	2.687465 1.839292 1 $Z_{\phi 3}$ 1.414215 24, 83.0741	3.44488 2.393052 1 $Z_{\phi 12}$ 2.190067 24, 50.560	3.58656 2.651759 1 $Z_{\phi 3}$ 2.1363 36, 69.966	4.386112 3.191479 1 $Z_{\phi 12}$ 2.960972 35, 46.7825	5.34465 3.993 1 $Z_{\phi 12}$ 3.747891 48, 42.009

T12E (III): T-S and (IV): $(G\text{-}S_1)+(T\text{-}S_2)$					
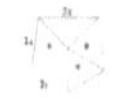		1. (1 3)	3. $(3^{-}3)$	8. $(3^{\sim}3^{-}3)$	16. $(\underline{6\ 3}^{-}3)$
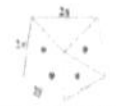	R x = y = z (III): N, Ds (IV): N, Ds	3.12084625 1.640726 24, 61.604 28, 71.871	4.8747473 2.5628063 48, 50.498 60, 63.123	6.6492512 3.4957185 84, 47.497 102, 61.069	8.4341 4.434032 120, 42.173 156, 54.826

T12E(IV): $(P\text{-}S_1)+(G\text{-}S_2)$	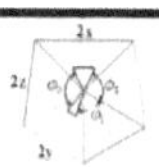	P1. (1 3)	P2. $(3^{\sim}3)$	P3. $(3^{-}3)$	P4. $(\underline{6\ 3})$	P8. $(3^{\sim}3^{-}3)$	P16. $(\underline{6\ 3}^{-}3)$
G1. (1 3)	R x y z N, Ds	2.80255 1.618037 1.618037 1 ← 20, 63.660	Same as P1-G2 (just switch x and y)				
G2. $(3^{\sim}3)$	R x y z N, Ds	2.98612 1.873688 1.632042 1 ← 28, 78.503	3.39782 1.90517 1.90517 $Z_{\phi 3}$ =1.38028 36, 77.955	Same as P2-G3.			
G3. $(3^{-}3)$	R x y z N, Ds	R =3.79191 z = 1 (has to) $\hat{\phi}_2 > \pi$ ×	3.51521 2.406498 1.911946 1 ← 36, 72.835	4.0313 2.484545 2.484545 1 ← 36, 5.380	Same as P3-G4		
G4. $(\underline{6\ 3})$	R x y z N, Ds	4.38701 1.686453 2.776358 $Z_{\phi 12}$ =2.58151 40, 51.959	4.22144 1.940549 2.756954 $Z_{\phi 3}$ =1.33527 48, 67.338	4.22522 2.506729 2.757401 1 ← 48, 67.224	4.966105 2.82839 2.82839 1 ← 60, 60.822	Same as P4-G8	
G8. $(3^{\sim}3^{-}3)$	R x y z N, Ds	$\|x-y\| > z$ ×	5.24222 1.962237 3.352656 $Z_{\phi 12}$ =3.11447 48, 43.667	R=4.38373 z = 1 $\hat{\phi}_2 > \pi$ ×	4.755705 2.811862 3.271645 1 ← 60, 66.323	5.2615625 3.35542 3.35542 1 ← 60, 54.183	Same as P8-G16
G16. $(\underline{6\ 3}^{-}3)$	R x y z N,Ds	$\|x-y\| > z$ ×	6.18315 1.973141 4.178907 $Z_{\phi 12}$ =3.84938 60, 39.235	$\|x-y\| > z$ ×	7.058565 2.917493 4.30708 $Z_{\phi 12}$ =4.08838 72, 36.128	R=5.59189 z = 1 $\hat{\phi}_2 > \pi$ ×	6.49797 4.230958 4.230958 1 84, 49.735

----------- End of [DATA 3-1] The T12E and Family ------------

[DATA 3-2] The T24E and Family:

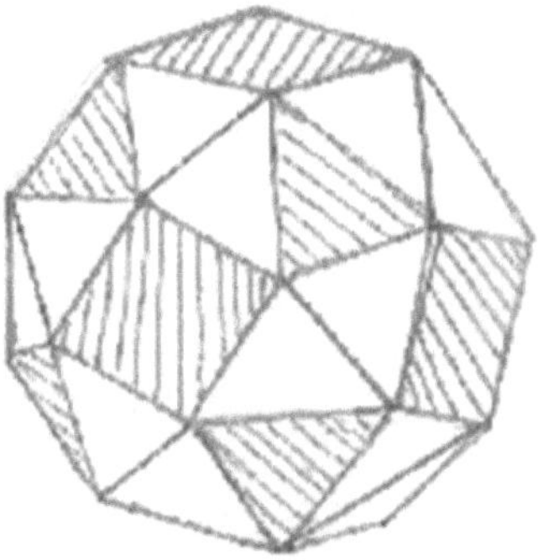

T24E:

$$R = \sqrt{\frac{1}{3}\left[\left(199+\sqrt{297}\right)^{\frac{1}{3}} + \left(199-\sqrt{297}\right)^{\frac{1}{3}} + 10\right]} \doteq 2.6874267474892....$$

$$Ds = \frac{2400}{4R^2} = 83.0764595\% \quad \theta_T = 4.9845665^{\circ}$$

T24E (I) P-S 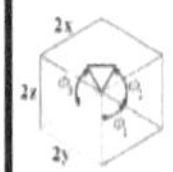		1. (1 ˇ4)e	3. (2 ~ˇ4)e	6. (4 ~ 4)	7. (4 ¯ 4)	9. ($\underline{8\ 4}$)	10. (1 4 ~4)e	23. (4 ~4 ¯4)e
	R x y z N,Ds	3.112792 1.33925 1 1 ← 30, 77. 404	3.47845 1.693024 1 1 ← 36, 74.383	3.77025 1.816146 1 $Z_{\phi 3}$ =1.216683 48, **84.419144**	4.69633 2.228963 1 $Z_{\phi 12}$ =1.972545 48, 54.408	5.246395 2.717776 1 $Z_{\phi 3}$ =2.040052 72, 65.396	4.09421 1.923877 1 $Z_{\phi 3}$ =1.522025 54, **80.53664**	6.072 3.039429 1 $Z_{\phi 12}$ =2.736286 72, 48.821
		30. (4 ~2~8 4)e	32. (4 −2~8 4)e [Rc<9.805]	35. ($\underline{8\ 4}$ ¯4)e	43 (1 4 ~4 ¯4)e	107 (4 ~2~8 ¯4 ¯4)e	111 (4 −2~8 ¯4 ¯4)e	153 (8 −2~8 2−8 4)e
	R x y z N,Ds	5.5758 2.833193 1 $Z_{\phi 3}$ =2.302421 96, 77.196	5.6351 2.850919 1 $Z_{\phi 3}$ =2.35932 96, 75.580	7.58544 3.925215 1 $Z_{\phi 12}$ =3.594484 96, 41.711	6.20115 3.115252 1 $Z_{\phi 12}$ =2.80903 78, 50.80	7.7309 4.010089 1 $Z_{\phi 12}$ =3.677445 120, 50.195	7.74454 4.018052 1 $Z_{\phi 12}$ =3.68523 120, 50.018	8.305075 4.518972 1 $Z_{\phi 3}$ =3.81269 168, 52.193

T24E (II): G-S 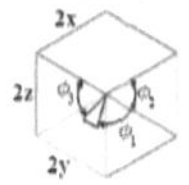		1. (1 3)	2. (3 ~ 3)	3. (3 ¯ 3)	4. ($\underline{6\ 3}$)	8. (3 ~3 ¯3)	16. ($\underline{6\ 3}$ ¯3)
	R x y z N, Ds	3.14637 1 1.642242 1 32, 80.811	3.77022 1 1.924326 $Z_{\phi 3}$ =1.58066 48, **84.42049**	4.48077 1 2.531725 $Z_{\phi 12}$ =2.033914 48, 59.769	4.93643 1 2.826196 $Z_{\phi 3}$ =2.38715 72, 72.866	5.6736 1 3.40788 $Z_{\phi 12}$ =2.89914 72, 55.918	6.8864 1 4.285666 $Z_{\phi 12}$ =3.77399 96, 50.609

T24E (III): T-S, (IV) T-S, G-S 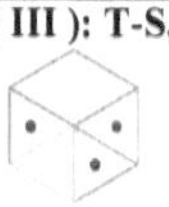 		1. (1 3)	3. (3 ¯3)	8. (3 ~3 ¯3)	16. ($\underline{6\ 3}$ ¯3)
	R	4.54058	7.129768	9.7329	12.3429
	x = y = z	1.689523	2.652945	3.621541	4.592696
	(III): N, Ds	48, 58.205	96, 47.213	168, 44.337	240, 39.384
	(IV): N, Ds	56, 67.906	120, 51.147	216, 46.448	312, 40.691

T24E (V): P4-S1, G3-S2	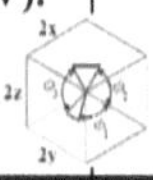	(1 ˇ4)	3. (2 ~ˇ4)	6. (4 ~4)	7. (4 ¯4)	9. ($\underline{8\ 4}$)	10. (1 4 ~4)	23. (4 ~4 ¯4)	30. ($\underline{8\ 4}$ ¯4)
1. (1 3)	R X Y Z N, Ds	3.69387 1.361405 1.667373 1 ← 38, 69.624	4.19174 1 ← 44, 62.604	4.32835 1.845716 1.685191 1 ← 56, 74.728	4.74332 2.230662 1.693122 1 ← 56, 62.225	6.34372 2.788936 1.710396 $Z_{\phi 12}$ 2.50299 80, 49.698	4.436572 1.937399 1.687478 1 ← 62, 78.750	$\|x-y\| > z$ ×	6.3052 2.816145 1.710129 $Z_{\phi 12}$ 2.22363 104, 65.40
2. (3 ~3)	R X Y Z N, Ds	4.4017 1.377234 1.945578 $Z_{\phi 12}$ 1.77915 54, 69.678	4.4006 1 ← 60, 77.458	4.68136 1.858844 1.952188 $Z_{\phi 3}$ 1.17655 72, **82.13504**	5.03617 2.251428 1.958954 1 ← 72, 70.970	6.05624 2.774263 1.971972 $Z_{\phi 3}$ 1.62177 96, 54.529	5.10119 1.95494 1.960035 $Z_{\phi 3}$ 1.647531 78, 74.936	7.1778 3.126253 1.980206 $Z_{\phi 12}$ 2.81875 96, 46.583	6.514 2.885349 1.975865 $Z_{\phi 3}$ 2.11867 120, 70.701
3. (3 ¯3)	R X Y Z N, Ds	$\|x-y\| > z$ ×	R = 4.5628 z = 1 $\bar{\phi}_{12} > \pi$ ×	4.82825 1.863425 2.55953 1 ← 72, 77.214	5.55365 2.280396 2.601664 1 ← 72, 58.36	6.23525 2.783655 2.628617 1 ← 96, 61.731	5.0055 1.93292 2.571535 1 ← 78, 77.829	6.560655 3.083178 2.638624 1 ← 96, 55.759	6.34535 2.878001 2.632176 1 ← 120, 74.509

4. (6 3)	R X Y Z N, D s	5.60785 1.391547 2.867106 $Z_{\phi 12}$ 2.5590 6 78, 62.007	6.3825 1.785446 2.898454 $Z_{\phi 12}$ 3.1808 1 84, 51.551	5.7182 1.883727 2.872436 $Z_{\phi 3}$ 1.7523 3 96, 73.384	6.99383 2.32988 2.915914 $Z_{\phi 12}$ 2.7612 1 96, 49.066	6.98917 2.875155 2.915799 $Z_{\phi 3}$ 1.67986 120, 61.415	6.19941 1.970995 2.892146 $Z_{\phi 3}$ 2.21222 102, 66.350	6.8497 3.104794 2.912233 1 ← 120, 63.941	7.54845 2.918542 2.928107 $Z_{\phi 3}$ 2.3390 3 144, 63.181
8. (3 ~3 ¯3)	R X Y Z N, D s	$\|x-y\| > z$ ×	$\|x-y\| > z$ ×	7.15715 1.901489 3.527973 $Z_{\phi 12}$ 3.1596 4 96, 46.852	$\|x-y\| > z$ ×	8.5945 2.855729 3.590393 $Z_{\phi 12}$ 3.15964 120, 40.614	6.95443 1.977452 3.51594 $Z_{\phi 12}$ 2.6935 102, 52.725	7.29325 3.133128 3.535495 1 ← 120, 56.4	8.3657 2.935401 3.582558 $Z_{\phi 12}$ 2.7808 3 144, 51.440
16. (6 3 ¯3)	R X Y Z N, D s	$\|x-y\| > z$ ×	$\|x-y\| > z$ ×	8.35397 1.909687 4.428304 $Z_{\phi 12}$ 3.9701 9 120, 42.987	$\|x-y\| > z$ ×	9.76454 2.873226 4.509553 $Z_{\phi 12}$ 4.09470 6 144 37.753	8.11715 1.983822 4.410379 $Z_{\phi 12}$ 3.4956 6 126 47.808	$\|x-y\| > z$ ×	9.47302 2.950911 4.495683 $Z_{\phi 12}$ 3.4240 7 168 46.803

(cont.)T24E (IV): P4-S1, G3-S2		32. (4 ¯8² ~4)	35. (8 4 ¯4)	43. (1 4 ~4 ¯4)	103 (4 ¯4 ~8² ¯4)	107. (4 ~8² ~4 ¯4)	111. (4 ¯8² ~4 ¯4)	153. (8 ¯8² ~8² ¯4)
1. (1 3)	R X Y Z N, Ds	6.2696 2.8864 1.709877 $Z_{\phi 12}$ = 2.147 104, 66.145	$\|x-y\| > z$ ×	$\|x-y\| > z$ ×	7.8725 3.705159 1.71802 $Z_{\phi 12}$ = 3.2941 128, 51.633	$\|x-y\| > z$ ×	$\|x-y\| > z$ ×	9.41866 4.620275 1.722261 $Z_{\phi 12}$ = 4.13817 176, 49.599
2. (3 ~3)	R X Y Z N, Ds	6.6004 2.899912 1.976506 $Z_{\phi 3}$ 2.23571 120, 68.862	8.66866 4.023864 1.986514 $Z_{\phi 12}$ 3.57888 120, 39.9225	7.2976 3.199014 1.980863 $Z_{\phi 12}$ = 2.87725 102, 47.883	7.51197 3.680787 1.981959 $Z_{\phi 3}$ 2.37372 144, 44.907	8.8039 4.104625 1.986929 $Z_{\phi 12}$ = 3.6505 144, 46.446	8.81315 4.110161 1.986959 $Z_{\phi 12}$ = 3.6554 144, 46.349	9.0146 4.588419 1.987542 $Z_{\phi 12}$ = 3.22099 192, 59.068
3. (3 ¯3)	R X Y Z N, Ds	6.35925 2.890337 2.632612 1 ← 120, 74.184	$\|x-y\| > z$ ×	6.62395 3.152562 2.640402 1 ← 102, 58.117	8.9535 2.759958 2.68189 $Z_{\phi 12}$ 3.4599 144, 44.907	$\|x-y\| > z$ ×	$\|x-y\| > z$ ×	10.47915 4.68526 2.695433 $Z_{\phi 12}$ 4.19414 192, 43.711
4. (6 3)	R X Y Z N, Ds	7.65748 2.929446 2.930187 $Z_{\phi 3}$ = 2.503741 144, 61.395	9.7891 4.093421 2.957668 $Z_{\phi 12}$ 3.7784 144, 37.568	6.93 3.175364 2.914313 1 ← 126, 65.591	8.36862 3.733202 2.941771 $Z_{\phi 3}$ = 2.08548 168, 59.971	9.916 4.171343 2.958758 $Z_{\phi 12}$ 3.83601 168, 42.714	9.92196 4.174958 2.958807 $Z_{\phi 12}$ 3.83871 168, 42.663	9.815425 4.647361 2.957896 $Z_{\phi 3}$ = 2.76871 216, 56.05
8. (3 ~3 ¯3)	R X Y Z N, Ds	8.26129 2.940433 3.57876 $Z_{\phi 12}$ = 2.60408 144, 52.748	8.43725 4.005942 3.585079 1 ← 144, 50.571	7.41075 3.205586 3.541661 1 ← 126, 57.357	8.1261 3.720199 3.573634 1 ← 168, 63.604	8.5063 4.082056 3.587448 1 ← 168, 58.0454	8.311117 4.087773 3.587614 1 ← 168, 57.980	11.67117 4.73289 3.653868 $Z_{\phi 12}$ = 4.440867 216, 40.308
16. (6 3 ¯3)	R X Y Z N, Ds	9.34402 2.954128 4.489124 $Z_{\phi 12}$ = 3.23547 168, 48.104	9.141073 4.056317 4.478244 1 ← 168, 50.264	$\|x-y\| > z$ ×	8.51995 3,740705 4.439994 1 ← 192, 66.125	9.2745 4.135866 4.485476 1 ← 192, 55.803	9.28211 4.140523 4.48588 1 ← 192, 55.7	10.0171 4.659769 4.520609 1 ← 240, 59.795

------------ End of [DATA 3-2] The T24E and Family ------------

[DATA 3-3] The T60E and Family

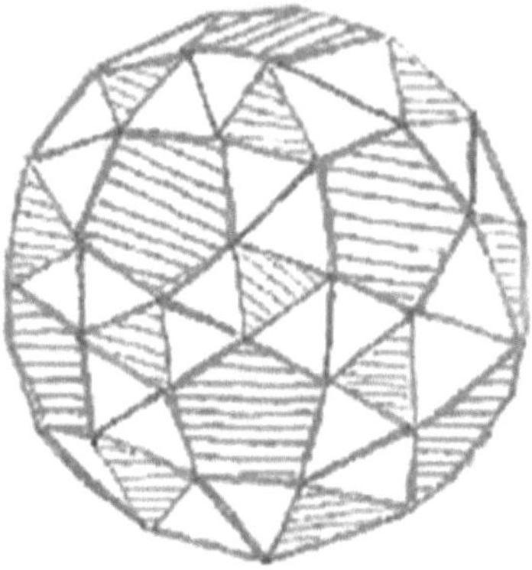

T60E:

$$R = \sqrt{A + B + \frac{1}{6}(27 + 7\sqrt{5})} \doteq \mathbf{4.31167475023127940...where}$$

$$\frac{A}{B} = \frac{1}{3}\sqrt[3]{\frac{1}{2}\left[\left(5112 + 2285\sqrt{5}\right) \pm \sqrt{64233 + 28728\sqrt{5}}\right]}$$

$$Ds = \frac{6000}{4R^2} \doteq 80.6862\%$$

$$\theta_T = 4.9845665^0$$

T60E (I): 12(P5-S)

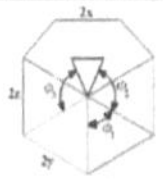

	1. (1 5)	2. (5 ~ 5)	3. (5 ¯ 5)	4. (<u>10 5</u>)	5. (1 5 ~ 5)	8. (5 ~ 5 ¯ 5)	11. (5 ~ $\overset{2}{10}$ ~ 5)	13. (5 ¯ $\overset{2}{10}$ ~ 5)
R	4.64998	5.95883	7.9314	8.66169	6.2666	9.43858	8.8365	8.88355
x	1.48064	1.770515	2.094739	2.730852	1.84539	2.869826	2.796605	2.807652
y	1	1	1	1	1	1	1	1
z	1 ←	$Z_{\phi 3}$ 1.08580	$Z_{\phi 12}$ 1.82775	$Z_{\phi 3}$ 1.97433	$Z_{\phi 3}$ 1.23566	$Z_{\phi 12}$ 2.54095	$Z_{\phi 3}$ 2.0955	$Z_{\phi 3}$ 2.117372
N, Ds	72 83.2473	3 120 84.48883	9 120 56.402	180 61.390	132 84.03302	180 50.513	240 76.846	240 76.030
	16. (<u>10 5</u> ¯ 5)	22. (1 5 ~ 5 ¯ 5)	25. (1 5 ~ $\overset{2}{10}$ ~ 5)	27. (1 5 ¯ $\overset{2}{10}$ ~ 5)	42. (5 ¯ 5 ~ $\overset{2}{10}$ ¯ 5)	46. (5 ~ $\overset{2}{10}$ ~ 5 ¯ 5)	50. (5 ¯ $\overset{2}{10}$ ~ 5 ¯ 5)	86. (10 ¯ $\overset{2}{10}$ ~ $\overset{2}{10}$ ¯ 5)
R	12.0813	9.6374	9.08335	9.18225	10.866	11.70268	12.3016	13.5092
x	3.819618	2.841352	2.853182	2.87487	3.5658	3.868373	3.898705	4.518645
y	1	1	1	1	1	1	1	1
z	$Z_{\phi 12}$ 3.4520	$Z_{\phi 12}$ 2.60857	$Z_{\phi 3}$ 2.2141	$Z_{\phi 3}$ 2.26529	$Z_{\phi 3}$ 2.78928	$Z_{\phi 3}$ 3.08762	$Z_{\phi 12}$ 3.52883	$Z_{\phi 12}$ 3.73148
N, Ds	3 240 41.108	192 51.680	252 76.357	252 74.721	300 63.545	300 54.763	300 49.561	420 57.535

T60E (II): G-S

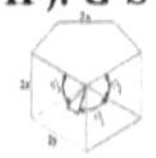

	1. (1 3)	2. (3 ~ 3)	3. (3 ¯ 3)	4. (<u>6 3</u>)	8. (3 ~ 3 ¯ 3)	16. (<u>6 3</u> ¯ 3)
R	4.9532	5.95883	6.70135	7.6758	8.4072	10.14567
x	1	1	1	1	1	1
y	1.696385	1.071022	2.642506	2.930627	3.584028	4.525928
z	1 ←	$Z_{\phi 3}$ 1.6373	$Z_{\phi 12}$ 1.96168	$Z_{\phi 3}$ 2.54136	$Z_{\phi 12}$ 2.891545	$Z_{\phi 12}$ 3.83009
N, Ds	80 81.519	120 84.48883	120 66.803	180 76.377	180 63.666	240 58.289

Group (III): T3-S (IV) T3-S, G3-S

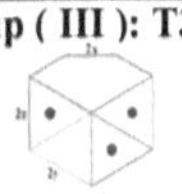

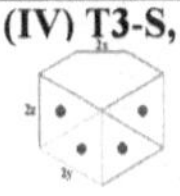

	1. (1 3)	3. (3 ¯ 3)	8. (3 ~ 3 ¯ 3)	16. (<u>6 3</u> ¯ 3)
R	7.3999	11.6526	15.9137	20.179
x = y = z	1.16163	2.70243	3.690675	4.679867
(III) N, Ds	120, 54.786	240, 44.188	420, 41.462	600, 36.838
(IV) N, Ds	140, 63.917	300, 55.235	540, 53.308	780, 47.889

T60E Group(V): P5-S1, G3-S2

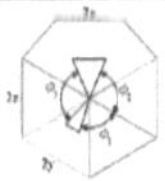

1. (1 3)

	1. (1 5)	2. (5 ~ 5)	3. (5 ¯ 5)	4. (<u>10 5</u>)	5. (1 5 ~ 5)	8. (5 ~ 5 ¯ 5)	11. (5 ~ $\overset{2}{10}$ ~ 5)	13. (5 ¯ $\overset{2}{10}$ ~ 5)
R	5.397328	6.93077	7.5771	10.2426	7.087	|x-y| > z ×	10.3193	10.3284
x	1.155217	1.785631	2.100698	2.760612	1.858208		2.825879	2.836693
y	1.702062	1.713925	1.7169	1.723775	1.714722		1.723899	1.723912
z	1 ←	1 ←	1 ←	$Z_{\phi 12}$ 2.43894	1 ←		$Z_{\phi 12}$ 2.34611	$Z_{\phi 12}$ 2.327550
$\bar{\phi}_3$	119.6844	90.60445	115.5578	118.4767	84.20444		109.0711	107.2583
$\hat{\phi}_{12}$	73.6747 163.1692	110.0736 97.38945	130.7505 112.0683	179.9995 ←	114.5653 88.17889		179.9997 ←	179.9994 ←
N,Ds	92, 78.953	140, 72.875	140, 60.962	200, 47.660	152, 75.659		40, 61.040	260, 60.932
	16. (<u>10 5</u> ¯ 5)	22. (1 5 ~ 5 ¯ 5)	25. (1 5 ~ $\overset{2}{10}$ ~ 5)	27. (1 5 ¯ $\overset{2}{10}$ ~ 5)	42. (5 ¯ 5 ~ $\overset{2}{10}$ ¯ 5)	46. (5 ~ $\overset{2}{10}$ ~ 5 ¯ 5)	50. (5 ¯ $\overset{2}{10}$ ~ 5 ¯ 5)	86. (10 ¯ $\overset{2}{10}$ ~ 10 ¯ 5)

	R	$\|x-y\| > z$ ×	$\|x-y\| > z$ ×	10.3467	10.344	12.57275	$\|x-y\| > z$ ×	$\|x-y\| > z$ ×	15.2195
	x			2.876496	2.896268	3.608153			4.564695
	y			1.723944	1.723937	1.726561			1.72831
	z			$Z_{\phi 12}$ 2.24749	$Z_{\phi 12}$ 2.19798	$Z_{\phi 12}$ 3.15439			$Z_{\phi 12}$ 4.02476
	$\bar{\phi}_3$			99.78778	95.40722	119.1667			119.4306
	$\hat{\phi}_{12}$			179.9997 ←	179.9995 ←	179.9997 ←			179.9995 ←
	N,Ds			272, 63.519	272, 63.552	320, 50.609			440, 47.489
2. $(3 \tilde{} 3)$		1. (1 5)	2. $(5 \tilde{} 5)$	3. $(5 \bar{} 5)$	4. $(\underline{10\ 5})$	5. $(1\ 5 \tilde{} 5)$	8. $(5 \tilde{} 5 \bar{} 5)$	11. $(5 \tilde{} \overset{2}{10} \tilde{} 5)$	13. $(5 \bar{} \overset{2}{10} \tilde{} 5)$
	R	6.41292	7.2896	7.9393	9.7539	7.6665	9.1066	10.121955	10.1864
	x	1.361191	1.789707	2.107388	2.753587	1.864791	2.859992	2.022749	2.834405
	y	1.975079	1.98082	1.983879	1.989377	1.982692	1.987794	1.990143	1.990267
	z	$Z_{\phi 12}$ 1.736135	$Z_{\phi 3}$ 1.0635	1 ←	$Z_{\phi 3}$ 1.49357	$Z_{\phi 3}$ 1.25357	1 ←	$Z_{\phi 3}$ 1.70783	$Z_{\phi 3}$ 1.74767
	$\bar{\phi}_3$	116.8311	60 ←	86.08444	60 ←	60 ←	69.28111	60 ←	60 ←
	$\hat{\phi}_{12}$	179.9997 ←	178.1025	85.25611 127.2647	178.9472	166.8578	150.145 79.68805	169.2572	167.4197
	N,Ds	132, 80.242	180, 84.68474	180, 71.392	240, 63.066	192, 81.66707	240, 72.350	300, 73.204	300, 72.280
		16. $(\underline{10\ 5} \bar{} 5)$	22. $(1\ 5 \tilde{} 5 \bar{} 5)$	25. $(1\ 5 \tilde{} \overset{2}{10} \tilde{} 5)$	27. $(1\ 5 \bar{} \overset{2}{10} \tilde{} 5)$	42. $(5 \bar{} 5 \tilde{} \overset{2}{10} \bar{} 5)$	46. $(5 \tilde{} \overset{2}{10} \tilde{} 5 \bar{} 5)$	50. $(5 \bar{} \overset{2}{10} \tilde{} 5 \bar{} 5)$	86. $(10 \bar{} \overset{2}{10} \tilde{} 10 \bar{} 5)$
	R	13.6765	9.04915	10.4371	10.57135	11.9984	13.866	13.8979	14.58761
	x	3.859925	2.923073	2.87782	2.899581	3.595987	3.92882	3.940345	4.549635
	y	1.994625	1.987638	1.990734	1.990972	1.993006	1.994772	1.994797	1.99528
	z	$Z_{\phi 12}$ 3.3759	1 ←	$Z_{\phi 3}$ 1.91039	$Z_{\phi 3}$ 2.0041	$Z_{\phi 3}$ 2.18034	$Z_{\phi 12}$ 3.43544	$Z_{\phi 12}$ 3.44543	$Z_{\phi 3}$ 3.05864
	$\bar{\phi}_3$	119.295	64.87833	60 ←	60 ←	60 ←	119.3139	119.3178	60 ←
	$\hat{\phi}_{12}$	179.9997 ←	163.3392 70.42472	159.7736	155.2517	179.3058	179.9997 ←	179.9995 ←	179.5308
	N,Ds	300, 40.097	252, 76.935	312, 71.604	312, 69.797	360, 62.517	360, 46.810	360, 46.596	480, 56.391
3. $(3 \bar{} 3)$		1. (1 5)	2. $(5 \tilde{} 5)$	3. $(5 \bar{} 5)$	4. $(\underline{10\ 5})$	5. $(1\ 5 \tilde{} 5)$	8. $(5 \tilde{} 5 \bar{} 5)$	11. $(5 \tilde{} \overset{2}{10} \tilde{} 5)$	13. $(5 \bar{} \overset{2}{10} \tilde{} 5)$
	R	$\|x-y\| > z$ ×	9.29733	8.516	10.15	9.34355	10.436	10.2925	10.31615
	x		1.804306	2.116325	2.759362	1.87724	2.893951	2.825465	2.836501
	y		2.685532	2.676604	2.693021	2.685994	2.69513	2.694094	2.694269
	z		$Z_{\phi 12}$ 2.3998	1 ←	1 ←	$Z_{\phi 12}$ 2.2322	1 ←	1 ←	1 ←
	$\bar{\phi}_3$		118.4811	107.815	79.72166	107.0989	108.7361	74.47389	73.48889
	$\hat{\phi}_{12}$		179.9994 ←	78.06555 173.0164	115.7136 103.7719	179.9995 ←	123.8739 126.6472	119.6761 95.28	120.3456 93.76139
	N,Ds		180, 52.059	180, 62.050	240, 58.240	192, 54.982	240, 55.091	300, 70.798	300, 70.474
		16. $(\underline{10\ 5} \bar{} 5)$	22. $(1\ 5 \tilde{} 5 \bar{} 5)$	25. $(1\ 5 \tilde{} \overset{2}{10} \tilde{} 5)$	27. $(1\ 5 \bar{} \overset{2}{10} \tilde{} 5)$	42. $(5 \bar{} 5 \tilde{} \overset{2}{10} \bar{} 5)$	46. $(5 \tilde{} \overset{2}{10} \tilde{} 5 \bar{} 5)$	50. $(5 \bar{} \overset{2}{10} \tilde{} 5 \bar{} 5)$	86. $(10 \bar{} \overset{2}{10} \tilde{} 10 \bar{} 5)$
	R	$\|x-y\| > z$ ×	10.5784	10.4015	10.444	14.19165	$\|x-y\| > z$ ×	$\|x-y\| > z$ ×	16.815
	x		2.964506	2.877302	2.897755	3.634676			4.595199
	y		2.696117	2.694884	2.695188	2.712088			2.717835
	z		1 ←	1 ←	1 ←	$Z_{\phi 12}$ 3.3101			$Z_{\phi 12}$ 4.0446
	$\bar{\phi}_3$		108.1672	69.43667	67.06445	119.345			119.5333
	$\hat{\phi}_{12}$		128.3208 122.7967	122.8436 87.75222	124.11 84.7075	179.9997 ←			179.9995 ←
	N,Ds		252, 56.299	312, 72.095	312, 71.509	360, 44.687			480, 42.441
4. $(\underline{6\ 3})$		1. (1 5)	2. $(5 \tilde{} 5)$	3. $(5 \bar{} 5)$	4. $(\underline{10\ 5})$	5. $(1\ 5 \tilde{} 5)$	8. $(5 \tilde{} 5 \bar{} 5)$	11. $(5 \tilde{} \overset{2}{10} \tilde{} 5)$	13. $(5 \bar{} \overset{2}{10} \tilde{} 5)$
	R	8.159	8.786	10.6388	11.10435	9.20495	10.74471	11.5468	11.62533
	x	1.166707	1.801535	2.137634	2.770751	1.876468	2.900098	2.841729	2.853685
	y	2.938676	2.947267	2.964236	2.967207	2.952035	2.964947	2.969695	2.970106
	z	$Z_{\phi 12}$ 2.59258	$Z_{\phi 3}$ 1.73073	$Z_{\phi 12}$ 2.68276	$Z_{\phi 3}$ 1.56902	$Z_{\phi 3}$ 1.87818	1 ←	$Z_{\phi 3}$ 1.8308	$Z_{\phi 3}$ 1.881334

	$\bar{\phi}_3$	118.0311	60 ←	118.8378	60 ←	60 ←	78.32722	60 ←	60 ←
	$\hat{\phi}_{12}$	179.9997 ←	178.6992	179.9995 ←	179.1892	167.2114	78.70416 142.2722	168.9633	167.0014
	N,Ds	192, 72.105	240, 58.295	240, 53.011	300, 60.824	252, 74.353	300, 64.964	360, 67.502	360, 66.594
		16. (10 5 ¯5)	22. (1 5 ~5 ¯5)	25. (1 5 ~10^2 ~5)	27. (1 5 ¯10^2 ~5)	42. (5 ¯5 ~10^2 ¯5)	46. (5 ~10^2 ~5 ¯5)	50. (5 ¯10^2 ~5 ¯5)	86. (10 ¯10^2 ~10 ¯5)
	R	12.2816	10.9078	11.9129	12.0755	13.2705	12.2457	15.53601	15.7868
	x	3.825564	2.971235	2.895745	2.916642	3.620812	3.886611	3.97038	4.576635
	y	2.973248	2.965999	2.971544	2.972325	2.977115	2.973004	2.983336	2.983863
	z	1 ←	1 ←	$Z_{\phi 3}$ 2.07811	$Z_{\phi 3}$ 2.1992	$Z_{\phi 3}$ 1.88766	1 ←	$Z_{\phi 12}$ 3.62241	$Z_{\phi 3}$ 2.56377
	$\bar{\phi}_3$	67.21333	78.15778	60 ←	60 ←	60 ←	64.66333	119.4533	60 ←
	$\hat{\phi}_{12}$	149.5797 82.73167	82.86 138.3375	159.3425	154.5892	179.4339	160.4008 74.47	179.9997 ←	179.6006
	N,Ds	360, 59.667	312, 65.557	372, 65.531	372, 63.778	420, 59.623	420, 70.02	420, 43.502	540, 54.168
8. (3 ~3 ¯3)		1. (1 5)	2. (5 ~5)	3. (5 ¯5)	4. (10 5)	5. (1 5 ~5)	8. (5 ~5 ¯5)	11. (5 ~10^2 ~5)	13 (5 ¯10^2 ~5)
	R	|x-y| > z ×	11.013	|x-y| > z ×	13.55915	11.0315	11.03125	10.7375	10.786
	x		1.810917		2.789532	1.884372	2.905346	2.831923	2.84343
	y		3.645713		3.675004	3.645996	3.645996	3.641232	3.642045
	z		$Z_{\phi 12}$ 3.19794		$Z_{\phi 12}$ 3.36597	$Z_{\phi 12}$ 3.01238	1	1 ←	1 ←
	$\bar{\phi}_3$		118.915		119.2822	107.3528	$\hat{\phi}_{12}$ =192.632 ×	63.58222	62.92833
	$\hat{\phi}_{12}$		179.9997 ←		179.9995 ←	179.9997 ←		60.35528 165.4453	61.42139 163.1445
	N,Ds		240, 49.470		300, 40.837	252, 51.769		360, 78.061	360, 77.361
		16. (10 5 ¯5)	22. (1 5 ~5 ¯5)	25. (1 5 ~10^2 ~5)	27. (1 5 ¯10^2 ~5)	42. (5 ¯5 ~10^2 ¯5)	46. (5 ~10^2 ~5 ¯5)	50. (5 ¯10^2 ~5 ¯5)	86. (10 ¯10^2 ~10 ¯5)
	R	13.6033	11.294…	10.9525	13.41525	13.0716	13.7408	13.763	18.4456
	x	3.85838	2.978…	2.88473	2.926969	3.617411	3.926082	3.937382	4.618328
	y	3.675436	3.65…	3.644758	2.673835	3.670738	3.67656	3.676747	3.701253
	z	1 ←	1	1 ←	$Z_{\phi 12}$ 2.79122	1 ←	1 ←	1 ←	$Z_{\phi 12}$ 4.287
	$\bar{\phi}_3$	104.01	$\hat{\phi}_2$ =185.8575 ×	60.0561	94.19833	74.78667	103.66	103.5944	119.6122
	$\hat{\phi}_{12}$	125.5461 130.0442		65.01862 154.8056	179.9997 ←	111.3061 113.4542	129.7253 126.2258	130.4339 125.5842	179.9995 ←
	N,Ds	360, 48.636		372, 77.528	372, 51.676	420, 61.451	420, 55.612	420, 55.432	540, 39.678
16. (6 3 ¯3)		1. (1 5)	2. (5 ~5)	3. (5 ¯5)	4. (10 5)	5. (1 5 ~5)	8. (5 ~5 ¯5)	11. (5 ~10^2 ~5)	13 (5 ¯10^2 ~5)
	R	|x-y| > z ×	12.75115	|x-y| > z ×	15.255	12.74922	|x-y| > z ×	15.2157	15.19865
	x		18.15055		2.79754	1.888884		2.867828	2.879492
	y		4.601465		4.64078	4.601429		4.640306	4.640097
	z		$Z_{\phi 12}$ 406539		$Z_{\phi 12}$ 4.10265	$Z_{\phi 12}$ 3.87384		$Z_{\phi 12}$ 3.8436	$Z_{\phi 12}$ 3.79178
	$\bar{\phi}_3$		119.1894		119.4328	107.5756		198.4933	106.3806
	$\hat{\phi}_{12}$		179.9997 ←		179.9995 ←	179.9995 ←		179.9997 ←	179.9997 ←
	N,Ds		300, 46.128		360, 38.674	312, 47.987		420, 45.354	420, 45.455
		16. (10 5 ¯5)	22. (1 5 ~5 ¯5)	25. (1 5 ~10^2 ~5)	27. (1 5 ¯10^2 ~5)	42. (5 ¯5 ~10^2 ¯5)	46. (5 ~10^2 ~5 ¯5)	50. (5 ¯10^2 ~5 ¯5)	86. (10 ¯10^2 ~10 ¯5)
	R	14.1165	|x-y| > z ×	15.11165	15.03371	17.5292	R=14.38…	R=14.4245	16.2282
	x	3.8687		2.916149	2.935773	3.66764			4.585057
	y	4.625474		4.639037	4.638076	4.662922			4.651402
	z	1		$Z_{\phi 12}$ 3.60225	$Z_{\phi 12}$ 3.47543	$Z_{\phi 12}$ 4.31174	Z = 1	Z=1	1 ←
	$\bar{\phi}_3$	$\hat{\phi}_2$ =195.9953 ×		98.80167	93.83111	119.5706	$\hat{\phi}_2$ =189… ×	$\hat{\phi}_2$ =188.2792 ×	71.67667
	$\hat{\phi}_{12}$			179.9997 ←	179.9997 ←	179.9994 ←			112.0883 115.9506
	N,Ds			432, 47.293	432, 47.785	480, 39.053			600, 56.957

------------ End of [DATA 3-3] The T60E and Family ----------

[DATA 4–1] The Basic Sq-SnE(k1-k2, 1) and Equivalent Configurations

	R	Eq. Conf. #1	R	Eq. Conf. #2	R
Sq- $12E'_1$ (4-3, 1)	$\sqrt{9+\sqrt{40}}$ = 3.9146589	24E'2[8(6S)] $(3\ \overset{\sim 2}{6})$ 6-e	3.914659	24E'3[6(8S)] $(4\ \overset{\sim 2}{8})$ 8-e	3.914659
Sq- $12E'_2$ (3-6, 1)	3.7229425	24E'2[4(6S)] $(3\ \overset{\sim 2}{6})$ 4-e	3.7229425	12E'2[4(6S)] $(6\ \overset{-2}{6})$ 6-e	3.7229424
Sq- $12E'_2$ (6-6, 1)	$\sqrt{10}$ = 3.16227766	24E'	$\sqrt{10}$		
Sq- $24E'_1$ (4-4, 1)	$\sqrt{2(2+\sqrt{2})(2+\sqrt{2+\sqrt{2}})+1}$ =5.2224652	48E'[6(8S)] $(4\ \overset{\sim 2}{8})$ 6 -e	5.222465		
Sq- $24E'_1$ (3-4, 1)	** $\sqrt{A+B+\frac{1}{3}(31+6\sqrt{2})}$ = 6.0751586	48E'[8(6S)] $(3\ \overset{\sim 2}{6})$ 8-e	5.9230146 Error = 2.504%		
Sq- $24E'_2$ (4-6, 1)	$\sqrt{\sqrt{525+\sqrt{1000}}+\sqrt{525-\sqrt{1000}}+10}$ = 5.1115333884	48E'[6(8S)] $(4\ \overset{\sim 2}{8})$ 4-e	5.111334	24E'2[8(6S)] $(6\ \overset{-2}{6})$ 6-e	5.1115338
Sq- $24E'_2$ (6-6, 1)	$\sqrt{13+6\sqrt{2}}$ = 4.6352218	48E'	$\sqrt{13+6\sqrt{2}}$		
Sq- $24E'_3$ (3-8, 1)	5.4269755	48E'[8(6S)] $(3\ \overset{\sim 2}{6})$ 4-e	5.4269762	24E'3[6(8S)] $(8\ \overset{-2}{8})$ 8-e	5.426976
Sq $24E'_3$ (8-8, 1)	$\sqrt{13+6\sqrt{2}}$ = 4.6352218	48E'	$\sqrt{13+6\sqrt{2}}$		
Sq- 30E' (5-3, 1)	6.175794	60E'3[12(10S)] $(5\ \overset{\sim 2}{10})$ 10-e	6.175792	60E'2[20(6S)] $(3\ \overset{\sim 2}{6})$ 6-e	6.175793
Sq- 48E' (8-6, 1)	6.601719	48E'[6(8S)] $(8\ \overset{-2}{8})$ 4-e	6.601718	48E'[8(6S)] $(6\ \overset{-2}{6})$ 4-e	6.601718
Sq- 48E' (8-4, 1)	* 7.04317	48E'[6(8S)] $(8\ \overset{-2}{8})$ 6-e	7.0431683		
Sq- 48E' (6-4, 1)	* 7.790735	48E'[8(6S)]	7.790672		
Sq- $60E'_1$ (5-6, 1)	7.935094	120E'[12(10S)] $(5\ \overset{\sim 2}{10})$ 4-e	7.935095	60E'1[20(6S)] $(6\ \overset{-2}{6})$ 6-e	7.935095
Sq- $60E'_1$ (6-6, 1)	$\sqrt{31+12\sqrt{5}}$ = 7.604789	120E'	$\sqrt{31+12\sqrt{5}}$		
Sq $60E'_2$ (5-4, 1)	* 7.995593	120E'[12(10S)] $(5\ \overset{\sim 2}{10})$ 6-e	7.995566		
Sq $60E'_2$ (3-4, 1)	* 9.8708	120E'[20(6S)] $(3\ \overset{\sim 2}{6})$ 10-e	9.870597		
Sq $60E'_3$ (3-10, 1)	8.835113	120E'[20(6S)] $(3\ \overset{\sim 2}{6})$ 4-e	8.835113	60E'3[12(10S)] $(10\ \overset{-2}{10})$ 10-e	8.835113
Sq $60E'_3$ (10-10, 1)	$\sqrt{31+12\sqrt{5}}$ = 7.604789	120E'	$\sqrt{31+12\sqrt{5}}$		
Sq 120E' (10-6, 1)	10.57276	120E'[12(10S)] $(10\ \overset{-2}{10})$ 4-e	10.57276	120E'[20(6S)] $(6\ \overset{-2}{6})$ 4-e	10.57276
Sq 120E' (10-4, 1)	* 11.1061	120E'[12(10S)] $(10\ \overset{-2}{10})$ 6-e	11.10587		
Sq 120E' (6-4, 1)	* 13.03042	120E'[20(6S)] $(6\ \overset{-2}{6})$ 10-e	13.030105		

** $A, B = \frac{1}{3}\left[\frac{1}{2}(51329 + 27612\sqrt{2}) \pm \frac{1}{2}\sqrt{5697+8856\sqrt{2}}\right]^{\frac{1}{3}}$ * The virtual value of R is possible to find theoretically.

End of [DATA 4-1] The basic Sq-SnE(k1-k2, 1) and their Equivalent Configurations

[DATA 4–2] The Basic Sq-SnE(k1-k2, m) while m = 1, 2, 3, 4, 5, 10, 50, 100

The Sq-SnE(k1-k2, m), m = 1, 2, 3, 4, 5, 10, 50, 100 — P. 1 of 2

Sq-SnE(k1-k2,m)	M	N	R	Ds	Equi.	Sq-SnE(k1-k2,m)	m	N	R	Ds	Equi.
Sq-S4E(3-3, m) $R_0 = \sqrt{\frac{3}{2}}$ = 1.2247448 L = 6 strings	1	12	2	75	12 E'_1	Sq-S12 E'_2 (6-6,m) $R_0 = \sqrt{\frac{11}{2}}$ =2.34520788 L = 6	1	24	3.16227766	60	24 E'_2
	2	24	2.92758	70.006			2	36	4.114857	53.154	
	3	36	3.91467	58.729			3	48	5.11155	45.928	
	4	48	4.9261	49.451			4	60	6.127478	39.951	
	5	60	5.949495	42.377			5	72	7.153416	35.176	
	10	120	11.131724	24.21			10	132	12.339146	21.674	
	50	600	52.962	53.476			50	612	54.16428	5.215	
	100	1200	105.2972	2.7058			100	1212	106.548	2.669	
Sq-S6E(3-3,m) $R_0 = \sqrt{2}$ L = 12	1	24	2.79793265	76.644	24 E'_1	Sq-S24 E'_1 (3-4,m) $R_0 = \sqrt{5+2\sqrt{2}}$ =2.79793265 L = 24	1	72	5.923028	51.308	
	2	48	4.33852	63.753			2	120	9.133939	35.959	
	3	72	592303	51.308			3	168	12.365	27.47	
	4	96	7.524452	42.39			4	216	15.604	22.178	
	5	120	9.133939	35.959			5	264	18.846602	18.581	
	10	240	17.22462	20.223			10	504	35.08017	10.239	
	50	1200	82.18533	4.442			50	2424	165.1480	2.222	
	100	2400	163.44688	2.246			100	4848	327.92698	1.121	
Sq-S8E(4-4,m) $R_0 = \sqrt{3}$ L = 12	1	24	2.79793265	76.644	24 E'_1	Sq-S12 E'_2 (6-6,m) $R_0 = \sqrt{\frac{11}{2}}$ =2.34520788 L = 6	1	72	5.2225	65.996	
	2	48	3.9910155	75.338			2	120	7.7264	50.254	
	3	72	5.2224652	65.997			3	168	10.247691	39.994	
	4	96	6.4701975	57.329			4	216	12.78	33.062	
	5	120	7.7262853	50.255			5	264	15.323	28.11	
	10	240	14.05343	30.38			10	504	28.026083	16.042	
	50	1200	64.955104	7.110			50	2424	129.89481	3.592	
	100	2400	128.67942	3.624			100	4848	257.19063	1.823	
Sq-S12E(3-3,m) $R_0 = \sqrt{\frac{1}{2}(5+\sqrt{5})}$ = 1.902113 L = 30	1	60	4.465901	75.2096	60 E'_2	Sq-S24 E'_1 (3-4,m) $R_0 = \sqrt{5+2\sqrt{2}}$ =2.79793265 L = 24	1	72	5.11152	68.893	
	2	120	7.1543	58.612			2	120	7.15335	58.628	
	3	180	9.8708	46.186			3	168	9.220467	49.402	
	4	240	12.5975	37.808			4	216	11.298245	42.303	
	5	300	15.329229	31.917			5	264	13.381401	36.859	
	10	600	29.01316	17.82			10	504	23.825596	22.196	
	50	3000	138.65	3.901			50	2424	107.54701	5.2303	
	100	6000	275.7575	1.973			100	4824	212.42827	2.6725	
Sq-S20E(5-5,m) $R_0 = \frac{\sqrt{3}}{2}(1+\sqrt{5})$ = 2.80251701 L = 30	1	60	4.465901	75.21	60 E'_2	Sq-S24 E'_1 (4-4,m) $R_0 = \sqrt{5+2\sqrt{2}}$ L = 24	1	48	4.635222	55.852	48E'
	2	120	6.217699	77.6			2	72	6.198546	46.848	
	3	180	7.995593	70.39			3	96	7.790735	39.542	
	4	240	9.7845912	62.671			4	120	9.395446	33.985	
	5	300	11.57931	55.937			5	144	11.006677	29.716	
	10	600	20.584906	35.399			10	264	19.09994	18.092	
	50	3000	9.2835629	8.7023			50	1224	84.06043	4.3305	
	100	6000	183.19731	4.4699			100	2424	165.41284	2.2148	
Sq-S12 E'_1 (3-4,m) $R_0 = 2$ L = 24	1	48	3.914652	78.306		Sq-S24 E'_3 (3-8,m) $R_0 = \sqrt{7+4\sqrt{2}}$ L= 24	1	72	5.4269383	61.117	
	2	96	5.94955	67.802			2	120	7.44699	54.095	
	3	144	8.014505	56.046			3	168	9.505297	46.486	
	4	192	10.091097	47.137			4	216	11.578671	40.279	
	5	240	12.173526	40.487			5	264	13.65924	35.375	
	10	480	22.6165343	23.460			10	504	24.0987	21.696	
	50	960	106.38316	5.302			50	2424	107.82887	5.212	
	100	4800	211.12314	2.692			100	4824	212.64727	2.667	
Sq-S12 E'_2 (3-6,m) $R_0 = \sqrt{\frac{11}{2}}$ =2.34520788 L = 12	1	36	3.722926	64.934		Sq-S24 E'_3 (8-8,m) $R_0 = \sqrt{7+4\sqrt{2}}$ L = 12	1	48	4.635222	55.852	48E'
	2	60	5.2705755	53.998			2	72	5.819	53.159	
	3	84	6.85781	44.653			3	96	7.04317	48.381	
	4	108	8.46055	37.72			4	120	8.285958	43.695	
	5	132	10.070565	32.539			5	144	9.528824	39.566	
	10	252	18.16197	19.099			10	264	15.858743	26.243	
	50	1212	83.14578	4.383			50	1224	66.762348	6.865	
	100	2412	164.3584	2.232			100	2424	130.46988	3.560	

End of [DATA 4-2, P. 1 of 2] The basic Sq-SnE(k1-k2, m), m = 1, 2, 3, 4, 5, 10, 50, 100

(continue) **(DATA 4-2)The basic Sq-SnE(k_1-k_2, m), m = 1, 2, 3, 4, 5, 10, 50, 100** P. 2 of 2

Sq-SnE(k1-k2, m)	m	N	R	Ds	
	1	120	6.17577	78.657	
Sq-S30E' (3-5, m)	2	240	9.20196	70.858	
	3	360	12.248396	59.991	
$R_0 = 1 + \sqrt{5}$	4	480	15.303	51.242	
=3.2360679775	5	600	18.361226	44.493	
	10	1200	33.67372	26.457	
L = 60	50	6000	156.38394	6.1335	
	100	12000	309.70907	3.1276	
Sq-S48E' (4-6, m)	1	96	7.79075	39.541	
	2	144	11.006675	29.716	
	3	192	14.23926	23.674	
$R_0 = \sqrt{13+6\sqrt{2}}$	4	240	17.478376	19.640	
=4.63522183	5	288	20.721	16.769	
	10	528	36.956586	8.6647	
L = 24	50	2448	167.02002	2.1939	
	100	4848	329.49852	1.1163	
	1	96	6.60174	55.067	
Sq-S48E' (6-8, m)	2	144	8.642078	48.202	
	3	192	10.707295	41.868	
$R_0 = \sqrt{13+6\sqrt{2}}$	4	240	12.78348	36.716	
=4.63522183	5	288	14.865219	32.583	
	10	528	25.307006	20.611	
L = 24	50	2448	109.02557	5.1487	
	100	4848	213.74904	2.6527	
Sq-S48E'(8-4, m)	1	96	7.04319	48.381	
	2	144	9.5388	39.565	
	3	192	12.0595	33.005	
$R_0 = \sqrt{13+6\sqrt{2}}$	4	240	14.590926	28.183	
	5	288	17.1275	24.544	
L = 24	10	528	29.838026	14.826	
	50	2448	131.719505	3.5274	
	100	4848	259.12923	1.805	
Sq-S60 E_1' (5-6,	1	180	7.93515	71.466	
m)	2	300	10.967816	62.348	
	3	420	14.016559	53.445	
	4	540	17.071704	46.321	
$R_0 = \sqrt{\frac{1}{2}(29+9\sqrt{5})}$	5	660	20.131054	40.715	
	10	1260	35.447324	25.069	
	50	6060	158.06155	6.064	
L = 60	100	12060	311.5804	3.106	
	1	120	7.604789	51.874	120E'
Sq-S60 E_1' (6-6,	2	180	10.30898	42.343	
m)	3	240	13.03042	35.337	
	4	300	15.759	30.2	
$R_0 = \sqrt{\frac{1}{2}(29+9\sqrt{5})}$	5	360	18.492141	26.319	
	10	660	32.17925	15.934	
L = 30	50	3060	141.86495	3.801	
	100	6060	279.11579	1.945	
	1	180	9.870569	46.188	
Sq-S60 E_2' (3-4,	2	300	15.329204	31.917	
m)	3	420	20.7998	24.27	
	4	540	26.27451	19.56	
$R_0 = \sqrt{11+4\sqrt{5}}$	5	660	31.75406	16.363	
	10	1260	59.16	9.0002	
L = 60	50	6060	278.73863	1.95	
	100	12060	552.98996	0.986	

Sq-SnE(k1-k2, m)	m	N	R	Ds	
	1	180	8.8352	57.647	
Sq-S60 E_3' (3-10, m)	2	300	11.8399	53.501	
	3	420	14.876	47.448	
	4	540	17.925	42.016	
$R_0 = \sqrt{\frac{1}{2}(37+15\sqrt{5})}$	5	660	20.9797	37.487	
	10	1260	36.287904	23.921	
L = 60	50	6060	158.91392	5.999	
	100	12060	312.54255	3.0867	
Sq-S60 E_3' (10-	1	120	7.604789	51.874	120E'
10,m)	2	180	9.340562	51.578	
	3	240	11.1061	48.644	
	4	300	12.886451	45.164	
$R_0 = \sqrt{\frac{1}{2}(37+15\sqrt{5})}$	5	360	14.675258	41,79	
	10	660	23.668	29.455	
	50	3060	95.89948	8.318	
L = 30	100	6060	186.33132	4.364	
	1	240	10.57277	53.675	
Sq-S120E' (10-6,	2	360	13.5957	48.69	
m)	3	480	16.638241	43.348	
	4	600	19.69	38.690	
$R_0 = \sqrt{31+12\sqrt{5}}$	5	720	22.74605	34.791	
	10	1320	38.05625	22.786	
L = 60	50	6120	160.77173	5.919	
	100	12120	314.0	3.073	
	1	240	13.030038	35.34	
Sq-S120E' (6-4, m)	2	360	18.492	26.319	
	3	480	23.963346	20.897	
$R_0 = \sqrt{31+12\sqrt{5}}$	4	600	29.4387	17.308	
	5	720	34.917	14.764	
L = 60	10	1320	62.326347	8.495	
	50	6120	281.78489	1.927	
	100	12120	555.97104	0.980	
	1	240	11.105783	48.647	
Sq-S120E' (4-10,	2	360	14.675	41.791	
m)	3	480	18.2658	35.967	
	4	600	21.865	31.376	
$R_0 = \sqrt{31+12\sqrt{5}}$	5	720	25.469102	27.749	
	10	1320	43.518	17.425	
L = 60	50	6120	188.02981	4.328	
	100	12120	368.99073	2.225	

Sq-S60 E'_2 (5-4, m) $Ro = \sqrt{11+4\sqrt{5}}$ L = 60	1	180	7.99555	70.391							
	2	300	11.5793	55.937							
	3	420	15.17718	45.583							
	4	540	18.78176	38.27							
	5	660	22.38908	32.916							
	10	1260	40.4445	19.257							
	50	6060	184.99447	4.427							
	100	12060	365.71956	2.254							

End of [DATA 4-2] The basic Sq-SnE(k1-k2, m), m = 1, 2, 3, 4, 5, 10, 50, 100

[DATA 4-3] The One-pearl and Special One-pearl-Ep of Sq-SEs

Sq-SnE(k1-2)	2. (1 4) 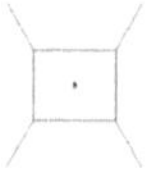 x1 = x2	5. (2 4) 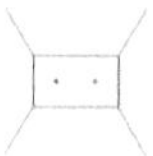 x1 = x2	4. (2 4) 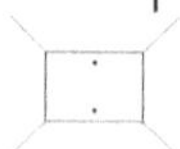 x1 = x2	Sq(.)-SnE(k1-k2) 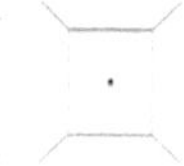 x1 = x2 = 1	Sq(:)-SnE(k1-k2) 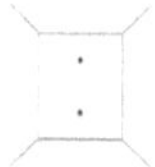 x1 = x2 = 1	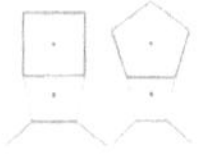	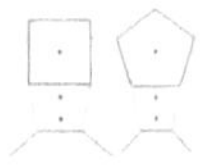
S4E(3-3)	R=2.57793 x1=x2=1.548583 N=18, Ds=67.713	3.55765 2.414214 24, 47.405	2.927575 1.863236 24, 70.006	2.577935 1 18, 67.712	3.55766 1 24, 47.405		
S6E(3-3)	3.609357 1.641023 36, 69.085	4.83168 2.559775 48, 51.403	3.991026 1.931852 48, 75.338	3.855791 1 60.536	5.436955 1 48, 40.595		
S8E(4-4)	3.85575 1.652557 36, 60.538	5.436909 2.596004 48, 40.595	4.33849 1.943081 48, 63.754	3.60935 1 36, 69.085	4.83167 1 48, 51.403	S8E(4̇ - 4̇) R=3.8188 x1=x2=1.364865 42, 77.000	S8E(4̇ - 4̇) 5.00678 x1=x2=1.385718 54, 53.854
S12E(3-3)	5.71285 1.696302 120, 53.575	7.4831 2.66024 120, 53.575	6.21775 1.973269 120, 77.60	6.391 1 90, 55.087	9.1059 1 120, 36.182		
S20E(5-5)	6.39095 1.703545 90, 55.087	9.10577 2.683557 120, 36.182	7.15423 1.979972 120, 58.613	5.71284 1 90, 68.941	7.483162 1 120, 53.574	S20E(5̇ - 5̇) 5.950856 x1=x2=1.158854 102, 72.008	S20E(5̇ - 5̇) 7.71947 x1=x2=1.165655 142, 59.574
S12E'1(4-3)	5.22109 1.689161 72, 66.031	7.09142 2.652087 96, 47.725	5.76035 1.968681 96, 72.329	5.357807 1 72, 62.704	7.418624 1 96, 43.608	S12E'1(4̇ -3) 5.4941 1.390591 1 78, 64.601	S12E'1(4̇ -3) 7.538798 1.401717 1 102, 44.868
S12E'2(3-6)	4.55776 1.675543 48, 57.767	5.77537 2.611485 60, 44.971	4.9175 1.956387 60, 62.03	4.8099 1 48, 51.869	6.38777 1 60, 36.761		
S12E'2(6-6)	3.813066 1.650723 30, 51.587	4.79361 2.557027 36, 39.167	4.11486 1.936216 36, 53.154	3.81307 1 30, 51.584	4.79362 1 36, 39.167		
S24E'1(4-4)	7.160434 1.709381 96, 46.809	9.8723 2.690794 120, 30.781	7.916106 1.983717 120, 47.874				
S24E'1(3-4)	7.86667 1.71329 96, 38.782	10.57608 2.696102 120, 26.821	8.61672 1.986302 120, 40.405				
S24E'2(4-6)	6.42508 1.70385 96, 58.137	8.2818 2.673424 120, 43.739	6.94728 1.97873 120, 62.157	65.7479 1 96, 55.520	8.63507 1 120, 40.234	S24E'2(4̇ -6) 6.698665 1.398366 1 102, 56.828	S24E'2(4̇ -6) 8.74986 1.404947 1 126, 41.144
S24E'2(6-6)	5.472714 1.693058 60, 50.082	6.6827 2.642006 72, 40.306	5.819032 1.969334 72, 53.158	5.74755 1 60, 45.407	7.33007 1 72, 44.668		
S24E'3(3-8)	6.7615 1.706607 96, 52.496	8.63721 2.67815 120, 40.214	7.2873 1.980715 120, 56.495	6.87495 1 50.778	8.92376 1 120, 37.673		

S24E'3(8-8)	5.747546 1.696736 60, 45.407	7.330068 2.65721 72, 33.501	6.19853 1.973097 72, 46.849	5.47275 1 60, 50.082	6.6827 1 72, 53.741		
S30E'(5-3)	8.44285 1.715775 180, 63.130	11.5981 2.702159 240, 44.604	9.31215 1.988298 240, 69.191	8.362806 1 180, 64.344	11.405606 1 240, 46.123	S30E'($\dot{5}$-3) 8.5624 1.167526 1 192, 65.471	S30E'($\dot{5}$-3) 11.60381 1.171199 1 252, 46.789
S48E'(8-6)	7.93119 1.713596 120, 47.692	9.7947 2.690138 144, 37.525	8.4465 1.985732 144, 50.46	8.07355 1 120, 46.025	10.13056 1 144, 35.078		
S48E'(8-4)	8.99507 1.71772 120, 37.078	11.70381 2.702697 144, 26.281	9.73948 1.989319 144, 37.952				
S48E'(6-4)	9.7374 1.719829 120, 31.640	12.4374 2.706058 144, 23.273	10.47628 1.990783 144, 32.801				
S60E'1(5-6)	10.20126 1.720919 240, 57.656	13.34285 2.709466 300, 42.127	11.0589 1.991741 300, 61.325	10.13575 1 240, 58.404	13.1798 1 43.176	S60E'1($\dot{5}$-6) 10.33375 1.170053 1 252, 58.996	S60E'1($\dot{5}$-6) 13.37694 1.172281 1 312, 43.589
S60E'1(6-6)	8.85855 1.717273 150, 47.787	10.61394 2.696358 180, 39.945	9.3404 1.988372 180, 51.58	9.5647 1 150, 40.991	12.2806 1 180, 29.838		
S60E'2(5-4)	11.19315 17.2281 180, 35.918	15.6051 2.715539 240, 24.639	12.3944 1.993439 240, 39.057				
S60E'2(3-4)	13.0767 1.725285 180, 26.316	17.4885 2.718904 240, 19.618	14.2729 1.995059 240, 29.453				
S60E'3(3-10)	11.12596 1.722697 240, 48.47	14.28823 2.712366 300, 36.737	11.98475 1.992878 300, 52.192	11.01583 1 240, 49.444	14.04394 1 300, 38.026		
S60E'3(10-10)	9.56463 1.719382 150, 40.992	12.280486 2.705389 180, 29.839	10.3088 1.990477 180, 42.344	8.8585 1 150, 47.787	10.61405 1 180, 39.944		
S12E'(10-6)	12.8571 1.725051 300, 45.371	16.009 2.716362 360, 29.264	13.71172 1.994847 360, 47.869	12.7723 1 300, 45.975	15.8088 1 360, 36.012		
S120E'(10-4)	14.31761 1.726409 300, 36.586	18.73085 2.720591 360, 25.652	15.512 1.995822 360, 37.403				
S120E'(6-4)	16.2376 17.27666 300, 28.446	20.6435 2.722616 360, 21.119	17.4271 1.996689 360, 29.634				

The $24E'_1$(4-4), $24E'_1$(3-4), 48E'(10-4), 48E'(6-4), $60E'_2$(5-4), $60E'_2$(3-4), 120E'(10-4), and 120E'(6-4) are having their 4-gon P-faces in the third set which can not connected with a string, otherwise the angle $\phi_0 > \pi$.

-- End of [4-3] The One-pearl and Special One-pearl-Ep of Sq-SEs --

At first glance on the above table, the R of **Sq-S6E(3-3)** is greater than that of **Sq-S8E(4-4).** It seems controversial due to the fact that the R of 6E should be smaller than that of 8E. To explain this, think about the open faces of **Sq(.)-** or **Sq(:)-SnE(k1-k2):** The **Sq(.)-S6E(3-3)** has six open faces of uneven 8-gons each with four inner surface angles of 180°_{-} and another four of 120°_{-}, so these 8-gons are very close to large squares of every edge-length of 4 units. The **Sq(.)-S8E(4-4)** has eight open faces of uneven 6-gons each with three inner surface angle of 150°_{-} and the other three of 120°_{-}. The ratio of these two areas is roughly **151:107**. Therefore, the R of **Sq(.)-S6E** is greater than that of **Sq(.)-S8E**.

For that same reason the R of **Sq(.)-S12E(3-3)** is greater than that of **Sq(.)-S20E(5-5),** and as well as some other **Sq(.) and Sq(:)**

configurations.

(4.12.2013.F)

[DATA 4A] The Five Sq-SnE(k-k, m) and Their Families

The edge-length x of P-face is limited:

While k = 3, 4, 5: $\frac{R^2 z + 2CR\sqrt{R^2-1}}{R^2 - 4} > x \geq \max[1, \frac{R^2-4}{R^2} z]$

While k = 6, 8, 10: $\frac{R^2 (z + 2c)}{R^2 - 4} > x \geq \max[1, \frac{R^2-4}{R^2} z]$

Every Sq-SnE(k, m) has 3 family groups: (T = Sq-pearl face)

(1) P-S (2) T-S (3) (T-S1) + (P-S2)

Every Sq-SnE'(k1-k2, m) has 7 family groups:

(1) P1-S (2) P2-S (3) T4-S

(4) (P1-S1) + (P2-S2) (5) (T-S1) + (P-S2) (6) (T-S1) + (P2-S2)

(7) (T-S1) + (P2-S2) + (P1-S3)

Sometimes, say, the P1-face is a 3-gon, it may be expressed as P1-face or P3-face.

[DATA 4A-1] The Sq-S4E(3-3, m) and Family

The R of Sq-S4E(3-3, 1) is: R = 2

Sq-S4E(3-3, m) (I): 4(P3-S)			m = 1	m = 3	m = 5	m = 10	m = 50	m = 100
1. (1 3)		R	2.290452	4.2890865	6.282398	11.439831	53.255888	105.620663
2. $(3^{\sim} 3)$		R	2.933527	4.37045	6.333813	11.473672	53.28339	105.6207
Sq-S4E(3-3, m) (II): 6(T-S)			m = 1	m = 3	m = 5	M = 10	m = 50	m = 100
2. $(1\ \overset{2}{4})$		R	2.57793	4.514074	6.51749	11.652334	53.42617	105.79
5. $(2^{-}\ \overset{2}{4})$		R	3.55765	5.42105	7.368766	12.418344	54.068113	106.41353
Sq-S4E(3-3, m) (III) 6(T-S1)+4(P3-S2)			m = 1	m = 3	m = 5	m = 10	m = 50	m = 100
2. $(1\ \overset{2}{4})$	1. (1 3)	R	--	4.5252	6.52225	11.653604	53.426172	105.79
	2. $(3^{\sim} 3)$	R	--	4.708204	6.683933	11.80221	53.564673	105.90117
	3. $(3^{-} 3)$	R	--	5.01173	6.9287	12.004394	53.7458	106.07
5. $(2^{-}\ \overset{2}{4})$	3. $(3^{-} 3)$	R	--	5.4210505	7.36867	12.418344	5.406811	106.41353
	4. $(\underline{6\ 3})$	R	--	5.63053	7.5454	12.57245	54.20965	106.52334
	8. $(3^{\sim} 3^{-} 3)$	R	--	5.954165	7.80923	12.783846	54.394754	106.72

----- End of [DATA 4A-1] The Sq-S4E(3-3,m) and Family -----

[DATA 4A-2] The Sq-S6E(3-3, m) and Family

The R of Sq-S6E(3-3, 1) is: $R = \sqrt{5+2\sqrt{2}} \doteq 2.7993265$

Sq-S6E(3-3,m) (I) 8(P3-S)		m = 1	m = 3	m = 5	m = 10	m = 50	m = 100
1. (1 3)	R	3.641162	6.43694	9.618721	17.692849	82.646221	163.96654
2. (1 $\overset{1}{4}$)	R	3.992632	6.5141	9.675856	17.73844	82.71247	163.96654

Sq-S6E(3-3,m) (II) 12(T4-S)		m = 1	m = 3	m = 5	m = 10	m = 50	m = 100
2. (1 $\overset{1}{4}$)	R	3.60935	6.720731	9.9053	17.961944	82.89066	164.2337
5. (2 $\overset{-1}{4}$)	R	4.831655	7.874965	11.01228	1.900725	83.85991	165.16756

Sq-S6E(3-3,m) (III) 12(T4-S1)+8(P3-S2)			m = 1	m = 3	m = 5	m = 10	m = 50	m = 100
2. (1 $\overset{1}{4}$)	1. (1 3)	R	--	6.72741	9.908109	17.96348	82.89066	164.23372
	2. (3 $\tilde{}$3)	R	--	6.9762	10.14231	18.1886	8.312432	164.3646
	3. (3 $^{-}$3)	R	--	7.34418	10.46602	18.483586	83.39299	164.6269
5. (2 $\overset{-1}{4}$)	3. (3 $^{-}$3)	R	--	7.874965	11.01231	19.00725	83.859906	165.16756
	4. ($\underline{6\ 3}$)	R	--	8.142114	11.256868	19.23574	84.09884	165.29988
	8. (3 $\tilde{}$3 $^{-}$3)	R	--	8.529755	11.5955	19.53763	84.3736	165.567

----- End of [DATA 4A-2] The Sq-S6E(3-3,m) and Family -----

[DATA 4A-3] The Sq-S8E(4-4, m) and Family

The R of Sq-S8E(4-4, 1) is: $R = \sqrt{5+2\sqrt{2}} \doteqdot 2.7993265$

Sq-S8E(4-4,m) (I) 6(P4-S)			m = 1	m = 3	m = 5	m = 10	m = 50	m = 100
1. $(1\ \overset{\vee}{4})e$		R	3.359605	5.71408	8.208425	14.52883	65,42877	129.082
3. $(2\ {}^{\sim}\overset{\vee}{4})e$		R	3.974698	6.116953	8.585754	14.88832	65.78335	129.48712
6. $(4\ {}^{\sim}4)e$		R	4.16386	6.211957	8.569142	14.96325	65.84633	129.56845
7. $(4\ {}^{-}4)e$		R	4.831636	6.479891	8.8832	15.145419	66.01483	129.73143
10. $(1\ 4\ {}^{\sim}4)e$		R	4.322195	6.276045	8.72052	15.006405	65.88838	129.56846
Sq-S8E(4-4,m) (II) 12(T4-S)			m = 1	m = 3	m = 5	m = 10	m = 50	m = 100
2. $(1\ \overset{2}{4})$		R	3.85575	6.270382	8.751	15.04542	65.90353	129.57618
5. $(2\ {}^{-}\overset{2}{4})$		R	5.436909	7.77695	10.21247	16.4432	67.015	130.9084
Sq-S8E(4-4,m) (III) 12(T4-S1) + 6(P4-S2)			m = 1	m = 3	m = 5	m = 10	m = 50	m = 100
2. $(1\ \overset{2}{4})$	3. $(2\ {}^{\sim}\overset{\vee}{4})e$	R	--	6.386206	8.8663	15.159014	66.00887	129.6576
	6. $(4\ {}^{\sim}4)e$	R	--	6.53384	9.004672	15.291	66.3572	129.82078
	7. $(4\ {}^{-}4)e$	R	--	7.055079	9.505787	15.77554	66.62649	130.3127
	9. $(\underline{8\ 4})e$	R	--	7.496402	9886445	16.10651	66.92882	130.56688
	10. $(1\ 4\ {}^{\sim}4)e$	R	--	6.637118	9.095089	15.37243	66.22056	129.9025
	30. $(4\ {}^{\sim}\overset{2}{8}{}^{\sim}\ 4)e$	R	--	7.550356	9.923969	16.132335	66.95052	130.6427
	32. $(4\ {}^{-}\overset{2}{8}{}^{\sim}\ 4)e$ [Rc<9.805]	R	--	7.555211	×	×	×	×
5. $(2\ {}^{-}\overset{2}{4})$	9. $(\underline{8\ 4})e$	R	--	8.033	10.45357	16.6883	67.44258	131.0748
	23. $(4\ {}^{\sim}4\ {}^{-}4)e$	R	--	8.47589	10.89845	17.11334	67.86311	131.4923
	30. $(4\ {}^{\sim}\overset{2}{8}{}^{\sim}\ 4)e$	R	--	8.151567	10.56506	16.77514	67.53066	131.15802
	32. $(4\ {}^{-}\overset{2}{8}{}^{\sim}\ 4)e$ [Rc<9.805]	R	--	8.16285	×	×	×	×
	35. $(\underline{8\ 4}\ {}^{-}4)e$	R	--	9.207147	×	×	×	×
	43. $(1\ 4\ {}^{\sim}4\ {}^{-}4)e$	R	--	8.557385	10.97012	17.17634	67.92994	131.57606
	103. $(4\ {}^{-}4\ {}^{\sim}\overset{2}{8}{}^{-}\ 4)e$	R	--	9.042745	11.38454	17.5245	68.22133	131.9124

----- End of [DATA 4A-3] The Sq-S8E(4-4,m) and Family -----

[DATA 4A-4] The Sq-S12E(3-3, m) and Family

The R of Sq-S12E(3-3, 1) is: $R = \sqrt{11+4\sqrt{5}} \doteq 4.465901$

Sq-S12E(3-3, m) (I) 20(P3-S)			m = 1	m = 3	m = 5	m = 10	m = 50	m = 100
1. $(1\ 3)$		R	5.729451	10.6807	16.12158	29.79503	139.4671	276.4968
2. $(3\ \tilde{3})$		R	6.21803	10.772	16.201	29.8682	139.55	276.867
Sq-S12E(3-3, m) (II) 30(Sq-S)			m = 1	M = 3	m = 5	m = 10	m = 50	m = 100
2. $(1\ \overset{2}{4})$		R	5.71285	11.09798	16.53871	30.2021	139.8524	276.8716
5. $(2\ \overset{-2}{4})$		R	7.4831	12.81605	18.22462	31.85047	141.40167	278.7502
Sq-S12E(3-3, m) (III) 30(Sq-S1) + 20(P3-S2)			m = 1	m = 3	m = 5	m = 10	m = 50	m = 100
2. $(1\ \overset{2}{4})$	1. $(1\ 3)$	R	--	11.1019	16.5407	30.2031	×	×
2. $(1\ \overset{2}{4})$	2. $(3\ \tilde{3})$	R	--	11.4928	16.92257	30.58058	140.23268	277.2437
2. $(1\ \overset{2}{4})$	3. $(3\ \bar{3})$	R	--	12.0205	17.42193	31.06158	140.71089	277.99092
5. $(2\ \overset{-2}{4})$	3. $(3\ \bar{3})$	R	--	12.81605	18.22462	31.85047	141.40197	278.7502
5. $(2\ \overset{-2}{4})$	4. $(\underline{6\ 3})$	R	--	13.2181	18.61293	32.22933	141.79025	279.1273
5. $(2\ \overset{-2}{4})$	8. $(3\ \tilde{3}\ \bar{3})$	R	--	13.7605	191.2205	32.71465	142.279	279.5055

----- End of [DATA 4A-4] The Sq-S12E(3-3,m) and Family -----

[DATA 4A-5] The Sq-S20E(5-5, m) and Family

The R of Sq-S20E(5-5, 1) is: $R = \sqrt{11+4\sqrt{5}} \doteq 4.465901$

Sq-S20E(5-5, m) (I): 12(P5-S)		m = 1	m = 3	m = 5	m = 10	m = 50	m = 100
1. (1 5) [x→2s (1.17557)]	R	4.874724	6.629226	11.999222	21.00643	93.25527	183.6766
2. (5 ~ 5) [x→c+s $\sqrt{3}$ (1.8271)]	R	6.605468	8.030193	13.342899	22.32878	94.58055	184.9944
3. (5 ¯ 5) [x→1+2s(2.17557)]	R	7.483054	8.585192	13.8669	22.838546	95.05993	185.49352
5. (1 5 ~ 5) [x→4cs(1.902113)]	R	6.807089	8.16699	13.47044	22.4525	94.71081	184.9945

Sq-S20E(5-5, m) (II): 30(T4-S)		m = 1	m = 3	m = 5	m = 10	m = 50	m = 100
2. (1 $\overset{2}{4}$) [z→ $\sqrt{3}$ (1.7320508)]	R	6.39095	8.133345	13.46538	22.448	94.67791	184.9999
5. (2 ¯ $\overset{2}{4}$) [z→1+ $\sqrt{3}$ (2.732051)]	R	9.10577	10.81666	16.08796	25.02176	97.19334	187.5343

Sq-S20E(5-5, m) (III): 30(T4-S_1) + 12(P5-S_2)			m = 1	m = 3	m = 5	m = 10	m = 50	m = 100
2. (1 $\overset{2}{4}$)	2. (5 ~ 5)	R	--	10.12527	13.69422	22.67781	94.89559	185.3322
	3. (5 ¯ 5)	R	--	10.92075	14.48373	23.46234	95.68755	186.00068
	4. (<u>10 5</u>)	R	--	12.1642	15.67802	24.61235	96.80971	187.18206
	5. (1 5 ~ 5)	R	--	10.31035	13.87619	22.85701	95.07045	185.49892
	8. (5 ~ 5 ¯ 5)	R	--	12.37623	15.88912	24.81989	96.99171	1.8735208
	11. (5 ~ $\overset{2}{10}$ ~ 5)	R	--	12.2683	15.77909	24.70962	96.90061	187.28606
	13. (5 ¯ $\overset{2}{10}$ ~ 5)	R	--	12.28502	15.795	24.72544	96.90062	187.35205
	22. (1 5 ~ 5 ¯ 5)	R	--	12.46721	15.97207	24.895	97.08295	187.5223
	25. (1 5 ~ $\overset{2}{10}$ ~ 5)	R	--	12.3401	15.8437	24.76668	96.94615	187.35205
	27. (1 5 ¯ $\overset{2}{10}$ ~ 5)	R	--	12.36795	15.86892	24.79029	96.99	187.35205
5. (2 ¯ $\overset{2}{4}$)	4. (<u>10 5</u>)	R	--	12.76282	16.30448	25.24761	97.42661	187.70486
	8. (5 ~ 5 ¯ 5)	R	--	13.1403	16.69448	25.64808	97.7957	188.2185
	11. (5 ~ $\overset{2}{10}$ ~ 5)	R	--	12.94131	16.141778	25.42674	97.151616	187.8758
	13. (5 ¯ $\overset{2}{10}$ ~ 5)	R	--	12.97075	16.11225	25.45532	97.105946	1.880469
	16. (<u>10 5</u> ¯ 5)	R	--	15.04341	18.52817	27.40779	99.53	189.9512
	22. (1 5 ~ 5 ¯ 5)	R	--	13.31602	16.866057	25.81647	97.98129	188.3903
	25. (1 5 ~ $\overset{2}{10}$ ~ 5)	R	--	13.06744	16.6027	25.5389	97.70315	188.047
	27. (1 5 ¯ $\overset{2}{10}$ ~ 5)	R	--	13.1179	16.651	25.58558	97.749401	188.047
	42. (5 ¯ 5 ~ $\overset{2}{10}$ ¯ 5)	R	--	14.6516	18.152	27.05131	99.1577	189.6021
	46. (5 ~ $\overset{2}{10}$ ~ 5 ¯ 5)	R	--	15.13648	18.613915	27.48601	99.5878	189.9512
	50. (5 ¯ $\overset{2}{10}$ ~ 5 ¯ 5)	R	--	15.15125	18.627	27.49769	99.58781	189.9513

----- End of [DATA 4A-5] The Sq-S20E(5-5,m) and Family -----

[DATA 4B] The Sq-SE' and Families

[DATA 4B-1] The Sq-S12E'1(4-3, m) and Family

The R of Sq-S12E'1(4-3, 1) is: $\boxed{R = \sqrt{9+\sqrt{40}} \doteq 3.9146589}$

Family groups:

(1) 6(P4-S) **(2) 8(P2-S)** **(3) 24(T4-S)**
(4) 6(P4-S1) + 8(P3-S2)
(5) 24(T4-S1) + 6(P4-S2)
(6) 24(T4-S1) + 8(P3-S2)
(7) 24(T4-S1) + 8(P3-S2) + 6(P4-S3) Sq-SnE'(4-3, 1)

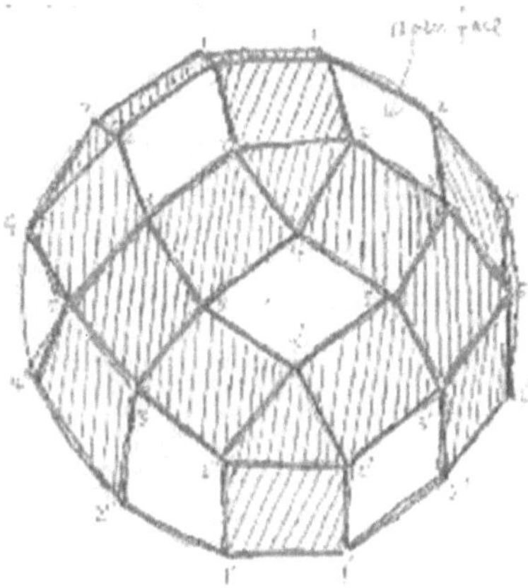

Sq-SnE'(4-3, 1)

[dd1 = half distance between two corner balls of two P1 (4-gon) faces across the open face, and should be >= 1.

dd2 = half distance between two corner balls of two P2 (3-gon) faces across the open face, and should be >= 1.]

Sq-S12E'1(4-3, m) Group (I): 6(P4-S) x1 = x1, z = 1, x2 = 1		m = 1	m = 2	m = 5	m = 10	m = 100
1. (1 4̌)e	R	4.33011	6.351305	12.56545	23.00604	211.5568
3. (2 ˜4̌)e	R	4.69202	6.676451	12.86343	23.29729	211.774
6. (4 ˜4)e	R	×	6.751795	12.92634	23.35641	211.774
7. (4 ¯4)e	R	×	6.962517	13.0819	23.49692	211.9916
10. (1 4 ˜4)e	R	×	6.80153	12.96289	23.39071	211.8898

Sq-S12E'1(4-3, m) Group (II): 8(P3-S) x1 = 1, x2 = x2, z = 1		m = 1	m = 2	m = 5	m = 10	m = 100
1. (1 3)	R	×	6.282426	12.47989	22.916905	211.34005
2. (3 ˜3)	R	×	6.333808	12.512711	22.94485	211.34005

Sq-S12E'1(4-3, m) (III): 24(T4-S) x1 = x2 = z		m = 1	m = 2	m = 5	m = 10	m = 100
2. (1 4)	R	5.221091	7.250713	13.4446	23.86319	212.4329
5. (2 ¯4)	R	7.091453	9.087774	15.22142	25.59493	213.9859

Sq-S12E'1(4-3,m) Group (IV): 6(P4-S1) + 8(P3-S2)							
6(P1-S1)	8(P2-S2)		m = 1	m = 2	m = 5	m = 10	m = 100
1. (1 4̌)e	1. (1 3)	R	×	6.678078	12.870565	23.30644	211.7741
	2. (3 ˜3)	R	×	6.7263	12.90317	23.335347	211.774
3. (2 ˜4̌)e	1. (1 3)	R	5.32577	6.9967	3.16775	23.59601	211.9917
	2. (3 ˜3)	R	5.50973	7.042424	13.20012	23.62563	212.2097
6. (4 ˜4)e	1. (1 3)	R	5.45572	7.069875	13.23036	23.65585	212.2097
	2. (3 ˜3)	R	5.6523	7.11496	13.26232	23.6851	212.2097
7. (4 ¯4)e	1. (1 3)	R	5.90395	7.272575	13.38486	23.79673	212.2098
	2. (3 ˜3)	R	6.15532	7.315805	13.416706	23.82413	212.2098
10. (1 4 ˜4)e	1. (1 3)	R	5.55185	7.117435	13.26671	23.69052	212.2097
	2. (3 ˜3)	R	5.75693	7.162087	13.29867	23.71767	212.2098

Sq-S12E'1(4-3,m) Group (V): 24(T4-S1) + 6(P3-S2) [x2 = z]							
24 (S-S1)	6(P1-S2)		m = 2	m = 3	m = 5	m = 10	m = 100
2. (1 4)	3. (2 ˜4̌)e	R	7.345681	9.398147	13.53781	23.9566	212.433
	6. (4 ˜4)e	R	7.463595	9.511181	13.64726	24.06465	212.433
	7. (4 ¯4)e	R	7.886643	9.922276	14.04772	24.45793	212.8715
	9. (8 4)e	R	8.253077	10.24352	14.33167	24.71665	213.0914
	10. (1 4 ˜4)e	R	7.543651	9.584616	13.71507	24.128845	212.6519

	30. (4 $\tilde{\overset{2}{8}}$ $\tilde{4}$)e R<9.805	R	8.21794	×	×	×	×
2. (1 $\overset{2}{4}$)	9. ($\underline{8\ 4}$)e	R	9.335533	11.3464	15.43605	25.79936	214.208
	23. (4 $\tilde{4}$ $\bar{4}$)e	R	9.659312	11.67875	15.77477	26.13999	214.4307
	30. (4 $\tilde{\overset{2}{8}}$ $\tilde{4}$)e R<9.805	R	9.418442	×	×	×	×
	43. (1 4 $\tilde{4}$ $\bar{4}$)e	R	9.721978	11.73634	15.8273	26.18893	214.6537

Sq-S12E'1(4-3,m) Group (VI): 24(T4-S1) + 8(P3-S2) [x1 = z]

24(S-S1)	8(G3-S2)		m = 2	m = 3	m = 5	m = 10	m = 100
2. (1 $\overset{2}{4}$)	1. (1 3)	R	7.254328	9.306132	13.44531	23.86377	212.4329
	2. (3 $\tilde{3}$)	R	7.41285	9.458375	13.593	24.00911	212.433
	3. (3 $\bar{3}$)	R	7.646656	9.672157	13.78962	24.19467	212.652
5. (2 $\overset{-2}{4}$)	3. (3 $\bar{3}$)	R	9.087774	11.11589	15.22105	25.59493	213.9859
	4. ($\underline{6\ 3}$)	R	9.267041	11.28249	15.37616	25.74242	214.208
	8. (3 $\tilde{3}$ $\bar{3}$)	R	9.4897	11.49248	15.57324	25.93	214.2258

Sq-S12E'1(4-3,m) Group (VII) 24(T4-S1) + 8(P3-S2) + 6(P4-S3)

24(S-S1)	8(P3-S2)	6(P4-S3)		m = 3	M = 4	m = 5	m = 10	m = 100
2. (1 $\overset{2}{4}$)	1. (1 3)	3. (2 $\tilde{\check{4}}$)e	R	9.400363	11.4656	13.53901	2.39566	212.4329
		6. (4 $\tilde{4}$)e	R	9.51332	11.57607	13.64833	24.06465	212.433
		7. (4 $\bar{4}$)e	R	9.92412	11.98048	14.04864	24.45793	212.8715
		9. ($\underline{8\ 4}$)e	R	10.24508	12.27796	14.33243	24.7171	213.0914
		10. (1 4 $\tilde{4}$)e	R	9.58657	11.646	13.7161	24.12884	212.6519
	2. (3 $\tilde{3}$)	3. (2 $\tilde{\check{4}}$)e	R	9.552442	11.61452	13.6862	24.1009	212.652
		6. (4 $\tilde{4}$)e	R	9.664838	11.72488	13.79518	24.20838	212.652
		7. (4 $\bar{4}$)e	R	10.07422	12.12846	14.19527	24.60259	213.0913
		9. ($\underline{8\ 4}$)e	R	10.39232	12.42463	14.47847	24.86143	213.3116
		10. (1 4 $\tilde{4}$)e	R	9.73759	11.79436	13.86308	24.27243	212.652
	3. (3 $\bar{3}$)	3. (2 $\tilde{\check{4}}$)e	R	9.76534	11.81723	13.88276	24.2873	212.652
		6. (4 $\tilde{4}$)e	R	9.8767	11.92704	13.99158	24.39482	212.8714
		7. (4 $\bar{4}$)e	R	10.28255	12.3285	14.3902	24.78734	213.3116
		9. ($\underline{8\ 4}$)e	R	10.59599	12.62206	14.66729	25.047102	213.5324
		10. (1 4 $\tilde{4}$)e	R	9.948543	11.99615	14.05892	24.4583	212.8715

(continue) **Sq-S12E'1(4-3,m) Group (VII) 24(T4-S1) + 8(P3-S2) + 6(P4-S3)**

24(Sq-S1)	8(P3-S2)	6(P4-S3)		m = 3	M = 4	m = 5	m = 10	m = 100
5. (2 $\overset{-2}{4}$)	3. (3 $\bar{3}$)	9. ($\underline{8\ 4}$)e	R	11.32175	13.366515	15.42373	25.79597	214.208
		23. (4 $\tilde{4}$ $\bar{4}$)e	R	11.67875	13.71988	15.77477	26.13999	214.4307

		43. (1 4 ˜4 ¯4)e	R	11.736345	13.7745	15.82715	26.18893	214.6537
	4. ($\underline{6\ 3}$)	9. ($\underline{8\ 4}$)e	R	11.4784	13.51929	15.573728	25.93967	214.4306
		23. (4 ˜4 ¯4)e	R	11.83388	13.87165	15.92828	26.28738	214.6538
		43. (1 4 ˜4 ¯4)e	R	11.891025	13.92597	15.9766	26.33555	214.6538
	8. (3 ˜3 ¯3)	9. ($\underline{8\ 4}$)e	R	11.69569	13.72636	15.774771	26.12904	214.4307
		23. (4 ˜4 ¯4)e	R	12.0485	14.07733	16.124	26.4738	214.87732
		43. (1 4 ˜4 ¯4)e	R	12.10471	14.131	16.17617	26.523385	214.8774

---- End of [DATA 4B-1] The Sq-S12E’1(4-3, m) and Family ----

[DATA 4B-2] The Sq-S30E'(5-3, m) and Family

The R of Sq-S30E'(5-3,1) is: R = 6.175794

There are three cases, each case has four groups:

Case (I): [z = 1] (I - 1) The basic Sq-S30E'(5-3, m)

(I - 2) 12(P5-S)
(I - 3) 20(P3-S)
(I - 4) 12(P5-S1) + 20(P3-S2)

Case (II): $[z = (1\ \overset{2}{4})]$, $z = \frac{1}{R}\sqrt{3R^2 - 4}$

(II - 1) 60(T4-S)
(II - 2) 60(T4-S1) + 12(P5-S2)
(II - 3) 60(T4-S1) + 20(P3-S2)
(II - 4) 60(T4-S1) + 12(P5-S2) + 20(P3-S3)

Case (III): $[z = 5.(2\ \overset{-2}{4})]$, $z = \frac{1}{R^2}[(R^2 - 2) + \sqrt{(R^2 - 1)(3R^2 - 4)}]$

(III - 1) 60(T4-S)
(III - 2) 60(T4-M) + 12(P5-A)
(III - 3) 60(T4-M) + 20(P3-A)
(III - 4) 60(T4-S) + 12(P5-S) + 20(P3-S)

Sq-S30E'(5-3, m)----Case (I): z = 1

Case (I - 1): the original Sq-S30E'(5-3, 1): z = 1, x1 = 1, x2 = 1

Z = 1 X1 = 1 X2 = 1

m	N	R	X1	X2	Ds	Remark
1	120	6.175795	1	1	78.657	
2	240	9.201964	1	1	70.858	
3	360	12.248395	1	1	59.991	
4	480	15.303	1	1	51.242	

Case (I - 2): z = 1, 12(P5-S), x2 = 1

Z = 1, X1 = 1. (1 5) X2 = 1

m	N	R	X1	X2	Ds	Remark
1	132	6.52573	1.161683	1	77.492	
2	252	9.5567	1.169116	1	68.98	
3	372	12.604884	1.171866	1	58.834	
4	492	15.6604	1.173172	1	50.153	

Z = 1 X1 = 2. (5 ~ 5) X2 = 1

M	N	R	X1	X2	Ds	Remark
1	180	7.721567	1.793854	1	75.475	
2	300	10.7128	1.809987	1	66.351	
3	420	13.343898	1.81674	1	55.587	
4	540	16.790196	1.820166	1	47.888	

Z = 1 X1 = 3. (5 ‾ 5) X2 = 1

m	N	R	X1	X2	Ds	Remark
1	180	8.18645	2.111449	1		$\phi_{02} < 60$ ×
2	300	11.868058	2.141149	1	60.132	
3	420	14.188	2.154252	1	52.161	
4	540	17.2286	2.161114	1	45.481	

Z = 1 X1 = 5. (1 5 ~ 5) X2 = 1

M	N	R	X1	X2	Ds	Remark
1	192	7.83925	1.866465	1	78.107	
2	312	10.82408	1.883675	1	66.575	
3	432	13.85214	1.890926	1	56.285	
4	552	16.896719	1.894615	1	48.336	

Case (I - 3): z = 1, 20(P3-S), x1 = 1

Z = 1 X1 = 1 X2 = 1. (1 3)

m	N	R	X1	X2	Ds	Remark
1						$\phi_{01} < 60$ ×
2	260	9.65876	1	1.722343	69.674	
3	380	12.69606	1	1.72667	58.937	
4	500	15.745948	1	1.728054	50.416	

Z = 1 X1 = 1 X2 = 2. (3 ~ 3)

m	N	R	X1	X2	Ds	Remark
1						$\phi_{01} < 60$ ×
2	300	9.71266	1	1.989286	79.504	
3	420	12.7439	1	1.993804	64.652	
4	540	15.791	1	1.995973	54.140	

Case (I - 4): z = 1, 12(P5-S1) + 20(P3-S2)

Z = 1 X1 = 1. (1 5) X2 = 1.(1 3)

m	N	R	X1	X2	Ds	Remark
1	152	7.07305	1.1763762	1.714653		$\phi_{01} < 60$ ×
2	272	10.01203	1.169692	1.72339	67.837	
3	392	13.052	1.172115	1.72696	57.527	

Z = 1 X1 = 1. (1 5) X2 = 2. (3 ~ 3)

m	N	R	X1	X2	Ds	Remark
1						$\phi_{01} < 60$ ×
2	312	10.06495	1.169755	1.99003	76.997	
3	432	13.09914	1.172138	1.994138	62.942	

m	N	R	X1	X2	Ds	Remark
4	512	16.103196	1.173301	1.728108	49.361	

m	N	R	X1	X2	Ds	Remark
4	552	16.14765	1.173317	1.99615	52.925	

Z = 1 X1 = 2. (5 ~ 5) X2 = 1.(1 3)

m	N	R	X1	X2	Ds	Remark
1	200	8.609175	1.800458	1.720327	67.460	
2	320	11.1603	1.811345	1.725684	64.230	
3	440	14.1872	1.817383	1.727743	54.651	
4	560	17.2315	1.820519	1.729232	47.150	

Z = 1 X1 = 2. (5 ~ 5) X2 = 2. (3 ~ 3)

m	N	R	X1	X2	Ds	Remark
1						$\phi_{01} < 60$ ×
2	360	11.2099	1.811484	1.991978	71.621	
3	480	14.232954	1.817444	1.995039	59.237	
4	600	17.274776	1.820552	1.996638	50.265	

Z = 1 X1 = 3. (5 ~ 5) X2 = 1.(1 3)

m	N	R	X1	X2	Ds	Remark
1	200	9.28506	2.125747	1.721976	57.996	
2	320	11.6118	2.14373	1.725616	59.332	
3	440	14.629	2.155621	1.728	51.4	
4	560	17.6684	2.161822	1.729274	44.847	

Z = 1 X1 = 3. (5 ~ 5) X2 = 2. (3 ~ 3)

m	N	R	X1	X2	Ds	Remark
1	240	9.5863	2.128833	1.988999	65.29	
2	360	11.6605	2.143995	1.992591	66.192	
3	480	14.6748	2.155644	1.995335	55.723	
4	600	1.77117	2.161889	1.996802	47.816	

Z = 1 X1 = 5. (1 5 ~ 5) X2 = 1.(1 3)

m	N	R	X1	X2	Ds	Remark
1	212	8.76878	1.873792	1.720751	68.928	
2	332	11.2705	1.885127	1.725219	65.342	
3	452	14.295	1.891611	1.727808	55.298	
4	572	17.3378	1.894997	1.729167	47.572	

Z = 1 X1 = 5. (1 5 ~ 5) X2 = 2. (3 ~ 3)

m	N	R	X1	X2	Ds	Remark
1						$\phi_{01} < 60$ ×
2	372	11.31987	1.885278	1.992135	72.577	
3	492	14.34059	1.891678	1.995114	59.810	
4	612	17.38115	1.895036	1.996679	50.645	

End of Sq-S30E'(5-3, m)----Case (I): z = 1

Sq-S30E'(5-3, m)----Case (II): $[z = (1\ \overset{2}{4})]$ $z = \frac{1}{R}\sqrt{3R^2-4}$

Sq-S30E'(5-3, m)----Case (II - 1): 60(T4-S)

$Z = (1\ \overset{2}{4})$, X1 = Z, X2 = Z

m	N	R	Z	X_1	X_2	Ds
1	180	8.44285	1.715775	= Z	= Z	63.13
2	360	11.4599	1.723236	= Z	= Z	68.53
3	540	14.496799	1.726548	= Z	= Z	64.238
4	720	17.543906	1.728295	= Z	= Z	58.482

Sq-S30E'(5-3, m)----Case (II - 2): 60(T4-S_1) +12(P5-S_2)

$Z = (1\ \overset{2}{4})$, X1 = 2. $(5 \sim 5)$ X2 = Z

m	N	R	Z	X_1	X_2	Ds
2	360	11.65305	1.723526	1.812661	= Z	66.477
3	540	14.691592	1.726693	1.818041	= Z	62.546
4	720	17.7392	1.728377	1.820896	= Z	57.201

$Z = (1\ \overset{2}{4})$, X1 = 3. $(5\ \bar{}\ 5)$ X2 = z

m	N	R	Z	X_1	X_2	Ds
2	360	12.32054	1.724434	2.147319	= Z	59.29
3	540	15.361227	1.72715	2.157383	= Z	52.211
4	720	18.4069	1.728639	2.162903	= Z	53.126

$Z = (1\ \overset{2}{4})$, X1 = 4. $(\underline{10\ 5})$, X2 = Z

m	N	R	Z	X_1	X_2	Ds
2	420	13.3654	1.725575	2.788461	= Z	58.779
3	600	16.37055	1.727737	2.801472	= Z	55.971
4	780	19.399021	1.72898	2.808895	= Z	51.817

$Z = (1\ \overset{2}{4})$, X1 = 5. $(1\ 5 \sim 5)$, X2 = Z

m	N	R	Z	X_1	X_2	Ds
2	372	11.80876	1.72375	1.88666	= Z	66.692
3	552	14.8455	1.726803	1.892383	= Z	62.625
4	732	17.8922	1.72844	1.895436	= Z	57.164

$Z = (1\ \overset{2}{4})$, X1 = 8.$(5 \sim 5\ \bar{}\ 5)$, X2 = Z

m	N	R	Z	X_1	X_2	Ds
2	420	13.546	1.726747	2.938079	= Z	57.223
3	600	16.54965	1.72783	2.959374	= Z	54.766
4	780	19.577099	1.729035	2.971731	= Z	50.363

Z = Z, X1 = 11. $(5 \sim \overset{2}{10} \sim 5)$, X2 = Z

m	N	R	Z	X_1	X_2	Ds
2	480	13.4531	1.725659	2.85802	= Z	66.304
3	660	16.455798	1.727781	2.872878	= Z	60.932
4	840	19.48273	1.729006	2.881355	= Z	55.325

$Z = (1\ \overset{2}{4})$, X1 = 13. $(5\ \bar{}\ \overset{2}{10} \sim 5)$, X2 = Z

m	N	R	Z	X_1	X_2	Ds
2	480	13.46707	1.725672	2.869622	= Z	66.166
3	660	16.46935	1.727788	2.884804	= Z	60.832
4	840	19.496	1.72961	2.893444	= Z	55.249

$Z = (1\ \overset{2}{4})$, X1 = 22.$(1\ 5 \sim 5\ \bar{}\ 5)$, X2 = Z

m	N	R	Z	X_1	X_2	Ds
2	432	13.62075	1.725816	3.009399	= Z	58.213
3	612	16.61955	1.727865	3.031806	= Z	55.393
4	792	19.64538			= Z	51.303

$Z = (1\ \overset{2}{4})$, X1 = 25. $(1\ 5 \sim \overset{2}{10} \sim 5)$, X2 = Z

m	N	R	Z	X_1	X_2	Ds
2	492	13.5117	1.725714	2.907592	= Z	67.373
3	672	16.510112	1.727809	2.921616	= Z	61.632
4	852	19.5346			= Z	55.818

$Z = (1\ \overset{2}{4})$, X1 = 27. $(1\ 5\ \bar{}\ \overset{2}{10} \sim 5)$, X2 = Z

m	N	R	Z	X_1	X_2	Ds
2	492	13.5345	1.725736	2.927736	= Z	67.146
3	672	16.5312	1.72782		= Z	61.475
4	852	19.5551			= Z	55.7

Sq-S30E'(5-3, m)----Case (II - 3): 60(S4-S_1) + 10(P3-S_2)

Z = (1 $\overset{2}{4}$), X1 = Z, X2 = 1.(1 3)

m	N	R	Z	X_1	X_2	Ds
2	492	11.461941	1.723239	= Z	1.725446	60.894
3	672	14.4982		= Z		59.468
4	852	17.5447		= Z		55.228

Z = (1 $\overset{2}{4}$), X1 = Z, X2 = 2. (3 ~ 3)

m	N	R	Z	X_1	X_2	Ds
2	360	11.6799	1.723566	= Z	1.992616	65.973
3	540	14.71297		= Z		62.364
4	720	17.757944		= Z		57.080

Z = (1 $\overset{2}{4}$), X1 = Z, X2 = 3. (3 ‾ 3)

m	N	R	Z	X_1	X_2	Ds
2	360	11.9755	1.72398	= Z	2.704014	62.756
3	540	14.9975		= Z		60.02
4	720	18.036179		= Z		55.333

Sq-S30E'(5-3, m)----Case (II - 4): 60(S4-S_1)+12(P5-S_2)+20(P3-S_3)

Z = (1 $\overset{2}{4}$), X1 =2. (5 ~ 5), X2 = 1. (1 3)

m	N	R	Z	X_1	X_2	Ds
3	380	14.692817	1.726694	1.878043	1.728034	44.006
4	560	17.73998	1.728378	1.820896	1.729297	44.486
5	740	20.793	1.729378	1.822586	1.730047	42.790

Z = (1 $\overset{2}{4}$), X1 = 2. (5 ~ 5), X2 = 2. (3 ~ 3)

m	N	R	Z	X_1	X_2	Ds
3	420	14.9075	1.726847	1.818301	1.99548	47.248
4	600	17.95307				46.539
5	780	21.005002	1.729432	1.822671	1.997728	44.197

Z = (1 $\overset{2}{4}$), X1 = 2. (5 ~ 5), X2 = 3. (3 ‾ 3)

m	N	R	Z	X_1	X_2	Ds
3	420	15.1917	1.72704	1.818629	2.714629	45.496
4	600	18.231	1.728573	1.821228	2.719954	45.131
5	780	21.279203	1.729499	1.822787	2.723171	43.065

Z = (1 $\overset{2}{4}$), X1 = 3.(5 ‾ 5) X2 = 1.(1 3)

m	N	R	Z	X_1	X_2	Ds
3	380	15.3624	1.727151	2.157385	1.728377	40.254
4	560	18.40725				41.319
5	740	21.458848				40.175

19. Z = (1 $\overset{2}{4}$), X1 = 3. (5 ‾ 5), X2 = 2. (3 ~ 3)

m	N	R	Z	X_1	X_2	Ds
3	420	15.5762	1.727285	2.157882	1.995861	55.643
4	600	18.620082	1.728717	2.163194	1.997107	51.917
5	780	21.67074	1.72959	2.166437	1.997866	47.911

20 Z = (1 $\overset{2}{4}$), X1 = 3. (5 ‾ 5), X2 = 3. (3 ‾ 3)

m	N	R	Z	X_1	X_2	Ds
3	420	15.8583	1.727453	2.158507	2.716063	53.681
4	600	18.8973	1.728814	2.163557	2.720792	50.405
5	780	21.944512	1.729651	2.166661	2.723701	46.723

21 Z = (1 $\overset{2}{4}$), X1 = 4.(10 5) X2 = 1. (1 3)

m	N	R	Z	X_1	X_2	Ds
3	560	16.371373	1.727737	2.801474	1.728817	52.235
4	740	19.3999	1.72898	2.808897	1.729748	49.156

22. Z = (1 $\overset{2}{4}$), X1 = 4. (10 5), X2 = 2. (3 ~ 3)

m	N	R	Z	X_1	X_2	Ds
3	600	16.58314	1.727847	2.802126	1.99635	54.345
4	780	19.6111	1.729046	2.809289	1.997393	50.703

23. Z = (1 $\overset{2}{4}$), X1 = 4. (10 5), X2 = 3. (3 ‾ 3)

m	N	R	Z	X_1	X_2	Ds
3	600	16.8612	1.727984	2.802951	2.717908	52.761
4	780	19.8855	1.729128	2.809782	2.721883	49.313

24. Z = (1 $\overset{2}{4}$), X1 = 5.(1 5 ~ 5) X2 = 1. (1 3)

m	N	R	Z	X_1	X_2	Ds
3	512	14.84665	1.726804	1.892385	1.728117	58.07
4	692	17.89292	1.72844	1.895439	1.729344	54.036

25. Z = (1 $\overset{2}{4}$), X1 = 5. (1 5 ~ 5), X2 = 2. (3 ~ 3)

26. Z = (1 $\overset{2}{4}$), X1 = 5. (1 5 ~ 5), X2 = 3. (3 ‾ 3)

m	N	R	Z	X1	X2	Ds
3	552	15.061100	1.7269530	1.892663	1.995572	60.837
4	732	18.1056	1.728525	1.895584	1.99694	55.825

m	N	R	Z	X1	X2	Ds
3	552	15.3445	1.72714	1.893007	2.714974	58.61
4	732	18.38385	1.728631	1.895787	2.720154	54.147

27 $Z = (1\,\overset{2}{4})$, $X1 = 8.(5\,^{\sim}5\,^{-}5)$ $X2 = 1.(1\,3)$

m	N	R	Z	X1	X2	Ds
3	560	16.5505	1.72783	2.959391	1.728886	51.11
4	740	19.5774	1.729035	2.971725	1.72979	48.268

28. $Z = (1\,\overset{2}{4})$, $X1 = 8.(5\,^{\sim}5\,^{-}5)$, $X2 = 2.(3\,^{\sim}3)$

m	N	R	Z	X1	X2	Ds
3	600	16.762	1.727936	2.960478	1.996428	53.388
4	780	19.7886	1,7291	2.972388	1.99744	49.797

29 $Z = (1\,\overset{2}{4})$, $X1 = 8.(5\,^{\sim}5\,^{-}5)$, $X2 = 3.(3\,^{-}3)$

m	N	R	Z	X1	X2	Ds
3	600	17.0394	1.728069	2.96183	2.718202	51.663

30. $Z = (1\,\overset{2}{4})$, $X1 = 11.(5\,^{\sim}\overset{2}{10}\,^{\sim}5)$ $X2 = 1.(1\,3)$

m	N	R	Z	X1	X2	Ds
3	620	16.4566	1.727782	2.872881	1.72885	57.234

31. $Z = (1\,\overset{2}{4})$, $X1 = 11.(5\,^{\sim}\overset{2}{10}\,^{\sim}5)$, $X2 = 2.(3\,^{\sim}3)$

m	N	R	Z	X1	X2	Ds
3	660	16.6682				59.389

32. $Z = (1\,\overset{2}{4})$, $X1 = 11.(5\,^{\sim}\overset{2}{10}\,^{\sim}5)$, $X2 = 3.(3\,^{-}3)$

m	N	R	Z	X1	X2	Ds
3	660	16.94585	1.728025	2.874581	2.718049	57.459

33 $Z = Z$, $X1 = 13.(5\,^{-}\overset{2}{10}\,^{\sim}5)$, $X2 = 1.(1\,3)$

m	N	R	Z	X1	X2	Ds
3	620	16.4702	1.727789	2.884794	1.728855	57.139
4	800					

34. $Z = Z$, $X1 = 13.(5\,^{-}\overset{2}{10}\,^{\sim}5)$, $X2 = 2.(3\,^{\sim}3)$

m	N	R	Z	X1	X2	Ds
3	660	16.6817	1.727896	2.885562	1.996394	59.293
4	840					

35. $Z = (1\,\overset{2}{4})$, $X1 = 13.(5\,^{-}\overset{2}{10}\,^{\sim}5)$, $X2 = 3.(3\,^{-}3)$

m	N	R	Z	X1	X2	Ds
3	660	16.9593	1.728031	2.886523	2.718071	57.368
4	840					

36 $Z = (1\,\overset{2}{4})$, $X1 = 22.(1\,5\,^{\sim}5\,^{-}5)$, $X2 = 1.(1\,3)$

m	N	R	Z	X1	X2	Ds
3	572	16.62037	1.727866	3.031818	1.728913	51.767
4	752					

37 $Z = (1\,\overset{2}{4})$, $X1 = 22.(1\,5\,^{\sim}5\,^{-}5)$, $X2 = 2.(3\,^{\sim}3)$

m	N	R	Z	X1	X2	Ds
3	612	16.8316	1.72797	3.032959	1.996458	54.006

38 $Z = (1\,\overset{2}{4})$, $X1 = 22.(1\,5\,^{\sim}5\,^{-}5)$, $X2 = 3.(3\,^{-}3)$

m	N	R	Z	X1	X2	Ds
3	612	17.1085	1.728101	3.034403	2.718314	52.272

39. $Z = (1\,\overset{2}{4})$, $X1 = 25.(1\,5\,^{\sim}\overset{2}{10}\,^{\sim}5)$ $X2 = 1.(1\,3)$

m	N	R	Z	X1	X2	Ds
3	632	16.5108	1.72781	2.921618	1.728871	57.959
4	812					

40 $Z = (1\,\overset{2}{4})$, $X1 = 25.(1\,5\,^{\sim}\overset{2}{10}\,^{\sim}5)$, $X2 = 2.(3\,^{\sim}3)$

m	N	R	Z	X1	X2	Ds
3	672	16.72215	1.727916	2.92232	1.996411	60.079
4	852					

41. $Z = (1\,\overset{2}{4})$, $X1 = 25.(1\,5\,^{\sim}\overset{2}{10}\,^{\sim}5)$, $X2 = 3.(3\,^{-}3)$

m	N	R	Z	X1	X2	Ds
3	672	16.9994	1.72805	2.923207	2.718137	58.136

42. $Z = (1\,\overset{2}{4})$, $X1 = 27.(1\,5\,^{-}\overset{2}{10}\,^{\sim}5)$, $X2 = 1.(1\,3)$

m	N	R	Z	X1	X2	Ds
3	632	16.53222	1.727821	2.941623	1.728879	57.809

43 $Z = (1\,\overset{2}{4})$, $X1 = 27.(1\,5\,^{-}\overset{2}{10}\,^{\sim}5)$, $X2 = 2.(3\,^{\sim}3)$

m	N	R	Z	X1	X2	Ds
3	672	16.7433	1.727927	2.942323	1.99642	59.928

44. $Z = (1\,\overset{2}{4})$, $X1 = 27.(1\,5\,^{-}\overset{2}{10}\,^{\sim}5)$, $X2 = 3.(3\,^{-}3)$

m	N	R	Z	X1	X2	Ds
3	672	17.0205	1.72806	2.943198	2.718171	57.992

[End of Sq-S30E'(5-3, m) ---- Case (II)]

Sq-S30E'(5-3, m)----Case (III):

$$z = 5.\ (2\ \overset{-2}{}\ 4),\quad z = \frac{1}{R^2}\left[(R^2 - 2) + \sqrt{(R^2 - 1)(3R^2 - 4)}\right]$$

Sq-S30E'(5-3, m)----Case (III - 1): $z = 5.\,(2\ \overset{-2}{}\ 4)$

$z = 5.\ (2\ \overset{-2}{}\ 4)$, X1 = Z, X2 = Z

m	N	R	Z	X1	X2	Ds
1	240	11.5981	2.702159	= Z	= Z	44.604
2	480	14.5885	2.713158	= Z	= Z	56.385
3	720	17.6058	2.719079	= Z	= Z	58.071
4	960	20.6376	2.72261	= Z	= Z	56.35

Sq-S30E'(5-3, m)----Case (III - 2): 60(T4-S) + 12(P5-S)

1. $z = 5.\,(2\ \overset{-2}{}\ 4)$, X1 = 4. (10 5), X2 = Z

m	N	R	Z	X1	X2	Ds
2	480	14.7693	2.713618	2.795543	= Z	55.013
3	720	17.7922	2.719349	2.805435	= Z	56.861
4	960	20.827	2.722781	2.81132	= Z	55.33

2. $z = 5.\ (2\ \overset{-2}{}\ 4)$, X1 = 8. $(5\ \overset{\sim}{}\ 5\ \overset{-}{}\ 5)$, X2 = Z

m	N	R	Z	X1	X2	Ds
2	480	15.097	2.714409	2.950655	= Z	52.650
3	720	18.235634	2.719812	2.96657	= Z	54.788
4	960	21.163871	2.723074	2.976195	= Z	53.582

3. $z = 5.\ (2\ \overset{-2}{}\ 4)$, X1 = 11. $(5\ \overset{\sim}{}\ \overset{2}{10}\ \overset{\sim}{}\ 5)$, X2 = Z

m	N	R	Z	X1	X2	Ds
2	540	14.9212	2.713991	2.866937	= Z	60.645
3	780	17.9437	2.719563	2.87759	= Z	60.564
4	1020	20.9785	2.722915	2.884246	= Z	57.942

4. $z = 5.\ (2\ \overset{-2}{}\ 4)$, X1 = 13. $(5\ \overset{-2}{}\ 10\ \overset{\sim}{}\ 5)$, X2 = Z

m	N	R	Z	X1	X2	Ds
2	540	14.9463	2.714052	2.87827	= Z	60.432
3	780	17.9687	2.719598	2.88963	= Z	60.395
4	1020				= Z	

5. $z = 5.\ (2\ \overset{-2}{}\ 4)$, X1 = 16. $(\underline{10\ 5}\ \overset{-}{}\ 5)$, X2 = Z

m	N	R	Z	X1	X2	Ds
2	540	16.6822	2.717603	3.906799	= Z	48.510
3	780	19.6701	2.721659	3.933734	= Z	50.399

6. $z = 5.\ (2\ \overset{-2}{}\ 4)$, X1 = 22. $(1\ 5\ \overset{\sim}{}\ 5\ \overset{-}{}\ 5)$, X2 = Z

m	N	R	Z	X1	X2	Ds
2	492	15.2443	2.714749	3.023169	= Z	53.927
3	732	18.2707	2.720006	3.039736	= Z	54.828

7. $z = 5.\ (2\ \overset{-2}{}\ 4)$, X1 = 25. $(1\ 5\ \overset{\sim}{}\ \overset{2}{10}\ \overset{\sim}{}\ 5)$, X2 = Z

m	N	R	Z	X1	X2	Ds
2	552	15.0248	2.714239	2.915756	= Z	61.131
3	792	18.04382	2.719701	2.926182	= Z	60.815

8. $z = 5.\ (2\ \overset{-2}{}\ 4)$, X1 = 27. $(1\ 5\ \overset{-2}{}\ 10\ \overset{\sim}{}\ 5)$, X2 = Z

m	N	R	Z	X1	X2	Ds
2	552	15.0666	2.714338	2.935923	= Z	60.792
3	792	18.0843	2.719756	2.946167	= Z	60.543

9. $z = 5.\ (2\ \overset{-2}{}\ 4)$, X1 = 42. $(5\ \overset{-}{}\ 5\ \overset{\sim}{}\ \overset{2}{10}\ \overset{-}{}\ 5)$, X2 = Z

m	N	R	Z	X1	X2	Ds
2	600	15.35565	2.71702	3.658357	= Z	56.073
3	840	19.3536	2.721316	3.678801	= Z	56.066

(10) $z = 5.\ (2\ \overset{-2}{}\ 4)$, X1 = 46. $(5\ \overset{\sim}{}\ \overset{2}{10}\ \overset{\sim}{}\ 5\ \overset{-}{}\ 5)$, X2 = Z

m	N	R	Z	X1	X2	Ds
2	600	16.7584	2.717734	3.975904	= Z	53.410
3	840	19.74172	2.721734	4.004405	= Z	53.883

(11) $z = 5.\ (2\ \overset{-2}{}\ 4)$, X1 = 50. $(10\ \overset{-2}{}\ 10\ \overset{\sim 2}{}\ 10\ \overset{-}{}\ 5)$ X2 = Z

m	N	R	Z	X1	X2	Ds
2	600	16.77024	2.717754	3.987422	= Z	53.335
3	840	19.75276	2.721746	4.016166	= Z	53.823

Sq-S30E'(5-3, m)----Case (III-3): 60(T4-S) + 20(P3-S)

$z = 5.\,(2\ \overset{-2}{}\ 4)$ (possible stacks for 3-gon as Single are: ##3, 4, and 8)

$z = 5.\,(2\ \overset{-2}{}\ 4)$, X1 = Z, X2 = 3. $(3\ \overset{-}{}\ 3)$

m	N	R	Z	X1	X2	Ds
2	420	14.5885	2.713158	= Z	2.713158	
3	660	17.6057	2.719079	= Z	2.719019	49.337

$z = 5.\,(2\ \overset{-2}{}\ 4)$, X1 = Z X2 = 4. $(\underline{6\ 3})$

m	N	R	Z	X1	X2	Ds
2	480	14.80892	2.713716	= Z	2.981649	54.719
3	720	17.822795	2.719393	= Z	2.987355	56.666

$z = 5.\ (2\ \overset{-2}{}\ 4)$, X1 = Z X2 = 8. $(3\ \overset{\sim}{}\ 3\ \overset{-}{}\ 3)$

m	N	R	Z	X1	X2	Ds
2	480	15.1049	2.714428	= Z	3.686126	52.595
3	720	18.10958	2.719791	= Z	3.700095	54.885

Sq-S30E'(5-3, m)----Case (III-4): 60(S4-S), 12(P5-S), 20(P3-S) $z = 5.(2^{-2}4)$

[5-gon as single, possible stacks: ## 4, 8, 11, 13, 16, 22, 25, 27, 42, 46, and 50] [3-gon S, possible stacks: ## 3, 4, and 8]

(1) $Z = 5.(2^{-2}4)$, $X1 = 4.(10\ 5)$, $X2 = 3.(3^{-}3)$

m	N	R	Z	X1	X2	Ds
3	660	17.7922	2.719349	2.805435	2.719349	52.122
4	900	20.827	2.722781	2.81132	2.722781	51.872

(2) $z = 5.(2^{-2}4)$ $X1 = 4.(10\ 5)$, $X2 = 4.(\underline{6\ 3})$

m	N	R	Z	X1	X2	Ds
3	720	18.00925	2.719654	2.805951	2.987616	55.499
4	960	21.0423	2.72297	2.811645	2.990939	54.203

(3) $Z = 5.(2^{-2}4)$ $X1 = 4.(10\ 5)$, $X2 = 8.(3^{\sim}3^{-}3)$

m	N	R	Z	X1	X2	Ds
3	720	18.2958	2.720039	2.806618	3.700742	53.774
4	960	21.323	2.723207	2.812052	3.708998	52.786

(4) $z = 5.(2^{-2}4)$ $X1 = 8.(5^{\sim}5^{-}5)$, $X2 = 3.(3^{-}3)$

m	N	R	Z	X1	X2	Ds
3	660	18.12566	2.719812	2.966573	2.719812	50.222
4	900	21.16388	2.723074	2.976187	2.723074	50.233

(5) $z = 5.(2^{-2}4)$ $X1 = 8.(5^{\sim}5^{-}5)$, $X2 = 4.(\underline{6\ 3})$

m	N	R	Z	X1	X2	Ds
3	720	18.3428	2.7201	2.967425	2.988064	53.499
4	960	21.3791	2.723254	2.976718	2.991223	52.509

(6) $z = 5.(2^{-2}4)$ $X1 = 8.(5^{\sim}5^{-}5)$, $X2 = 8.(3^{\sim}3^{-}3)$

m	N	R	Z	X1	X2	Ds
3	720	18.6286	2.720464	2.968488	3.70185	51.870
4	960	21.6597	2.72348	2.977392	3.709709	51.157

(7) $z = 5.(2^{-2}4)$ $X1 = 11.(5^{\sim}\overset{2}{10}{}^{\sim}5)$ $X2 = 3.(3^{-}3)$

m	N	R	Z	X1	X2	Ds
3	720	17.9437	2.719563	2.87759	2.719563	55.905
4	960	20.9785	2.722915	2.884246	2.722915	54.533

(8) $z = 5.(2^{-2}4)$ $X1 = 11.(5^{\sim}\overset{2}{10}{}^{\sim}5)$ $X2 = 4.(\underline{6\ 3})$

m	N	R	Z	X1	X2	Ds
3	780	18.1607	2.71986	2.878193	2.987823	59.125
4	1020	21.1936	2.723099	2.884607	2.991068	56.772

(9) $z = 5.(2^{-2}4)$ $X1 = 11.(5^{\sim}\overset{2}{10}{}^{\sim}5)$ $X2 = 8.(3^{\sim}3^{-}3)$

m	N	R	Z	X1	X2	Ds
3	780	18.4469	2.720235	2.8278938	3.701253	57.304
4	1020	21.474165	2.723332	2.885062	3.709321	55.298

(10) $z = 5.(2^{-2}4)$ $X1 = 13.(5^{-}\overset{2}{10}{}^{\sim}5)$, $X2 = 3.(3^{-}3)$

m	N	R	Z	X1	X2	Ds
3	720	17.9687	2.719598	2.88963	2.719598	55.749
4	960	21.00335	2.722936	2.896401	2.722936	54.404

(11) $z = 5.(2^{-2}4)$ $X1 = 13.(5^{-}\overset{2}{10}{}^{\sim}5)$ $X2 = 4.(\underline{6\ 3})$

m	N	R	Z	X1	X2	Ds
3	780	18.1856	2.719893	2.890234	2.987856	58.963
4	1020	21.2184	2.72312	2.896774	2.991089	56.639

(12) $z = 5.(2^{-2}4)$ $X1 = 13.(5^{-}\overset{2}{10}{}^{\sim}5)$ $X2 = 8.(3^{\sim}3^{-}3)$

m	N	R	Z	X1	X2	Ds
3	780	18.4718	2.720267	2.890993	3.701336	57.15
4	1020	21.49898	2.723352	2.897244	3.709314	55.170

(13) $z = 5.(2^{-2}4)$ $X1 = 16.(\underline{10\ 5}^{-}5)$ $X2 = 3.(3^{-}3)$

m	N	R	Z	X1	X2	Ds
3	720	19.6701	2.721659	3.933734	2.721659	46.522
4	960	22.6823	2.724236	3.950844	2.724236	46.648

(14) $z = 5.(2^{-2}4)$ $X1 = 16.(\underline{10\ 5}^{-}5)$ $X2 = 4.(\underline{6\ 3})$

m	N	R	Z	X1	X2	Ds
3	780	19.8835	2.721881	3.9352-6	2.989848	49.323
4	1020	22.89518	2.72438	3.951792	2.99235	48.647

(15) $z = 5.(2^{-2}4)$ $X1 = 16.(\underline{10\ 5}^{-}5)$ $X2 = 8.(3^{\sim}3^{-}3)$

m	N	R	Z	X1	X2	Ds
3	780	20.1633	2.722161	3.937068	3.706271	47.964
4	102	23.17177	2.72456	3.953023	3.71252	47.49

(16) $z = 5.(2^{-2}4)$ $X1 = 22.(1\ 5^{\sim}5^{-}5)$ $X2 = 3.(3^{-}3)$

m	N	R	Z	X1	X2	Ds
3	672	18.2708	2.720006	3.039731	2.720006	50.326
4	912	21.3078	2.72319	3.04978	2.72319	50.21

	0		2		9	2				5	4	5	8

(17) $z = 5.\ (2\,{}^{-2}4)$ $X1 = 22.(1\ 5\,\tilde{}\,5\,\bar{}\,5)$ $X2 = 4.(\underline{6\ 3})$

m	N	R	Z	X1	X2	Ds
3	732	18.4878	2.720287	3.040614	2.988251	53.540
4	972	21.5227	2.723371	3.050324	2.99134	52.458

(18) $z = 5.\ (2\,{}^{-2}4)$ $X1 = 22.(1\ 5\,\tilde{}\,5\,\bar{}\,5)$ $X2 = 8.(3\,\tilde{}\,3\,\bar{}\,3)$

m	N	R	Z	X1	X2	Ds
3	732	18.773	2.720642	3.04172	3.702313	51.926
4	972	21.8029	2.723593	3.051032	3.710001	51.119

(19) $z = 5.\ (2\,{}^{-2}4)$ $X1 = 25.(1\ 5\,\tilde{}\,\overset{2}{10}\,\tilde{}\,5)$ $X2 = 3.\ (3\,\bar{}\,3)$

m	N	R	Z	X1	X2	Ds
3	732	18.04382	2.719701	2.926182	2.719701	56.208
4	972	21.07643	2.722999	2.932371	2.722999	54.703

(20) $z = 5.\ (2\,{}^{-2}4)$ $X1 = 25.(1\ 5\,\tilde{}\,\overset{2}{10}\,\tilde{}\,5)$ $X2 = 4.(\underline{6\ 3})$

m	N	R	Z	X1	X2	Ds
3	792	18.26045	2.719992	2.926727	2.987956	59.380
4	1032	21.291	2.723181	2.932712	2.99115	56.915

(21) $z = 5.\ (2\,{}^{-2}4)$ $X1 = 25.(1\ 5\,\tilde{}\,\overset{2}{10}\,\tilde{}\,5)$ $X2 = 8.(3\,\tilde{}\,3\,\bar{}\,3)$

m	N	R	Z	X1	X2	Ds
3	792	18.54596	2.720361	2.927424	3.701581	57.566
4	1032	21.57152	2.72341	2.933139	3.709526	55.445

(22) $z = 5.\ (2\,{}^{-2}4)$ $X1 = 27.(1\ 5\,\bar{}\,\overset{2}{10}\,\tilde{}\,5)$ $X2 = 3.\ (3\,\bar{}\,3)$

m	N	R	Z	X1	X2	Ds
3	732	18.0843	2.719756	2.946167	2.719756	55.956
4	972	21.1435	2.723034	2.952243	2.723034	54.497

(23) $z = 5.\ (2\,{}^{-2}4)$ $X1 = 27.(1\ 5\,\bar{}\,\overset{2}{10}\,\tilde{}\,5)$ $X2 = 4.(\underline{6\ 3})$

m	N	R	Z	X1	X2	Ds
3	792	18.301	2.720046	2.946698	2.988009	59.117
4	1032	21.3312	2.723214	2.95257	2.991183	56.701

(24) $z = 5.\ (2\,{}^{-2}4)$ $X1 = 27.(1\ 5\,\bar{}\,\overset{2}{10}\,\tilde{}\,5)$ $X2 = 8.(3\,\tilde{}\,3\,\bar{}\,3)$

m	N	R	Z	X1	X2	Ds
3	792	18.58632	2.720412	2.947386	3.701713	57.316
4	1032	21.6112	2.723442	2.952992	3.709609	55.241

(25) $z = 5.\ (2\,{}^{-2}4)$ $X1 = 42.(5\,\bar{}\,5\,\tilde{}\,\overset{2}{10}\,\bar{}\,5)$ $X2 = 3.(3\,\bar{}\,3)$

m	N	R	Z	X1	X2	Ds
3	780	19.3536	2.721316	3.678801	2.721316	52.061
4	1020	22.373	2.724018	3.691589	2.724018	50.944

(26) $z = 5.\ (2\,{}^{-2}4)$ $X1 = 42.(5\,\bar{}\,5\,\tilde{}\,\overset{2}{10}\,\bar{}\,5)$ $X2 = 4.(\underline{6\ 3})$

m	N	R	Z	X1	X2	Ds
3	840	19.568	2.72155	3.679923	2.989517	54.844
4	1080	22.58633	2.724169	3.692297	2.992138	52.926

(27) $z = 5.\ (2\,{}^{-2}4)$ $X1 = 42.(5\,\bar{}\,5\,\tilde{}\,\overset{2}{10}\,\bar{}\,5)$ $X2 = 8.(3\,\tilde{}\,3\,\bar{}\,3)$

m	N	R	Z	X1	X2	Ds
3	840	19.849	2.721845	3.681315	3.705448	53.302
4	1080	22.8635	2.724359	3.693176	3.711999	51.651

(28) $z = 5.\ (2\,{}^{-2}4)$ $X1 = 46.(5\,\tilde{}\,\overset{2}{10}\,\tilde{}\,5\,\bar{}\,5)$ $X2 = 3.(3\,\bar{}\,3)$

m	N	R	Z	X1	X2	Ds
3	780	19.74172	2.721734	4.004405	2.721734	50.034
4	1020	22.751	2.724283	4.022532	2.724283	49.265

(29) $z = 5.\ (2\,{}^{-2}4)$ $X1 = 46.(5\,\tilde{}\,\overset{2}{10}\,\tilde{}\,5\,\bar{}\,5)$ $X2 = 4.(\underline{6\ 3})$

m	N	R	Z	X1	X2	Ds
3	840	19.9549	2.721953	4.005959	2.989921	52.738
4	1080	22.9637	2.724426	4.023551	2.992345	51.201

(30) $z = 5.\ (2\,{}^{-2}4)$ $X1 = 46.(5\,\tilde{}\,\overset{2}{10}\,\tilde{}\,5\,\bar{}\,5)$ $X2 = 8.(3\,\tilde{}\,3\,\bar{}\,3)$

m	N	R	Z	X1	X2	Ds
3	840	20.234	2.72223	4.007918	3.706451	51.293
4	1080	23.23967	2.724606	4.024811	3.712643	49.992

(31) $z = 5.\ (2\,{}^{-2}4)$, $X1 = 50.(5\,\bar{}\,\overset{2}{10}\,\tilde{}\,5\,\bar{}\,5)$ $X2 = 3.\ (3\,\bar{}\,3)$

m	N	R	Z	X1	X2	Ds
3	780	19.73277	2.721746	4.016165	2.721746	49.978

(32) $z = 5.\ (2\,{}^{-2}4)$ $X1 = 50.(5\,\bar{}\,\overset{2}{10}\,\tilde{}\,5\,\bar{}\,5)$ $X2 = 4.(\underline{6\ 3})$

m	N	R	Z	X1	„X2	Ds
3	840	19.966	2.721965	4.017741	2.989932	52.679

4	1020	22.76167	2.72429	4.034482	2.72429	49.219	4	1080	22.9742	2.724433	4.03549	2.992402	51.154
(33) $z = 5.\,(2\,^{-2}\,4)$ $X1 = 50.(5\,^{-}\,10^{2}\sim 5\,^{-}\,5)$ $X2 = 8.(3\sim 3\,^{-}\,3)$													
m	N	R	Z	X1	X2	Ds							
3	840	20.2451	2.722241	4.019731	3.706479	51.237							
4	1080	23.25	2.724613	4.036788	3.71266	47.173							

[End of Sq-S30E'(5-3, m)----Case (III-4)]

------- End of [DATA 4B-2] Sq-S30E'(5-3, m) and Family ------

The families of Sq-S12E1'(4-3, m) and of Sq-S30E'(5-3, m) are the only two completed data in all eleven Sq-SnE' families.

[DATA 5-1]
The Basic Tr-SE and Equivalent Configurations

Basic Tr-SnE(k1-k2, 1)			
Tr-SnE(k1-k2, 1)	**R**	**N, Ds**	**Equi. Conf.**
Tr-4E (3-3,1)	$\sqrt{\frac{1}{2}(5+\sqrt{5})}$ =1.902113	12, 82.91796	T12E
Tr-6E(3-3, 1)	$\sqrt{\frac{1}{3}[(199+\sqrt{297})^{\frac{1}{3}}+(199-\sqrt{297})^{\frac{1}{3}}]+10}$ = 2.68742675	24, 83.07646	T24E
Tr-8E(4-4, 1)	$\sqrt{\frac{1}{3}[(199+\sqrt{297})^{\frac{1}{3}}+(199-\sqrt{297})^{\frac{1}{3}}]+10}$ = 2.68742675	24, 83.07646	T24E
Tr-12E(3-3, 1)	$\sqrt{A+B+\frac{1}{6}(27+7\sqrt{5})}$ = 4. 31167475	60, 0.68624	T60E
Tr-20E(5-5, 1)	$\sqrt{A+B+\frac{1}{6}(27+7\sqrt{5})}$ = 4.31167475 $\frac{A}{B}=\frac{1}{3}\sqrt[3]{\frac{1}{2}\left[\left(5112+2285\sqrt{5}\right)\pm\sqrt{64233+28728\sqrt{5}}\right]}$	60, 80.68624	T60E
Tr- $12E_1'$ (4-3, 1)	3.7701264	48, 84.42468	
Tr- $12E_2'$ (3-6, 1)	3.586498	36, 69.968	
Tr- $12E_2'$ (6-6, 1)	3.060468	24, 64.058	
Tr- $24E_1'$ (4-4, 1)	×		
Tr- $24E_1'$ (3-4, 1)	×		
Tr- $24E_2'$ (4-6, 1)	4.9362681	72, 73.871	
Tr- $24E_2'$ (6-6, 1)	4.4863475	48, 59.62	
Tr- $24E_3'$ (3-8, 1)	5.2461605	72, 65.402	
Tr- $24E_3'$ (8-8, 1)	4.5702751	48, 58.99	
Tr- $30E'$ (5-3, 1)	5.9581525	120, 84.50804	
Tr- $48E'$ (8-6, 1)	6.405194	96, 58.499	
Tr- $48E'$ (8-4, 1)	×		
Tr- $48E'$ (6-4, 1)	×		
Tr- $60E_1'$ (5-6, 1)	7.6750843	180, 76.392	
Tr- $60E_1'$ (6-6, 1)	7.3627843	120, 55.34	
Tr- $60E_2'$ (5-4, 1)	×		
Tr- $60E_2'$ (3-4, 1)	×		
Tr- $60E_3'$ (3-10, 1)	8.561015	180, 61.399	
Tr- $60E_3'$ (10-10, 1)	7.420845	120, 54.477	
Tr- $120E'$ (10-6, 1)	10.276787	240, 56.812	
Tr- $120E'$ (10-4,k 1)	×		
Tr- $120E'$ (6-4, 1)	×		

× The 4-gon in the third set of the mother E' can not connect with a tr-string. It can connect with a sq-string.

--- End of [DATA 5-1] The basic Tr-SnE(k1-k2, 1) and some Equivalent Configurations ---

[DATA 5-2] The R and Ds of Basic Sq-SE and Tr-SE (for Comparison) and Their Long-String Resemblance Configurations

Basic Sq-SnE(k1-k2, 1)				Basic Tr-SnE(k1-k2, 1)				Long-string Resemblance Configuration	
	R	N	Ds		R	N	Ds	R (m = 1000)	
Sq-4E(3-3)	2	12	75	Tr-4E(3-3)	1.902113	12	82.91796	1046.7733	4E
Sq-6E(3-3)	2.79793265	24	76.644	Tr-6E(3-3)	2.68742675	24	83.07646	1624.7489	8E
Sq-8E(4-4)	2.79793265	24	76.644	Tr-8E(4-4)	2.68742675	24	83.07646	1273.2395	6E
Sq-12E(3-3)	4.4659	60	75.21	Tr-12E(3-3)	4. 31167475	60	80.68624	1902.1174	20E
Sq-20E(5-5)	4.46590	60	75.21	Tr-20E(5-5)	4.31167475	60	80.68624	2740.7687	12E
Sq- $12E'_1$ (4-3)	3.9146589	48	78.306	Tr $12E'_1$ (4-3)	3.7701264	48	84.42468	2093.5467	A14E
Sq- $12E'_2$ (3-6)	3.7229425	36	64.934	Tr $12E'_2$ (3-6)	3.586498	36	69.968	1624.7489	8E
Sq- $12E'_2$ (6-6)	3.16227766	24	60	Tr- $12E'_2$ (6-6)	3.060468	24	64.058	1044.7733	4E
Sq- $24E'_1$ (4-4)	5.2224652	72	65.997	Tr $24E'_1$ (4-4)	×			2547.7528	6E
Sq- $24E'_1$ (3-4)	6.0751586	72	48.771	Tr $24E'_1$ (3-4)	×			3251.1226	8E
Sq- $24E'_2$ (4-6)	5.1115333884	72	68.892	Tr $24E'_2$ (4-6)	4.9362681	72	73.871	2093.5467	A14E
Sq- $24E'_2$ (6-6)	4.6352218	48	55.852	Tr $24E'_2$ (6-6)	4.4863475	48	59.62	1624.7489	8E
Sq- $24E'_3$ (3-8)	5.4269755	72	61.116	Tr $24E'_3$ (3-8)	5.2461605	72	65.402	2093.5467	A14E
Sq- $24E'_3$ (8-8)	4.6352218	48	55.852	Tr $24E'_3$ (8-8)	4.5702751	48	58.99	1273.2395	6E
Sq- 30E' (5-3)	6.175794	120	78.657	Tr 30E' (5-3)	5.9581525	120	84.50804	3065.8006	A32E
Sq- 48E' (8-6)	6.601719	96	55.068	Tr 48E' (8-6)	6.405194	96	58.499	1624.7489	A14E
Sq- 48E' (8-4)	7.04317	96	48.381	Tr 48E' (8-4)	×			2547.7528	6E
Sq- 48E' (6-4)	7.790735	96	39.542	Tr 48E' (6-4)	×			3251.1226	8E
Sq- $60E'_1$ (5-6)	7.935094	180	71.468	Tr $60E'_1$ (5-6)	7.6750843	180	76.392	3065.8006	A32E
Sq- $60E'_1$ (6-6)	7.604789	120	51.874	Tr $60E'_1$ (6-6)	7.3627843	120	55.34	2740.7487	20E
Sq- $60E'_2$ (5-4)	7.995593	180	70.39	Tr $60E'_2$ (5-4)	×			3806.1288	12E
Sq- $60E'_2$ (3-4)	9.8708	180	46.186	Tr $60E'_2$ (3-4)	×			5482.2390	20E
Sq- $60E'_3$ (3-10)	8.835113	180	57.649	Tr $60E'_3$ (3-10)	8.561015	180	61.399	3065.8006	A32E
Sq- $60E'_3$ (10-10)	7.604789	120	51.874	Tr $60E'_3$ (10-10)	7.420845	120	54.477	1273.2396	12E
Sq- 120E' (10-6)	10.57276	240	53.675	Tr 120E' (10-6)	10.276787	240	56.812	3065.8006	A32E
Sq- 120E' (10-4)	11.1061	240	48.644	Tr 120E' (10-4)	×			3806.1282	12E
Sq- 120E' (6-4)	13.03042	240	35.337	Tr 120E' (6-4)	×			5481.4975	20E

× The 4-gon of the mother E' is in the third set, so that it can not connect with a tr-string. It can connect with a sq-string.

End of [DATA 5-2] The R and Ds of basic Sq-SE and Tr-SE (for comparison) And Their Long-String Resemblance Configurations

[DATA 5-3] The Five Tr-SnE(k-k, m), m = 1, 2, 3, 10, 100, and Families

The longer possible z-value would yield a more efficient pack. In the following chart, z = 1 and z = min(z_d , z_x) are provided to compare the efficiency of the pack. That is while z = min(z_d , z_x) provides shorter R-value than while z = 1 does. (But due to the computer's long procedure of calculation, sometimes the error made the solution in the opposite direction.)

The value of z_d and z_x are: $z_d = \sqrt{\frac{3R^2-4}{R}}$ [= z(d)] $z_x = \frac{R^2x}{\sqrt{(R^2-2)^2+R^2x^2}}$ [= z(x)]

		m = 1	m = 2	m = 3	m = 10	m = 100
Tr-S4E (3-3, m)	N	12	24	36	120	1200
	R	1.902113	2.8277565	3.8169841	11.044125	105.24245
	x	1	1	1	1	1
	z	1	1	1	1	1
	Ds	82.91796	7.504	61.773	44.596	2.7086
	R	Non-Existing	2.738664	3.76988	11.037649	105.24247
	x		1	1	1	1
	z(x)		1.220674	1.11994	1.012405	1.000135
	Ds		79.997	63.327	24.625	2.709
Tr-S6E (3-3, m)	N	24	48	72	240	2400
	R	2.68742675	4.2162764	5.7978199	17.098169	163.4464
	x	1	1	1	1	1
	z	1	1	1	1	1
	Ds	83.07646	67.503	53.548	20.524	2.246
	R	Non-Existing	4.1776216	5.7733113	17.095488	163.44626
	x		1	1	1	1
	z(x)		1.090287	1.046221	1.005149	1.000056
	Ds		68.798	54.004	20.530	22.460
Tr-S8E (4-4,m)	N	24	48	72	240	2400
	R	2.68742675	3.8507039	5.07063	13.887962	128.4386
	x	1	1	1	1	1
	z	1	1	1	1	1
	Ds	83.07646	80.92841	70.008	31.108	3.637
	R	Non-Existing	3.7942255	5.0290668	13.880403	128.43844
	x		1	1	1	1
	z(x)		1.110487	1.061408	1.07823	1.000091
	DS		83.35564	71.170	31.142	3.637
Tr-S12E (3-3, m)	N	60	120	180	600	6000
	R	4.31167475	6.9738538	9.6803011	28.810702	275.7576
	x	1	1	1	1	1
	z	1	1	1	1	1
	Ds	80.68624	61.684	48.021	18.071	1.973
	R	Non-Existing	6.9547302	9.6668754	28.810784	275.75795
	x		1	1	1	1
	z(x)		1.031594	1.016208	1.001809	1.00002
	Ds		62.024	48.155	18.071	1.973
Tr-S20E (5-5, m)	N	60	120	180	600	6000
	R	4.31167475	6.0018061	7.748933	20.2819	182.8622
	x	1	1	1	1	1
	z	1	1	1	1	1
	Ds	80.68624	83.08318	74.943	36.465	4.486
	R	Non-Existing	5.969884	7.72088	20.276665	182.8625
	x		1	1	1	1
	z(x)		1.043154	1.025547	1.003657	1.000043
	Ds		84.17623	75.488	36.484	4.486

		m = 1	m = 2	m = 3	m = 10	M= 100
Tr-S4E (3̇-3̇, m)	N	16	28	40	124	1204
	R	2.284633	3.0246852	3.899048	11.016191	105.2424
	x	1.557317	1.634652	1.674116	1.7249	1.731973
	z	1	1	1	1	1
	Ds	76.635	76.513	6.778	25.545	2.718
	R	Non-existing	2.9411375	3.8711644	11.06094	105.2426
	x		1.628862	1.673264	1.724958	1.731973
	z(d)		1.693897	1.711286	1.72967	1.732025
	Ds		80.92214	66.773	25.338	2.718
Tr-S6E (3̇-3̇, m)	N	32	56	80	248	2408
	R	3.1462785	4.3492205	5.847632	1.7084109	164.4464
	x	1.642237	1.685646	1.706537	1.729081	1.732018
	z	1	1	1	1	1
	Ds	80.81569	74.01253	58.48843	21.24247	2.25344
	R	Non-existing	4.4210918	5.9280292	17.140576	163.4466
	x		1.687162	1.707229	1.729101	1.732018
	z(d)		1.716415	1.723575	1.731065	1.73204
	Ds		71.626	56.913	21.103	2.253
Tr-S8E (4̇-4̇, m)	N	30	54	78	246	2406
	R	3.1127017	4.293146	5.520018	14.343927	128.9202
	x	1.339245	1.375314	1.390814	1.410773	1.414171
	z	1	1	1	1	1
	Ds	77.408	73.246	63.996	29.891	3.619
	R	Non-existing	4.1210057	5.3054302	14.088649	128.7603
	x		1.371945	1.388865	1.410647	1.414771
	z(x)		1.454939	1.439049	1.417771	1.414256
	Ds		79.493	69.278	30.984	3.628
Tr-S12E (3̇-3̇, m)	N	80	140	200	620	6020
	R	4.9527915	7.031144	9.690987	28.794628	275.7576
	x	1.696379	1.714444	1.722805	1.731006	1.732039
	z	1	1	1	1	1
	Ds	81.53233	70.797	53.240	18.694	1.979
	R	Non-existing	7.3269503	9.9253326	28.891278	275.7574
	x		1.715843	1.723237	1.731013	1.732039
	z(d)		1.726563	1.729088	1.731704	1.732047
	Ds		65.196	50.755	18.569	1.9792
Tr-S20E (5̇-5̇, m)	N	72	132	192	612	5012
	R	4.6495649	6.381981	8.1542601	20.7355	183.5129
	x	1.148059	1.161049	1.166697	1.174203	1.175553
	z	1	1	1	1	1
	Ds	83.26221	81.022	72.189	35.585	4.463
	R	Non-existing	6.277958	8.0086007	20.517492	183.1874
	x		1.160561	1.16637	1.174173	1.175553
	z(x)		1.200057	1.190535	1.177833	1.175599
	Ds		83.72924	74.839	36.345	4.4789

------ End of [DATA 5-3] The Tr-SnE(k, m) and Families ------

[DATA 5-4] The eleven Tr-SnE'(k1-k2, m), m = 1, 2, 3, 10, 100, and Families

P. 1 of 4

		m = 1	m = 2	M = 3	m = 10	m = 100
Tr-S12E'1 (4-3, m)	N	48	96	144	480	4800
	R	3.7701264	5.7629725	7.816318	22.408388	210.9076
	X1	1	1	1	1	1
	X2	1	1	1	1	1
	z	1	1	1	1	1
	Ds	84.42468	72.263	58.925	23.904	2.698
	R	Non-Existing	57.432325	7.801939	22.40065	210.9011
	X1		1	1	1	1
	X2		1	1	1	1
	z(x)		1.046719	1.025011	1.002995	1.000034
	Ds		72.761	59.142	23.914	2.698
Tr-S12E'1 (3-4̇, m)	N	54	102	150	486	4806
	R	4.094064	6.1095559	8.1706605	22.770269	211.3398
	X1	1	1	1	1	1
	X2	1.371378	1.395148	1.303582	1.41285	1.414198
	z	1	1	1	1	1
	Ds	80.54238	68.443	56.772	23.423	2.690
	R	Non-Existing	6.1038797	8.16475	22.775422	211.3399
	X1		1	1	1	1
	X2		1.395105	1.403573	1.41285	1.414198
	z(x)		1.041237	1.022793	1.002897	1.000034
	Ds		68.443	56.215	23.423	2.690
Tr-S12E'1 (3̇-4, m)	N	56	104	152	488	4808
	R	Non-Existing	5.93537	7.95877	22.43937	210.9076
	X1		1.707291	1.718245	1.73033	1.732031
	X2		1	1	1	1
	z		1	1	1	1
	Da		73.804	60.338	24.229	2.702
	R	Non-Existing	5.9480785	7.946023	22.43947	210.9077
	X1		1.707397	1.71828	1.73033	1.732031
	X2		1	1	1	1
	z(x)		1.043479	1.024099	1.002984	1.000034
	Ds		73.489	60.184	24.229	2.702
Tr-S12E'1 (3̇-4̇, m)	N	62	110	158	494	4814
	R	Non-Existing	6.216584	8.236043	22.78798	211.3398
	X1		1.709494	1.719238	1.730382	1.732031
	X2		1.395796	1.403752	1.412851	1.414198
	z		1	1	1	1
	Ds		71.160	58.223	23.782	2.695
	R	Non-Existing	6.1803801	8.2189259	22.79591	211.3411
	X1		1.709228	1.719183	1.730584	1.732031
	X2		1.39538	1.403707	1.412852	1.414198
	Z(x)		1.43258	1.424638	1.415574	1.414229
	Ds		71.995	58.475	23.766	2.945
Tr-S12E'2 (3-6,m)	N	36	60	84	252	2412
	R	3.586498	5.074332	6.6455213	17.92278	164.2273
	X1	1	1	1	1	1
	X2	1	1	1	1	1
	z	1	1	1	1	1
	Ds	69.968	58.255	47.598	19.612	2.236
	R	Non-Existing	5.0469087	6.620549	17.919854	164.227
	X1		1	1	1	1
	X2		1	1	1	1
	z(x)		1.060961	1.034929	1.004684	1.000056
	Ds		58.890	47.911	10.2765	2.236

		m = 1	m = 2	m = 3	m = 10	m = 100
Tr-S12E'2 (6-6,m)		24	36	48	132	1212
	R	3.060468	3.9596742	4.9295356	12.114171	105.3273
	X1	1	1	1	1	1
	X2	1	1	1	1	1
	z	1	1	1	1	1
	Ds	64.058	57.402	49.382	22.487	2.680
	R	Non-existing	3.9004902	4.874161	12.100543	106.3272
	X1		1	1	1	1
	X2		1	1	1	1
	z(x)		1.104255	1.065513	1.010308	1.000013
	Ds		59.158	50.510	22.537	2.680
Tr-S24E'2 (4-6,m)	N	72	120	168	504	4824
	R	4.9362681	6.89916	8.9317145	23.47832	212.2094
	X1	1	1	1	1	1
	X2	1	1	1	1	1
	z	1	1	1	1	1
	Ds	73.871	63.027	52.648	22.858	3.6180
	R	Non-existing	6.8808183	8.915466	23.47302	212.2094
	X1		1	1	1	1
	X2		1	1	1	1
	z(x)		1.032289	1.019088	1.002727	1.000033
	Ds		63.364	52.840	22.868	2.6180
Tr-S24E'2 (4̇-6, m)	N	78	126	174	510	4830
	R	5.1763775	7.198657	9.256863	2.304085	212.6007
	X1	1.387573	1.400502	1.405937	1.412969	1.414198
	X2	1	1	1	1	1
	z	1	1	1	1	1
	Ds	72.775	6.0787	50.765	22.432	2.6715
	R	Non-existing	7.171082	9.23774	23.83639	212.599
	X1		1.400396	1.405903	1.412969	1.414198
	X2		1	1	1	1
	z(x)		1.029684	1.017765	1.002644	1.000033
	Ds		61.255	5.098	22.440	2.672
Tr-S24E'2 (6-6,m)	N	48	72	96	264	2424
	R	4.4863475	5.978204	7.531588	18.766608	165.0156
	X1	1	1	1	1	1
	X2	1	1	1	1	1
	z	1	1	1	1	1
	Ds	59.620	5.0365	42.310	18.740	2.2255
	R	Non-existing	5.945231	7.5003065	18.75658	165.0157
	X1		1	1	1	1
	X2		1	1	1	1
	z(x)		1.043522	1.027095	1.004275	1.000055
	Ds		50.925	42.663	18.760	2.225
Tr-S24E'3 (3-8,m)	N	72	120	168	504	4824
	R	5.2461605	7.176737	9.194233	23.72055	212.2095
	X1	1	1	1	1	1
	X2	1	1	1	1	1
	z	1	1	1	1	1
	Ds	65.402	58.246	49.682	22.393	2.6780
	R		7.1585825	9.178109	23.71515	212.2095
	X1	Non-Existing	1	1	1	1
	X2		1	1	1	1
	z		1.02979	1.018	1.002671	1.000033
	Ds		58.542	49.859	22.404	2.6780

Tr-S12E'2 (3̇-6, m)		40	64	88	256	2416
	R	Non-Existing	5.2364947	6.7624577	17.960082	164.2273
	X1		1.700175	1.713009	1.729364	1.732019
	X2		1	1	1	1
	z		1	1	1	1
	Ds		58.350	48.108	19.841	2.239
	R	Non-Existing	5.23101	6.76281	17.960152	164.2271
	X1		1.700107	1.713011	1.729364	1.732019
	X2		1	1	1	1
	z(x)		1.056616	1.033447	1.004663	1.000056
	Ds		58.472	48.103	19.841	2.239

Tr-S24E'3 (3-8̃, m)	N	96	144	192	528	4848
	R	5.5755435	7.574767	9.627121	24.214892	212.6471
	X1	1	1	1	1	1
	X2	1.320204	1.364044	1.383356	1.409382	1.414151
	z	1	1	1	1	1
	Ds	77.203	62.743	51.801	22.512	2.688
	R	Non-Existing	7.567274	9.619913	24.2094	212.6469
	X1		1	1	1	1
	X2		1.363943	1.383316	1.409379	1.414151
	z		1.02661	1.026238	1.002563	1.000033
	Ds		62.867	51.868	22.522	2.688

End of [DATA 5-4] The eleven Tr-SnE'(k1-k2, m) and Families (P. 1 of 4)

P. 2 of 4

		m=1	m=2	M=3	m=10	m=100
Tr-	N		128	176	512	4832
S24E'3	R		7.3692475	9.343538	23.76971	212.2095
(3̇ -8,	X1	Non-	1.71603	1.722102	1.730517	1.732032
m)	X2	Existing	1	1	1	1
	z		1	1	1	1
	Ds		58.926	50.40	22.655	2.6825
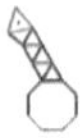	R		7.3669255	9.341653	23.76973	212.2095
	X1	Non-	1.71602	1.722098	1.730517	1.732032
	X2	Existing	1	1	1	1
	z		1.028101	1.017368	1.002659	1.000033
	Ds		58.963	50.420	22.655	2.6825
Tr-	N	48	72	96	264	2424
S24 E'_3	R	4.5102751	5.615589	6.791984	15.50538	131.1584
(8-8,	X1	1	1	1	1	1
m)	X2	1	1	1	1	1
	z	1	1	1	1	1
	Ds	58.990	57.080	52.026	27.449	3.583
	R		5.580754	6.753675	15.49277	131.058
	X1	Non-	1	1	1	1
	X2	Existing	1	1	1	1
	z		1.049555	1.03354	1.006273	1.000089
	Ds		57.795	52.618	27.497	2.583
Tr-	N	72	96	120	288	2448
S24 E'_3	R	4.8692202	6.048037	7.271686	16.089585	130.71695
(8̃ - 8̃ ,	X1	1.289664	1.334716	1.359692	1.403245	1.414048
m)	X2	1.289664	1.334716	1.359692	1.403245	1.414048
	z	1	1	1	1	1
	Ds	75.920	65.612	56.735	27.813	3.582
	R		5.8608825	6.9613895	15.50292	129.895
	X1	Non-	1.329403	1.354619	1.402396	1.414046
	X2	existing	1.329403	1.354619	1.402396	1.414046
	z		1.372349	1.384697	1.400317	1.41413
	Ds		69.869	61.906	29.196	3.6271
Tr-	N	120	240	360	1200	12000
S30E'	R	5.9581525	3.910702	11.927163	33.30634	309.7096
(3-5,m)	X1	1	1	1	1	1
	X2	1	1	1	1	1
	z	1	1	1	1	1
	Ds	84.50804	75.566	63.266	27.044	3.128
	R		8.898403	11.91670	33.30495	309.7094
	X1	Non-	1	1	1	1
	X2	Existing	1	1	1	1
	z		1.019162	1.010631	1.001353	1.000016
	Ds		75.775	63.377	27.046	3.128
Tr-	N	132	252	372	1212	12012
S30E'	R	6.265894	9.256925	12.289761	33.6974	309.7097
(3- 5̇ ,	X1	1	1	1	1	1
m)	X2	1.160503	1.168691	1.171672	1.175053	1.175564
	z	1	1	1	1	1
	Ds	84.05196	73.520	61.574	26.684	3.131
	R		9.2482762	12.282959	33.69754	309.7095
	X1	Non-	1	1	1	1
	X2	Existing	1.168678	1.171668	1.175053	1.175564
	z		1.017725	1.010002	1.001322	1.000016
	Ds		73.658	61.643	26.684	3.131

		m=1	m=2	m=3	m=10	m=100
Tr-S30E'	N	152	272	392	1232	12032
(3̇ - 5̇ , m)	R		9.442499	12.423559	33.74154	309.7090
	X1	Non-	1.70231	1.729431	1.73129	1.732042
	X2	existing	1.168959	1.171756	1.175054	1.175564
	z		1	1	1	1
	Ds		76.267	63.494	27.053	3.126
	R		9.471121	12.453093	33.75486	309.7090
	X1	Non-	1.722369	1.726457	1.731291	1.732042
	X2	existing	1.168999	1.171774	1.175035	1.175564
	z		1.186143	1.181727	1.176406	1.175581
	Ds		75.806	63.193	27.032	3.136
Tr-S48E'	N	96	144	192	528	4848
(6-8, m)	R	6.405194	8.333866	10.337362	24.80801	213.5275
	X1	1	1	1	1	1
	X2	1	1	1	1	1
	z	1	1	1	1	1
	Ds	58.499	51.834	44.918	21.448	2.658
	R		8.318796	10.322155	24.80688	213.5276
	X1	Non-	1	1	1	1
	X2	existing	1	1	1	1
	z(x)		1.021961	1.014199	1.002441	1.000033
	Ds		52.021	45.051	21.450	2.658
Tr-S48E'	N	120	168	216	552	4872
(6- 8̃ , m)	R	6.6796945	8.6709227	10.7131	25.27428	213.9705
	X1	1	1	1	1	1
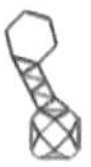	X2	1.349369	1.376087	1.389353	1.409779	1.414152
	z	1	1	1	1	1
	Ds	67.237	55.862	46.050	21.603	2.660
	R		8.662897	10.705305	25.27431	213.9704
	X1	Non-	1	1	1	1
	X2	existing	1.376015	1.389316	1.409779	1.414152
	z(x)		1.02023	1.013193	1.002342	1.000033
	Ds		55.966	47.119	21.603	2.660
Tr-S60E'1	N	180	300	420	1260	12060
(5-6,m)	R	7.6750843	10.58833	13.575497	34.88113	311.5810
	X1	1	1	1	1	1
	X2	1	1	1	1	1
	z	1	1	1	1	1
	Ds	76.392	66.897	56.974	25.890	3.106
	R		10.57689	13.56479	34.88128	311.5808
	X1	Non-	1	1	1	1
	X2	existing	1	1	1	1
	z(x)		1.013518	1.008193	1.001234	1.000015
	Ds		67.042	57.064	25.890	3.106
Tr-S60E'1	N	192	312	432	1272	12072
(5-6, m)	R	7.954668	10.907624	13.9158	35.26229	311.5810
	X1	1.166224	1.17062	1.172531	1.175098	1.175564
	X2	1	1	1	1	1
	z	1	1	1	1	1
	Ds	75.857	65.559	55.771	2.557	3.109
	R		10.39857	13.90689	35.26242	311.5808
	X1	Non-	1.170612	1.172527	1.175098	1.175564
	X2	existing	1	1	1	1
	z(x)		1.012726	1.007793	1.001207	1.000015
	Ds		65.668	55.842	25.574	3.109

Tr-S30E' (3̇ -5, m) 	N	140	260	380	1220	12020
	R	Non-Existing	9.135694	12.0916605	33.36019	309.7096
	X1		1.721643	1.726122	1.731273	1.732042
	X2		1	1	1	1
	z		1	1	1	1
	Ds		77.881	64.976	27.406	3.133
	R	Non-existing	9.14412	12.096297	33.36014	309.7094
	X1		1.721662	1.726122	1.741273	1.732042
	X2		1	1	1	1
	z		1.018135	1.010316	1.001349	1.000016
	Ds		77.737	64.927	27.406	3.132

Tr-S60E'1 (6-6, m) 	N	120	180	240	660	6060
	R	7.3627843	9.953203	12.610968	31.62387	278.7386
	X1	1	1	1	1	1
	X2	1	1	1	1	1
	z	1	1	1	1	1
	Ds	55.340	45.424	37.727	16.499	1.950
	R	Non-existing	9.934989	12.59322	31.623858	278.7390
	X1		1	1	1	1
	X2		1	1	1	1
	z(x)		1.015357	1.009513	1.001501	1.000019
	Ds		45.591	37.834	16.499	1.950

End of the eleven Tr-SnE'(k1-k2, m) and Families P. 2 of 4

P. 3 of 4

		m = 1	m = 2	m = 3	m = 10	m = 100
Tr-S60 E'_3 (3-10, m)	N	180	300	420	1260	12060
	R	8.561015	11.423669	14.382313	35.63955	311.5811
	X1	1	1	1	1	1
	X2	1	1	1	1	1
	z	1	1	1	1	1
	Ds	61.399	57.471	50.761	2.480	3.106
	R		11.413029	14.37192	35.634618	311.5809
	X1	Non-	1	1	1	1
	X2	Existing	1	1	1	1
	z(x)		1.011596	1.007294	1.001182	1.000015
	Ds		51.578	50.835	24.807	3.106
Tr-S60 E'_3 (3-10, m)	N	240	360	480	1320	12120
	R	8.835507	11.777353	14.780474	36.12589	312.5252
	X1	1	1	1	1	1
	X2	1.132479	1.15146	1.160304	1.173025	1.175537
	z	1	1	1	1	1
	Ds	76.858	24.885	54.929	2.5286	3.102
	R		11.769098	14.77231	36.11735	312.5250
	X1	Non-	1	1	1	1
	X2	Existing	1.151426	1.160287	1.173024	1.175537
	z(x)		1.010901	1.006903	1.001155	1.000015
(5 $\overset{\sim 2}{10}$)	Ds		64.976	54.990	25.298	3.182
Tr-S60 E'_3 (3-10, m)	N	252	372	492	1332	12132
	R	9.08247	12.05441	15.07431	36.45745	312.5253
	X1	1	1	1	1	1
	X2	1.261987	1.284237	1.294683	1.30994	1.313029
	z	1	1	1	1	1
	Ds	76.372	64.002	54.129	25.054	3.105
	R		12.048805	15.068788	36.45487	312.5251
	X1	Non-	1	1	1	1
	X2	Existing	1.28421	1.294669	1.30994	1.313009
	z(x)		1.010398	1.006633	1.00113	1.000015
(1 5 $\overset{\sim 2}{10}$)	Ds		64.061	54.169	25.057	3.105
Tr-S60 E'_3 (3-$\overline{10}$, m)	N	240	360	480	1320	12120
	R	8.882804	11.83837	14.84924	36.20179	312.52515
	X1	1	1	1	1	1
	X2	1.156384	1.179414	1.19022	1.205889	1.209014
	z	1	1	1	1	1
	Ds	76.042	64.218	54.422	25.180	3.102
	R		11.83077	14.8419	36.20187	312.52505
	X1	Non-	1	1	1	1
	X2	Existing	1.179376	1.190202	1.205889	1.209014
	z(x)		1.010787	1.006838	1.001145	1.000015
(5 $\overset{- 2}{10}$)	Ds		64.301	54.476	25.180	3.102
Tr-S60 E'_3 (3-$\overline{10}$, m)	N	252	372	492	1332	12132
	R	9.181419	12.17213	15.20225	36.59957	312.5253
	X1	1	1	1	1	1
	X2	1.317864	1.346566	1.368204	1.380408	1.38457
	z	1	1	1	1	1
	Ds	74.734	62.770	53.222	24.8595	3.105
	R		12.167081	15.19801	36.5997	312.5251
	X1	Non-	1	1	1	1
	X2	Existing	1.346535	1.360191	1.380408	1.38457
	z(x)		1.010195	1.00652	1.001122	1.000015
(1 5 $\overset{- 2}{10}$)	Ds		62.822	53.257	24.859	3.105

		m = 1	m = 2	m = 3	M = 10	m = 100
Tr-S60 E'_3 ($\dot{3}$-10, m)	N	200	320	440	1280	12080
	R		11.6946	14.59789	35.72592	311.5811
	X1	Non-	1.725707	1.727982	1.731372	1.732042
	X2	existing	1	1	1	1
	z		1	1	1	1
	Ds		58.495	51.619	35.072	3.111
	R	Non-	11.6937	14.59594	35.72355	311.5809
	X1	existing	1.725706	1.727981	1.731372	1.732042
	X2		1	1	1	1
	z(x)		1.011043	1.007071	1.001176	1.000015
	Ds		58.504	51.633	25.075	3.111
Tr-S60 E'_3 ($\dot{3}$-10,m)	N		380	500	1340	12140
	R	Non-	12.004444	14.95755	36.18908	312.5252
	X1	existing	1.726031	1.728176	1.73139	1.732942
	X2		1.15237	1.160665	1.173034	1.175537
	z		1	1	1	1
	Ds		65.923	55.871	25.579	3.107
	R	Non-	11.935301	14.84754	35.97739	312.5250
	X1	existing	1.725961	1.728118	1.731382	1.732042
	X2		1.152098	1.160442	1.173004	1.174437
	z(x)		1.162944	1.167441	1.174194	1.175552
(5 $\overset{\sim 2}{10}$)	Ds		66.689	56.702	25.881	3.107
Tr-S60 E'_3 ($\dot{3}$-10, m)	N		392	512	1352	12152
	R	Non-	12.23752	15.21561	36.509	312.5253
	X1	existing	1.726258	1.728306	1.731401	1.732042
	X2		1.2851	1.295024	1.309949	1.313029
	z		1	1	1	1
	Ds		65.439	55.288	25.358	3.110
	R	Non-	12.010592	14.91314	36.0283	312.5251
	X1	existing	1.726037	1.728153	1.731384	1.732042
	X2		1.284024	1.294281	1.309865	1.313029
	z(x)		1.294492	1.301046	1.311018	1.313044
(1 5 $\overset{\sim 2}{10}$)	Ds		67.936	57.5535	2.604	3.110
Tr-S60 E'_3 ($\dot{3}$-10, m)	N		380	500	1340	12140
	R	Non-	12.05679	15.019225	36.2656	312.52525
	X1	existing	1.726083	1.728207	17.31392	1.732042
	X2		1.180479	1.190644	1.2059	1.209014
	z		1	1	1	1
	Ds		65.352	55.413	25.472	3.107
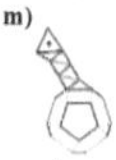	R	Non-	11.96277	14.87502	36.005105	312.5244
	X1	existing	1.725989	1.728132	1.731383	1.732042
	X2		1.180028	1.190285	1.205854	1.209014
	z(x)		1.190809	1.197245	1.207038	1.20903
(5 $\overset{- 2}{10}$)	Ds		66.383	56.493	25.184	3.107
Tr-S60 E'_3 ($\dot{3}$-$\bar{10}$, m)	N		392	512	1352	12152
	R	Non-	12.23752	15.21561	36.50906	312.5253
	X1	existing	1.726258	1.728306	1.731401	1.732042
	X2		1.2851	1.295024	1.309949	1.31029
	z		1	1	1	1
	Ds		65.439	55.288	25.358	3.110
	R	Non-	12.010592	14.91314	36.02821	312.5251
	X1	existing	1.726037	1.728163	1.731384	1.732042
	X2		1.284024	1.294281	1.309865	1.313029
	z(x)		1.294492	1.301046	1.311018	1.313044
(1 5 $\overset{- 2}{10}$)	Ds		67.936	57.5535	26.039	3.110

End of the eleven Tr-SnE'(k1-k2, m) and Families P. 3 of 4

		m = 1	m = 2	m = 3	m = 10	m = 100
Tr- S60 E'_3 (10-10, m)	N	120	180	240	660	6060
	R	7.420845	9.035749	10.720265	23.07509	185.4933
	X1	1	1	1	1	1
	X2	1	1	1	1	1
	z	1	1	1	1	1
	Ds	54.477	55.117	52.208	30.988	4.4839
	R	Non-Existing	9.016626	10.697639	23.065015	185.4935
	X1		1	1	1	1
	X2		1	1	1	1
	z(x)		1.018657	1.013212	1.002824	1.000044
	Ds		55.351	52.429	31.025	4.403
Tr- S60 E'_3 (10-10, m)	N	180	240	300	720	6120
	R	7.6882025	9.377666	11.115766	23.61225	186.1629
	X1	1.118412	1.137375	1.148479	1.169606	1.175475
	X2	1.118412	1.137375	1.148479	1.169606	1.175475
	z	1	1	1	1	1
	Ds	76.131	68.228	60.699	32.285	4.415
	R	Non-Existing	9.26844	10.92607	23.19733	185.8282
	X1		1.136459	1.147522	1.16939	1.175474
	X2		1.136459	1.147522	1.16939	1.175474
(5 $\tilde{10}^2$)	z(x)		1.154487	1.160473	1.172253	1.175519
	Ds		69.845	62.825	33.450	4.431
Tr- S60 E'_3 (10-10, m)	N	192	252	312	732	6132
	R	7.9673535	9.688216	11.44631	23.99072	186.8374
	X1	1.246465	1.268248	1.281065	1.305833	1.312953
	X2	1.246465	1.268248	1.281065	1.305833	1.312953
	z	1	1	1	1	1
	Ds	75.616	67.120	59.534	31.795	4.392
	R	Non-Existing	9.486544	11.09751	23.2354	185.4939
	X1		1.266298	1.279004	1.305354	1.312951
	X2		1.266298	1.279004	1.305354	1.312951
	z(x)		1.283177	1.291286	1.308129	1.312994
(1 5 $\tilde{10}^2$)	Ds		70.004	63.335	33.896	4.455
Tr- S60 E'_3 (10-10, m)	N	180	240	300	720	6720
	R	7.735104	9.438696	11.18664	23.70432	186.4995
	X1	1.139571	1.162411	1.175857	1.201666	1.208938
	X2	1.139571	1.162411	1.175857	1.201666	1.208938
	z	1	1	1	1	1
	Ds	75.211	67.348	59.932	32.034	4.399
	R	Non-Existing	9.311547	10.964082	23.21343	185.8279
	X1		1.161127	1.174495	1.201351	1.208937
	X2		1.161127	1.174495	1.201351	1.208937
	z(x)		1.178977	1.187342	1.204202	1
(5 $\bar{10}^4$)	Ds		69.200	62.390	33.404	4.4361
Tr- S60 E'_3 (10-10, m)	N	192	252	312	732	6132
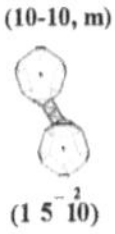	R	8.077418	9.81799	11.58859	24.16398	186.8375
	X1	1.298481	1.326207	1.342648	1.374951	1.384465
	X2	1.298481	1.326207	1.342638	1.374951	1.384465
	z	1	1	1	1	1
	Ds	73.569	65.357	58.081	31.341	4.3915
	R	Non-Existing	9.574368	11.16383	23.221265	185.4937
	X1		1.323209	1.339403	1.37415	1.384463
	X2		1.323209	1.339403	1.37415	1.384463
(1 5 $\tilde{10}^2$)	z(x)		1.331419	1.351239	1.376839	1.384505
	Ds		68.726	62.585	33.937	4.455

		m = 1	m = 2	m = 3	m = 10	M = 100
Tr- S120E' (6-10,m)	N	240	360	480	1320	12120
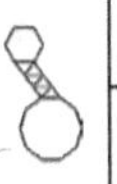	R	10.276787	13.128423	16.067025	37.24647	313.47513
	X1	1	1	1	1	1
	X2	1	1	1	1	1
	z	1	1	1	1	1
	Ds	56.81154	52.21766	41.48476	23.78722	3.083446
	R	Non-existing	13.119554	16.05724	37.24015	313.47493
	X1		1	1	1	1
	X2		1	1	1	1
	z(x)		1.008761	1.005838	1.001082	1.000015
	Ds		52.288	46.541	23.795	3.083
Tr- S120E' (6- $\tilde{10}$,m)	N	300	420	540	1380	12180
	R	10.546776	13.463752	16.443235	37.709	314.4308
	X1	1	1	1	1	1
	X2	1.145449	1.157154	1.163427	1.173234	1.175537
	z	1	1	1	1	1
	Ds	6.743	57.924	49.930	24.262	3.080
	R	Non-existing	13.4566555	16.435408	37.70915	314.4305
	X1		1	1	1	1
	X2		1.157134	1.163235	1.173234	1.175537
(5 $\tilde{10}^2$)	z(x)		1	1	1.001056	1.000015
	Ds		57.985	49.977	24.262	3.080
Tr- S120E' (6- $\tilde{10}$,m)	N	312	432	552	1392	12192
	R	10.758439	13.708572	16.709515	38.0285	314.4308
	X1	1	1	1	1	1
	X2	1.276802	1.290814	1.298119	1.310194	1.31303
	z	1	1	1	1	1
	Ds	67.398	57.470	49.425	24.064	3.083
	R	Non-existing	13.702965	16.703005	38.02845	314.4305
	X1		1	1	1	1
	X2		1.290796	1.298108	1.310194	1.31303
(1 5 $\tilde{10}^2$)	z(x)		1.008027	1.005394	1.001038	1.000015
	Ds		57.517	49.464	24.064	3.083
Tr- S120E' (6- $\tilde{10}$,m)	N	300	420	540	1380	12180
	R	10.592266	13.520693	16.50771	37.79187	314.4308
	X1	1	1	1	1	1
	X2	1.172024	1.186335	1.193816	1.20615	1.209015
	z	1	1	1	1	1
	Ds	66.847	57.437	49.540	24.156	3.08
	R	Non-existing	13.513782	16.49999	37.791905	314.4305
	X1		1	1	1	1
	X2		1.186312	1.193802	1.20615	1.209015
(5 $\bar{10}^2$)	z(x)		1.008255	1.005528	1.001057	1.000015
	Ds		57.496	49.587	24.156	3.08
Tr- S120E' (6- $\bar{10}$,m)	N	312	432	552	1392	12192
	R	10.842992	13.810348	16.82162	38.16902	314.4309
	X1	1	1	1	1	1
	X2	1.336695	1.355043	1.364674	1.380748	1.38457
	z	1	1	1	1	1
	Ds	66.343	56.626	48.769	23.887	3.083
	R	Non-existing	13.805296	16.816408	38.16915	314.4306
	X1		1	1	1	1
	X2		1.355022	1.364662	1.380748	138475
(1 5 $\bar{10}^2$)	z(x)		1.007908	1.005321	1.00103	1.000015
	Ds		56.667	48.799	23.887	3.083

P. 4 of 4

--- End of [DATA 5-4] Tr-SnE'(k1-k2, m) and Families [completed for all eleven mother E's] ---

(2.9.12. done typing table of Tr-SE)
(3.12.12.done sketchup of Tr-SE)

[DATA 5-5] The High Ds Configurations

- **The highest surface density configurations are among the TEs or Tr-SEs (or Tr-SE's).**
- **The following chart is a list of the high Ds configurations with Ds ≥ 80 % .**
- **Some configurations in different families are of the same constructions and certainly have the same Ds, and hence, they are listed in the same box.**

High Ds Configurations (Ds > 80%)				
Configurations	**R**	**N**	**Ds %**	**Order of Density**
T60E: (5 ~ 5) + (3 ~ 3)	7.2896	180	84.68473	1
Tr-S30E'(5-3, 1)	5.9581525	120	84.50804	2
T60E: (5 ~ 5) or (3 ~ 3)	5.95883	120	84.49843	3
Tr-S12E'1(4-3, 1) T24E: (4 ~ 4)e or (3 ~ 3)	3.7701988	48	84.421436	4
Tr-S20E(5-5, 2): (z = 1.043154)	5.969884	120	84.17623	5
Tr-S30E'($\dot{5}$ -3,1) T60E: (1 5 ~ 5)	6.266247	132	84.04249	6
Tr-S20E($\dot{5}$ - $\dot{5}$, 2): (z = 1.200057)	6.277958	132	83.72924	7
TR-S8E(4-4, 2): (z = 1.110487)	3.7942255	48	83.35564	8
Tr-S20E($\dot{5}$ - $\dot{5}$, 1) T60E: (1 5)	4.6497724	72	83.254767	9
Tr-S20E(5-5, 2)	6.001806	120	83.28319	10
T24E T12E : (3 ~ 3) Tr-S6E(3-3, 1) Tr-S8E(4-4, 1)	2.68742675	24	83.07646	11
12E T12E Tr-S4E(3-3, 1)	1.902113	12	82.91796	12
A32E	3.1208482	32	82.138105	13
T60E: (1 5 ~ 5) + (3 ~ 3)	7.6665	192	81.66707	14
Tr-S12E'1($\dot{3}$ - $\dot{3}$, 1) T60E: (1 3)	4.952996	80	81.5256	15
T24E: (4 ~ 4)e + (3 ~ 3)	4.68136	72	81.13504	16
Tr-S20E($\dot{5}$ - $\dot{5}$, 2)	6.381981	132	81.022	17
Tr-S8E(4-4, 2)	3.850704	48	80.92841	18
Tr-S4E($\dot{3}$ - $\dot{3}$, 2): (z = 1.693897)	2.9411375	28	80.922143	19
Tr-S6E($\dot{3}$ - $\dot{3}$, 1)	3.1462785	32	80.81569	20
T60E Tr-S12E(3-3, 1) Tr-S20E(5-5, 1)	4.31167475	60	80.6862	21
Tr-S12E'1(3- $\dot{4}$, 1) T24E: (1 ‾ 4 ~ 4)e	4.094137	54	80.539517	22

End of [DATA 5-6] The High Ds Configurations

INDEX

F

G

H

I

K

L

M

N

O

P

R

S

www.ingramcontent.com/pod-product-compliance
Lightning Source LLC
Chambersburg PA
CBHW020726180526
45163CB00001B/131
9781514451175